Strength of Materials and Structures

Fourth edition

JOHN CASE
M.A., F.R.Ae.S.
Formerly Head of the Department of Applied Mechanics, Royal Naval Engineering College, Plymouth

LORD CHILVER of Cranfield
M.A., D.Sc., F.Eng., F.R.S.
Formerly Vice Chancellor, Cranfield Institute of Technology, and Professor of Civil Engineering, University College, London

CARL T.F. ROSS
B.S.c., Ph.D., D.Sc., C. Eng., F.R.I.N.A., M.S.N.A.M.E.
Professor of Structural Dynamics, University of Portsmouth, Portsmouth

ARNOLD
A member of the Hodder Headline Group
LONDON • SYDNEY • AUCKLAND
Co-published in North, Central and South America by
John Wiley & Sons Inc., New York • Toronto

First published in Great Britain in 1959 as *Strength of Materials*
Reprinted 1961, 1964
Second edition 1971
Reprinted 1985, 1986
Third edition 1993
Reprinted 1992, 1994, 1995, 1997, 1998
Fourth edition published in 1999 by
Arnold, a member of the Hodder Headline Group,
338 Euston Road, London NW1 3BH

http://www.arnoldpublishers.com

Co-published in North, Central and South America by
John Wiley & Sons Inc., 605 Third Avenue,
New York, NY 10158-0012

British Library Cataloguing in Publication Data
A catalogue record for this book is available from the British Library

Library of Congress Cataloging-in-Publication Data
A catalog record for this book is available from the Library of Congress

ISBN 0 340 71920 6
ISBN 0 470 37980 4 (Wiley)

1 2 3 4 5 6 7 8 9 10

Commissioning Editor: Matthew Flynn
Cover Designer: Terry Griffiths

Printed and bound in Great Britain by J.W. Arrowsmith Ltd, Bristol

What do you think about this book? Or any other Arnold title?
Please send your comments to feedback.arnold@hodder.co.uk

Strei
and

Contents

Preface

This new edition is updated by Professor Ross, and while it retains much of the basic and traditional work in Case & Chilver's *Strength of Materials and Structures*, it introduces modern numerical techniques, such as matrix and finite element methods.

Additionally, because of the difficulties experienced by many of today's students with basic traditional mathematics, the book includes an introductory chapter which covers in some detail the application of elementary mathematics to some problems involving simple statics.

The 1971 edition was begun by Mr. John Case and Lord Chilver but, because of the death of Mr. John Case, it was completed by Lord Chilver.

Whereas many of the chapters are retained in their 1971 version, much tuning has been applied to some chapters, plus the inclusion of other important topics, such as the plastic theory of rigid jointed frames, the torsion of non-circular sections, thick shells, flat plates and the stress analysis of composites.

The book covers most of the requirements for an engineering undergraduate course on strength of materials and structures.

The introductory chapter presents much of the mathematics required for solving simple problems in statics.

Chapter 1 provides a simple introduction to direct stresses and discusses some of the fundamental features under the title: Strength of materials and structures.

Chapter 2 is on pin-jointed frames and shows how to calculate the internal forces in some simple pin-jointed trusses. Chapter 3 introduces shearing stresses and Chapter 4 discusses the modes of failure of some structural joints.

Chapter 5 is on two-dimensional stress and strain systems and Chapter 6 is on thin walled circular cylindrical and spherical pressure vessels.

Chapter 7 deals with bending moments and shearing forces in beams, which are extended in Chapters 13 and 14 to include beam deflections. Chapter 8 is on geometrical properties.

Chapters 9 and 10 cover direct and shear stresses due to the bending of beams, which are extended in Chapter 13. Chapter 11 is on beam theory for beams made from two dissimilar materials. Chapter 15 introduces the plastic hinge theory and Chapter 16 introduces stresses due to torsion. Chapter 17 is on energy methods and, among other applications, introduces the plastic design of rigid-jointed plane frames.

Chapter 18 is on elastic buckling.

Chapter 19 is on flat plate theory and Chapter 20 is on the torsion of non-circular sections. Chapter 21 is on thick cylinders and spheres.

Chapter 22 introduces matrix algebra and Chapter 23 introduces the matrix displacement method.

Chapter 24 introduces the finite element method and in Chapter 25 this method is extended to cover the vibrations of complex structures.

CTFR, 1999

Acknowledgements

I would like to thank my wife, Anne, and my children, Nicolette and Jonathan, who have suffered my nebulous number-crunching world of eigenvalue economisers and matrix manipulators over many years.

My thanks are extended to Mrs. Joanna Russell and Mrs. Helen Facey for the considerable care and devotion they showed in typing this manuscript.

CTFR, 1999

❑❑❑❑❑❑❑❑❑

"Only when you climb the highest mountain, will you be aware of the vastness that lies around you."

Oscar Wilde, 1854—1900.

❑❑❑❑❑❑❑❑❑

Chinese Proverb — *It is better to ask a question and look a fool for five minutes, than not to ask a question at all and be a fool for the rest of your life.*

Heaven and Hell — *In heaven you are faced with an infinite number of solvable problems and in hell you are faced with an infinite number of unsolvable problems.*

Principal notation

a	length	A	area
b	breadth	C	complementary energy
c	wave velocity, distance	D	diameter
d	diameter	E	young's modulus
h	depth	F	shearing force
j	number of joints	G	shearing modulus
l	length	H	force
m	mass, modular ratio, number of numbers	I	second moment of area
n	frequency, load factor, distance	J	torsion constant
p	pressure	K	bulk modulus
q	shearing force per unit length	L	length
r	radius	M	bending moment
s	distance	P	force
t	thickness	Q	force
u	displacement	R	force, radius
v	displacement, velocity	S	force
w	displacement, load intensity, force	T	torque
x	coordinate	U	strain energy
y	coordinate	V	force, volume, velocity
z	coordinate	W	work done, force
		X	force
		Y	force
		Z	section modulus, force

α	coefficient of linear expansion	ρ	density
γ	shearing strain	σ	direct stress
δ	deflection	τ	shearing stress
ε	direct strain	ω	angular velocity
η	efficiency	Δ	deflection
θ	temperature, angle of twist	Φ	step-function
ν	Poisson's ratio		

[k]	element stiffness matrix	**[K]**	system stiffness matrix
[m]	elemental mass matrix	**[M]**	system mass matrix

Note on SI units

The units used throughout the book are those of the Système Internationale d'Unités; this is usually referred to as the SI system. In the field of the strength of materials and structures we are concerned with the following basic units of the SI system:

length	metre (m)
mass	kilogramme (kg)
time	second (s)
temperature	kelvin (K)

There are two further basic units of the SI system – electric current and luminous intensity – which we need not consider for our present purposes, since these do not enter the field of the strength of materials and structures. For temperatures we shall use conventional degrees centigrade (°C), since we shall be concerned with temperature changes rather than absolute temperatures. The units which we derive from the basic SI units, and which are relevant to out fielf of study, are:

force	newton (N)	$kg.m.s^{-2}$
work, energy	joule (J)	$kg.m^2.s^{-2}$ = Nm
power	watt (W)	$kg.m^2.s^{-3} = Js^{-1}$
frequency	hertz (Hz)	cycle per second
pressure	Pascal (Pa)	$N.m^{-2} = 10^{-5}$ bar

The acceleration due to gravity is taken as:

$$g = 9.81 ms^{-2}$$

Linear distances are expressed in metres and multiples or divisions of 10^3 of metres, i.e.

Kilometre (km)	10^3 m
metre (m)	1 m
millimetre (mm	10^{-3} m

In many problems of stress analysis these are not convenient units, and others, such as the centimetre (cm), which is 10^{-2} m, are more appropriate.

The unit of force, the newton (N), is the force required to give unit acceleration (ms^{-2}) to unit mass kg). In terms of newtons the common force units in the foot-pound-second-system (with g = 9.81 ms^{-2}) are

1 lb.wt = 4.45 newtons (N)

1 ton.wt = 9.96×10^3 newtons (N)

In general, decimal multiples in the SI system are taken in units of 10^3. The prefixes we make most use of are:

kilo	k	10^3
mega	M	10^6
giga	G	10^9

Thus:

$$1 \text{ ton.wt} = 9.96 \text{ kN}$$

The unit of force, the newton (N), is used for external loads and internal forces, such as shearing forces. Torques and bending of moments are expressed in newton-metres (Nm).

An important unit in the strength of materials and structures is stress. In the foot-pound-second system, stresses are commonly expressed in lb.wt/in^2, and tons/in^2. In the SI system these take the values:

$$1 \text{ lb.wt/in}^2 = 6.89 \times 103 \text{ N/m}^2 = 6.89 \text{ kN/m}^2$$

$$1 \text{ ton.wt/in}^2 = 15.42 \times 106 \text{ N/m}^2 = 15.42 \text{ MN/m}^2$$

Yield stresses of the common metallic materials are in the range:

$$200 \text{ MN/m}^2 \text{ to } 750 \text{ MN/m}^2$$

Again, Young's modulus for steel becomes:

$$E_{\text{steel}} = 30 \times 106 \text{ lb.wt/in}^2 = 207 \text{ GN/m}^2$$

Thus, working and yield stresses will usually be expressed in MN/m^2 units, while Young's modulus will usually be given in GN/m^2 units.

Introduction

I.1 Introduction

Stress analysis is an important part of engineering science, as failure of most engineering components is usually due to stress. The component under a stress investigation can vary from the legs of an integrated circuit to the legs of an offshore drilling rig, or from a submarine pressure hull to the fuselage of a jumbo jet aircraft.

The present chapter will commence with elementary trigonometric definitions and show how elementary trigonometry can be used for analysing simple pin-jointed frameworks (or trusses). The chapter will then be extended to define couples and show the reader how to take moments.

I.2 Trigonometrical definitions

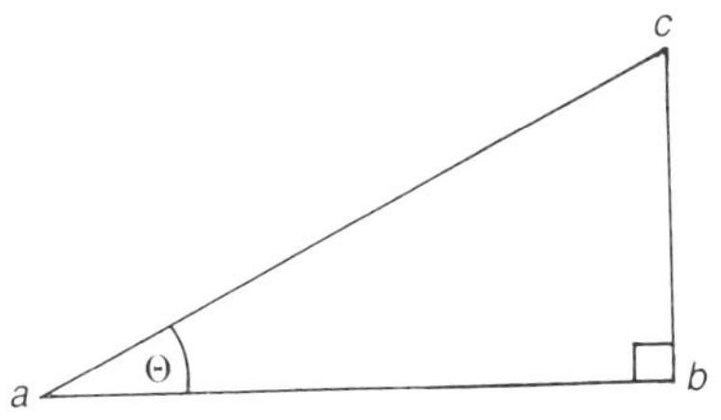

Figure I.1 Right-angled triangle.

With reference to Figure I.1,

$$\sin\theta = bc/ac$$

$$\cos\theta = ab/ac \qquad \text{(I.1)}$$

$$\tan\theta = bc/ab$$

For a triangle without a right angle in it, as shown in Figure I.2, the *sine* and *cosine* rules can be used to determine the lengths of unknown sides or the value of unknown angles.

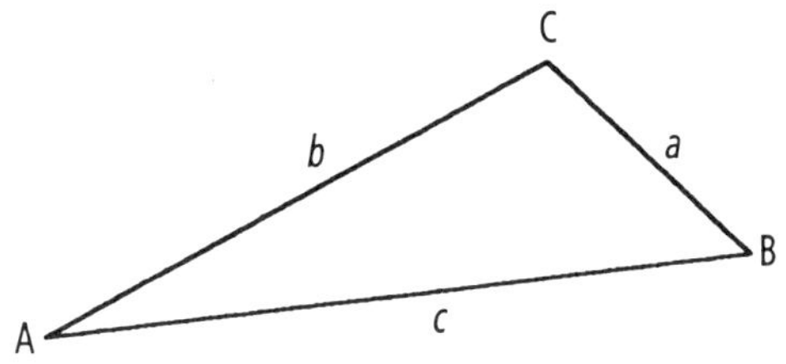

Figure I.2. Triangle with no right angle.

The *sine rule* states that:

$$\frac{a}{\sin A} = \frac{b}{\sin B} = \frac{c}{\sin C} \qquad (I.2)$$

where

a = length of side BC; opposite the angle A

b = length of side AC; opposite the angle B

c = length of side AB; opposite the angle C

The cosine rule states that:

$$a^2 = b^2 + c^2 - 2bc \cos A$$

I.3 Vectors and scalars

A scalar is a quantity which has magnitude but no direction, such as a mass, length and time. A vector is a quantity which has magnitude and direction, such as weight, force, velocity and acceleration.

NB It is interesting to note that the moment of a couple, (Section I.6) and energy (Chapter 17), have the same units; but a moment of a couple is a vector quantity and energy is a scalar quantity.

I.4 Newton's laws of motion

These are very important in engineering mechanics, as they form the very fundamentals of this topic.

Newton's three laws of motion were first published by Sir Isaac Newton in *The Principia* in 1687, and they can be expressed as follows:

(1) Every body continues in its state of rest or uniform motion in a straight line, unless it is compelled by an external force to change that state.

(2) The rate of change of momentum of a body with respect to time, is proportional to the resultant force, and takes place in a direction of which the resultant force acts.

(3) Action and reaction are equal and opposite.

I.5 Elementary statics

The trigonometrical formulae of I.2 can be used in statics. Consider the force F acting on an angle θ to the horizontal, as shown by Figure I.3(a). Now as the force F is a vector, (i.e. it has magnitude and direction), it can be represented as being equivalent to its horizontal and vertical components, namely F_H and F_V, respectively, as shown by Figure I.3(b). These horizontal and vertical components are also vectors, as they have magnitude and direction.

NB If F is drawn to scale, it is possible to obtain F_H and F_V from the scaled drawing.

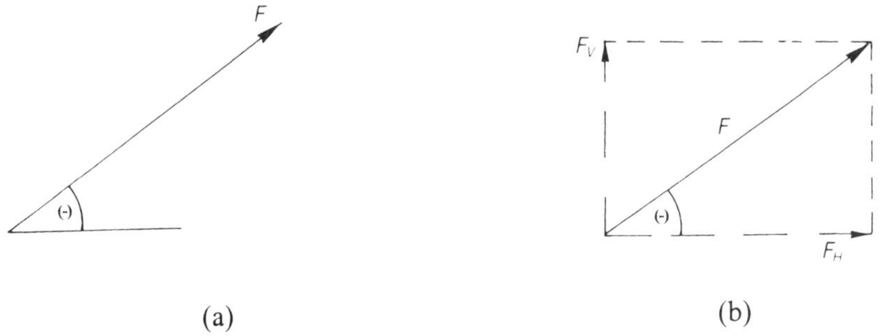

Figure I.3 Resolving a force.

From elementary trigonometry

$$\frac{F_H}{F} = \cos\theta$$

$$\therefore F_H = F\cos\theta\text{—horizontal component of } F$$

Similarly,

$$\frac{F_V}{F} = \sin\theta$$

$$\therefore F_V = F\sin\theta\text{—vertical component of } F$$

Problem I.1 Determine the forces in the plane pin-jointed framework shown below.

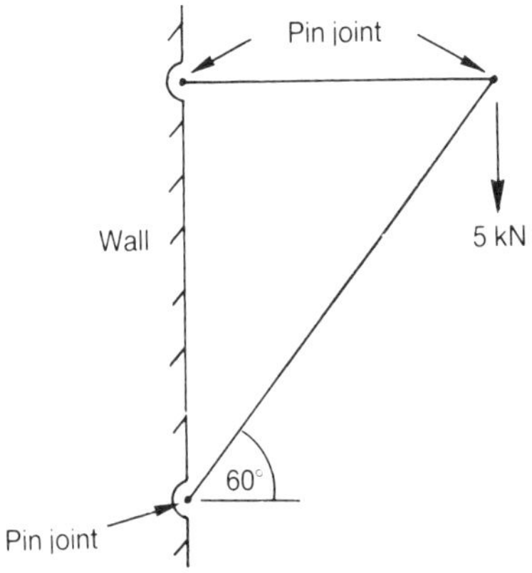

Solution

Assume all unknown forces in each member are in tension, i.e. the internal force in each member is pulling away from its nearest joint, as shown below.

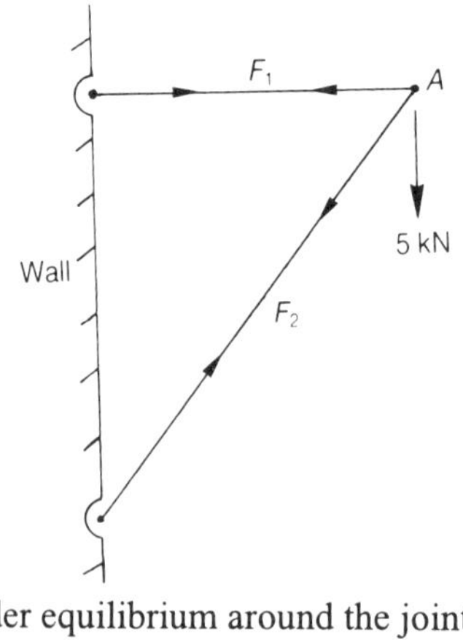

Isolate joint A and consider equilibrium around the joint,

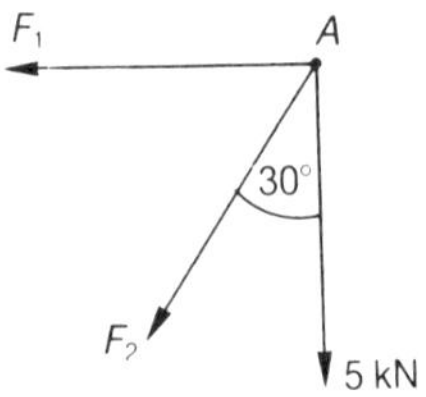

Resolving forces vertically
From Section I.7

$$\text{upward forces} = \text{downward forces}$$
$$0 = 5 + F_2 \cos 30$$

or
$$F_2 = -\frac{5}{\cos 30} = -5.77 \text{ kN (compression)}$$

The negative sign for F_2 indicates that this member is in compression.

Resolving forces horizontally
From Section I.7

$$\text{forces to the left} = \text{forces to the right}$$
$$F_1 + F_2 \sin 30 = 0$$
$$F_1 = -F_2 \sin 30 = 5.77 \sin 30$$
$$F_1 = 2.887 \text{ kN (tension)}$$

The force diagram is as follows:

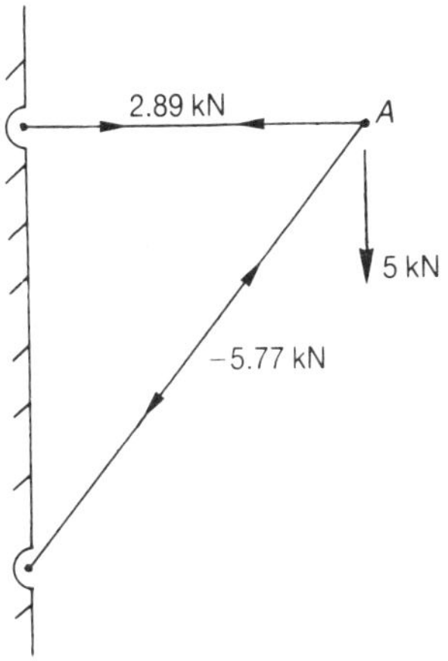

Another method of determining the internal forces in the truss shown on page 4 is through the use of the triangle of forces. For this method, the magnitude and the direction of the known force, namely the 5kN load in this case, must be drawn to scale.

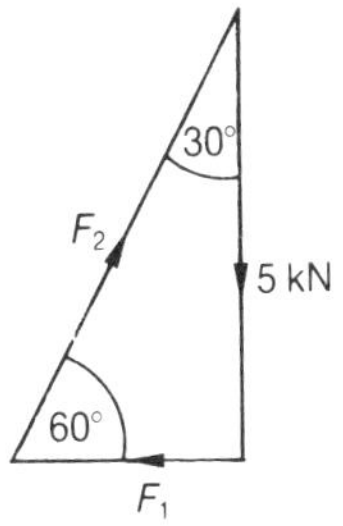

To complete the triangle, the directions of the unknown forces, namely F_1 and F_2 must be drawn, as shown above. The directions of these forces can then be drawn by adding the arrowheads to the triangle so that the arrowheads are either all in a clockwise direction or, alternatively, all in a counter-clockwise direction.

Applying the sine rule to the triangle of forces above,

$$\frac{5}{\sin 60} = \frac{F_1}{\sin 30}$$

$$\therefore F_1 = \frac{5 \times 0.5}{0.866} = 2.887 \text{ kN}$$

Similarly by applying the sine rule:

$$\frac{5}{\sin 60} = \frac{F_2}{\sin 90}$$

$$\therefore F_2 = \frac{5}{0.866} = 5.77 \text{ kN}$$

These forces can now be transferred to the joint A of the pin-jointed truss below, where it can be seen that the member with the load F_1 is in tension, and that the member with the load F_2 is in compression.

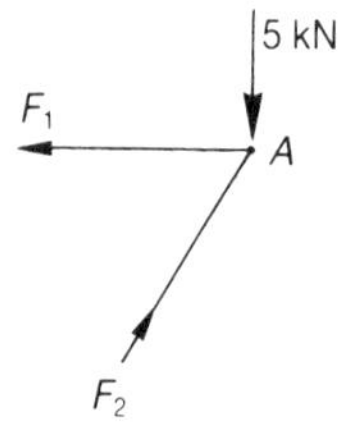

This is known as a free body diagram.

I.6 Couples

A couple can be described as the moment produced by two equal and opposite forces acting together, as shown in Figure I.4 where,

the moment at the couple = $M = F \times l$ (N.m)

F = force (N)

l = lever length (m)

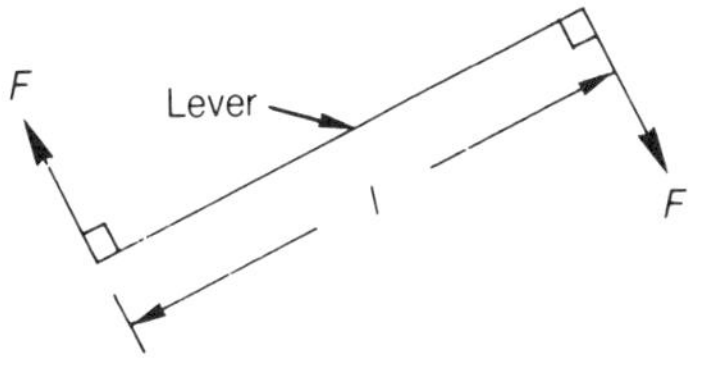

Figure I.4 A clockwise couple.

For the counter-clockwise couple of Figure I.5,

$$M = F \cos \theta \times l$$

where $F \cos \theta$ = the force acting perpendicularly to the lever of length l.

NB The components of force $F \sin \theta$ will simply place the lever in tension, and will not cause a moment.

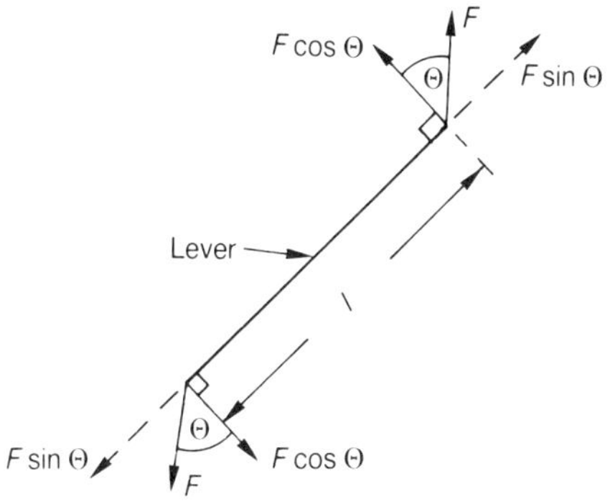

Figure I.5 A counter-clockwise couple.

It should be noted from Figure I.4 that the lever can be described as the perpendicular distance between the line of action of the two forces causing the couple.

Furthermore, in Figure I.5, although the above definition still applies, the same value of couple can be calculated, if the lever is chosen as the perpendicular distance between the components of the force that are perpendicular to the lever, and the forces acting on this lever are in fact those components of force.

I.7 Equilibrium

This section will be limited to one- or two-dimensional systems, where all the forces and couples will be acting in on plane; such a system of forces is called a coplanar system.

In two dimensions, equilibrium is achieved when the following laws are satisfied:

(1) upward forces = downward forces

(2) forces to the left = forces to the right

(3) clockwise couples = counter-clockwise couples.

To demonstrate the use of these two-dimensional laws of equilibrium, the following problems will be considered.

Problem I.2 Determine the values of the reactions R_A and R_B, when a beam is simply-supported at its ends and subjected to a downward force of 5 kN.

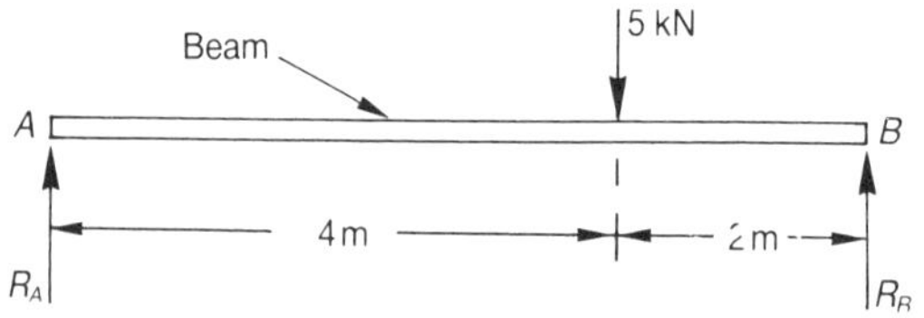

<u>*Solution*</u>

For this problem, it will be necessary to take moments. By taking moments, it is meant that the values of the moments must be considered about a suitable position.

Suitable positions for taking moments on this beam are A and B. This is because, if moments are taken about A, the unknown section R_A will have no lever and hence, no moment about A, thereby simplifying the arithmetic. Similarly, by taking moments about B, the unknown R_B will have no lever and hence, no moment about B, thereby simplifying the arithmetic.

Taking moments about B

clockwise moments = counter-clockwise moments

$$R_A \times (4 + 2) = 5 \times 2$$

or

$$R_A = 10/6$$

$$R_A = 1.667 \text{ kN}$$

Resolving forces vertically

$$\text{upward forces} = \text{downward forces}$$

$$R_A + R_B = 5$$

or

$$R_B = 5 - R_A = 5 - 1.667$$

$$R_B = 3.333 \text{ kN}$$

Problem I.3 Determine the values of the reactions of R_A and R_B for the simply-supported beam shown.

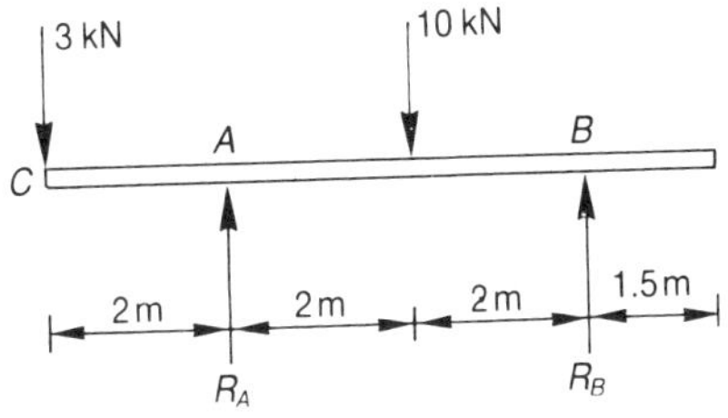

<u>*Solution*</u>

Taking moments about B

$$\text{clockwise couples} = \text{counter-clockwise couples}$$

$$R_A \times 4 = 3 \times 6 + 10 \times 2$$

$$R_A = \frac{18 + 20}{4}$$

$$R_A = 9.5 \text{ kN}$$

Resolving forces vertically

$$R_A + R_B = 3 + 10$$

or

$$R_B = 13 - 9.5 = 3.5 \text{ kN}$$

Further problems ***(answers on page 691)***

Problem I.4 Determine the reactions R_A and R_B for the simply-supported beams.

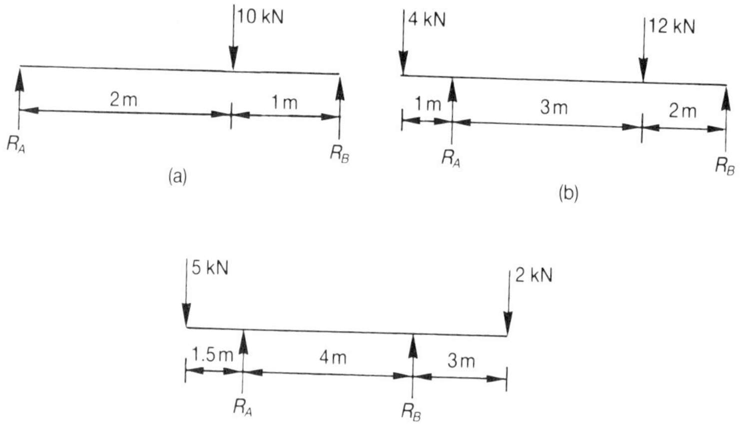

Problem I.5 Determine the forces the pin-jointed trusses shown.

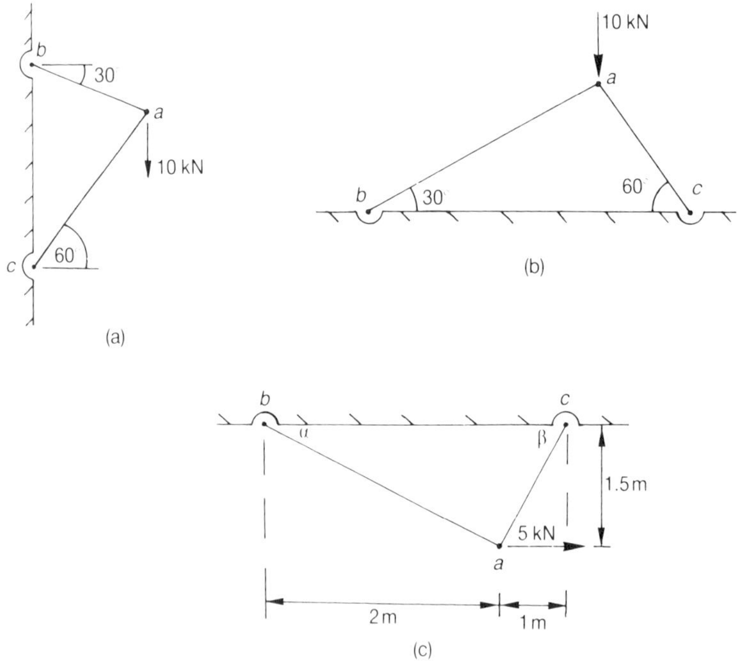

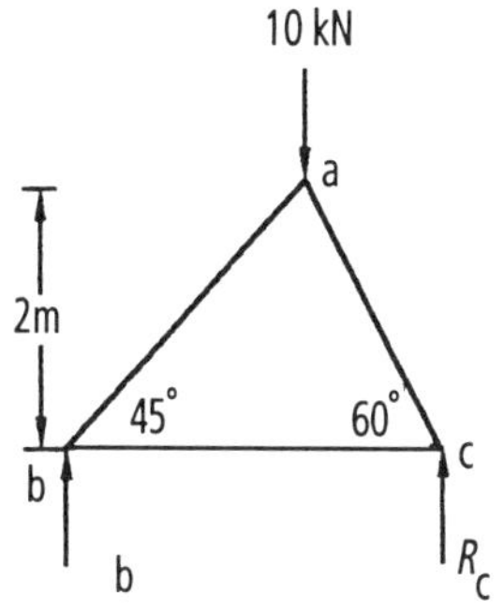

(d)

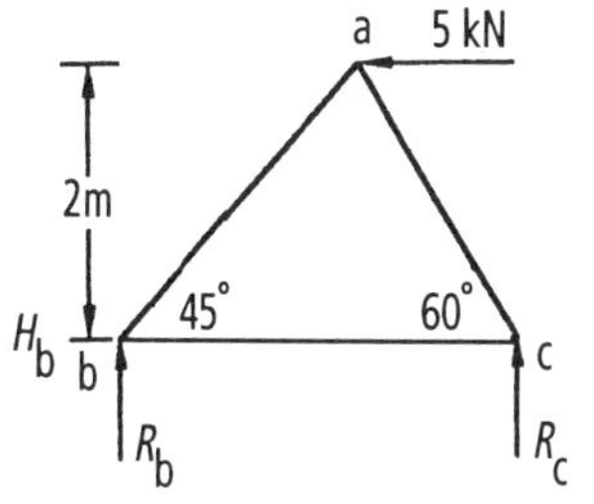

(e)

1 Tension and compression: direct stresses

1.1 Introduction

The strength of a material, whatever its nature, is defined largely by the internal stresses, or intensities of force, in the material. A knowledge of these stresses is essential to the safe design of a machine, aircraft, or any type of structure. Most practical structures consist of complex arrangements of many component members; an aircraft fuselage, for example, usually consists of an elaborate system of interconnected sheeting, longitudinal stringers, and transverse rings. The detailed stress analysis of such a structure is a difficult task, even when the loading conditions are simple. The problem is complicated further because the loads experienced by a structure are variable and sometimes unpredictable. We shall be concerned mainly with stresses in materials under relatively simple loading conditions; we begin with a discussion of the behaviour of a stretched wire, and introduce the concepts of direct stress and strain.

1.2 Stretching of a steel wire

One of the simplest loading conditions of a material is that of *tension*, in which the fibres of the material are stretched. Consider, for example, a long steel wire held rigidly at its upper end, Figure 1.1, and loaded by a mass hung from the lower end. If vertical movements of the lower end are observed during loading it will be found that the wire is stretched by a small, but measurable, amount from its original unloaded length. The material of the wire is composed of a large number of small crystals which are only visible under a microscopic study; these crystals have irregularly shaped boundaries, and largely random orientations with respect to each other; as loads are applied to the wire, the crystal structure of the metal is distorted.

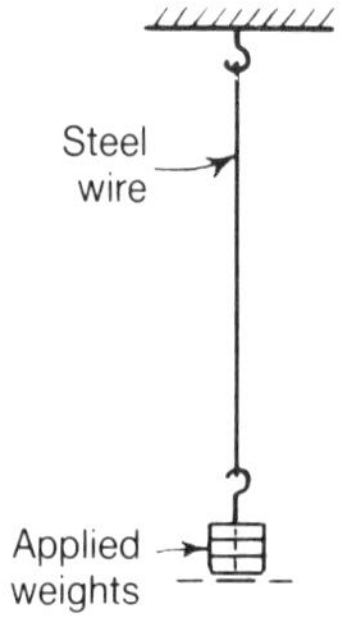

Figure 1.1 Stretching of a steel wire under end load.

For small loads it is found that the extension of the wire is roughly proportional to the applied load, Figure 1.2. This linear relationship between load and extension was discovered by Robert Hooke in 1678; a material showing this characteristic is said to obey *Hooke's law*.

As the tensile load in the wire is increased, a stage is reached where the material ceases to show this linear characteristic; the corresponding point on the load–extension curve of Figure 1.2 is known as the *limit of proportionality*. If the wire is made from a high-strength steel then the load–extension curve up to the *breaking point* has the form shown in Figure 1.2. Beyond the limit of proportionality the extension of the wire increases non-linearly up to the elastic limit and, eventually, the breaking point.

The elastic limit is important because it divides the load–extension curve into two regions. For loads up to the elastic limit, the wire returns to its original unstretched length on removal of the loads; this property of a material to recover its original form on removal of the loads is known as *elasticity*; the steel wire behaves, in fact, as a still elastic spring. When loads are applied above the elastic limit, and are then removed, it is found that the wire recovers only part of its extension and is stretched permanently; in this condition the wire is said to have undergone an *inelastic*, or *plastic*, extension. For most materials, the limit of proportionality and the elastic limit are assumed to have the same value.

In the case of elastic extensions, work performed in stretching the wire is stored as *strain energy* in the material; this energy is recovered when the loads are removed. During inelastic extensions, work is performed in making permanent changes in the internal structure of the material; not all the work performed during an inelastic extension is recoverable on removal of the loads; this energy reappears in other forms, mainly as heat.

The load–extension curve of Figure 1.2 is not typical of all materials; it is reasonably typical, however, of the behaviour of *brittle* materials, which are discussed more fully in Section 1.5. An important feature of most engineering materials is that they behave elastically up to the limit of proportionality, that is, all extensions are recoverable for loads up to this limit. The concepts of linearity and elasticity[1] form the basis of the theory of small deformations in stressed materials.

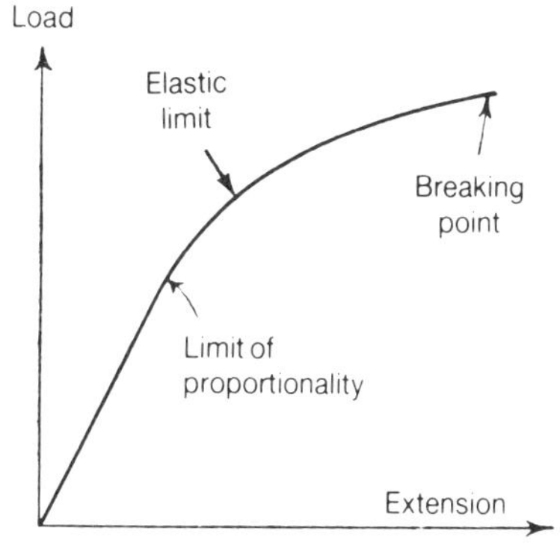

Figure 1.2 Load–extension curve for a steel wire, showing the limit of linear-elastic behaviour (or limit of proportionality) and the breaking point.

[1]The definition of elasticity requires only that the extensions are recoverable on removal of the loads; this does not preclude the possibility of a non-linear relation between load and extension .

1.3 Tensile and compressive stresses

The wire of Figure 1.1 was pulled by the action of a mass attached to the lower end; in this condition the wire is in *tension*. Consider a cylindrical bar *ab*, Figure 1.3, which has a uniform cross-section throughout its length. Suppose that at each end of the bar the cross-section is divided into small elements of equal area; the cross-sections are taken normal to the longitudinal axis of the bar. To each of these elemental areas an equal tensile load is applied normal to the cross-section and parallel to the longitudinal axis of the bar. The bar is then uniformly stressed in tension.

Suppose the total load on the end cross-sections is P; if an imaginary break is made perpendicular to the axis of the bar at the section c, Figure 1.3, then equal forces P are required at the section c to maintain equilibrium of the lengths ac and cb. This is equally true for any section across the bar, and hence on any imaginary section perpendicular to the axis of the bar there is a total force P.

When tensile tests are carried out on steel wires of the same material, but of different cross-sectional area, the breaking loads are found to be proportional approximately to the respective cross-sectional areas of the wires. This is so because the tensile strength is governed by the intensity of force on a normal cross-section of a wire, and not by the total force. This intensity of force is known as *stress*; in Figure 1.3 the *tensile stress* σ at any normal cross-section of the bar is

$$\sigma = \frac{P}{A} \tag{1.1}$$

where P is the total force on a cross-section and A is the area of the cross-section.

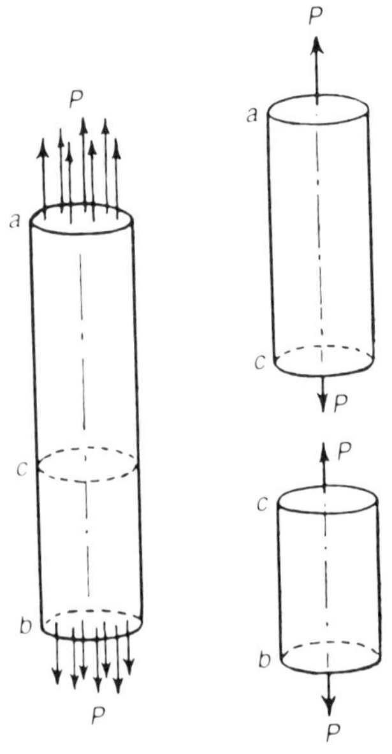

Figure 1.3 Cylindrical bar under uniform tensile stress; there is a similar state of tensile stress over any imaginary normal cross-section.

In Figure 1.3 uniform stressing of the bar was ensured by applying equal loads to equal small areas at the ends of the bar. In general we are not dealing with equal force intensities of this type, and a more precise definition of stress is required. Suppose δA is an element of area of the cross-section of the bar, Figure 1.4; if the normal force acting on this element is δP, then the tensile stress at this point of the cross-section is defined as the limiting value of the ratio $(\delta P/\delta A)$ as δA becomes infinitesimally small. Thus

$$\sigma = \operatorname*{Limit}_{\delta A \to 0} \frac{\delta P}{\delta A} = \frac{\mathrm{d}P}{\mathrm{d}A} \tag{1.2}$$

This definition of stress is used in studying problems of non-uniform stress distribution in materials.

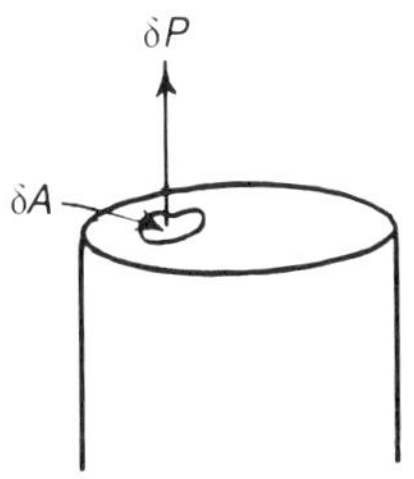

Figure 1.4 Normal load on an element of area of the cross-section.

When the forces P in Figure 1.3 are reversed in direction at each end of the bar they tend to *compress* the bar; the loads then give rise to *compressive stresses*. Tensile and compressive stresses are together referred to as *direct* (or *normal*) *stresses*, because they act perpendicularly to the surface.

Problem 1.1 A steel bar of rectangular cross-section, 3 cm by 2 cm, carries an axial load of 30 kN. Estimate the average tensile stress over a normal cross-section of the bar.

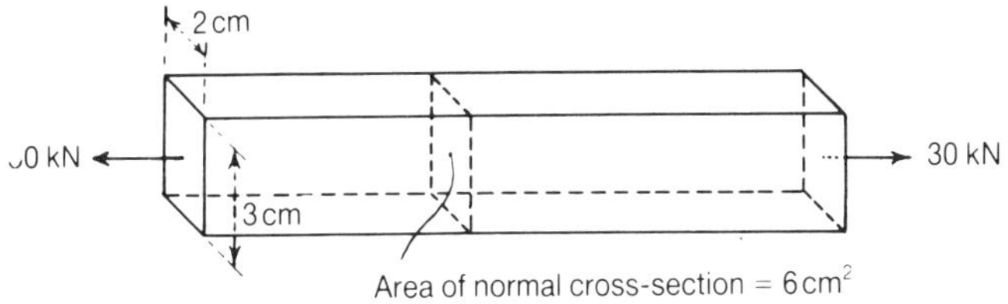

Solution

The area of a normal cross-section of the bar is

$$A = 0.03 \times 0.02 = 0.6 \times 10^{-3}\ \text{m}^2$$

The average tensile stress over this cross-section is then

$$\sigma = \frac{P}{A} = \frac{30 \times 10^3}{0.6 \times 10^{-3}} = 50\text{MN/m}^2$$

Problem 1.2 A steel bolt, 2.50 cm in diameter, carries a tensile load of 40 kN. Estimate the average tensile stress at the section *a* and at the screwed section *b*, where the diameter at the root of the thread is 2.10 cm.

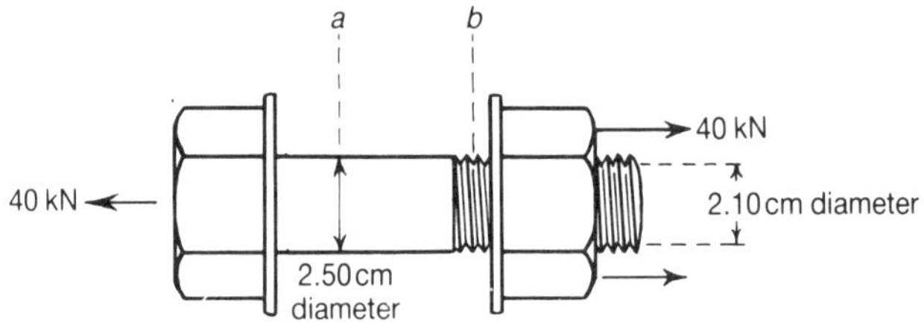

Solution

The cross-sectional area of the bolt at the section *a* is

$$A_a = \frac{\pi}{4}(0.025)^2 = 0.491 \times 10^{-3}\ \text{m}^2$$

The average tensile stress at A is then

$$\sigma_a = \frac{P}{A_a} = \frac{40 \times 10^3}{0.491 \times 10^{-3}} = 81.4\ \text{MN/m}^2$$

The cross-sectional area at the root of the thread, section *b*, is

$$A_b = \frac{\pi}{4}(0.021)^2 = 0.346 \times 10^{-3}\ \text{m}^2$$

The average tensile stress over this section is

$$\sigma_b = \frac{P}{A_b} = \frac{40 \times 10^3}{0.346 \times 10^{-3}} = 115.6\ \text{MN/m}^2$$

1.4 Tensile and compressive strains

In the steel wire experiment of Figure 1.1 we discussed the extension of the whole wire. If we measure the extension of, say, the lowest quarter-length of the wire we find that for a given load it is equal to a quarter of the extension of the whole wire. In general we find that, at a given load, the ratio of the extension of any length to that length is constant for all parts of the wire; this ratio is known as the *tensile strain*.

Suppose the initial unstrained length of the wire is L_0, and the e is the extension due to straining; the tensile strain ε is defined as

$$\varepsilon = \frac{e}{L_0} \tag{1.3}$$

This definition of strain is useful only for small distortions, in which the extension e is small compared with the original length L_0; this definition is adequate for the study of most engineering problems, where we are concerned with values of ε of the order 0.001, or so.

If a material is compressed the resulting strain is defined in a similar way, except that e is the contraction of a length.

We note that strain is a *non-dimensional* quantity, being the ratio of the extension, or contraction, of a bar to its original length.

Problem 1.3 A cylindrical block is 30 cm long and has a circular cross-section 10 cm in diameter. It carries a total compressive load of 70 kN, and under this load it contracts by 0.02 cm. Estimate the average compressive stress over a normal cross-section and the compressive strain.

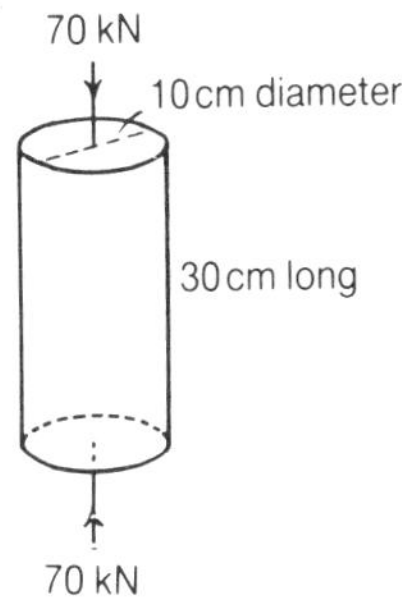

<u>*Solution*</u>

The area of a normal cross-section is

$$A = \frac{\pi}{4}(0.10)^2 = 7.85 \times 10^{-3}\ \mathrm{m}^2$$

The average compressive stress over this cross-section is then

$$\sigma = \frac{P}{A} = \frac{70 \times 10^3}{7.85 \times 10^{-3}} = 8.92\ \mathrm{MN/m^2}$$

The average compressive strain over the length of the cylinder is

$$\varepsilon = \frac{0.02 \times 10^{-2}}{30 \times 10^{-2}} = 0.67 \times 10^{-3}$$

1.5 Stress–strain curves for brittle materials

Many of the characteristics of a material can be deduced from the tensile test. In the experiment of Figure 1.1 we measured the extensions of the wire for increasing loads; it is more convenient to compare materials in terms of stresses and strains, rather than loads and extensions of a particular specimen of a material.

The tensile *stress–strain* curve for a high-strength steel has the form shown in Figure 1.5. The stress at any stage is the ratio of the load of the *original* cross-sectional area of the test specimen; the strain is the elongation of a unit length of the test specimen. For stresses up to about 750 $\mathrm{MN/m^2}$ the stress–strain curve is linear, showing that the material obeys Hooke's law in this range; the material is also elastic in this range, and no permanent extensions remain after removal of the stresses. The ratio of stress to strain for this linear region is usually about 200 $\mathrm{GN/m^2}$ for steels; this ratio is known as *Young's modulus* and is denoted by *E*. The strain at the limit of proportionality is of the order 0.003, and is small compared with strains of the order 0.100 at fracture.

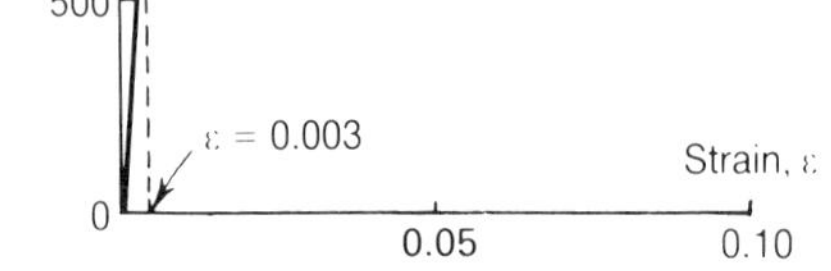

Figure 1.5 Tensile stress–strain curve for a high-strength steel.

We note that *Young's modulus* has the units of a stress; the value of E defines the constant in the linear relation between stress and strain in the elastic range of the material. We have

$$E = \frac{\sigma}{\varepsilon} \tag{1.4}$$

for the linear-elastic range. If P is the total tensile load in a bar, A its cross-sectional area, and L_0 its length, then

$$E = \frac{\sigma}{\varepsilon} = \frac{P / A}{e / L_0} \tag{1.5}$$

where e is the extension of the length L_0. Thus the expansion is given by

$$e = \frac{PL_0}{EA} \tag{1.6}$$

If the material is stressed beyond the linear-elastic range the limit of proportionality is exceeded, and the strains increase non-linearly with the stresses. Moreover, removal of the stress leaves the material with some permanent extension; this range is then both non-linear and inelastic. The maximum stress attained may be of the order of 1500 MN/m^2, and the total extension, or *elongation*, at this stage may be of the order of 10%.

The curve of Figure 1.5 is typical of the behaviour of *brittle* materials—as, for example, area characterized by small permanent elongation at the breaking point; in the case of metals this is usually 10%, or less.

When a material is stressed beyond the limit of proportionality and is then unloaded, permanent deformations of the material take place. Suppose the tensile test-specimen of Figure 1.5 is stressed beyond the limit of proportionality, (point a in Figure 1.6), to a point b on the stress–strain diagram. If the stress is now removed, the stress–strain relation follows the curve bc; when the stress is completely removed there is a residual strain given by the intercept $0c$ on the ε-axis. If the stress is applied again, the stress–strain relation follows the curve cd initially, and finally the curve df to the breaking point. Both the unloading curve bc and the reloading curve cd are approximately parallel to the elastic line $0a$; they are curved slightly in opposite directions. The process of unloading and reloading, bcd, had little or no effect on the stress at the breaking point, the stress–strain curve being interrupted by only a small amount bd, Figure 1.6.

The stress–strain curves of brittle materials for tension and compression are usually similar in form, although the stresses at the limit of proportionality and at fracture may be very different for the two loading conditions. Typical tensile and compressive stress–strain curves for concrete are shown in Figure 1.7; the maximum stress attainable in tension is only about one-tenth of that in compression, although the slopes of the stress–strain curves in the region of zero stress are nearly equal.

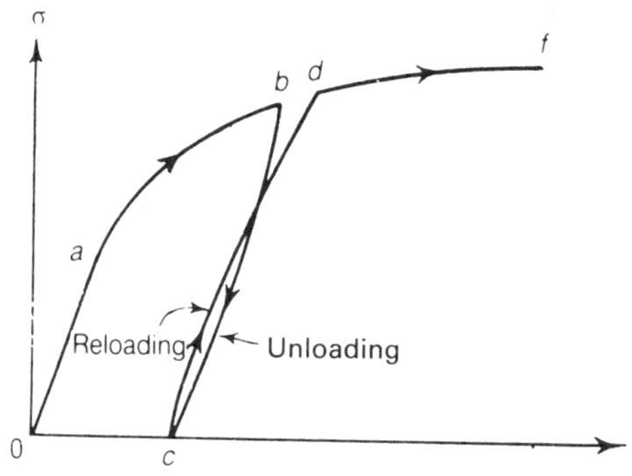

Figure 1.6 Unloading and reloading of a material in the inelastic range; the paths *bc* and *cd* are approximately parallel to the linear-elastic line *oa*.

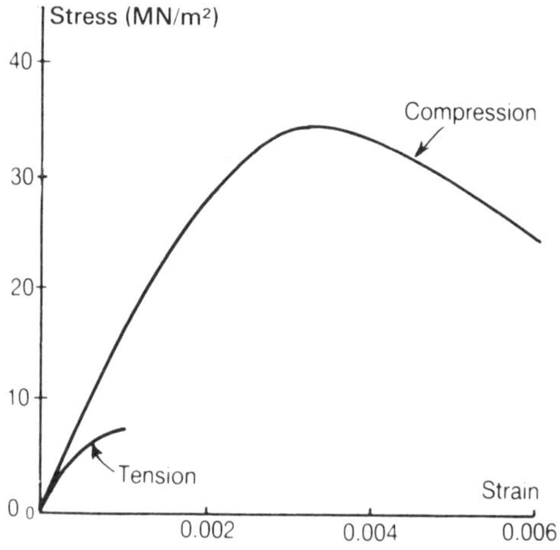

Figure 1.7 Typical compressive and tensile stress–strain curves for concrete, showing the comparative weakness of concrete in tension.

1.6 Ductile materials *(see Section 1.8)*

A brittle material is one showing relatively little elongation at fracture in the tensile test; by contrast some materials, such as mild steel, copper, and synthetic polymers, may be stretched appreciably before breaking. These latter materials are *ductile* in character.

If tensile and compressive tests are made on a mild steel, the resulting stress–strain curves are different in form from those of a brittle material, such as a high-strength steel. If a tensile test specimen of mild steel is loaded axially, the stress–strain curve is linear and elastic up to a point *a*, Figure 1.8; the small strain region of Figure 1.8. is reproduced to a larger scale in Figure 1.9. The ratio of stress to strain, or Young's modulus, for the linear portion 0*a* is usually about 200 GN/m^2, ie, 200 $\times 10^9$ N/m^2. The tensile stress at the point *a* is of order 300 MN/m^2, i.e. 300 $\times$ 10^6 N/m^2. If the test specimen is strained beyond the point *a*, Figures 1.8 and 1.9, the stress must be reduced almost immediately to maintain equilibrium; the reduction of stress, *ab*, takes place rapidly, and the form of the curve *ab* is difficult to define precisely. Continued straining proceeds at a roughly constant stress along *bc*. In the range of strains from *a* to *c* the material is said to *yield*; *a* is the *upper yield point*, and *b* the *lower yield point*. Yielding at constant stress along *bc* proceeds usually to a strain about 40 times greater than that at *a*; beyond the point *c* the material *strain-hardens*, and stress again increases with strain where the slope from *c* to *d* is about 1/50th that from 0 to *a*. The stress for a tensile specimen attains a maximum value at *d* if the stress is evaluated on the basis of the original cross-sectional area of the bar; the stress corresponding to the point *d* is known as the *ultimate stress*, σ_{ult}, of the material. From *d* to *f* there is a reduction in the nominal stress until fracture occurs at *f*. The ultimate stress in tension is attained at a stage when *necking* begins; this is a reduction of area at a relatively weak cross-section of the test specimen. It is usual to measure the diameter of the neck after fracture, and to evaluate a true stress at fracture, based on the breaking load and the reduced cross-sectional area at the neck. Necking and considerable elongation before fracture are characteristics of ductile materials; there is little or no necking at fracture for brittle materials.

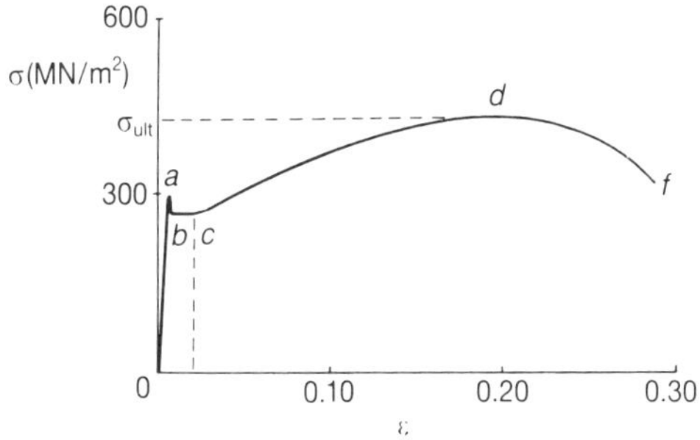

Figure 1.8 Tensile stress–strain curve for an annealed mild steel, showing the drop in stress at yielding from the upper yield point *a* to the lower yield point *b*.

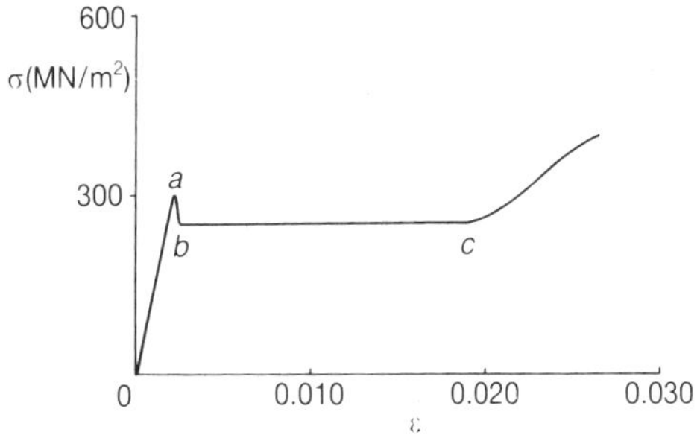

Figure 1.9 Upper and lower yield points of a mild steel.

Compressive tests of mild steel give stress–strain curves similar to those for tension. If we consider tensile stresses and strains as positive, and compressive stresses and strains as negative, we can plot the tensile and compressive stress–strain curves on the same diagram; Figure 1.10 shows the stress–strain curves for an annealed mild steel. In determining the stress–strain curves experimentally, it is important to ensure that the bar is loaded axially; with even small eccentricities of loading the stress distribution over any cross-section of the bar is non-uniform, and the upper yield point stress is not attained in all fibres of the material simultaneously. For this reason the lower yield point stress is taken usually as a more realistic definition of yielding of the material.

Some ductile materials show no clearly defined upper yield stress; for these materials the limit of proportionality may be lower than the stress for continuous yielding. The term *yield stress* refers to the stress for continuous yielding of a material; this implies the lower yield stress for a material in which an upper yield point exists; the yield stress is denoted by σ_Y.

Tensile failures of some steel bars are shown in Figure 1.11; specimen (ii) is a brittle material, showing little or no necking at the fractured section; specimens (i) and (iii) are ductile steels showing a characteristic necking at the fractured sections. The tensile specimens of Figure 1.12 show the forms of failure in a ductile steel and a ductile light-alloy material; the steel specimen (i) fails at a necked section in the form of a 'cup and cone'; in the case of the light-alloy bar, two 'cups' are formed. The compressive failure of a brittle cast iron is shown in Figure 1.13. In the case of a mild steel, failure in compression occurs in a 'barrel-like' fashion, as shown in Figure 1.14.

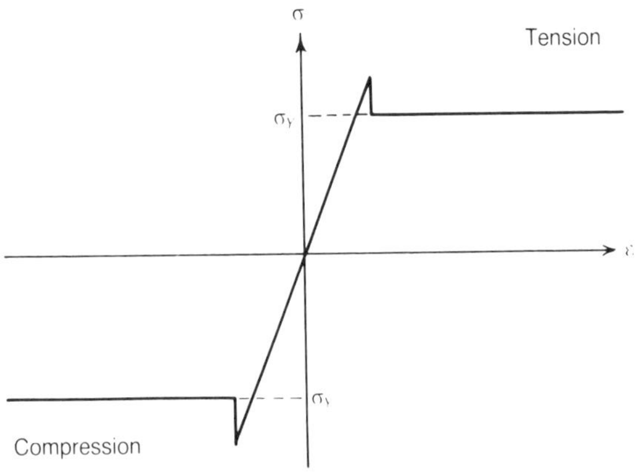

Figure 1.10 Tensile and compressive stress–strain curves for an annealed mild steel; in the annealed condition the yield stresses in tension and compression are approximately equal.

The stress–strain curves discussed in the preceding paragraph refer to static tests carried out at negligible speed. When stresses are applied rapidly the yield stress and ultimate stresses of metallic materials are usually raised. At a strain rate of 100 per second the yield stress of a mild steel may be twice that at negligible speed.

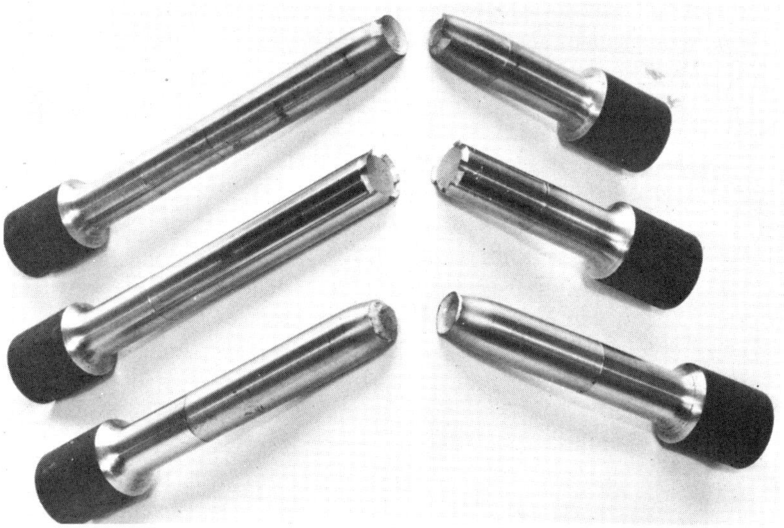

(i)

(ii)

(iii)

Figure 1.11 Tensile failures in steel specimens showing necking in mild steel, (i) and (iii), and brittle fracture in high-strength steel, (ii).

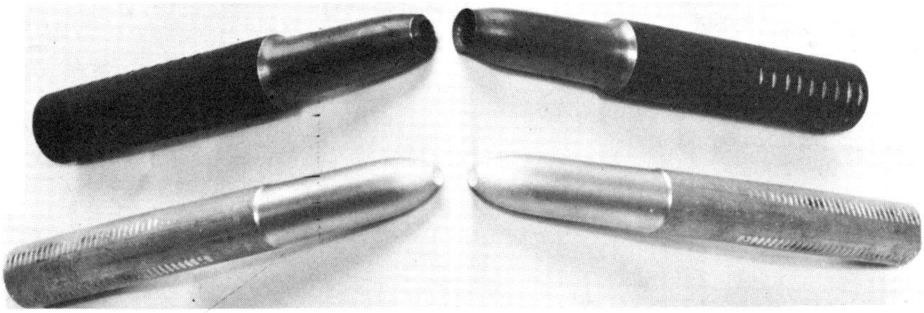

(i)

(ii)

Figure 1.12 Necking in tensile failures of ductile materials.
(i) Mild-steel specimen showing 'cup and cone' at the broken section.
(ii) Aluminium-alloy specimen showing double 'cup' type of failure.

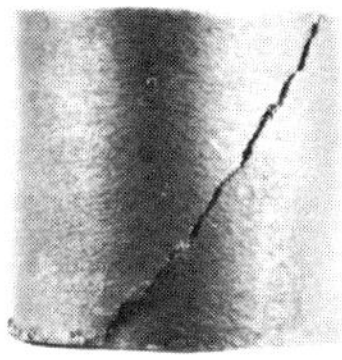

Figure 1.13 Failure in compression of a circular specimen of cast iron, showing fracture on a diagonal plane.

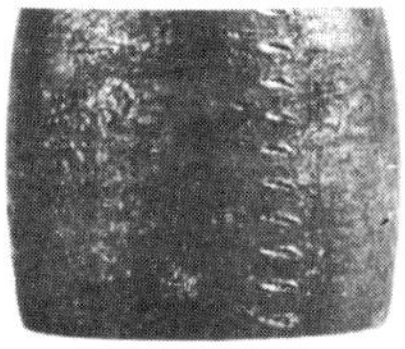

Figure 1.14 Barrel-like failure in a compressed specimen of mild steel.

Problem 1.4 A tensile test is carried out on a bar of mild steel of diameter 2 cm. The bar yields under a load of 80 kN. It reaches a maximum load of 150 kN, and breaks finally at a load of 70 kN.

Estimate:

(i) the tensile stress at the yield point;
(ii) the ultimate tensile stress;
(iii) the average stress at the breaking point, if the diameter of the fractured neck is 1 cm.

Solution

The original cross-section of the bar is

$$A_0 = \frac{\pi}{4}(0.020)^2 = 0.314 \times 10^{-3}\ \mathrm{m}^2$$

(i) The average tensile stress at yielding is then

$$\sigma_Y = \frac{P_Y}{A_0} = \frac{80 \times 10^3}{0.314 \times 10^{-3}} = 254\ \mathrm{MN/m^2},$$

where P_Y = load at the yield point

(ii) The ultimate stress is the nominal stress at the maximum load, i.e.,

$$\sigma_{ult} = \frac{P_{max}}{A_0} = \frac{150 \times 10^3}{0.314 \times 10^{-3}} = 477\ \mathrm{MN/m^2}$$

where P_{max} = maximum load

(iii) The cross-sectional area in the fractured neck is

$$A_f = \frac{\pi}{4}(0.010)^2 = 0.0785 \times 10^{-3}\ \mathrm{m}^2$$

The average stress at the breaking point is then

$$\sigma_f = \frac{P_f}{A_f} = \frac{70 \times 10^3}{0.0785 \times 10^{-3}} = 892\ \mathrm{MN/m^2},$$

where P_f = final breaking load.

Problem 1.5 A circular bar of diameter 2.50 cm is subjected to an axial tension of 20 kN. If the material is elastic with a Young's modulus $E = 70\ \text{GN/m}^2$, estimate the percentage elongation.

Solution

The cross-sectional area of the bar is

$$A = \frac{\pi}{4}(0.025)^2 = 0.491 \times 10^{-3}\ \text{m}^2$$

The average tensile stress is then

$$\sigma = \frac{P}{A} = \frac{20 \times 10^3}{0.491 \times 10^{-3}} = 40.7\ \text{MN/m}^2$$

The longitudinal tensile strain will therefore be

$$\varepsilon = \frac{\sigma}{E} = \frac{40.7 \times 10^6}{70 \times 10^9} = 0.582 \times 10^{-3}$$

The percentage elongation will therefore be

$$(0.582 \times 10^{-3})\,100 = 0.058\%$$

Problem 1.6 The piston of a hydraulic ram is 40 cm diameter, and the piston rod 6 cm diameter. The water pressure is $1\ \text{MN/m}^2$. Estimate the stress in the piston rod and the elongation of a length of 1 m of the rod when the piston is under pressure from the piston-rod side. Take Young's modulus as $E = 200\ \text{GN/m}^2$.

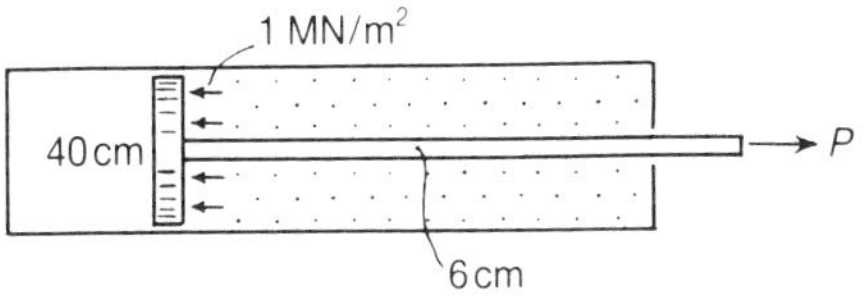

Solution

The pressure on the back of the piston acts on a net area

$$\frac{\pi}{4}\left[(0.40)^2 - (0.06)^2\right] = \frac{\pi}{4}(0.46)(0.34) = 0.123 \text{ m}^2$$

The load on the piston is then

$$P = (1)(0.123) = 0.123 \text{ MN}$$

Area of the piston rod is

$$A = \frac{\pi}{4}(0.060)^2 = 0.283 \times 10^{-2} \text{ m}^2$$

The average tensile stress in the rod is then

$$\sigma = \frac{P}{A} = \frac{0.123 \times 10^6}{0.283 \times 10^{-2}} = 43.5 \text{ MN/m}^2$$

From equation (1.6), the elongation of a length $L = 1$ m is

$$e = \frac{PL}{EA} = \frac{P}{A}\left(\frac{L}{E}\right) = \frac{\sigma L}{E}$$

$$= \frac{(43.5 \times 10^6)(1)}{200 \times 10^9}$$

$$= 0.218 \times 10^{-3} \text{ m}$$

$$= 0.0218 \text{ cm}$$

Problem 1.7 The steel wire working a signal is 750 m long and 0.5 cm diameter. Assuming a pull on the wire of 1.5 kN, find the movement which must be given to the signal-box end of the wire if the movement at the signal end is to be 17.5 cm. Take Young's modulus as 200 GN/m^2.

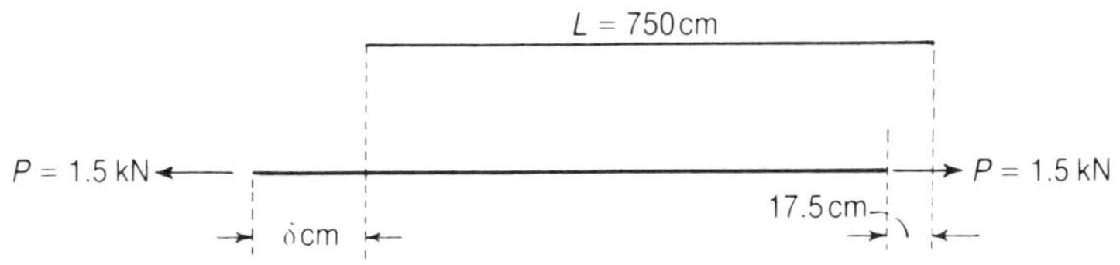

Solution

If δ(cm) is the movement at the signal-box end, the actual stretch of the wire is $e = (\delta - 17.5)$cm

The longitudinal strain is then

$$\varepsilon = \frac{(\delta - 17.5)\ 10^{-2}}{750}$$

Now the cross-sectional area of the wire is

$$A = \frac{\pi}{4}(0.005)^2 = 0.0196 \times 10^{-3}\ \text{m}^2$$

The longitudinal strain can also be defined in terms of the tensile load, namely,

$$\varepsilon = \frac{e}{L} = \frac{P}{EA} = \frac{1.5 \times 10^3}{(200 \times 10^9)(0.0196 \times 10^{-3})}$$

$$= 0.383 \times 10^{-3}$$

On equating these two values of ε,

$$\frac{(\delta - 17.5)\ 10^{-2}}{750} = 0.383 \times 10^{-3}$$

The equation gives

$$\delta = 46.2\ \text{cm}$$

Problem 1.8 A circular, metal rod of diameter 1 cm is loaded in tension. When the tensile load is 5kN, the extension of a 25 cm length is measured accurately and found to be 0.0227 cm. Estimate the value of Young's modulus, E, of the metal.

Solution

The cross-sectional area is

$$A = \frac{\pi}{4}(0.01)^2 = 0.0785 \times 10^{-3} \text{ m}^2$$

The tensile stress is then

$$\sigma = \frac{P}{A} = \frac{5 \times 10^3}{0.0785 \times 10^{-3}} = 63.7 \text{ MN/m}^2$$

The measured tensile strain is

$$\varepsilon = \frac{e}{L} = \frac{0.0227 \times 10^{-2}}{25 \times 10^{-2}} = 0.910 \times 10^{-3}$$

Then Young's modulus is defined by

$$E = \frac{\sigma}{\varepsilon} = \frac{63.7 \times 10^6}{0.91 \times 10^{-3}} = 70 \text{ GN/m}^2$$

Problem 1.9 A straight, uniform rod of length L rotates at uniform angular speed ω about an axis through one end and perpendicular to its length. Estimate the maximum tensile stress generated in the rod and the elongation of the rod at this speed. The density of the material is ρ and Young's modulus is E.

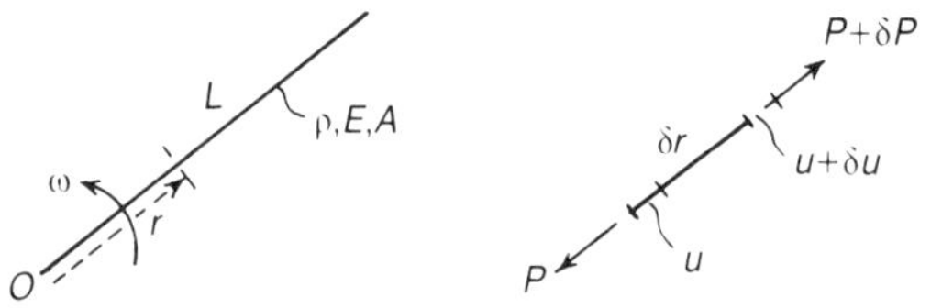

<u>*Solution*</u>

Suppose the radial displacement of any point a distance r from the axis of rotation is u. The radial displacement a distance $r + \delta r$) from 0 is then $(u + \delta u)$, and the elemental length δr of the rod is stretched therefore an amount δu. The longitudinal strain of this element is therefore

$$\varepsilon = \lim_{\delta r \to 0} \frac{\delta u}{\delta r} = \frac{\mathrm{du}}{\mathrm{dr}}$$

The longitudinal stress in the elemental length is then

$$\sigma = E\varepsilon = E \frac{\mathrm{du}}{\mathrm{dr}}$$

If A is the cross-sectional area of the rod, the longitudinal load at any radius r is then

$$P = \sigma A = EA \frac{\mathrm{du}}{\mathrm{dr}}$$

The centrifugal force acting on the elemental length δr is

$$(\rho A \, \delta r) \, \omega^2 r$$

Then, for radial equilibrium of the elemental length,

$$\delta P + \rho A \omega^2 \, r \, \delta r = 0$$

This gives

$$\frac{dP}{dr} = -\rho A \omega^2 r$$

On integrating, we have

$$P = -\frac{1}{2} \rho A \omega^2 r^2 + C$$

where C is an arbitrary constant; if $P = 0$ at the remote end, $r = L$, of the rod, then

$$C = \frac{1}{2} \rho A \omega^2 L^2$$

and

$$P = \frac{1}{2}\rho A \omega^2 L^2 \left(1 - \frac{r^2}{L^2}\right)$$

The tensile stress at any radius is then

$$\sigma = \frac{P}{A} = \frac{1}{2}\rho \omega^2 L^2 \left(1 - \frac{r^2}{L^2}\right)$$

This is greatest at the axis of rotation, $r = 0$, so that

$$\sigma_{max} = \frac{1}{2}\rho \omega^2 L^2$$

The longitudinal stress, σ, is defined by

$$\sigma = E \frac{du}{dr}$$

so

$$\frac{du}{dr} = \frac{\sigma}{E} = \frac{\rho\, \omega^2 L^2}{2E}\left(1 - \frac{r^2}{L^2}\right)$$

On integrating,

$$u = \frac{\rho\, \omega^2 L^2}{2E}\left(r - \frac{r^3}{3L^2} + D\right)$$

where D is an arbitrary constant; if there is no radial movement at 0, then $u = 0$ at $= r = 0$, and we have $D = 0$.

Thus

$$u = \frac{\rho\, \omega^2 L^2}{2E}\left[r\left(1 - \frac{r^2}{3L^2}\right)\right]$$

At the remote end, $r = L$,

$$u_L = \frac{\rho\, \omega^2 L^2}{2E}\left[L\left(\frac{2}{3}\right)\right] = \frac{\rho\, \omega^2 L^3}{3E}$$

1.7 Proof stresses

Many materials show no well-defined yield stresses when tested in tension or compression. A typical stress–strain curve for an aluminium alloy is shown in Figure 1.15.

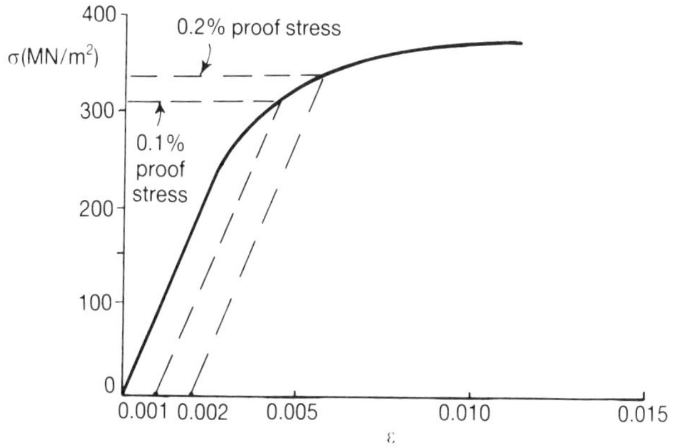

Figure 1.15 Proof stresses of an aluminium-alloy material; the proof stress is found by drawing the line parallel to the linear-elastic line at the appropriate proof strain.

The limit of proportionality is in the region of 300 MN/m^2, but the exact position of this limit is difficult to determine experimentally. To overcome this problem a *proof stress* is defined; the 0.1% proof stress required to produce a permanent strain of 0.001 (or 0.1%) on removal of the stress. Suppose we draw a line from the point 0.001 on the strain axis, Figure 1.15, parallel to the elastic line of the material; the point where this line cuts the stress–strain curve defines the proof stress. The 0.2% proof stress is defined in a similar way.

1.8 Ductility measurement

The Ductility value of a material can be described as the ability of the material to suffer plastic deformation while still being able to resist applied loading. The more ductile a material is the more it is said to have the ability to deform under applied loading.

The ductility of a metal is usually measured by its percentage reduction in cross-sectional area or by its percentage increase in length, i.e.

$$\text{percentage reduction in area} = \frac{(A_I - A_F)}{A_I} \times 100\%$$

and

$$\text{percentage increase in length} = \frac{(L_I - L_F)}{L_I} \times 100\%$$

where

A_I = initial cross-sectional area of the tensile specimen

A_F = final cross-sectional area of the tensile specimen

L_I = initial gauge length of the tensile specimen

L_F = final gauge length of the tensile specimen

It should be emphasised that the shape of the tensile specimen plays a major role on the measurement of the ductility and some typical relationships between length and character for tensile specimens i.e. given in Table 1.1

Materials such as copper and mild steel have high ductility and brittle materials such as bronze and cast iron have low ductility.

Table 1.1 Circular cylindrical tensile specimens

Place	L_I	L_I/D_I*
UK	$4\sqrt{\text{area}}$	3.54
USA	$4.51\sqrt{\text{area}}$	4.0
Europe	$5.65\sqrt{\text{area}}$	5.0

area = cross-sectional area

* D_I = initial diameter of the tensile specimen

1.9 Working stresses

In many engineering problems the loads sustained by a component of a machine or structure are reasonably well-defined; for example, the lower stanchions of a tall building support the weight of material forming the upper storeys. The stresses which are present in a component, under normal working conditions, are called the *working stresses*; the ratio of the yield stress, σ_Y, of a material to the largest working stress, σ_W, in the component is the *stress factor* against yielding. The stress factor on yielding is then

$$\frac{\sigma_Y}{\sigma_W} \tag{1.7}$$

If the material has no well-defined yield point, it is more convenient to use the *proof stress*, σ_p; the stress factor on proof stress is then

$$\frac{\sigma_P}{\sigma_W} \tag{1.8}$$

Some writers refer to the stress factor defined above as a 'safety factor'. It is preferable, however, to avoid any reference to 'safe' stresses, as the degree of safety in any practical problem is difficult to define. The present writers prefer the term 'stress factor' as this defines more precisely that the working stress is compared with the yield, or proof stress of the material. Another reason for using 'stress factor' will become more evident after the reader has studied Section 1.10.

1.10 Load factors

The *stress factor* in a component gives an indication of the working stresses in relation to the yield, or proof, stress of the material. In practical problems working stresses can only be estimated approximately in stress calculations. For this reason the stress factor may give little indication of the degree of safety of a component.

A more realistic estimate of safety can be made by finding the extent to which the working loads on a component may be increased before collapse or fracture occurs. Consider, for example, the continuous beam in Figure 1.16, resting on three supports. Under working conditions the beam carries lateral loads P_1, P_2 and P_3, Figure 1.16(i). If all these loads can be increased simultaneously by a factor n before collapse occurs, the load factor against collapse is n. In some complex structural systems, as for example continuous beams, the collapse loads, such as nP_1, nP_2 and nP_3, can be estimated reasonably accurately; the value of the load factor can then be deduced to give working loads P_1, P_2 and P_3.

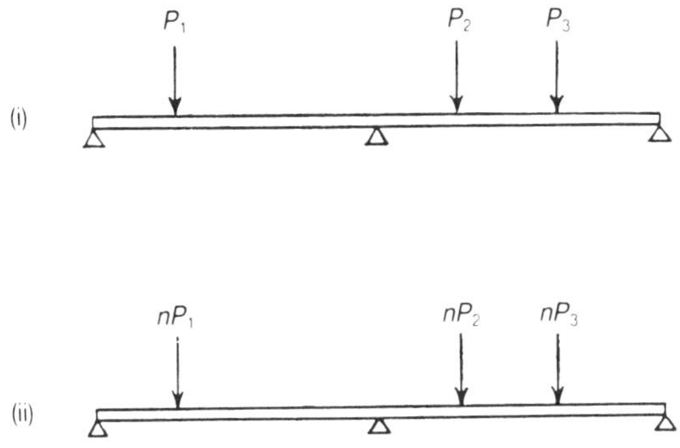

Figure 1.16 Factored loads on a continuous beam.
(i) Working loads. (ii) Factored working loads leading to collapse.

1.11 Lateral strains due to direct stresses

When a bar of a material is stretched longitudinally—as in a tensile test—the bar extends in the direction of the applied load. This longitudinal extension is accompanied by a lateral contraction of the bar, as shown in Figure 1.17. In the linear-elastic range of a material the lateral strain is proportional to the longitudinal strain; if ε_x is the longitudinal strain of the bar, then the lateral strain is

$$\varepsilon_y = \nu\varepsilon_x \qquad (1.9)$$

The constant ν in this relationship is known as *Poisson's ratio*, and for most metals it has a value of about 0.3 in the linear-elastic range; it cannot exceed a value of 0.5. For concrete it has a value of about 0.1. If the longitudinal strain is tensile, the lateral strain is a contraction; for a compressed bar there is a lateral expansion.

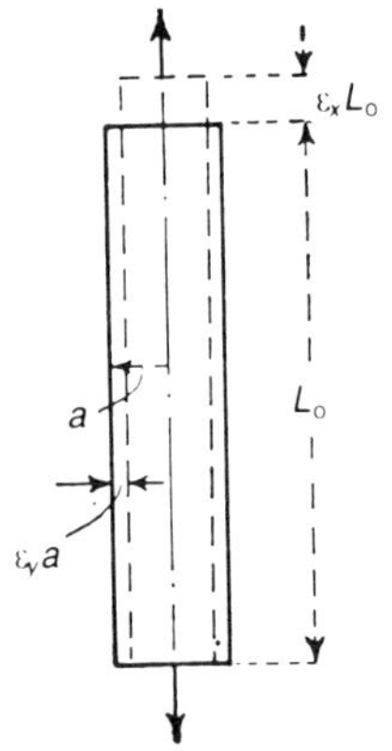

Figure 1.17 The Poisson ratio effect leading to lateral contraction of a bar in tension.

With a knowledge of the lateral contraction of a stretched bar it is possible to calculate the change in volume due to straining. The bar of Figure 1.17 is assumed to have a square cross-section of side a; L_0 is the unstrained length of the bar. When strained longitudinally an amount ε_x, the corresponding lateral strain of contractions is ε_y. The bar extends therefore an amount $\varepsilon_x L_0$, and each side of the cross-section contracts an amount $\varepsilon_y a$. The volume of the bar before stretching is

$$V_0 = a^2 L_0$$

After straining the volume is

$$V = (a - \varepsilon_y a)^2 \, (L_0 + \varepsilon_x L_0)$$

which may be written

$$V = a^2 L_0 (1 - \varepsilon_y)^2 \, (1 + \varepsilon_x) = V_0 (1 - \varepsilon_y)^2 \, (1 + \varepsilon_y)$$

If ε_x and ε_y are small quantities compared to unit, we may write

$$(1 - \varepsilon_y)^2 \, (1 + \varepsilon_x) = (1 - 2\,\varepsilon_y)\,(1 + \varepsilon_x) = 1 + \varepsilon_x - 2\,\varepsilon_y$$

ignoring squares and products of ε_x and ε_y. The volume after straining is then

$$V = V_0 \, (1 + \varepsilon_x - 2\,\varepsilon_y)$$

The *volumetric strain* is defined as the ratio of the change of volume to the original volume, and is therefore

$$\frac{V - V_0}{V_0} = \varepsilon_x - 2\,\varepsilon_y \tag{1.10}$$

If $\varepsilon_y = v\,\varepsilon_{xy}$ then the volumetric strain is $\varepsilon_x(1 - 2v)$. Equation (1.10) shows why v cannot be greater than 0.5; if it were, then under *compressive* hydrostatic stress a *positive* volumetric strain will result, which is impossible.

Problem 1.10 A bar of steel, having a rectangular cross-section 7.5 cm by 2.5 cm, carries an axial tensile load of 180 kN. Estimate the decrease in the length of the sides of the cross-section if Young's modulus, E, is 200 GN/m^2 and Poisson's ratio, v, is 0.3.

Solution

The cross-sectional area is

$$A = (0.075)(0.025) = 1.875 \times 10^{-3}\ \text{m}^2$$

The average longitudinal tensile stress is

$$\sigma = \frac{P}{A} = \frac{180 \times 10^3}{1.875 \times 10^{-3}} = 96.0\ \text{MN/m}^2$$

The longitudinal tensile strain is therefore

$$\varepsilon = \frac{\sigma}{E} = \frac{96.0 \times 10^6}{200 \times 10^9} = 0.48 \times 10^{-3}$$

The lateral strain is therefore

$$v\varepsilon = 0.3(0.48 \times 10^{-3}) = 0.144 \times 10^{-3}$$

The 7.5 cm side then contracts by an amount

$$(0.075)(0.144 \times 10^{-3}) = 0.0108 \times 10^{-3}\ \text{m} = 0.00108\ \text{cm}$$

The 2.5 cm side contracts by an amount

$$(0.025)(0.144 \times 10^{-3}) = 0.0036 \times 10^{-3}\ \text{m} = 0.00036\ \text{cm}$$

1.12 Strength properties of some engineering materials

The mechanical properties of some engineering materials are given in Table 1.2. Most of the materials are in common engineering use, including a number of relatively new and important materials; namely glass-fibre composites, carbon-fibre composites and boron composites. In the case of some brittle materials, such as cast iron and concrete, the ultimate stress in tension is considerably smaller than in compression.

Composite materials, such as glass fibre reinforced plastics, (GRP), carbon-fibre reinforced plastics (CFRP), boron-fibre reinforced plastics, 'Kevlar' and metal-matrix composites are likely to revolutionise the design and construction of many structures in the 21st century. The glass fibres used in GRP are usually made from a borosilicate glass, similar to the glass used for cooking utensils. Borosilicate glass fibres are usually produced in 'E' glass or glass that has good electrical resistance. A very strong form of borosilicate glass fibre appears in the form of 'S' glass which is much more expensive than 'E' glass.

Some carbon fibres, namely high modulus (HM) carbon fibres , have a tensile modulus much larger than high strength steels, whereas other carbon fibres have a very high tensile strength (HS) much larger than high tensile steels.

Currently 'S' glass is some eight times more expensive than 'E' glass and HS carbon is about 50 times more expensive than 'E' glass. HM carbon is some 250 times more expensive than 'E' glass while 'Kevlar' is some 15 times more expensive than 'E' glass.

1.13 Weight and stiffness economy of materials

In some machine components and structures it is important that the weight of material should be as small as possible. This is particularly true of aircraft, submarines and rockets, for example, in which less structural weight leads to a larger pay–load. If σ_{ult} is the ultimate stress of a material in tension and ρ is its density, then a measure of the strength economy is the ratio

$$\frac{\sigma_{ult}}{\rho}$$

The materials shown in Table 1.2 are compared on the basis of strength economy in Table 1.3 from which it is clear that the modern fibre-reinforced composites offer distinct savings in weight over the more common materials in engineering use.

In some engineering applications, stiffness rather than strength is required of materials; this is so in structures likely to buckle and components governed by deflection limitations. A measure of the stiffness economy of a material is the ratio

$$\frac{E}{\rho},$$

some values of which are shown in Table 1.2. Boron composites and carbon-fibre composites show outstanding stiffness properties, whereas glass-fibre composites fall more into line with the best materials already in common use.

Table 1.2 Approximate strength properties of some engineering materials

Material	Limit of proportionality (MN/m^2)	Ultimate stress σ_{ult} (MN/m^2)	Elongation at tensile fracture (as a fraction of the original length)	Young's modulus E (GN/m^2)	Density ρ (kg/m^3)	σ_{ult}/ρ $(m/s)^2$	E/ρ $(m/s)^2$	Coefficient of linear expansion α (per °C)
Medium-strength mild steel	280	370	0.30	200	7840	47×10^3	25×10^6	1.2×10^{-5}
High-strength steel	770	1550	0.10	200	7840	198×10^3	25×10^6	1.3×10^{-5}
Medium-strength aluminium alloy	230	430	0.10	70	2800	154×10^3	25×10^6	2.3×10^{-5}
Titanium alloy	385	690	0.15	120	4500	153×10^3	27×10^6	0.9×10^{-5}
Magnesium alloy	155	280	0.08	45	1800	156×10^3	25×10^6	2.7×10^{-5}
Wrought iron	185	310	—	190	7670	40×10^3	25×10^6	1.2×10^{-5}
Cast iron } tension	—	155	—	140	7200	—	20×10^6	1.1×10^{-5}
} compression	—	700	—	140	7200	$97^* \times 10^3$	20×10^6	1.1×10^{-5}
Concrete } tension	—	3.0	—	14	2410	—	6×10^6	1.2×10^{-5}
} compression	—	30.0	—	14	2410	$12^* \times 10^3$	6×10^6	1.2×10^{-5}
Nylon (polyamide)	77	90	1.00	2		79×10^3	1.8×10^6	10×10^{-5}
Polystyrene	46	60	0.03	3.5	1050	57×10^3	3.3×10^6	10×10^{-5}
Fluon (tetrafluoroethylene)	8	15	2.00	0.4	2220	7×10^3	0.2×10^6	11×10^{-5}
Polythene (ethylene)	6	12	5.00	0.2	915	13×10^3	0.2×10^6	28×10^{-5}
High-strength glass-fibre composite	—	1600	—	60	2000	800×10^3	30×10^6	—
Carbon-fibre composite	—	1400	—	170	1600	875×10^3	105×10^6	—
Boron composite	—	1300	—	270	2000	650×10^3	135×10^6	—

* Evaluate on the compressive value of σ_{ult}.

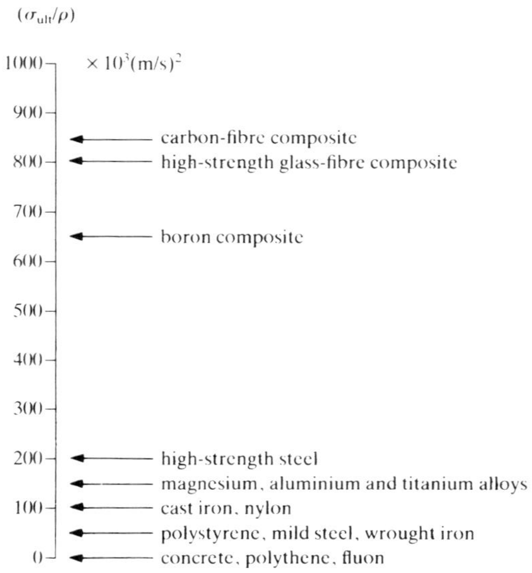

Table 1.3 Strength economy of some engineering materials

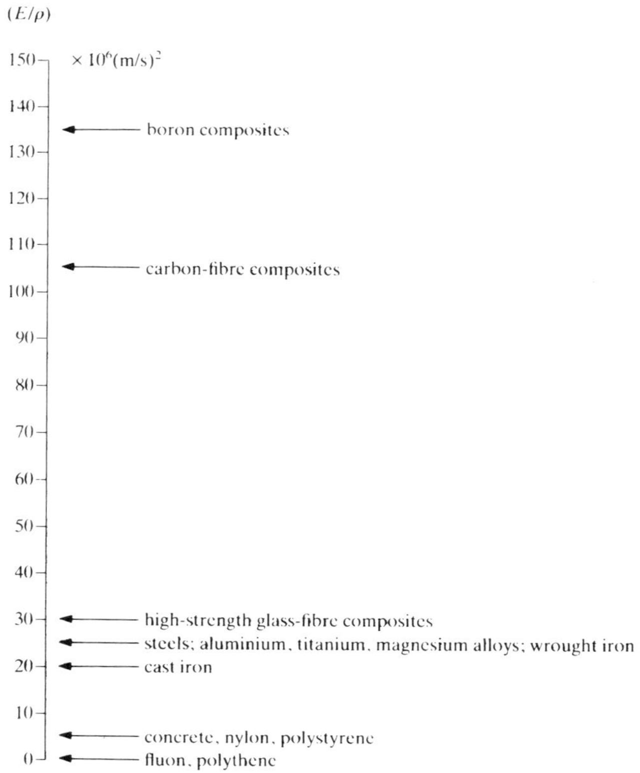

Table 1.4 Stiffness economy of some engineering materials

1.14 Strain energy and work done in the tensile test

As a tensile specimen extends under load, the forces applied to the ends of the test specimen move through small distances. These forces perform work in stretching the bar. If, at a tensile load P, the bar is stretched a small additional amount δe, Figure 1.18, then the work done on the bar is approximately $P\delta e$

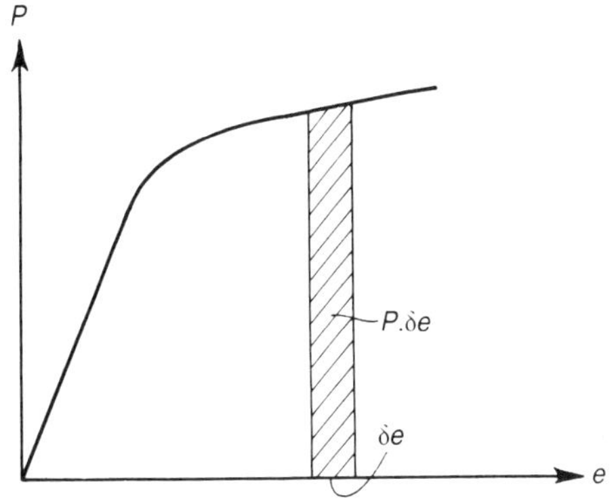

Figure 1.18 Work done in stretching a bar through a small extension, δe.

The total work done in extending the bar to the extension e is then

$$W = \int_0^e P\,de, \tag{1.11}$$

which is the area under the P–e curve up to the stretched condition. If the limit of proportionality is not exceeded, the work done in extending the bar is stored as *strain energy*, which is directly recoverable on removal of the load. For this case, the strain energy, U, is

$$U = W = \int_0^e P\,de \tag{1.12}$$

But in the linear-elastic range of the material, we have from equation (1.6) that

$$e = \frac{PL_0}{EA}$$

where L_0 is the initial length of the bar, A is its cross-sectional area and E is Young's modulus. Then equation (1.12) becomes

$$U = \int_0^e \frac{EA}{L_0}\, e\,de = \frac{EA}{2L_0}\left(e^2\right) \tag{1.13}$$

In terms of P

$$U = \frac{EA}{2L_0}\left(e^2\right) = \frac{L_0}{2EA}\left(P^2\right) \tag{1.14}$$

Now (P/A) is the tensile stress σ in the bar, and so we may write

$$U = \frac{AL_0}{2E}\left(\sigma^2\right) = \frac{\sigma^2}{2E} \times \text{ the volume} \tag{1.15}$$

Moreover, as AL_0 is the original volume of the bar, the strain energy per unit volume is

$$\frac{\sigma^2}{2E} \tag{1.16}$$

When the limit of proportionality of a material is exceeded, the work done in extending the bar is still given by equation (1.11); however, not all this work is stored as strain energy; some of the work done is used in producing permanent distortions in the material, the work reappearing largely in the form of heat. Suppose a mild-steel bar is stressed beyond the yield point, Figure 1.19, and up to the point where strain-hardening begins; the strain at the limit of proportionality is small compared with this large inelastic strain; the work done per unit volume in producing a strain ε is approximately

$$W = \sigma_Y \varepsilon \tag{1.17}$$

in which σ_Y is the yield stress of the material. This work is considerably greater than that required to reach the limit of proportionality. A ductile material of this type is useful in absorbing relatively large amounts of work before breaking.

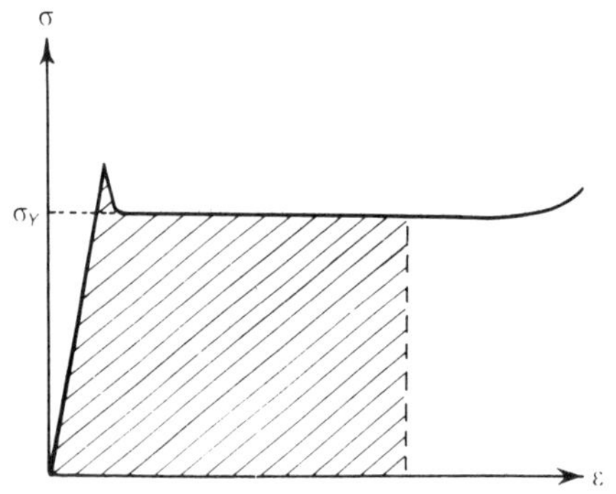

Figure 1.19 Work done in stretching a mild-steel bar; the work done during plastic deformation is very considerable compared with the elastic strain energy.

1.15 Initial stresses

It frequently happens that, before any load is applied to some part of a machine or structure, it is already in a state of stress. In other words, the component is *initially stressed* before external forces are applied. Bolted joints and connections, for example, involve bolts which are pre-tensioned; subsequent loading may, or may not, affect the tension in a bolt. Most forms of welded connections introduce initial stresses around the welds, unless the whole connection is stress relieved by a suitable heat treatment; in such cases, the initial stresses are not usually known with any real accuracy. Initial stresses can also be used to considerable effect in strengthening certain materials; for example, concrete can be made a more effective material by precompression in the form of prestressed concrete. The problems solved below are *statically indeterminate* (see Chapter 2) and therefore require *compatibility* considerations as well as *equilibrium* considerations.

Problem 1.11 A 2.5 cm diameter steel bolt passes through a steel tube 5 cm internal diameter, 6.25 cm external diameter, and 40 cm long. The bolt is then tightened up onto the tube through rigid end blocks until the tensile force in the bolts is 40 kN. The distance between the head of the bolt and the nut is 50 cm. If an external force of 30 kN is applied to the end blocks, tending to pull them apart, estimate the resulting tensile force in the bolt.

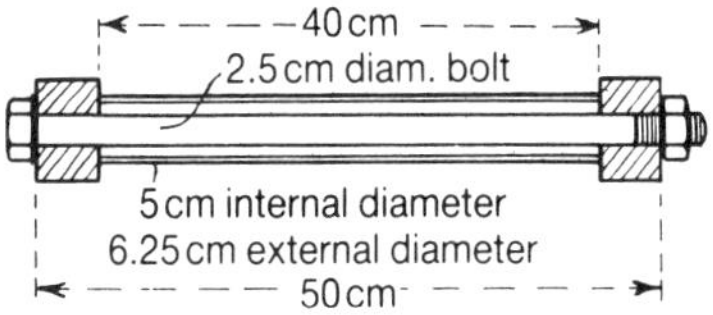

Solution:

The cross-sectional area of the bolt is

$$\frac{\pi}{4}(0.025)^2 = 0.491 \times 10^{-3}\ \mathrm{m}^2$$

The cross-sectional are of the tube is

$$\frac{\pi}{4}\left[(0.0625)^2 - (0.050)^2\right] = \frac{\pi}{4}(0.1125)(0.0125) = 0.110 \times 10^{-2}\ \mathrm{m}^2$$

Before the external load of 30 kN is applied, the bolt and tube carry internal loads of 40 kN. When the external load of 30 kN is applied, suppose the tube and bolt are each stretched by amounts δ; suppose further that the *change* of load in the bolt is $(\Delta P)_b$, tensile, and the *change* of load in the tube is $(\Delta P)_t$, tensile.

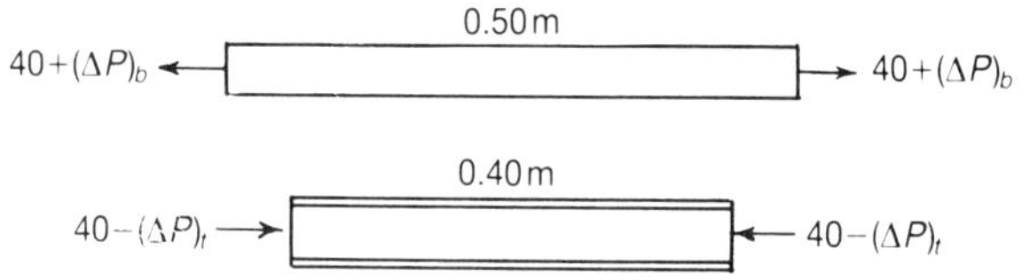

Then for *compatibility*, the elastic stretch of each component due to the additional external load of 30 kN is

$$\delta = \frac{(\Delta P)_b\,(0.50)}{(0.491 \times 10^{-3})\,E} = \frac{(\Delta P)_t\,(0.40)}{(0.110 \times 10^{-2})\,E}$$

where E is Young's modulus. Then

$$(\Delta P)_b = 0.357\,(\Delta P)_t$$

But for *equilibrium* of internal and external forces,

$$(\Delta P)_b + (\Delta P)_t = 30 \text{ kN}$$

These two equations give

$$(\Delta P)_b = 7.89 \text{ kN}, \qquad (\Delta P)_t = 22.11 \text{ kN}$$

The resulting tensile force in the bolt is

$$40 + (\Delta P)_b = 47.89 \text{ kN}$$

1.16 Composite bars in tension or compression

A composite bar is one made of two materials, such as steel rods embedded in concrete. The construction of the bar is such that constituent components extend or contract equally under load. To illustrate the behaviour of such bars consider a rod made of two materials, 1 and 2, Figure 1.20; A_1, A_2 are the cross-sectional areas of the bars, and E_1, E_2 are the values of Young's modulus. We imagine the bars to be rigidly connected together at the ends; then for *compatibility*, the longitudinal strains to be the same when the composite bar is stretched we must have

$$\varepsilon = \frac{\sigma_1}{E_1} = \frac{\sigma_2}{E_2} \tag{1.18}$$

Figure 1.20 Composite bar in tension; if the bars are connected rigidly at their ends, they suffer the same extensions.

where σ_1 and σ_2 are the stresses in the two bars. But from *equilibrium* considerations,

$$P = \sigma_1 A_1 + \sigma_2 A_2 \tag{1.19}$$

Equations (1.18) and (1.19) give

$$\sigma_1 = \frac{PE_1}{A_1 E_1 + A_2 E_2}, \quad \sigma_2 = \frac{PE_2}{A_1 E_1 + A_2 E_2} \tag{1.20}$$

Problem 1.12 A concrete column, 50 cm square, is reinforced with four steel rods, each 2.5 cm in diameter, embedded in the concrete near the corners of the square. If Young's modulus for steel is 200 GN/m^2 and that for concrete is 14 GN/m^2, estimate the compressive stresses in the steel and concrete when the total thrust on the column is 1 MN.

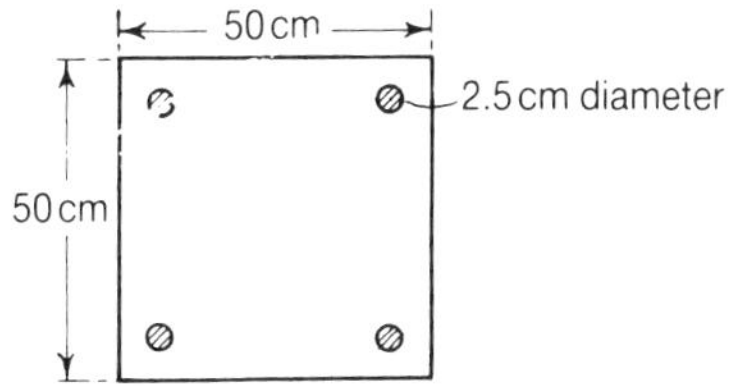

<u>*Solution*</u>

Suppose subscripts c and s refer to concrete and steel, respectively. The cross-sectional area of steel is

$$A_s = 4\left[\frac{\pi}{4}(0.025)^2\right] = 1.96 \times 10^{-3}\ \text{m}^2$$

and the cross-sectional area of concrete is

$$A_c = (0.50)^2 - A_s = 0.248 \text{ m}^2$$

Equations (1.20) then give

$$\sigma_c = \frac{10^6}{(0.248) + (1.96 \times 10^{-3})\left(\frac{200}{14}\right)} = 3.62 \text{ MN/m}^2$$

$$\sigma_s = \frac{10^6}{(0.248)\left(\frac{14}{200}\right) + (1.96 \times 10^{-3})} = 51.76 \text{ MN/m}^2$$

Problem 1.13 A uniform beam weighing 500 N is held in a horizontal position by three vertical wires, one attached to each end of the beam, and one at the mid-length. The outer wires are brass of diameter 0.125 cm, and the central wire is of steel of diameter 0.0625 cm. If the beam is rigid and the wires are of the same length, and unstressed before the beam is attached, estimate the stresses in the wires. Young's modulus for brass is 85 GN/m^2 and for steel is 200 GN/m^2.

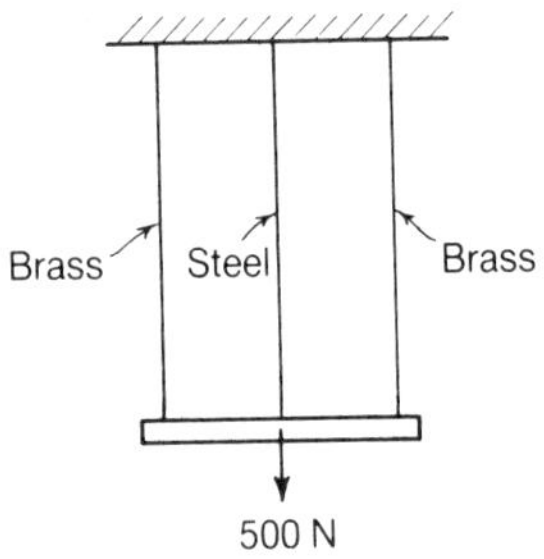

Solution

On considering the two outer brass wires together, we may take the system as a composite one consisting of a single brass member and a steel member. The area of the steel member is

$$A_s = \frac{\pi}{4}(0.625 \times 10^{-3})^2 = 0.306 \times 10^{-6} \text{ m}^2$$

The total area of the two brass members is

$$A_b = 2\left[\frac{\pi}{4}(1.25 \times 10^{-3})^2\right] = 2.45 \times 10^{-6} \text{ m}^2$$

Equations (1.20) then give, for the steel wire

$$\sigma_s = \frac{500}{(0.306 \times 10^{-6}) + (2.45 \times 10^{-6})\left(\frac{85}{200}\right)} = 370 \text{ MN/m}^2$$

and for the brass wires

$$\sigma_b = \frac{500}{(0.306 \times 10^{-6})\left(\frac{200}{85}\right) + (2.45 \times 10^{-6})} = 158 \text{ MN/m}^2$$

1.17 Temperature stresses

When the temperature of a body is raised, or lowered, the material expands, or contracts. If this expansion or contraction is wholly or partially resisted, stresses are set up in the body. Consider a long bar of a material; suppose L_0 is the length of the bar at a temperature θ_0, and that α is the coefficient of linear expansion of the material. The bar is now subjected to an increase θ in temperature. If the bar is completely free to expand, its length increases by $\alpha L_0 \theta$, and the length becomes $L_0(1 + \alpha\theta)$ were compressed to a length L_0; in this case the compressive strain is

$$\varepsilon = \frac{\alpha L_0 \theta}{L_0(1 + \alpha\theta)} = \alpha\theta$$

since $\alpha\theta$ is small compared with unity; the corresponding stress is

$$\sigma = E\varepsilon = \alpha\theta E \qquad (1.21)$$

By a similar argument the tensile stress set up in a constrained bar by a fall θ in temperature is $\alpha\theta E$. It is assumed that the material remains elastic.

In the case of steel $\alpha = 1.3 \times 10^{-5}$ per °C; the product αE is approximately 2.6 MN/m^2 per °C, so that a change in temperature of 4°C produces a stress of approximately 10 MN/m^2 if the bar is completely restrained.

1.18 Temperature stresses in composite bars

In a component or structure made wholly of one material, temperature stresses arise only if external restraints prevent thermal expansion or contraction. In composite bars made of materials with different rates of thermal expansion, internal stresses can be set up by temperature changes; these stresses occur independently of those due to external restraints.

Consider, for example, a simple composite bar consisting of two members—a solid circular bar, 1, contained inside a circular tube, 2, Figure 1.21. The materials of the bar and tube have different coefficients of linear expansion, α_1 and α_2, respectively. If the ends of the bar and tube are attached rigidly to each other, longitudinal stresses are set up by a change of temperature. Suppose firstly, however, that the bar and tube are quite free of each other; if L_0 is the original length of each bar, Figure 1.21, the extensions due to a temperature increase θ are $\alpha_1 \theta L_0$ and $\alpha_2 \theta L_0$, Figure 1.21(ii). The difference in lengths of the two members is $(\alpha_1 - \alpha_2) \theta L_0$; this is now eliminated by compressing the inner bar with a force P, and pulling the outer tube with an equal force P, Figure 1.21(iii).

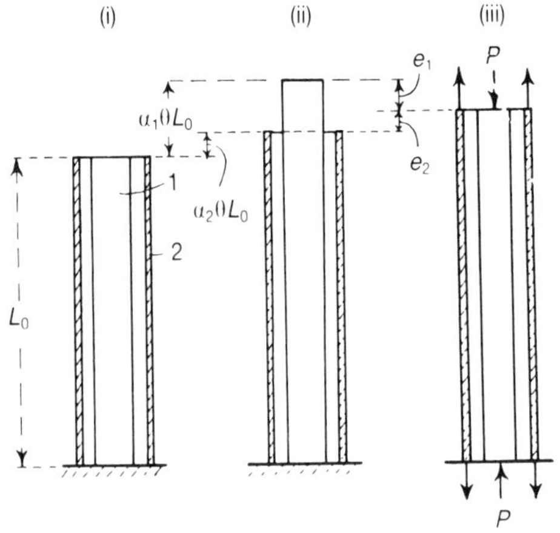

(i) (ii) (iii)

Figure 1.21 Temperature stress in a composite bar.

If A_1 and E_1 are the cross-sectional area and Young's modulus, respectively, of the inner bar, and A_2 and E_2 refer to the outer tube, then the contraction of the inner bar to P is

$$e_1 = \frac{PL_0}{E_1 A_1}$$

and the extension of the outer tube due to P is

$$e_2 = \frac{PL_0}{E_2 A_2}$$

Then from *compatibility* considerations, the difference in lengths $(\alpha_1 - \alpha_2) \theta L$, is eliminated completely when

$$(\alpha_1 - \alpha_2) \theta L_0 = e_1 + e_2$$

On substituting for $e_1 + e_2$, we have

$$(\alpha_1 - \alpha_2)\theta L_0 = PL\left(\frac{1}{E_1A_1} + \frac{1}{E_2A_2}\right) \tag{1.22}$$

The force P is induced by the temperature change θ if the ends of the two members are attached rigidly to each other; from equation (1.22), P has the value

$$P = \frac{(\alpha_1 - \alpha_2)\,\theta}{\left(\dfrac{1}{E_1\ A_1} + \dfrac{1}{E_2\ A_2}\right)} \tag{1.23}$$

An internal load is only set up if α_1 is different from α_2.

Problem 1.14 An aluminium rod 2.2 cm diameter is screwed at the ends, and passes through a steel tube 2.5 cm internal diameter and 0.3 cm thick. Both are heated to a temperature of 140°C, when the nuts on the rod are screwed lightly on to the ends of the tube. Estimate the stress in the rod when the common temperature has fallen to 20°C. For steel, E = 200 GN/m^2 and α = 1.2×10^{-5} per °C, and for aluminium, E = 70 GN/m^2 and α = 2.3×10^{-5} per °C, where E is Young's modulus and α is the coefficient of linear expansion.

Solution

Let subscript a refer to the aluminium rod and subscript s to the steel tube. The problem is similar to the one discussed in Section 1.17, except that the composite rod has its temperature lowered, in this case from 140°C to 20°C. From equation (1.23), the common force between the two components is

$$P = \frac{(\alpha_a - \alpha_s)\,\theta}{\dfrac{1}{(EA)_a} + \dfrac{1}{(EA)_s}}$$

The stress in the rod is therefore

$$\frac{P}{A_a} = \frac{(\alpha_a - \alpha_s)\,\theta}{\dfrac{1}{E_a} + \dfrac{A_a}{E_s\ A_s}} = \frac{(\alpha_a - \alpha_s)\,E_a\ \theta}{1 + \dfrac{E_a\ A_a}{E_s\ A_s}}$$

Now

$$(EA)_a = (70 \times 10^9)\left[\frac{\pi}{4}\ (0.022)^2\right] = 26.6\ \text{MN}$$

Again

$$(EA)_s = (200 \times 10^9)\left[\pi\,(0.028)\,(0.003)\right] = 52.8 \text{ MN}$$

Then

$$\frac{P}{A_a} = \frac{\left[(2.3 - 1.2)\,10^{-5}\right](70 \times 10^9)\,(120)}{1 + \left(\dfrac{26.6}{52.8}\right)} = 61.4 \text{ MN/m}^2$$

1.19 Circular ring under radial pressure

When a thin circular ring is loaded radially, a circumferential force is set up in the ring; this force extends the circumference of the ring, which in turn leads to an increase in the radius of the ring. Consider a thin ring of mean radius r, Figure 1.22(i), acted upon by an internal radial force of intensity p per unit length of the boundary. If the ring is cut across a diameter, Figure 1.22(ii), circumferential forces P are required at the cut sections of the ring to maintain equilibrium of the half-ring. For equilibrium

$$2P = 2pr$$

so that

$$P = pr \tag{1.24}$$

A section may be taken across any diameter, leading to the same result; we conclude, therefore, that P is the circumferential tension in all parts of the ring.

If A is the cross-sectional area of the ring at any point of the circumference, then the tensile circumferential stress in the ring is

$$\sigma = \frac{P}{A} = \frac{pr}{A} \tag{1.25}$$

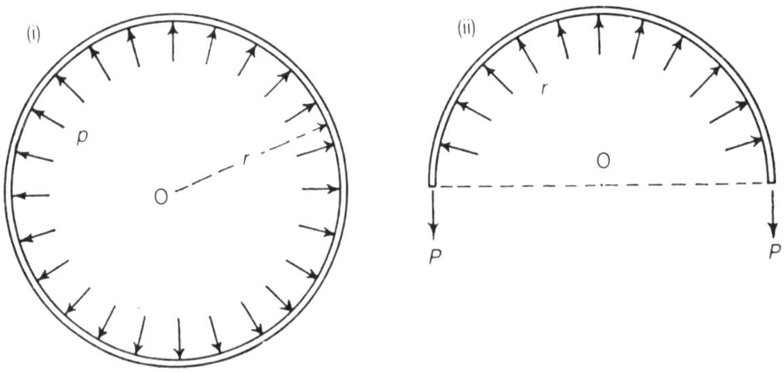

(i) (ii)

Figure 1.22 Thin circular ring under uniform radial loading, leading to a uniform circumferential tension.

If the cross-section is a rectangle of breadth b, (normal to the plane of Figure 1.22), and thickness t, (in the plane of Figure 1.22), then

$$\sigma = \frac{pr}{bt} \tag{1.26}$$

Circumferential stresses of a similar type are set up in a circular ring rotating about an axis through its centre. We suppose the ring is a uniform circular one, having a cross-sectional area A at any point, and that it is rotating about its central axis at uniform angular velocity ω. If ρ is the density of the material of the ring, then the centrifugal force on a unit length of the circumference is

$$\rho A \omega^2 r$$

In equation (1.25) we put this equal to p; thus, the circumferential tensile stress in the ring is

$$\sigma = \frac{pr}{A} = \rho \omega^2 r^2 \tag{1.27}$$

which we see is independent of the actual cross-sectional area. Now, ωr is the circumferential velocity, V (say), of the ring, so

$$\sigma = \rho V^2 \tag{1.28}$$

For steel we have $\rho = 7840 \text{ kg/m}^3$; to produce a tensile stress of 10 MN/m^2, the circumferential velocity must be

$$V = \sqrt{\frac{\sigma}{\rho}} = \sqrt{\frac{(10 \times 10^6)}{7840}} = 35.7 \text{ m/s}$$

Problem 1.15 A circular cylinder, containing oil, has an internal bore of 30 cm diameter. The cylinder is 1.25 cm thick. If the tensile stress in the cylinder must not exceed 75 MN/m^2, estimate the maximum load W which may be supported on a piston sliding in the cylinder.

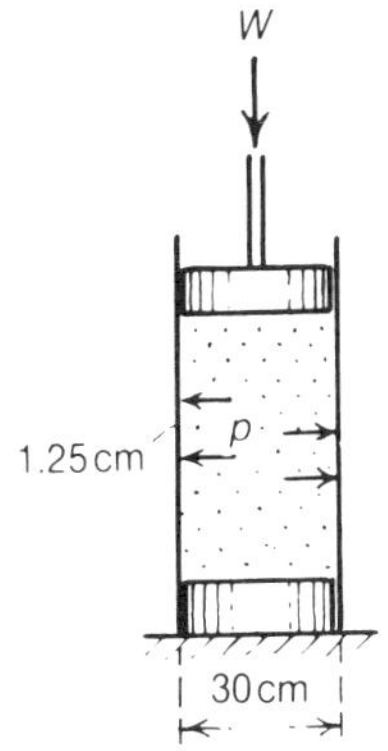

Solution

A load W on the piston generates an internal pressure p given by

$$W = \pi r^2 p$$

where r is the radius of the cylinder. In this case

$$p = \frac{W}{\pi r^2} = \frac{W}{\pi (0.150)^2}$$

A unit length of the cylinder is equivalent to a circular ring subjected to an internal load of p per unit length of circumference. The circumferential load set up by p in this ring is, from equation (1.24),

$$P = pr = p\,(0.150)$$

The circumferential stress is, therefore,

$$\sigma = \frac{P}{1 \times t} = \frac{P}{0.0125} = 80P$$

where t is the thickness of the wall of the cylinder. If σ is limited to 75 MN/m^2, then

$$80P = 75 \times 10^6$$

But

$$80P = 80\,[p\,(0.150)] = 12p = \frac{12W}{\pi (0.150)^2}$$

Then

$$\frac{12W}{\pi (0.150)^2} = 75 \times 10^6$$

giving

$$W = 441 \text{ kN}$$

Problem 1.16 An aluminium-alloy cylinder of internal diameter 10.000 cm and wall thickness 0.50 cm is shrunk onto a steel cylinder of external diameter 10.004 cm and wall thickness 0.50 cm. If the values of Young's modulus for the alloy and the steel are 70 GN/m^2 and 200 GN/m^2, respectively, estimate the circumferential stresses in the cylinders and the radial pressure between the cylinders.

<u>*Solution*</u>

We take unit lengths of the cylinders as behaving like thin circular rings. After the shrinking operation, we suppose p is the radial between the cylinders. The mean radius of the steel tube is

$$\tfrac{1}{2}\,[10.004\;-0.50] \;=\; 4.75\text{ cm}$$

The compressive circumferential stress in the steel tube is then

$$\sigma_s \;=\; \frac{pr}{t} \;=\; \frac{p\,(0.0475)}{0.0050} \;=\; 9.5p$$

The circumferential strain in the steel tube is then

$$\varepsilon_s \;=\; \frac{\sigma_s}{E_s} \;=\; \frac{9.50p}{200\;\times\;10^9}$$

The mean radius of the alloy tube is

$$\tfrac{1}{2}\,[10.000 + 0.50] \;=\; 5.25\text{ cm}$$

The tensile circumferential stress in the alloy tube is then

$$\sigma_a \;=\; \frac{pr}{t} \;=\; \frac{p(0.0525)}{(0.0050)} \;=\; 10.5p$$

The circumferential strain in the alloy tube is then

$$\varepsilon_a \;=\; \frac{\sigma_a}{E_a} \;=\; \frac{10.5p}{70\;\times\;10^9}$$

The circumferential expansion of the alloy tube is

$$2\pi\; r\; \varepsilon_a$$

so the mean radius increases effectively by an amount

$$\delta_a \;=\; r\,\varepsilon_a \;=\; 0.0525\;\varepsilon_a$$

Similarly, the mean radius of the steel tube contracts by an amount

$$\delta_s \;=\; r\,\varepsilon_s \;=\; 0.0475\;\varepsilon_s$$

For the shrinking operation to be carried out we must have that the initial lack of fit, δ, is given by

$$\delta = \delta_a + \delta_s$$

Then

$$\delta_a + \delta_s = 0.002 \times 10^{-2}$$

On substituting for δ_a and δ_s, we have

$$0.0525 \left[\frac{10.5p}{70 \times 10^9}\right] + 0.0475 \left[\frac{9.50p}{200 \times 10^9}\right] = 0.002 \times 10^{-2}$$

This gives

$$p = 1.97 \text{ MN/m}^2$$

The compressive circumferential stress in the steel cylinder is then

$$\sigma_s = 9.50p = 18.7 \text{ MN/m}^2$$

The tensile circumferential stress in the alloy cylinder is

$$\sigma_a = 10.5p = 20.7 \text{ MN/m}^2$$

1.20 Creep of materials under sustained stresses

At ordinary laboratory temperatures most metals will sustain stresses below the limit of proportionality for long periods without showing additional measurable strains. At these temperatures metals deform continuously when stressed above the elastic range. This process of continuous inelastic strain is called *creep*. At high temperatures metals lose some of their elastic properties, and creep under constant stress takes place more rapidly.

When a tensile specimen of a metal is tested at a high temperature under a constant load, the strain assumes instantaneously some value ε_0, Figure 1.23. If the initial strain is in the inelastic range of the material then creep takes place under constant stress. At first the creep rate is fairly rapid, but diminishes until a point *a* is reached on the strain–time curve, Figure 1.23; the point *a* is a point of inflection in this curve, and continued application of the load increases the creep rate until fracture of the specimen occurs at *b*.

At ordinary temperatures concrete shows creep properties; these may be important in prestressed members, where some of the initial stresses in the concrete may be lost after a long period due to creep. Composites are also vulnerable to creep and this must be considered when using them for construction.

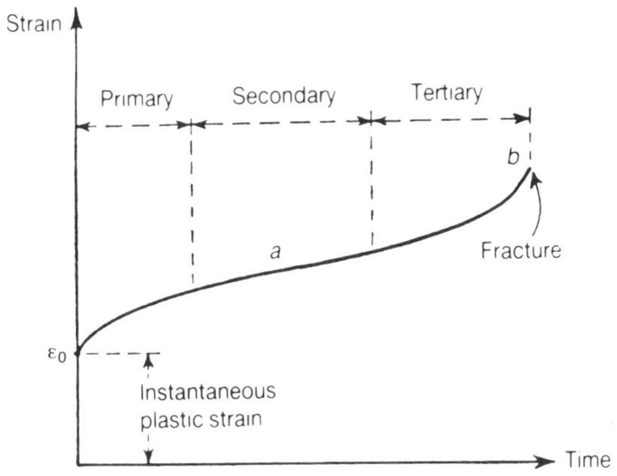

Figure 1.23 Creep curve for a material in the inelastic range; ε_0 is the instantaneous plastic strain.

1.21 Fatigue under repeated stresses

When a material is subjected to repeated cyclic loading, it can fail at a stress which may be much less than the material's yield stress. The problem that occurs here, is that the structure might have minute cracks in it or other stress raisers. Under repeated cyclic loading the large stresses that occur at these stress concentrations cause the cracks to grow, until fracture eventually occurs. Materials likely to suffer fatigue include aluminium alloys and composites; see Figure 1.24.

Failure of a material after a large number of cycles of tensile stress occurs with little, or no, permanent set; fractures show the characteristics of brittle materials. Fatigue is primarily a problem of repeated tensile stresses; this is due probably to the fact that microscopic cracks in a material can propagate more easily when the material is stressed in tension. In the case of steels it is found that there is a critical stress—called the *endurance limit*—below which fluctuating stresses cannot cause a fatigue failure; titanium alloys show a similar phenomenon. No such endurance limit has been found for other non-ferrous metals and other materials.

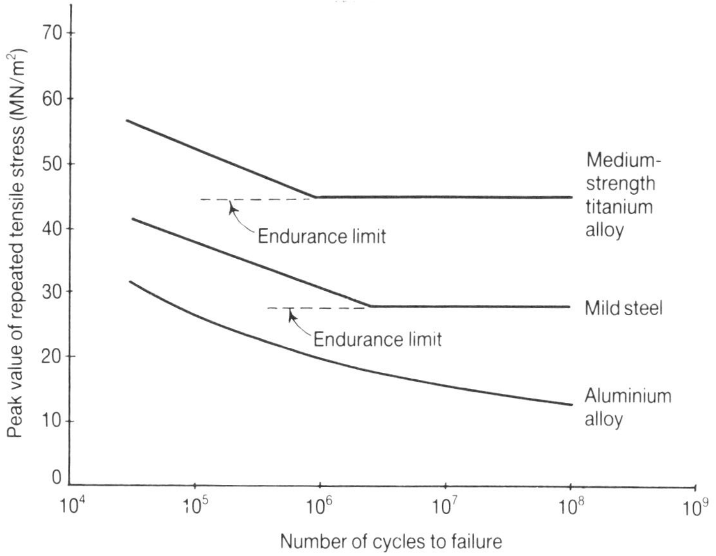

Figure 1.24 Comparison of the fatigue strengths of metals under repeated tensile stresses.

Further problems *(answers on page 691)*

1.17 The piston rod of a double-acting hydraulic cylinder is 20 cm diameter and 4 m long. The piston has a diameter of 40 cm, and is subjected to 10 MN/m^2 water pressure on one side and 3 MN/m^2 on the other. On the return stroke these pressures are interchanged. Estimate the maximum stress occurring in the piston-rod, and the change of length of the rod between two strokes, allowing for the area of piston-rod on one side of the piston. Take E = 200 GN/m^2. *(RNC)*

1.18 A uniform steel rope 250 m long hangs down a shaft. Find the elongation of the first 125 m at the top if the density of steel is 7840 kg/m^3 and Young's modulus is 200 GN/m^2. *(Cambridge)*

1.19 A steel wire, 150 m long, weighs 20 N per metre length. It is placed on a horizontal floor and pulled slowly along by a horizontal force applied to one end. If this force measures 600 N, estimate the increase in length of the wire due to its being towed, assuming a uniform coefficient of friction. Take the density of steel as 7840 kg/m^3 and Young's modulus as 200 GN/m^2. *(RNEC)*

1.20 The hoisting rope for a mine shaft is to lift a cage of weight W. The rope is of variable section so that the stress on every section is equal to σ when the rope is fully extended. If ρ is the density of the material of the rope, show that the cross-sectional area A at a height z above the cage is

$$A = \left(\frac{W}{\sigma}\right) e^{\rho g z/\sigma}$$

1.21 To enable two walls, 10 m apart, to give mutual support they are stayed together by a 2.5 cm diameter steel tension rod with screwed ends, plates and nuts. The rod is heated to 100°C when the nuts are screwed up. If the walls yield, relatively, by 0.5 cm when the rod cools to 15°C, find the pull of rod at that temperature. The coefficient of linear expansion of steel is $\alpha = 1.2 \times 10^{-5}$ per °C, and Young's modulus E = 200 GN/m^2. *(RNEC)*

1.22 A steel tube 3 cm diameter, 0.25 cm thick and 4 m long, is covered and lined throughout with copper tubes 0.2 cm thick. The three tubes are firmly joined together at their ends. The compound tube is then raised in temperature by 100°C. Find the stresses in the steel and copper, and the increase in length of the tube, will prevent its expansion? Assume E = 200 GN/m^2 for steel and E = 110 GN/m^2 for copper; the coefficients of linear expansion of steel and copper are 1.2×10^{-5} per °C and 1.9×10^{-5} per °C, respectively.

2 Pin-jointed frames or trusses

2.1 Introduction

In problems of stress analysis we discriminate between two types of structure; in the first, the forces in the structure can be determined by considering only its statical equilibrium. Such a structure is said to be *statically determinate*. The second type of structure is said to be *statically indeterminate*. In the case of the latter type of structure, the forces in the structure cannot be obtained by considerations of statical equilibrium alone. This is because there are more unknown forces than there are simultaneous equations obtained from considerations of statical equilibrium alone. For statically indeterminate structures, other methods have to be used to obtain the additional number of the required simultaneous equations; one such method is to consider compatibility, as was adopted in Chapter 1. In this chapter, we will consider statically determinate frames and one simple statically indeterminate frame.

Figure 2.1 shows a rigid beam BD supported by two vertical wires BF and DG; the beam carries a force of $4W$ at C. We suppose the wires extend by negligibly small amounts, so that the geometrical configuration of the structure is practically unaffected; then for equilibrium the forces in the wires must be $3W$ in BF and W in DG. As the forces in the wires are known, it is a simple matter to calculate their extensions and hence to determine the displacement of any point of the beam. The calculation of the forces in the wires and structure of Figure 2.1 is said to be *statically determinate*. If, however, the rigid beam be supported by three wires, with an additional wire, say, between H and J in Figure 2.1, then the forces in the three wires cannot be solved by considering statical equilibrium alone; this gives a second type of stress analysis problem, which is discussed more fully in Section 2.5; such a structure is *statically indeterminate*.

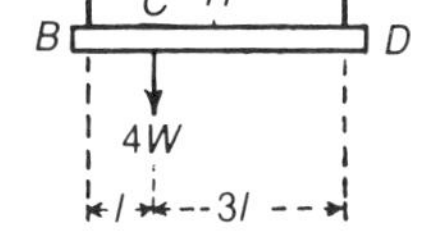

Figure 2.1 Statically determinate system of a beam supported by two wires.

2.2 Statically determinate pin-jointed frames

By a *frame* we mean a structure which is composed of straight bars joined together at their ends. A *pin-jointed frame or truss* is one in which no bending actions can be transmitted from one bar to another as described in the introductory chapter; ideally this could be achieved if the bars were joined together through pin-joints. If the frame has just sufficient bars or rods to prevent collapse without the application of external forces, it is said to be *simply-stiff*; when there are more bars or rods than this, the frame is said to be *redundant*. A redundant framework is said to contain one or more redundant members, where the latter are not required for the framework to be classified as a framework, as distinct from being a mechanism. It should be emphasised, however, that if a redundant member is removed from the framework, the stresses in the remaining members of the framework may become so large that the framework collapses. A redundant member of a framework does not necessarily have a zero internal force in it. Definite relations exist which must be satisfied by the numbers of bars and joints if a frame is said to be simply-stiff, or statically determinate.

In the plane frame of Figure 2.2, *BC* is one member. To locate the joint *D* relative to *BC* requires two members, namely, *BD* and *CD*; to locate another joint *F* requires two further members, namely, *CF* and *DF*. Obviously, for each new joint of the frame, two new members are required. If *m* be the total number of members, including *BC*, and *j* is the total number of joints, we must have

$$m = 2j - 3, \tag{2.1}$$

if the frame is to be simply-stiff or statically determinate.

When the frame is rigidly attached to a wall, say at *B* and *C*, *BC* is not part of the frame as such, and equation (2.1) becomes, omitting member *BC*, and joints *B* and *C*,

$$m = 2j \tag{2.2}$$

These conditions must be satisfied, but they may not necessarily ensure that the frame is simply-stiff. For example, the frames of Figures 2.2 and 2.3 have the same numbers of members and joints; the frame of Figure 2.2 is simply-stiff. The frame of Figure 2.3 is *not* simply-stiff, since a mechanism can be formed with pivots at *D*, *G*, *J*, *F*. Thus, although a frame having *j* joints must have at least $(2j - 3)$ members, the mode of arrangement of these members is important.

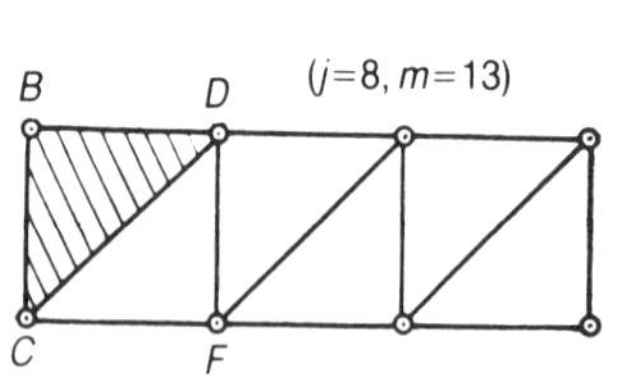

Figure 2.2 Simply-stiff plane frame built up from a basic triangle *BCD*.

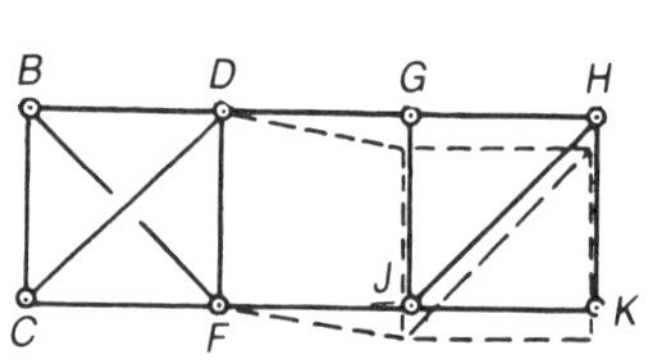

Figure 2.3 Rearrangement of the members of Figure 2.2 to give a mechanism.

For a pin-jointed space frame attached to three joints in a rigid wall, the condition for the frame to be simply-stiff is

$$m = 3j \qquad (2.3)$$

where m is the total number of members, and j is the total number of joints, exclusive of the three joints in the rigid wall. When a space frame is not rigidly attached to a wall, the condition becomes

$$m = 3j - 6, \qquad (2.4)$$

where m is the total number of members in the frame, and j the total number of joints.

2.3 The method of joints

This method can only be used to determine the internal forces in the members of statically determinate pin-jointed trusses. It consists of isolating each joint of the framework in the form of a *free-body diagram* and then by considering equilibrium at each of these joints, the forces in the members of the framework can be determined. Initially, all *unknown forces* in the members of the framework are assumed to be in *tension*, and before analysing each joint it should be ensured that each joint does not have more than two unknown forces.

To demonstrate the method, the following example will be considered.

Problem 2.1 Using the method of joints, determine the member forces of the plane pin-jointed truss of Figure 2.4.

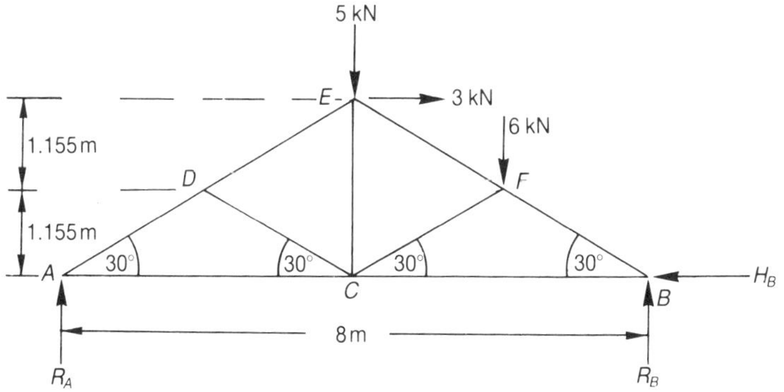

Figure 2.4 Pin-jointed truss.

Solution

Assume all unknown internal forces are in tension, because if they are in compression, their signs will be negative.

As each joint must only have two unknown forces acting on it, it will be necessary to determine the values of R_A, R_B and H_B, prior to using the method of joints.

Resolving the forces horizontally

$$\text{forces to the left} = \text{forces to the right}$$

$$3 = H_B$$

$$\therefore\ H_B = 3\ \text{kN}$$

Taking moments about B

$$\text{clockwise moments} = \text{counter-clockwise moments}$$

$$R_A \times 8 + 3 \times 2.311 = 5 \times 4 + 6 \times 2$$

$$\therefore\ R_A = 25.07/8 = 3.13\ \text{kN}$$

Resolving forces vertically

$$\text{upward forces} = \text{downward forces}$$

$$R_A + R_B = 5 + 6$$

or

$$R_B = 11 - 3.13 = 7.87\ \text{kN}$$

Isolate *joint A* and consider equilibrium, as shown by the following free-body diagram.

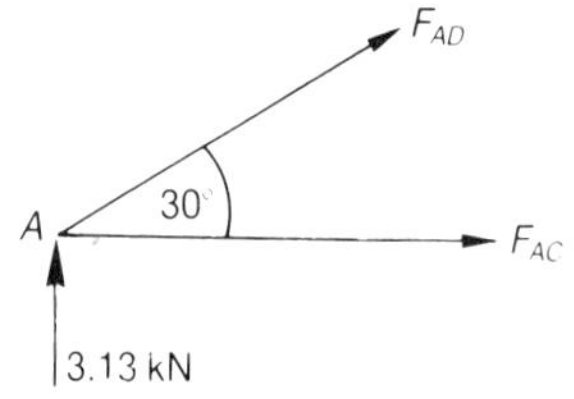

Resolving forces vertically

$$\text{upward forces} = \text{downward forces}$$

$$3.13 + F_{AD} \sin 30 = 0$$

or

$$F_{AD} = -6.26 \text{ kN (compression)}$$

NB The *negative sign* for this force denotes that this member is in *compression*, and such a member is called a *strut*.

Resolving forces horizontally

$$\text{forces to the right} = \text{forces to the left}$$

$$F_{AC} + F_{AD} \cos 30 = 0$$

or

$$F_{AC} = 6.26 \times 0.866$$

$$F_{AC} = 5.42 \text{ kN (tension)}$$

NB The *positive sign* for this force denotes that this member if in *tension*, and such a member is called a *tie*.

It is possible now to analyse *joint D*, because F_{AD} is known and therefore the joint has only two unknown forces acting on it, as shown by the free-body diagram.

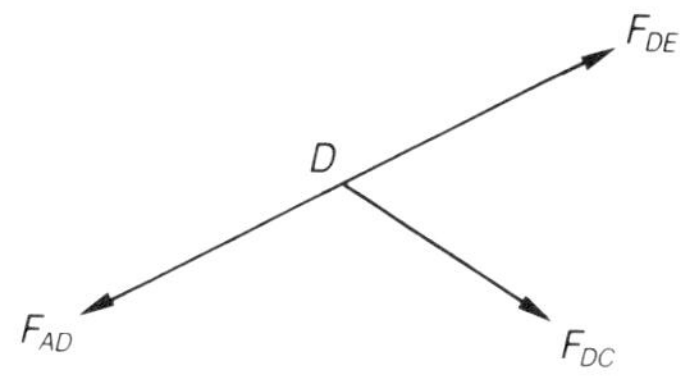

Resolving vertically

$$\text{upward forces} = \text{downward forces}$$

$$F_{DE} \sin 30 = F_{AD} \sin 30 + F_{DC} \sin 30$$

or

$$F_{DE} = -6.26 + F_{DC} \tag{2.5}$$

Resolving horizontally

$$\text{forces to the left} = \text{forces to the right}$$

$$F_{AD}\cos 30 = F_{DE}\cos 30 + F_{DC}\cos 30$$

or
$$F_{DE} = -6.26 - F_{DC} \tag{2.6}$$

Equating (2.5) and (2.6)

$$-6.26 + F_{DC} = -6.26 - F_{DC}$$

or
$$F_{DC} = 0 \tag{2.7}$$

Substituting equation (2.7) into equation (2.5)

$$F_{DE} = -6.26 \text{ kN (compression)}$$

It is now possible to examine *joint E*, as it has two unknown forces acting on it, as shown:

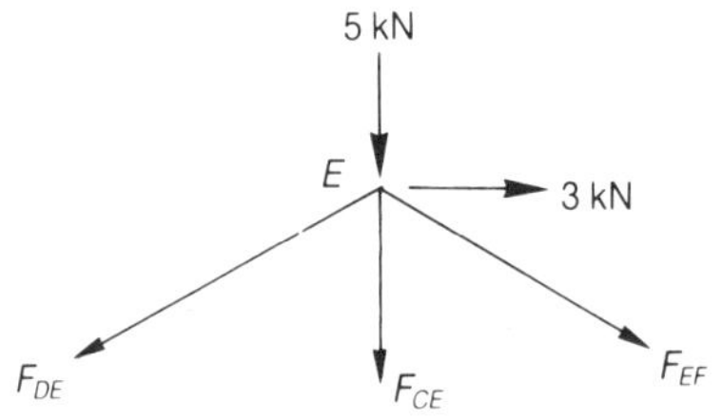

Resolving horizontally

$$\text{forces to the left} = \text{forces to the right}$$

$$F_{DE}\cos 30 = F_{EF}\cos 30 + 3$$

or
$$F_{EF} = -6.26 - 3/0.866$$

$$F_{EF} = -9.72 \text{ kN (compression)}$$

Resolving vertically

$$\text{upward forces} = \text{downward forces}$$

$$0 = 5 + F_{DE}\sin 30 + F_{CE} + F_{EF}\sin 30$$

$$F_{CE} = -5 + 6.26 \times 0.5 + 9.72 \times 0.5$$

$$F_{CE} = 3 \text{ kN (tension)}$$

It is now possible to analyse either *joint F* or *joint C*, as each of these joints has only got two unknown forces acting on it. Consider *joint F*,

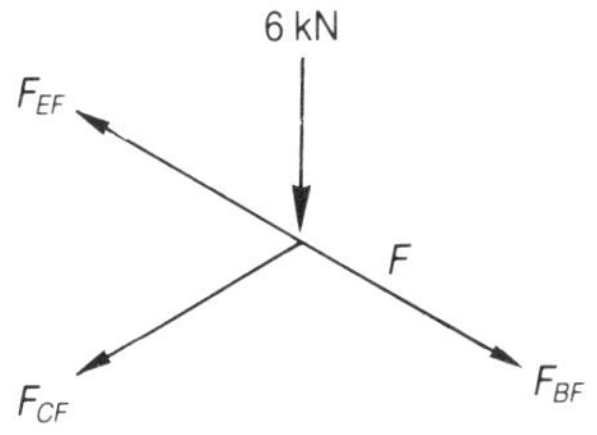

Resolving horizontally

$$\text{forces to the left} = \text{forces to the right}$$

$$F_{EF} \cos 30 + F_{CF} \cos 30 = F_{BF} \cos 30$$

$$\therefore F_{BF} = -9.72 + F_{CF} \quad (2.8)$$

Resolving vertically

$$\text{upward forces} = \text{downward forces}$$

$$F_{EF} \sin 30 = F_{CF} \sin 30 + F_{BF} \sin 30 + 6$$

or

$$F_{BF} \times 0.5 = -9.72 \times 0.5 - 0.5\, F_{CF} - 6$$

$$\therefore F_{BF} = -21.72 - F_{CF} \quad (2.9)$$

Equating (2.8) and (2.9)

$$-9.72 + F_{CF} = -21.72 - F_{CF}$$

$$\therefore F_{CF} = -6 \text{ kN (compression)} \quad (2.10)$$

Substituting equation (2.10) into equation (2.8)

$$F_{BF} = -9.72 - 6 = -15.72 \text{ kN (compression)}$$

Consider *joint B* to determine the remaining unknown force, namely F_{BC},

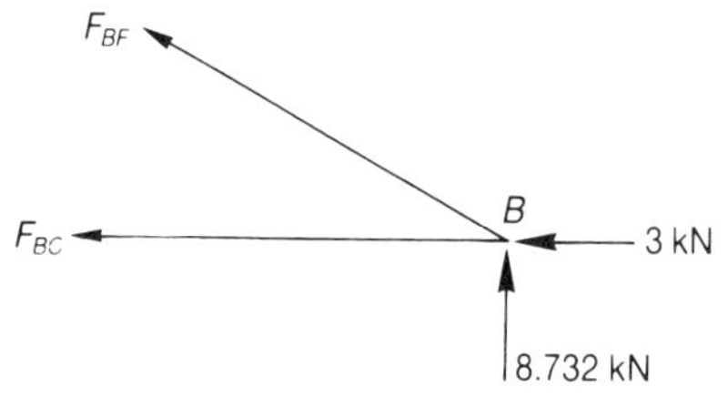

Resolving horizontally

forces to the left = forces to the right

$$F_{BF}\ \cos 30 + F_{BC} + 3 = 0$$

$$\therefore F_{BC} = -3 + 15.72 \times 0.866 = \text{kN (tension)}$$

Here are the magnitudes and 'directions' of the internal forces in this truss:

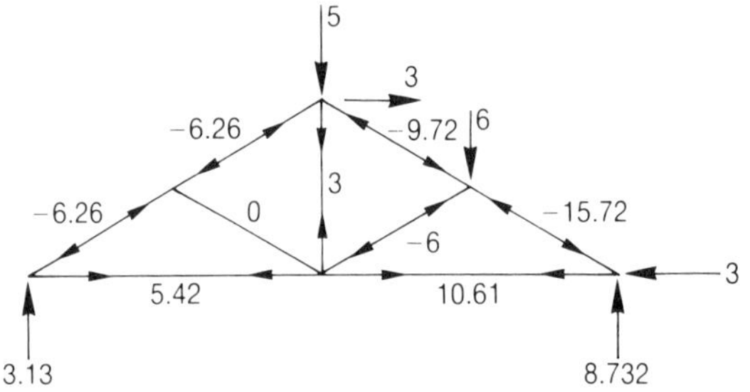

2.4 The method of sections

This method is useful if it is required to determine the internal forces in only a few members. The process is to make an imaginary cut across the framework, and then by considering equilibrium, to determine the internal forces in the members that lie across this path. In this method, it is only possible to examine a section that has a maximum of three unknown internal forces, and here again, it is convenient to assume that all unknown forces are in tension.

To demonstrate the method, an imaginary cut will be made through members DE, CD and AC of the truss of Figure 2.4, as shown by the free-body diagram of Figure 2.5

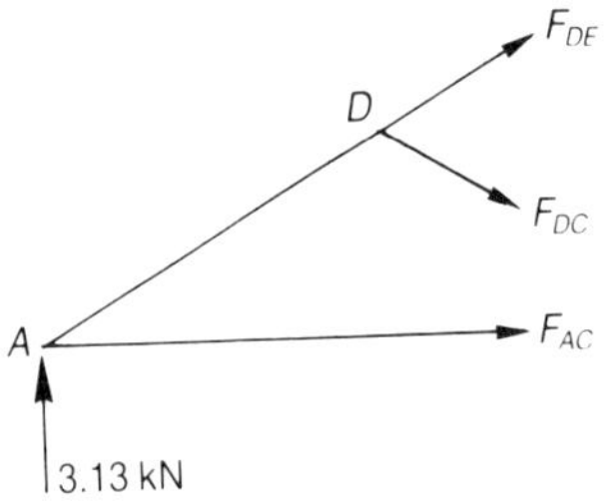

Figure 2.5 Free-body diagram.

Taking moments about D

counter-clockwise couples = clockwise couples

$$F_{AC} \times 1.55 = 3.13 \times 2$$

$$\therefore F_{AC} = 5.42 \text{ kN}$$

NB It was convenient to take moments about D, as there were two unknown forces acting through this point and therefore, the arithmetic was simplified.

Resolving vertically

upward forces = downward forces

$$F_{DE} \sin 30 + 3.13 = F_{DC} \sin 30$$

$$\therefore F_{DC} = 6.26 + F_{DE} \tag{2.11}$$

Resolving horizontally

forces to the right = forces to the left

$$F_{DE} \cos 30 + F_{DC} \cos 30 + F_{AC} = 0$$

$$\therefore F_{DC} = -5.42/0.866 - F_{DE}$$

or

$$F_{DC} = -6.26 - F_{DE} \tag{2.12}$$

Equating (2.11) and (2.12)

$$F_{DE} = -6.26 \text{ kN} \tag{2.13}$$

Substituting equation (2.13) into equation (2.11)

$$F_{DC} = 0 \text{ kN}$$

These values can be seen to be the same as those obtained by the method of joints.

2.5 A statically indeterminate problem

In Section 2.1 we mentioned a type of stress analysis problem in which internal stresses are not calculable on considering statical equilibrium alone; such problems are *statically indeterminate*. Consider the rigid beam BD of Figure 2.6 which is supported on three wires; suppose the tensions in the wires are T_1, T_2 and T_3. Then by resolving forces vertically, we have

$$T_1 + T_2 + T_3 = 4W \tag{2.14}$$

and by taking moments about the point C, we get

$$T_1 - T_2 - 3T_3 = 0 \tag{2.15}$$

From these equilibrium equations alone we cannot derive the values of the three tensile forces T_1, T_2, T_3; a third equation is found by discussing the extensions of the wires or considering compatibility. If the wires extend by amounts e_1, e_2, e_3, we must have from Figure 2.6(ii) that

$$e_1 + e_3 = 2e_2 \tag{2.16}$$

because the beam BD is rigid. Suppose the wires are all of the same material and cross-sectional area, and that they remain elastic. Then we may write

$$e_1 = \lambda T_1 , \qquad e_2 = \lambda T_2 , \qquad e_3 = \lambda T_3 , \tag{2.17}$$

where λ is a constant common to the three wires. Then equation (2.16) may be written

$$T_1 + T_3 = 2T_2 \tag{2.18}$$

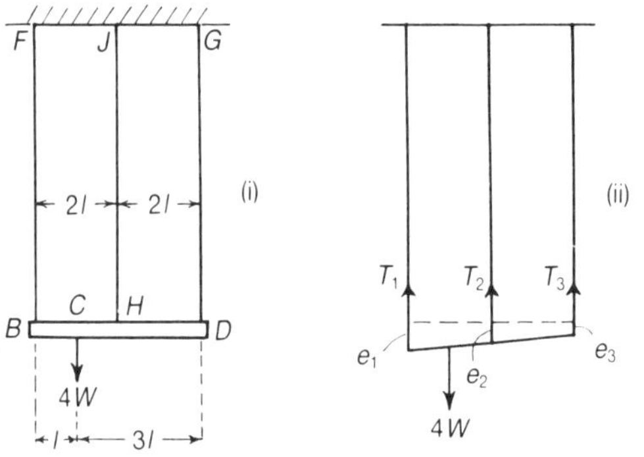

Figure 2.6 A simple statically indeterminate system consisting of a rigid beam supported by three extensible wires.

The three equations (2.14), (2.15) and (2.18) then give

$$T_1 = \frac{7W}{12} \qquad T_2 = \frac{4W}{12} \qquad T_3 = \frac{W}{12} \tag{2.19}$$

Equation (2.16) is a condition which the extensions of the wires must satisfy; it is called a *strain compatibility* condition. Statically indeterminate problems are soluble if strain compatibilities are considered as well as statical equilibrium.

Further problems *(answers on page 691)*

2.2 Determine the internal forces in the plane pin-jointed trusses shown below:

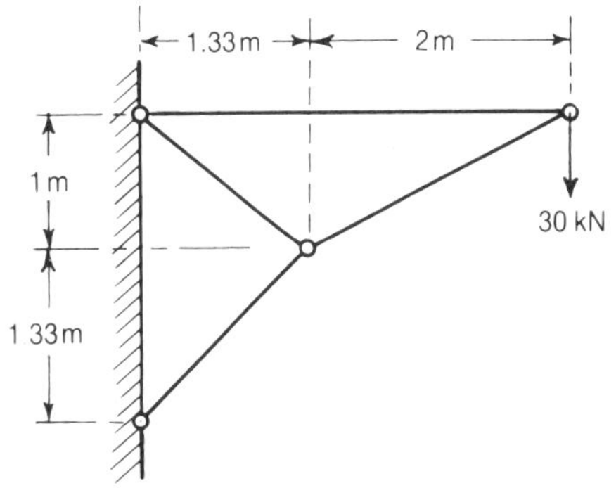

(a)

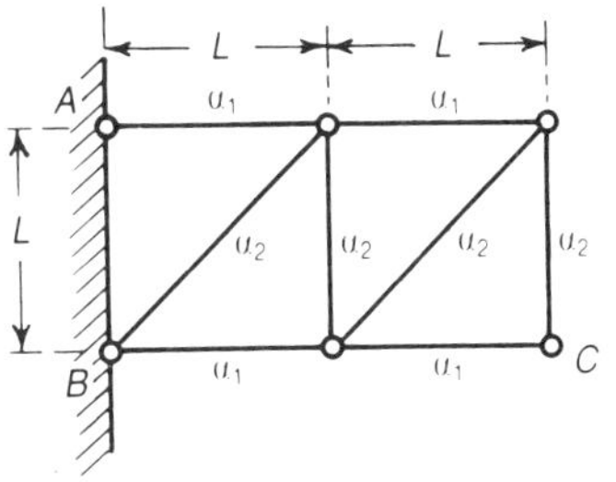

(b)

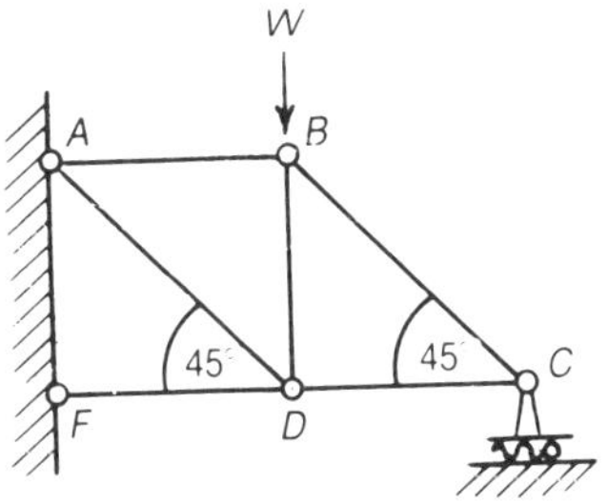

(c)

2.3 The plane pin-jointed truss below is firmly pinned at A and B and subjected to two point loads at the joint F.

Using any method, determine the forces in all the members, stating whether they are tensile or compressive. *(Portsmouth 1982)*

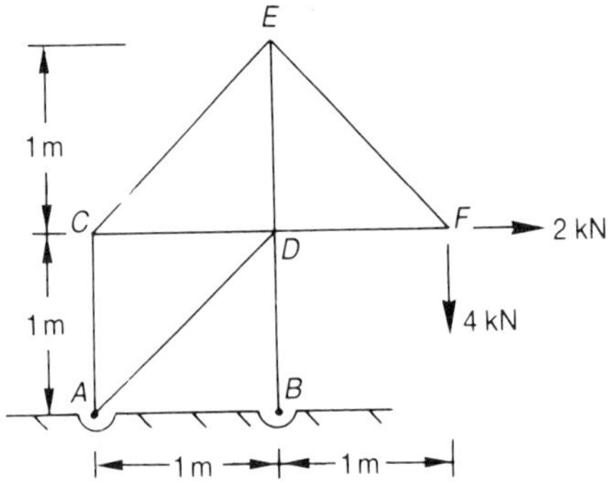

2.4 A plane pin-jointed truss is firmly pinned at its base, as shown below.

Determine the forces in the members of this truss, stating whether they are in tension or compression. *(Portsmouth 1980)*

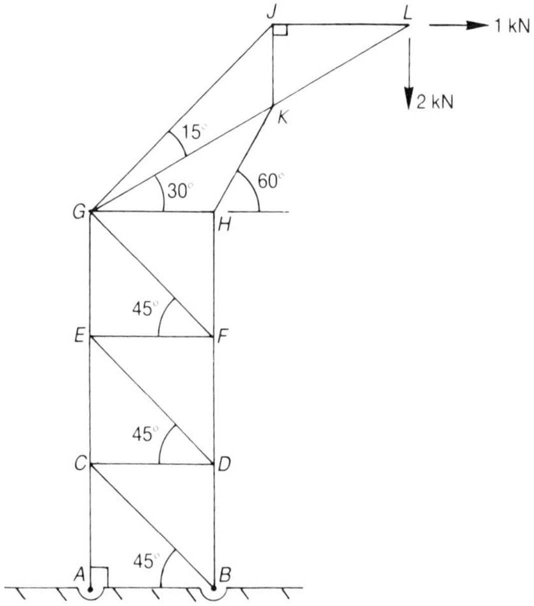

2.5 Determine the internal forces in the pin-jointed truss, below, which is known as a Warren girder.

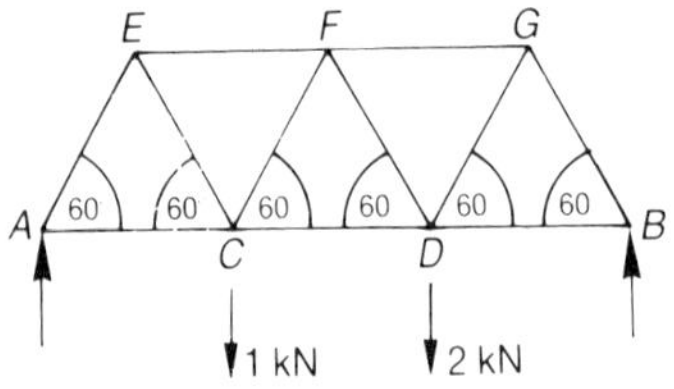

3 Shearing stress

3.1 Introduction

In Chapter 1 we made a study of tensile and compressive stresses, which we called direct stresses. There is another type of stress which plays a vital role in the behaviour of materials, especially metals.

Consider a thin block of material, Figure 3.1, which is glued to a table; suppose a thin plate is now glued to the upper surface of the block. If a horizontal force F is applied to the plate, the plate will tend to slide along the top of the block of material, and the block itself will tend to slide along the table. Provided the glued surfaces remain intact, the table resists the sliding of the block, and the block resists the sliding of the plate on its upper surface. If we consider the block to be divided by any imaginary horizontal plane, such as *ab*, the part of the block above this plane will be trying to slide over the part below the plane. The material on each side of this plane will be trying to slide over the part below the plane. The material on each side of this plane is said to be subjected to a *shearing* action; the stresses arising from these actions are called *shearing stresses*. Shearing stresses act *tangential* to the surface, unlike direct stresses which act perpendicular to the surface.

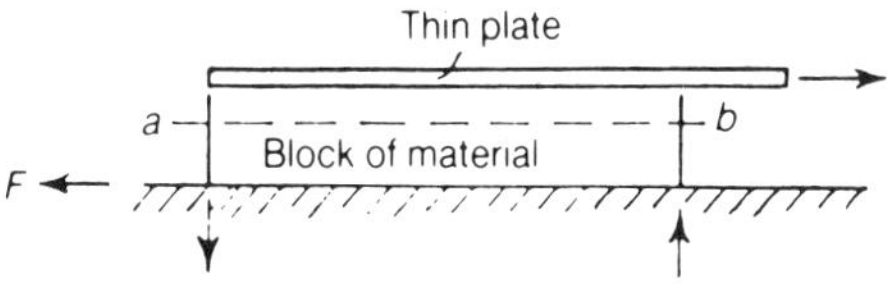

Figure 3.1 Shearing stresses caused by shearing forces.

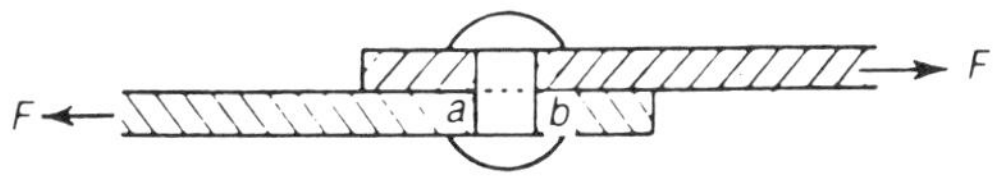

Figure 3.2 Shearing stresses in a rivet; shearing forces F is transmitted over the face *ab* of the rivet.

In general, a pair of garden shears cuts the stems of shrubs through shearing action and not bending action. Shearing stresses arise in many other practical problems. Figure 3.2 shows two flat plates held together by a single rivet, and carrying a tensile force F. We imagine the rivet divided into two portions by the plane *ab*; then the upper half of the rivet is tending to slide over the lower half, and a shearing stress is set up in the plane *ab*. Figure 3.3 shows a circular shaft *a*, with a collar *c*, held in bearing *b*, one end of the shaft being pushed with a force F; in this case there is, firstly, a tendency for the shaft to be pushed bodily through the collar, thereby inducing shearing stresses over the cylindrical surfaces *d* of the shaft and the collar; secondly, there is a tendency for the collar to push through the bearing, so that shearing stresses are set up on cylindrical surfaces such as *e* in the bearing. As a third example, consider the case of a steel bolt in the end of a bar of wood, Figure 3.4, the bolt being pulled by forces F; suppose the grain of the wood runs parallel to the length of the bar; then if the forces F are large enough the block *abcd* will be pushed out, shearing taking place along the planes *ab* and *cd*.

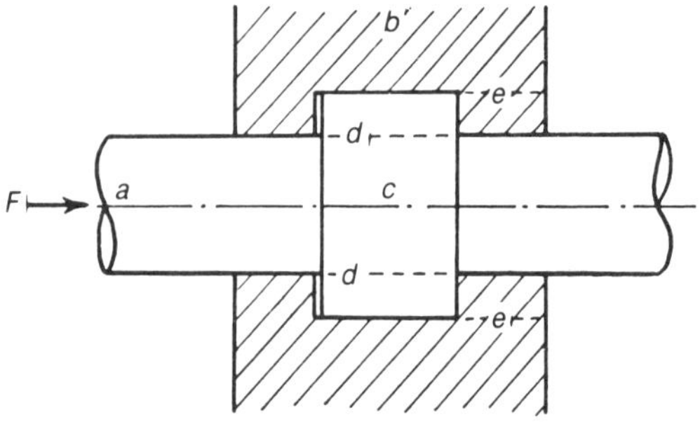

Figure 3.3 Thrust on the collar of a shaft, generating shearing stress over the planes *d*.

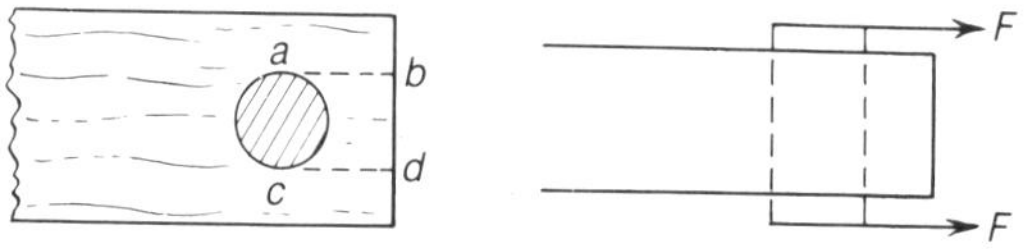

Figure 3.4 Tearing of the end of a timber member by a steel bolt, generating a shearing action on the planes *ab* and *cd*.

3.2 Measurement of shearing stress

Shearing stress on any surface is defined as the intensity of shearing force tangential to the surface. If the block of material of Figure 3.1 has an area A over any section such as *ab*, the average shearing stress τ over the section *ab* is

$$\tau = \frac{F}{A} \tag{3.1}$$

In many cases the shearing force is not distributed uniformly over any section; if δF is the shearing force on any elemental area δA of a section, the shearing stress on that elemental area is

$$\tau = \operatorname*{Limit}_{\delta A \to 0} \frac{\delta F}{\delta A} = \frac{dF}{dA} \tag{3.2}$$

Problem 3.1 Three steel plates are held together by a 1.5 cm diameter rivet. If the load transmitted is 50 kN, estimate the shearing stress in the rivet.

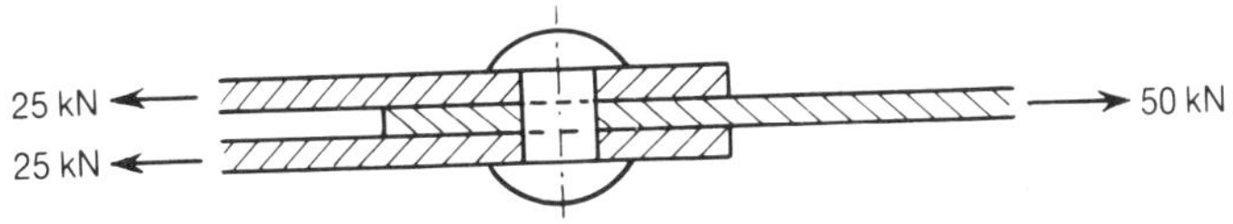

<u>*Solution*</u>

There is a tendency to shear across the planes in the rivet shown by broken lines. The area resisting shear is twice the cross-sectional area of the rivet; the cross-sectional area of the rivet is

$$A = \frac{\pi}{4}(0.015)^2 = 0.177 \times 10^{-3} \text{ m}^2$$

The average shearing stress in the rivet is then

$$\tau = \frac{F}{A} = \frac{25 \times 10^3}{0.177 \times 10^{-3}} = 141 \text{ MN/m}^2$$

Problem 3.2 Two steel rods are connected by a cotter joint. If the shearing strength of the steel used in the rods and the cotter is 150 MN/m^2, estimate which part of the joint is more prone to shearing failure.

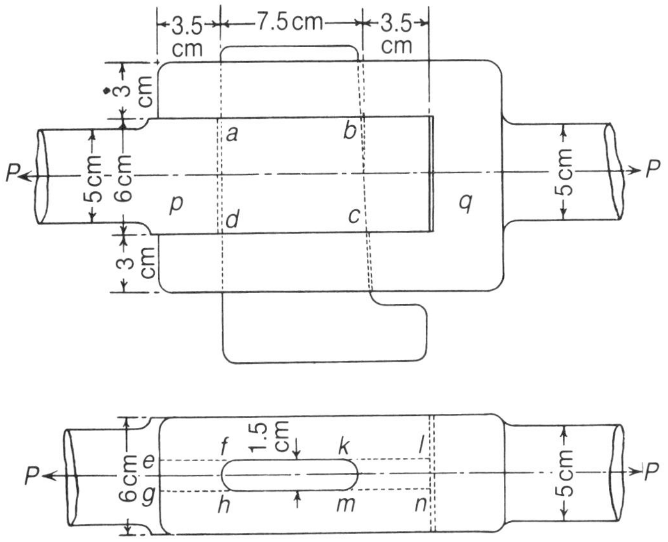

<u>*Solution*</u>

Shearing failure may occur in the following ways:

(i) Shearing of the cotter in the planes *ab* and *cd*.
The area resisting shear is 2(*fkmh*) = 2(0.075 (0.015) = 2.25×10^{-3} m^2

For a shearing failure on these planes, the tensile force is

$$P = \tau A = (150 \times 10^6)(2.25 \times 10^{-3}) = 338 \text{ kN}$$

(ii) By the cotter tearing through the ends of the socket q, i.e. by shearing the planes *ef* and *gh*. The total area resisting shear is

$$A = 4(0.030)(0.035) = 4.20 \times 10^{-3} \text{ m}^2$$

For a shearing failure on these planes

$$P = \tau A = (150 \times 10^6)(4.20 \times 10^{-3}) = 630 \text{ kN}$$

(iii) By the cotter tearing through the ends of the rod p, i.e. by shearing in the planes *kl* and *mn*. The total area resisting shear is

$$A = 2(0.035)(0.060) = 4.20 \times 10^{-3} \text{ m}^2$$

For a shearing failure on these planes

$$P = \tau A = (150 \times 10^6)(4.20 \times 10^{-3}) = 630 \text{ kN}$$

Thus, the connection is most vulnerable to shearing failure in the cotter itself, as discussed in (i); the tensile load for shearing failure is 338 kN.

Problem 3.3 A lever is keyed to a shaft 4 cm in diameter, the width of the key being 1.25 cm and its length 5 cm. What load P can be applied at an arm of $a = 1$ m if the average shearing stress in the key is not to exceed 60 MN/m^2?

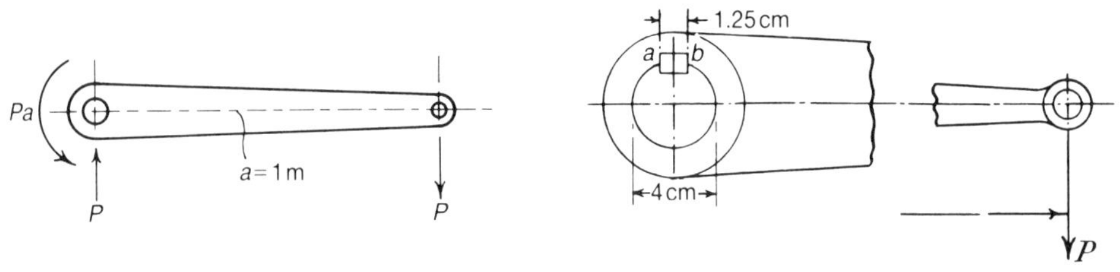

Solution

The torque applied to the shaft is Pa. If this is resisted by a shearing force F on the plane *ab* of the key, then

$$Fr = Pa$$

where r is the radius of the shaft. Then

$$F = \frac{Pa}{r} = \frac{P(1)}{(0.02)} = 50P$$

The area resisting shear in the key is

$$A = 0.0125 \times 0.050 = 0.625 \times 10^{-3}\ \text{m}^2$$

The permissible shearing force on the plane *ab* of the key is then

$$F = \tau A = (60 \times 10^6)(0.625 \times 10^{-3}) = 37.5\ \text{kN}$$

The permissible value of P is then

$$P = \frac{F}{50} = 750\ \text{N}$$

3.3 Complementary shearing stress

Let us return now to the consideration of the block shown in Figure 3.1. We have seen that horizontal planes, such as *ab*, are subjected to shearing stresses. In fact the state of stress is rather more complex than we have supposed, because for rotational equilibrium of the whole block an external couple is required to balance the couple due to the shearing forces F. Suppose the material of the block is divided into a number of rectangular elements, as shown by the full lines of Figure 3.5. Under the actions of the shearing forces F, which together constitute a couple, the elements will tend to take up the positions shown by the broken lines in Figure 3.5. It will be seen that there is a tendency for the vertical faces of the elements to slide over each other. Actually the ends of the elements do not slide over each other in this way, but the tendency to so do shows that the shearing stress in horizontal planes is accompanied by shearing stresses in vertical planes perpendicular to the applied shearing forces. This is true of all cases of shearing action: a given shearing stress acting on one plane is always accompanied by a *complementary shearing stress* on planes at right angles to the plane on which the given stress acts.

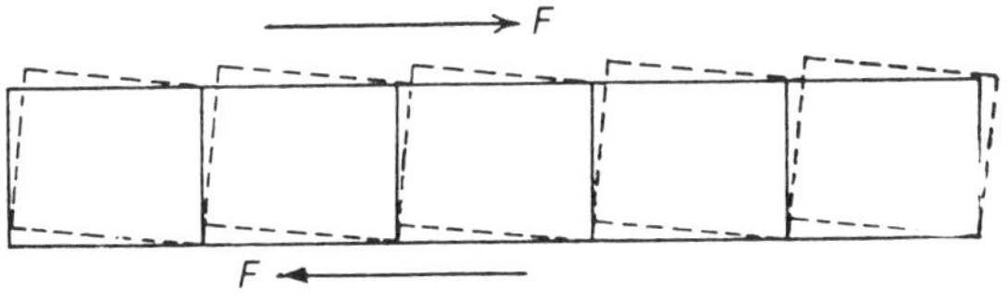

Figure 3.5 Tendency for a set of disconnected blocks to rotate when shearing forces are applied.

Consider now the equilibrium of one of the elementary blocks of Figure 3.5. Let τ_{xy} be the shearing stress on the horizontal faces of the element, and τ_{yx} the complementary shearing stress[2]

[2]Notice that the first suffix x shows the direction, the second the plane on which the stress acts; thus τ_{xy} acts in direction of x axis on planes y = constant.

on vertical faces of the element, Figure 3.6. Suppose a is the length of the element, b its height, and that it has unit thickness. The total shearing force on the upper and lower faces is then

$$\tau_{xy} \times a \times 1 = a\tau_{xy}$$

while the total shearing force on the end faces is

$$\tau_{yx} \times b \times 1 = b\tau_{yx}$$

For rotational equilibrium of the element we then have

$$(a\tau_{xy}) \times b = (b\tau_{yx}) \times a$$

and thus

$$\tau_{xy} = \tau_{yx}$$

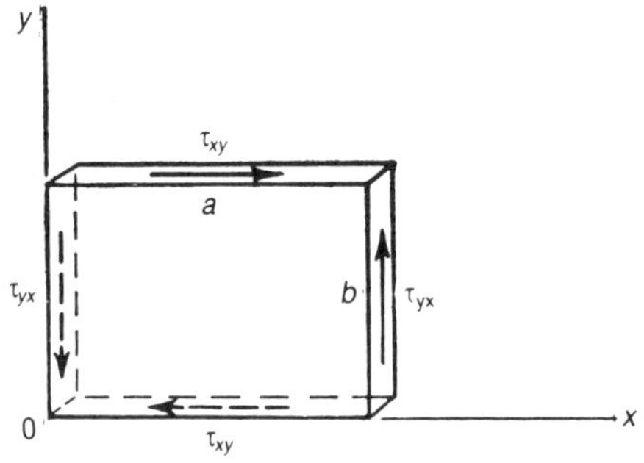

Figure 3.6 Complementary shearing stresses over the faces of a block when they are connected.

We see then that, whenever there is a shearing stress over a plane passing through a given line, there must be an *equal* complementary shearing stress on a plane perpendicular to the given plane, and passing through the given line. The directions of the two shearing stresses must be either both towards, or both away from, the line of intersection of the two planes in which they act.

It is extremely important to appreciate the existence of the complementary shearing stress, for its necessary presence has a direct effect on the maximum stress in the material, as we shall see later in Chapter 5.

3.4 Shearing strain

Shearing stresses in a material give rise to *shearing strains*. Consider a rectangular block of material, Figure 3.7, subjected to shearing stresses τ in one plane. The shearing stresses distort the rectangular face of the block into a parallelogram. If the right-angles at the corners of the face change by amounts γ, then γ is the shearing strain. The angle γ is measured in radians, and is non-dimensional therefore.

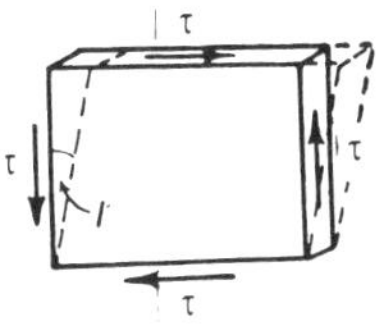

Figure 3.7 Shearing strain in a rectangular block; small values of γ lead to a negligible change of volume in shear straining.

For many materials shearing strain is linearly proportional to shearing stress within certain limits. This linear dependence is similar to the case of direct tension and compression. Within the limits of proportionality

$$\tau = G_\gamma \ , \tag{3.3}$$

where G is the *shearing modulus* or modulus of rigidity, and is similar to Young's modulus E, for direct tension and compression. For most materials E is about 2.5 times greater than G.

It should be noted that no volume changes occur as a result of shearing stresses acting alone. In Figure 3.7 the volume of the strained block is approximately equal to the volume of the original rectangular prism if the angular strain γ is small.

3.5 Strain energy due to shearing actions

In shearing the rectangular prism of Figure 3.7, the forces acting on the upper and lower faces undergo displacements. Work is done, therefore, during these displacements. If the strains are kept within the elastic limit the work done is recoverable, and is stored in the form of strain energy. Suppose all edges of the prism of Figure 3.7 are of unit length; then the prism has unit volume, and the shearing forces on the sheared faces are τ. Now suppose τ is increased by a small amount, causing a small increment of shearing strain $\delta\gamma$. The work done on the prism during this small change is $\tau\delta\gamma$, as the force τ moves through a distance $\delta\gamma$. The total work done in producing a shearing strain γ is then

$$\int_0^y \tau d\gamma$$

While the material remains elastic, we have from equation (3.3) that $\tau = G\gamma$, and the work done is stored as strain energy; the strain energy is therefore

$$\int_0^{\gamma} \tau \, d\gamma = \int_0^{\gamma} G\gamma \, d\gamma = \frac{1}{2} G\gamma^2 \tag{3.4}$$

per unit volume. In terms of τ this becomes

$$\frac{1}{2} G\gamma^2 = \frac{\tau^2}{2G} \quad = \text{ shear strain energy per unit volume} \tag{3.5}$$

Further problems *(answers on page 691)*

3.4 Rivet holes 2.5 cm diameter are punched in a steel plate 1 cm thick. The shearing strength of the plate is 300 MN/m^2. Find the average compressive stress in the punch at the time of punching.

3.5 The diameter of the bolt circle of a flanged coupling for a shaft 12.5 cm in diameter is 37.5 cm. There are six bolts 2.5 cm diameter. What power can be transmitted at 150 rev/min if the shearing stress in the bolts is not to exceed 60 MN/m^2?

3.6 A pellet carrying the striking needle of a fuse has a mass of 0.1 kg; it is prevented from moving longitudinally relative to the body of the fuse by a copper pin *A* of diameter 0.05 cm. It is prevented from turning relative to the body of the fuse by a steel stud *B*. *A* fits loosely in the pellet so that no stress comes on *A* due to rotation. If the copper shears at 150 MN/m^2, find the retardation of the shell necessary to shear *A*. *(RNC)*

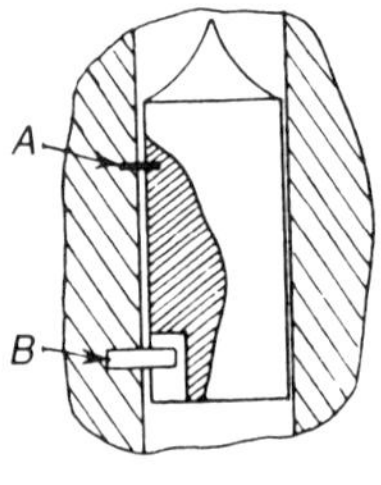

3.7 A lever is secured to a shaft by a taper pin through the boss of the lever. The shaft is 4 cm diameter and the mean diameter of the pin is 1 cm. What torque can be applied to the lever without causing the average shearing stress in the pin to exceed 60 MN/m^2.

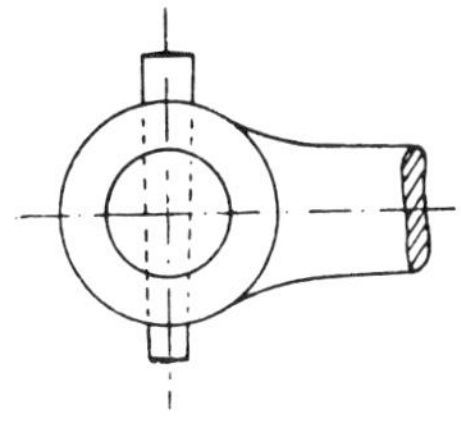

3.8 A cotter joint connects two circular rods in tension. Taking the tensile strength of the rods as 350 MN/m^2, the shearing strength of the cotter 275 MN/m^2, the permissible bearing pressure between surfaces in contact 700 MN/m^2, the shearing strength of the rod ends 185 MN/m^2, calculate suitable dimensions for the joint so that it may be equally strong against the possible types of failure. Take the thickness of the cotter = *d/4*, and the taper of the cotter 1 in 48.

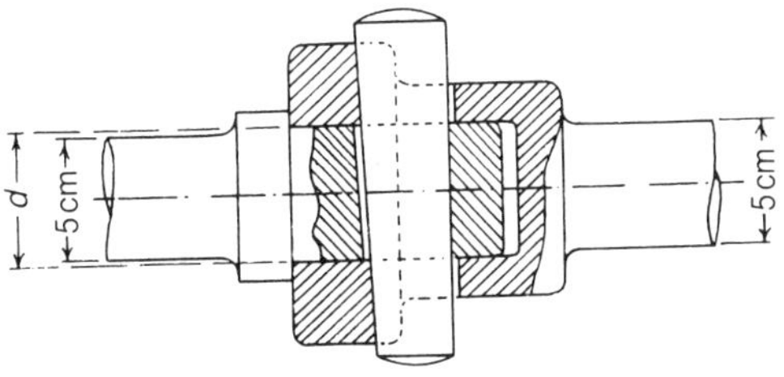

3.9 A horizontal arm, capable of rotation about a vertical shaft, carries a mass of 2.5 kg, bolted to it by a 1 cm bolt at a distance 50 cm from the axis of the shaft. The axis of the bolt is vertical. If the ultimate shearing strength of the bolt is 50 MN/m^2, at what speed will the bolt snap? *(RNEC)*

3.10 A copper disc 10 cm in diameter and 0.0125 cm thick, is fitted in the casing of an air compressor, so as to blow and safeguard the cast-iron case in the event of a serious compressed air leak. If pressure inside the case is suddenly built up by a burst cooling coil, calculate at what pressure the disc will blow out, assuming that failure occurs by shear round the edges of the disc, and that copper will normally fail under a shearing stress of 120 MN/m^2. *(RNEC)*

4 Joints and connections

4.1 Importance of connections

Many engineering structures and machines consist of components suitably connected through carefully designed joints. In metallic materials, these joints may take a number of different forms, as for example welded joints, bolted joints and riveted joints. In general, such joints are stressed in complex ways, and it is not usually possible to calculate stresses accurately because of the geometrical discontinuities in the region of a joint. For this reason, good design of connections is a mixture of stress analysis and experience of the behaviour of actual joints; this is particularly true of connections subjected to repeated loads.

Bolted joints are widely used in structural steel work and recently the performance of such joints has been greatly improved by the introduction of high-tensile, friction-grip bolts. Welded joints are widely used in steel structures, as for example, in ship construction. Riveted joints are still widely used in aircraft-skin construction in light-alloy materials. Epoxy resin glues are often used in the aeronautical field to bond metals.

4.2 Modes of failure of simple bolted and riveted joints

One of the simplest types of joint between two plates of material is a bolted or riveted lap joint, Figure 4.1.

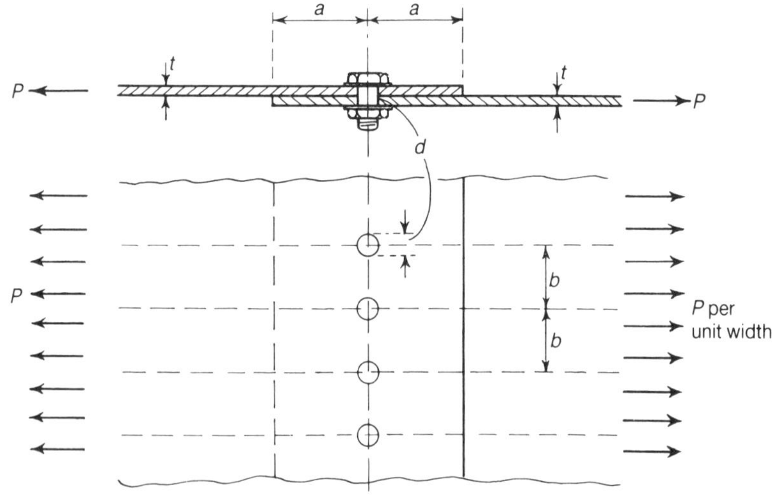

Figure 4.1 Single-bolted lap joint under tensile load.

We shall discuss the forms of failure of the joint assuming it is bolted, but the analysis can be extended in principle to the case of a riveted connection. Consider a joint between two wide plates, Figure 4.1; suppose the plates are each of thickness t, and that they are connected together with a single line of bolts, giving a total overlap of breadth $2a$. Suppose also that the bolts are each of diameter d, and that their centres are a distance b apart along the line of bolts; the line of bolts is a distance a from the edge of each plate. It is assumed that a bolt fills a hole, so that the holes in the plates are also of diameter d.

We consider all possible simple modes of failure when each plate carries a tensile load of P per unit width of plate:

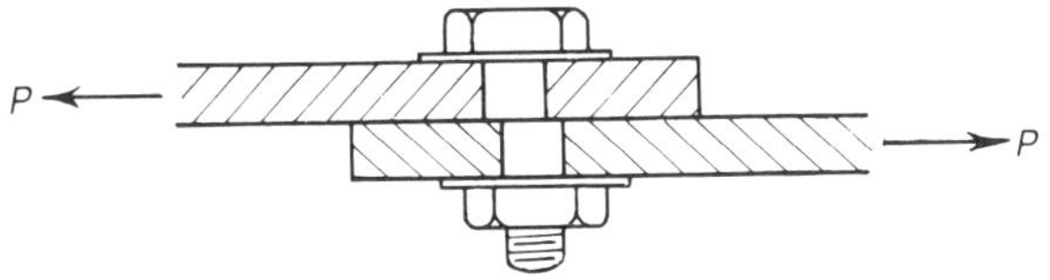

Figure 4.2 Failure by shearing of the bolts.

(1) The bolts may fail by shearing, as shown in Figure 4.2; if τ_1 is the maximum shearing stress the bolts will withstand, the total shearing force required to shear a bolt is

$$\tau_1 \times \left(\frac{\pi d^2}{4}\right)$$

Now, the load carried by a single bolt is Pb, so that a failure of this type occurs when

$$Pb = \tau_1 \left(\frac{\pi d^2}{4}\right)$$

This gives

$$P = \frac{\pi d^2 \tau_1}{4b} \tag{4.1}$$

(2) The bearing pressure between the bolts and the plates may become excessive; the total bearing load taken by a bolt is Pb, Figure 4.3, so that the average bearing pressure between a bolt and its surrounding hole is

$$\frac{Pb}{td}$$

If P_b is the pressure at which either the bolt or the hole fails in bearing, a failure of this type occurs when:

$$P = \frac{P_b td}{b} \tag{4.2}$$

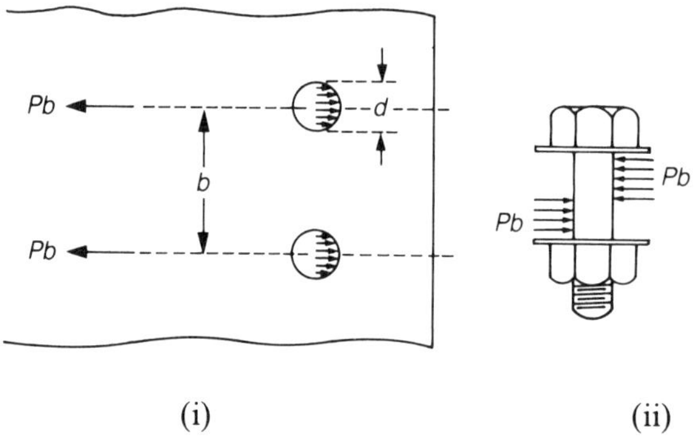

Figure 4.3 (i) Bearing pressure on the holes of the upper plate.
(ii) Bearing pressures on a bolt.

(3) Tensile failures may occur in the plates; clearly the most heavily stressed regions of the plates are on sections such as *ee*, Figure 4.4, through the line of bolts. The average tensile stress on the reduced area of plate through this section is

$$\frac{Pb}{(b-d)t}$$

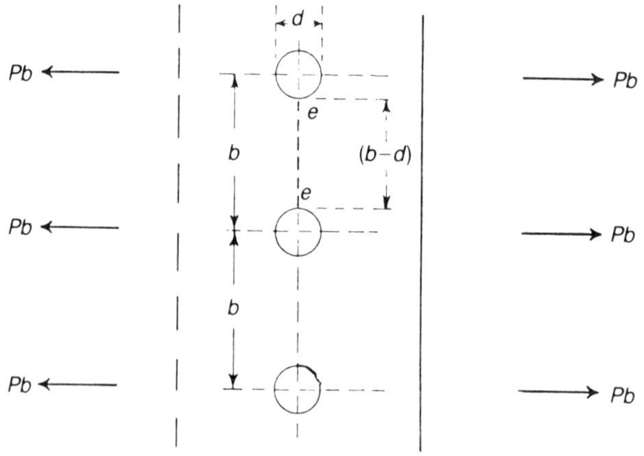

Figure 4.4 Tensile failures in the plates.

If the material of the plate has an ultimate tensile stress of σ_{ult}, then a tensile failure occurs when

$$P = \frac{\sigma_{ult}\, t(b - d)}{b} \tag{4.3}$$

(4) Shearing of the plates may occur on planes such as *cc*, Figure 4.5, with the result that the whole block of material *cccc* is sheared out of the plate. If τ_2 is the maximum shearing stress of the material of the plates, this mode of failure occurs when

$$Pb = \tau_2 \times 2at$$

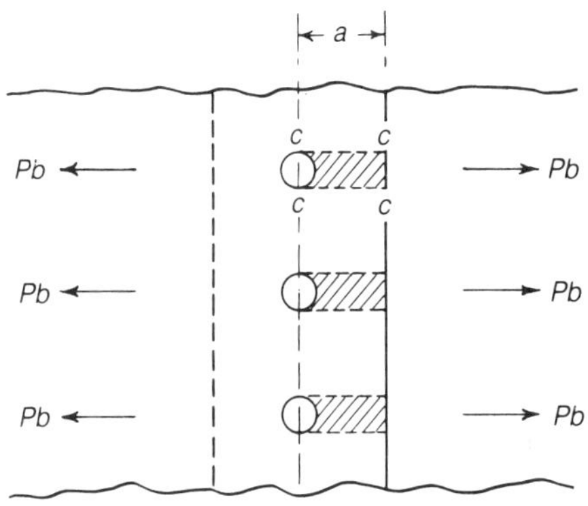

Figure 4.5 Shearing failure in the plates.

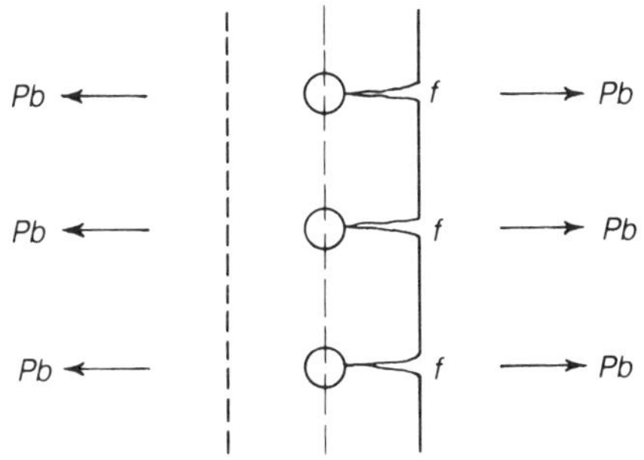

Figure 4.6 Tensile failures at the free edges of the plates.

This gives

$$P = \frac{2at\,\tau_2}{b} \tag{4.4}$$

(5) The plates may fail due to the development of large tensile stresses in the regions of points such as *f*, Figure 4.6. The failing load in this condition is difficult to estimate, and we do not attempt the calculation at this stage.

In riveted joints it is found from tests on mild-steel plates and rivets that if the centre of a rivet hole is not less than 1½ times the rivet hole diameter from the edge of the plate, then failure of the plate by shearing, as discussed in (4) and (5), does not occur. Thus, if for mild-steel plates and rivets,

$$a \geq 1.5d \tag{4.5}$$

we can disregard the modes of failure discussed in (4) and (5). In the case of wrought aluminium alloys, the corresponding value of a is

$$a \geq 2d \tag{4.6}$$

We have assumed, in discussing the modes of failure, that all load applied to the two plates of Figure 4.1 is transmitted in shear through the bolts or rivets. This is so only if there is a negligible frictional force between the two plates. If hot-driven rivets are used, appreciable frictional forces are set up on cooling; these forces play a vital part in the behaviour of the connection. With cold-driven rivets the frictional force is usually small, and may be neglected.

Problem 4.1 Two steel plates, each 1 cm thick, are connected by riveting them between cover plates each 0.6 cm thick. The rivets are 1.6 cm diameter. The tensile stress in the plates must not exceed 140 MN/m^2, and the shearing stress in the rivets must not exceed 75 MN/m^2. Find the proportions of the joint so that it shall be equally strong in shear and tension, and estimate the bearing pressure between the rivets and the plates.

Solution

Suppose b is the rivet pitch, and that P is the tensile load per metre carried by the connection. Then the tensile load on one rivet is Pb. The cover plates, taken together, are thicker than the main plates, and may be disregarded therefore, in the strength calculations. We imagine there is no restriction on the distance from the rivets to the extreme edges of the main plates and cover plates; we may disregard then any possibility of shearing or tensile failure on the free edges of the plates.

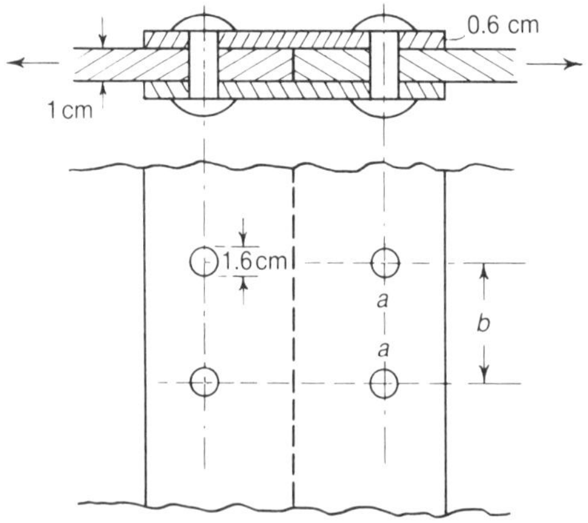

There are then two possible modes of failure:

(1) Tensile failure of the main plates may occur on sections such as aa. The area resisting tension is

$$0.010\,(b - 0.016)\ \mathrm{m}^2$$

The permissible tensile load is, therefore,

$$Pb = (140 \times 10^6)\,[0.010\,(b - 0.016)]\ \text{N per rivet}$$

(2) The rivets may fail by shearing. The area of each rivet is

$$\frac{\pi}{4}(0.016)^2 = 0.201 \times 10^{-3}\ \text{m}^2$$

The permissible load per rivet is then

$$Pb = 2(75 \times 10^6)\,(0.201 \times 10^{-3})\ \text{N}$$

as each rivet is in double shear.

If the joint is equally strong in tension and shear , we have, from (1) and (2),

$$(140 \times 10^6)\,[0.010\,(b - 0.016)] = 2(75 \times 10^6)\,(0.201 \times 10^{-3})$$

This gives

$$b = 0.038\ \text{m}$$

Now

$$Pb = 2(75 \times 10^6)\,(0.201 \times 10^{-3}) = 30.2\ \text{kN}$$

The average bearing pressure between the main plates and rivets is

$$\frac{30.2 \times 10^3}{(0.016)\,(0.010)} = 189\ \text{MN/m}^2$$

4.3 Efficiency of a connection

After analysing the connection of Figure 4.1, suppose we find that in the weakest mode of failure the carrying capacity of the joint is P_0. If the two plates were continuous through the connection, that is, if there were no overlap or bolts, the strength of the plates in tension would be

$$P_{ult} = \sigma_{ult} t$$

where σ_{ult} is the ultimate tensile stress of the material of the plates. The ratio

$$\eta = \frac{P_0}{P_{ult}} = \frac{P_0}{\sigma_{ult}\, t} \qquad (4.7)$$

is known as the *efficiency* of the connection; clearly, η defines the extent to which the strength of the connection attains the full strength of the continuous plates. Joint efficiencies are also described in Chapter 6.

Problem 4.2 What is the efficiency of the joint of Problem 4.1?

Solution

The permissible tensile load per rivet is 30.2 kN. For a continuous joint the tensile load which could be carried by a 3.8 cm width of main plate is

$$(0.038)\,(0.010)\,(140\times10^{6}) = 53.2\text{ kN}$$

Then

$$\eta = \frac{30.2}{53.2} = 0.57, \text{ or } 57\%$$

4.4 Group-bolted and -riveted joints

When two members are connected by cover plates bolted or riveted in the manner shown in Figure 4.7, the joint is said to be *group-bolted* or *-riveted.*

The greatest efficiency of the joint shown in Figure 4.7 is obtained when the bolts or rivets are re-arranged in the form shown in Figure 4.8, where it is supposed six bolts or rivets are required each side of the join. The loss of cross-section in the main members, on the line a, is that due to one bolt or rivet hole. If the load is assumed to be equally distributed among the bolts or rivets, the bolt or rivet on the line a will take one-sixth of the total load, so that the tension in the main plates, across b, will be 5/6ths of the total.

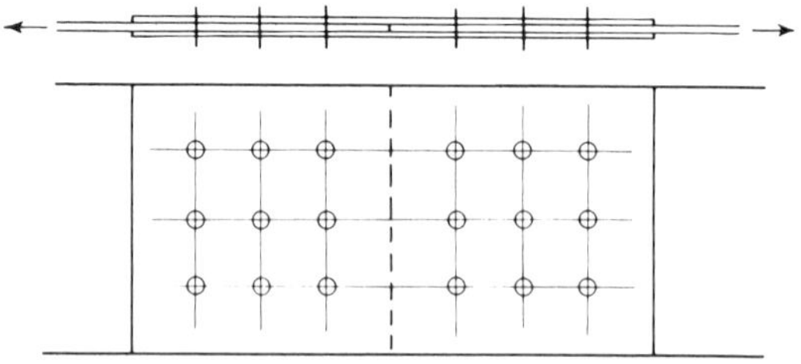

Figure 4.7 A group-bolted or -riveted joint.

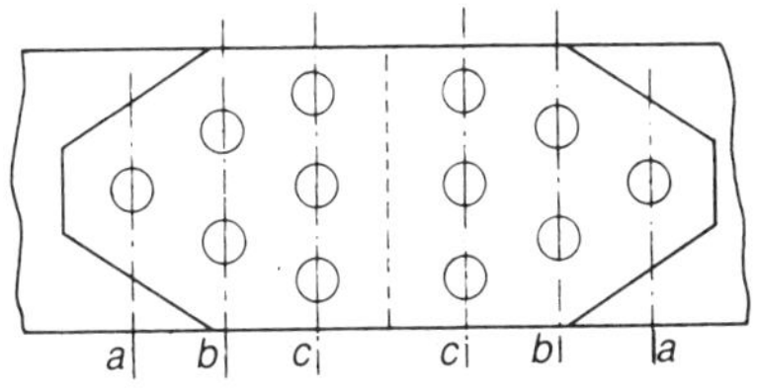

Figure 4.8 Joint with tapered cover plates.

But this section is reduced by two bolt or rivet holes, so that, relatively, it is as strong as the section a, and so on: the reduction of the nett cross-section of the main plates increases as the load carried by these plates decreases. Thus a more efficient joint is obtained than when the bolts or rivets are arranged as in Figure 4.7.

4.5 Eccentric loading of bolted and riveted connections

Structural connections are commonly required to transmit moments as well as axial forces. Figure 4.9 shows the connection between a bracket and a stanchion; the bracket is attached to the stanchion through a system of six bolts or rivets, a vertical load P is applied to the bracket. Suppose the bolts or rivets are all of the same diameter. The load P is then replaced by a parallel load P applied to the centroid C of the rivet system, together with a moment Pe about the centroid Figure 4.9(ii); e is the perpendicular distance from C onto the line of action of P.

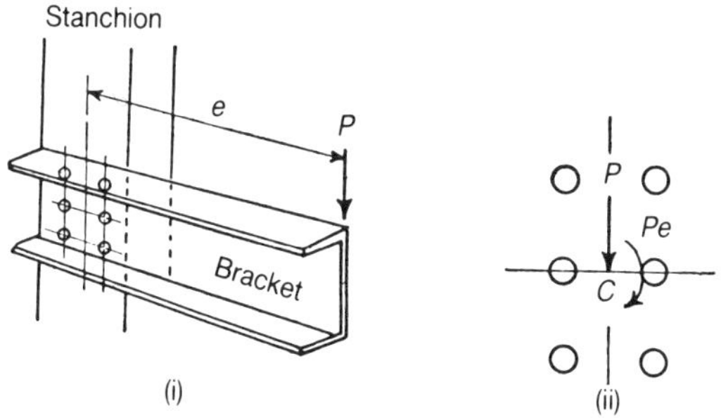

Figure 4.9 Eccentrically loaded connection leading to a bending action on the group of bolts, as well as a shearing action.

Consider separately the effects of the load P at C and the moment Pe. We assume that P is distributed equally amongst the bolts or rivets as a shearing force parallel to the line of action of P.

The moment Pe is assumed to induce a shearing force F in any bolt or rivet perpendicular to the line joining C to the bolt or rivet; moreover the force F is assumed to be proportional to the distance r from the bolt or rivet to C, (Figure 4.10).

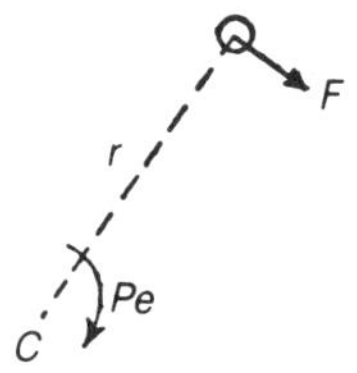

Figure 4.10 Assumed forces on the bolts.

For equilibrium we have

$$Pe = \Sigma Fr$$

If $F = kr$, where k is constant for all rivets, then

$$Pe = k \Sigma r^2$$

Thus, we have

$$k = \frac{Pe}{\Sigma r^2}$$

The force on a rivet is

$$F = kr = \frac{Pe}{\Sigma r^2} r \tag{4.8}$$

The resultant force on a bolt or rivet is then the vector sum of the forces due to P and Pe.

Problem 4.3 A bracket is bolted to a vertical stanchion and carries a vertical load of 50 kN. Assuming that the total shearing stress in a bolt is proportional to the relative displacement of the bracket and the stanchion in the neighbourhood of the bolt, find the load carried by each of the bolts. *(Cambridge)*

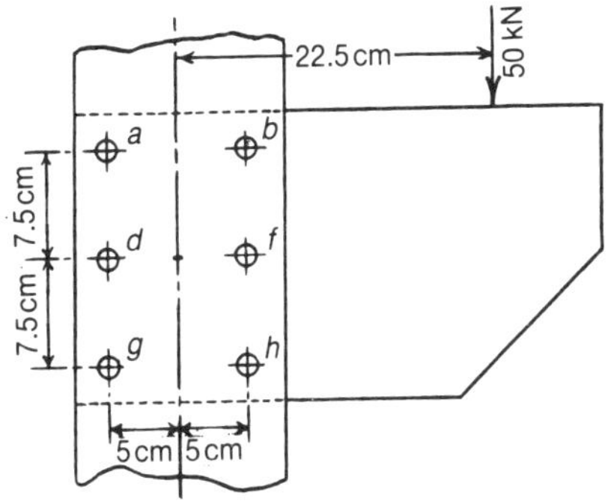

Solution

The centroid of the bolt system is at the point C. For bolt a

$$r = aC = [(0.050)^2 + (0.075)^2]^{1/2} = 0.0902 \text{ m}$$

For bolt b,

$$r = bC = aC = 0.0902 \text{ m}$$

For bolts d and f,

$$r = 0.050 \text{ m}$$

For bolts g and h,

$$r = gC = aC = 0.0902 \text{ m}$$

Then

$$\Sigma r^2 = 4(0.0902)^2 + 2(0.050)^2 = 0.0376 \text{ m}^2$$

Now

$$e = 0.225 \text{ m} \quad \text{and} \quad P = 50 \text{ kN}$$

Then

$$Pe = (0.225)(50 \times 10^3) = 11.25 \times 10^3 \text{ Nm}$$

The loads on the bolts *a, b g, h*, due to the couple *Pe* alone, are then

$$\frac{Pe}{\Sigma r^2} r = \frac{11.25 \times 10^3}{0.0376} (0.0902) = 28.0 \text{ kN}$$

These loads are at right-angles to *Ca, Cb, Cg* and *Ch*, respectively. The corresponding loads on the bolts *d* and *f* are

$$\frac{Pe}{\Sigma r^2} r = \frac{11.25 \times 10^3}{0.0376} (0.050) = 15.0 \text{ kN}$$

perpendicular to *Cd* and *Cf*, respectively.

The load on each bolt due to the vertical shearing force of 50 kN alone is

$$\frac{50 \times 10^3}{6} = 8.33 \times 10^3 \text{ N} = 8.33 \text{ kN}$$

This force acts vertically downwards on each bolt. The resultant loads on all the rivets are found by drawing parallelograms of forces as follows:

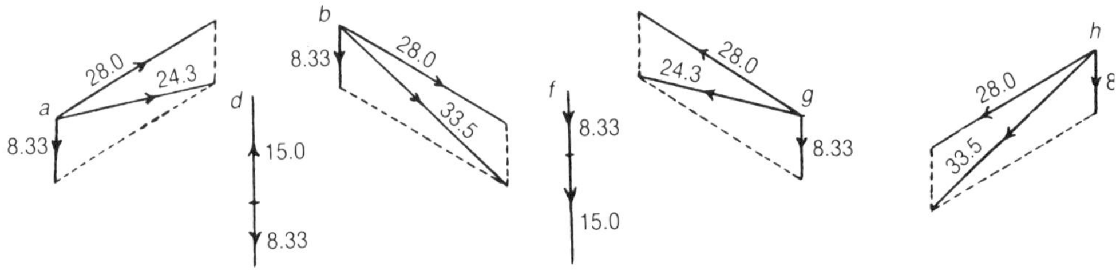

All force vectors are in kN. The resultant loads on the bolts are then as follows:

Bolts	Resultant Load
a and *g*	24.3 kN
b and *h*	33.5 kN
d	6.7 kN
f	23.3 kN

4.6 Welded connections

Some metals used in engineering—such as steel and aluminium—can be deposited in a molten state between two components to form a joint, which is then called a welded connection. The metal deposited to form the joint is called the weld. Two types of weld are in common use, the *butt weld* and the *fillet weld*; Figure 4.11 shows two plates connected by a butt weld; the plates are tapered at the joint to give sufficient space for the weld material. If the plates carry a tensile load the weld material carries largely tensile stresses. Figure 4.12 shows two plates connected by fillet welds; if the joint carries a tensile load the welds carry largely shearing stresses, although the state of stress in the welds is complex, and tensile stresses may also be present. Fillet welds of the type indicated in Figure 4.12 transmit force between the two plates by shearing actions within the welds; if the weld has the triangular cross-section shown in Figure 4.13(i), the shearing stress is greatest across the narrowest section of the weld, having a thickness $t/\sqrt{2}$. This section is called the *throat* of the weld. In Figure 4.13(ii), the weld has the same thickness t at all sections. To estimate approximately the strength of the welds in Figure 4.13 it is assumed that failure of the welds takes place by shearing across the throats of the welds.

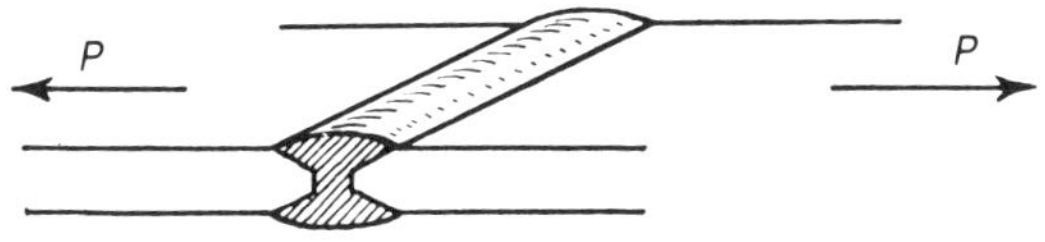

Figure 4.11 Butt weld between two plates.

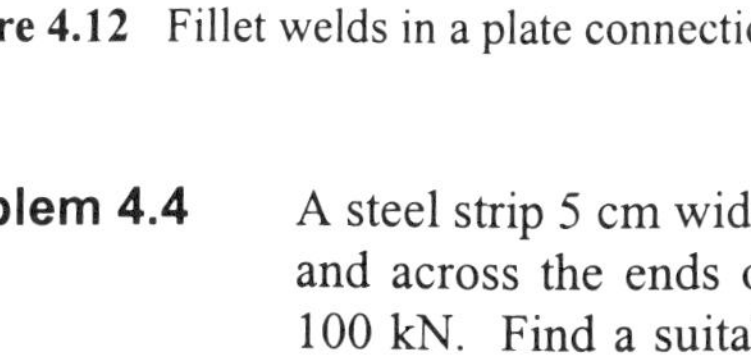

Figure 4.12 Fillet welds in a plate connection.

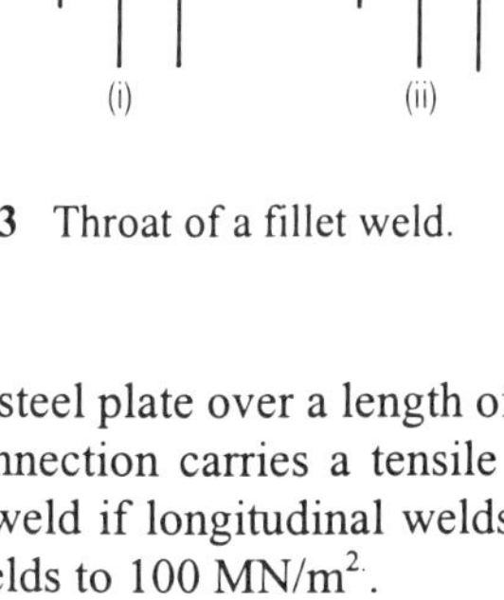

Figure 4.13 Throat of a fillet weld.

Problem 4.4 A steel strip 5 cm wide is fillet-welded to a steel plate over a length of 7.5 cm and across the ends of the strip. The connection carries a tensile load of 100 kN. Find a suitable size of the fillet weld if longitudinal welds can be stressed to 75 MN/m^2 and the transverse welds to 100 MN/m^2.

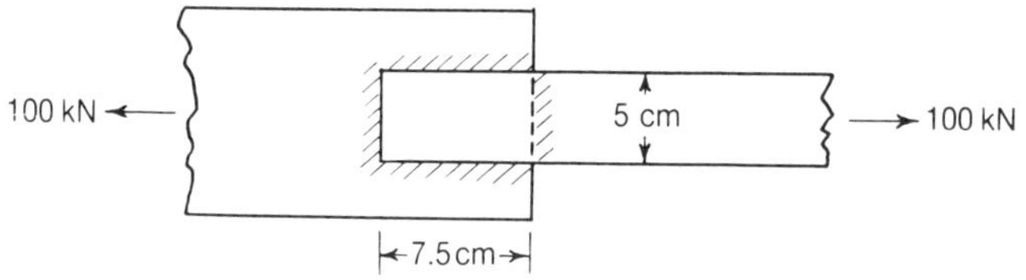

Solution

Suppose the throat thickness of the fillet-welds is t. Then the longitudinal welds carry a shearing force

$$\tau A = (75 \times 10^6)(0.075 \times 2t) = (11.25 \times 10^6)\, t \text{ N}$$

The transverse welds carry a shearing force

$$\tau A = (100 \times 10^6)(0.050 \times 2t) = (10 \times 10^6)\, t \text{ N}$$

Then

$$(11.25 \times 10^6)\, t + (10 \times 10^6)\, t = 100 \times 10^3$$

and therefore,

$$t = \frac{100}{21.25} \times 10^{-3} = 4.71 \times 10^{-3} \text{ m} = 0.471 \text{ cm}$$

The fillet size is then

$$t\sqrt{2} = 0.67 \text{ cm}$$

Problem 4.5 Two metal plates of the same material and of equal breadth are fillet welded at a lap joint. The one plate has a thickness t_1 and the other a thickness t_2. Compare the shearing forces transmitted through the welds, when the connection is under a tensile force P.

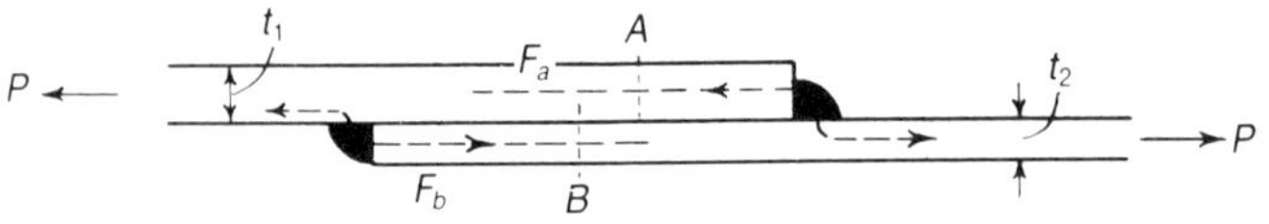

<u>*Solution*</u>

The sections of the plates between the welds will stretch by approximately the same amounts; thus, these sections will suffer the same strains and, as they are the same materials, they will also suffer the same stresses. If a shearing force F_a is transmitted by the one weld and a shearing force F_b by the other, then the tensile force over the section A in the one plate is F_a and over the section B in the other plate is F_b. If the plates have the same breadth and are to carry equal tensile stresses over the sections A and B, we have

$$\frac{F_a}{t_1} = \frac{F_b}{t_2}$$

and thus

$$\frac{F_a}{F_b} = \frac{t_1}{t_2}$$

We also have

$$F_a + F_b = P$$

and so

$$F_a = \frac{P}{1 + \frac{t_2}{t_1}} \quad \text{and} \quad F_b = \frac{P}{1 + \frac{t_1}{t_2}}$$

4.7 Welded connections under bending actions

Where a welded connection is required to transmit a bending moment we adopt a simple empirical method of analysis similar to that for bolted and riveted connections discussed in Section 4.5. We assume that the shearing stress in the weld is proportional to the distance of any part of the weld from the centroid of the weld. Consider, for example, a plate which is welded to a stanchion and which carries a bending moment M in the plane of the welds, Figure 4.14. We suppose the fillet-welds are of uniform thickness t around the parameter of a rectangle of sides a and b. At any point of the weld we take the shearing stress, τ, as acting normal to the line joining that point to the centroid C of the weld. If δA is an elemental area of weld at any point, then

$$M = \int \tau r \, dA$$

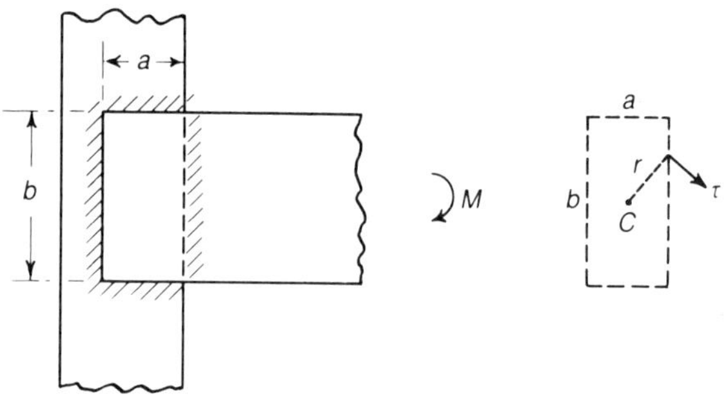

Figure 4.14 A plate fillet welded to a column, and transmitting a bending moment M.

If

$$\tau = kr$$

then

$$M = \int kr^2 \, dA = kJ$$

where J is the polar second moment of area of the weld about the axis through C and normal to the plane of the weld. Thus

$$k = \frac{M}{J}$$

and

$$\tau = \frac{Mr}{J} \tag{4.9}$$

According to this simple empirical theory, the greatest stresses occur at points of the weld most remote from the centroid C.

Problem 4.6 Two steel plates are connected together by 0.5 cm fillet welds. Estimate the maximum shearing stress in the welds if the joint carries a bending moment of 2500 Nm.

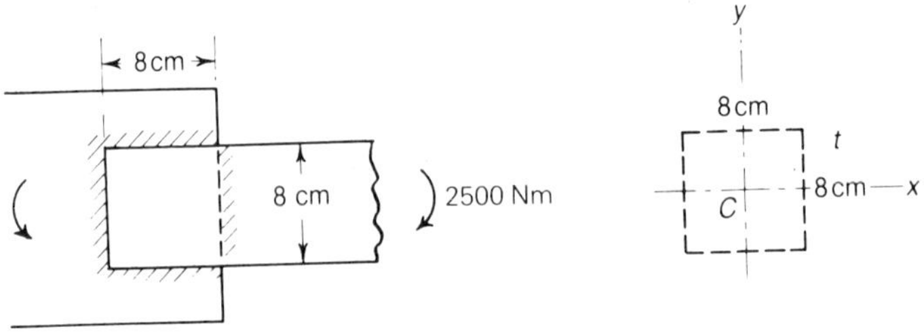

<u>Solution</u>

The centroid of the welds is at the centre of an 8 cm square. Suppose t is the throat or thickness of the welds. The second moment of area of the weld about Cx or Cy is

$$I_x = I_y = 2\left[\frac{1}{12}(t)(0.08)^3\right] + 2\left[(t)(0.08)(0.04)^2\right]$$

$$= (0.341 \times 10^{-3})\, t \text{ m}^4$$

The polar second moment of area about an axis through C is then

$$J = I_x + I_y = 2(0.341 \times 10^{-3})\ t = (0.682 \times 10^{-3})\ t \text{ m}^4$$

Now $t = 0.005/\sqrt{2}\,m$, and so

$$J = 2.41 \times 10^{-6} \text{ m}^4$$

The shearing stress in the weld at any radius r is

$$\tau = \frac{Mr}{J}$$

This is greatest at the corners of the square where it has the value

$$\tau_{max} = \frac{M}{J}\left(\frac{0.08}{\sqrt{2}}\right) = \frac{2500}{2.41 \times 10^{-6}}\left(\frac{0.08}{\sqrt{2}}\right)$$

$$= 58.6 \text{ MN/m}^2$$

Further problems *(answers on page 692)*

4.7 Two plates, each 1 cm thick are connected by riveting a single cover strap to the plates through two rows of rivets in each plate. The diameter of the rivets is 2 cm, and the distance between rivet centres along the breadth of the connection is 12.5 cm. Assuming the other unstated dimensions are adequate, calculate the strength of the joint per metre breadth, in tension, allowing 75 MN/m^2 shearing stress in the rivets and a tensile stress of 90 MN/m^2 in the plates. *(Cambridge)*

4.8 A flat steel bar is attached to a gusset plate by eight bolts. At the section AB the gusset plate exerts on the flat bar a vertical shearing force F and a counter-clockwise couple M.

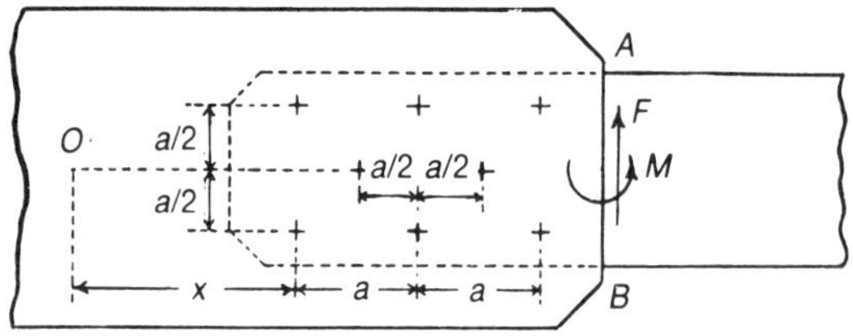

Assuming that the gusset plate, relative to the flat bar, undergoes a minute rotation about a point O on the line of the two middle rivets, also that the loads on the rivets are due to and proportional to the relative movement of the plates at the rivet holes, prove that

$$x = -a \times \frac{4M + 3aF}{4M + 6aF}$$

Prove also that the horizontal and vertical components of the load on the top right-hand rivet are

$$\frac{2M + 3aF}{24a} \quad \text{and} \quad \frac{4M + 9aF}{24a}$$

respectively.

4.9 A steel strip of cross-section 5 cm by 1.25 cm is bolted to two copper strips, each of cross-section 5 cm by 0.9375 cm, there being two bolts on the line of pull. Show that, neglecting friction and the deformation of the bolts, a pull applied to the joint will be shared by the bolts in the ratio 3 to 4. Assume that E for steel is twice E for copper.

4.10 Two flat bars are riveted together using cover plates, x being the pitch of the rivets in a direction at right angles to the plane of the figure. Assuming that the rivets themselves do not deform, show that the load taken by the rivets (1) is $tPx / (t + 2t^1)$ and that the rivets (2) are free from load.

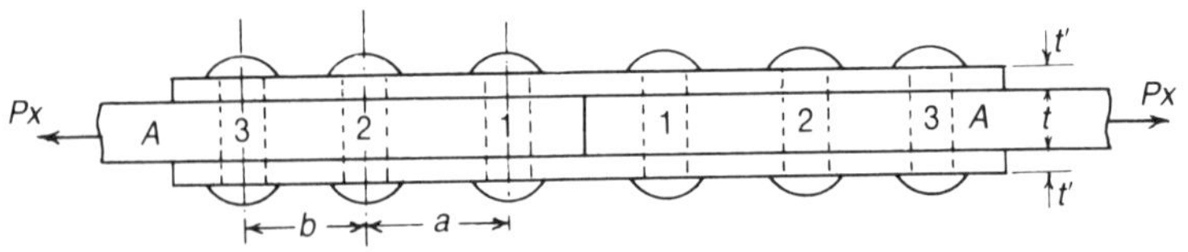

4.11 Two tie bars are connected together by 0.5 cm fillet welds around the end of one bar, and around the inside of a slot machined in the same bar. Estimate the strength of the connection in tension if the shearing stresses in the welds are limited to 75 MN/m^2.

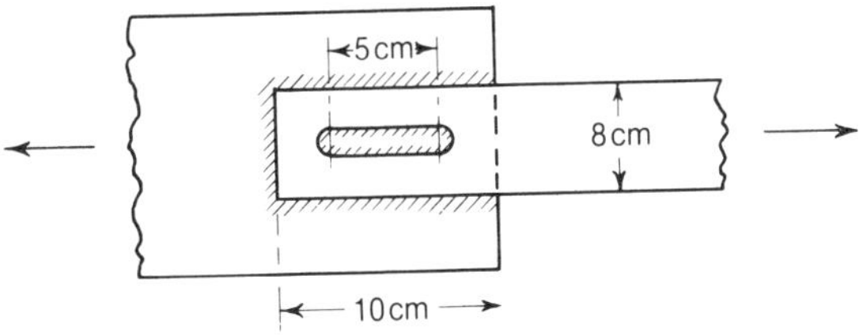

4.12 A bracket plate is welded to the face of a column and carries a vertical load P. Determine the value of P such that the maximum shearing stress in the 1 cm weld is 75 MN/m^2. *(Bristol)*

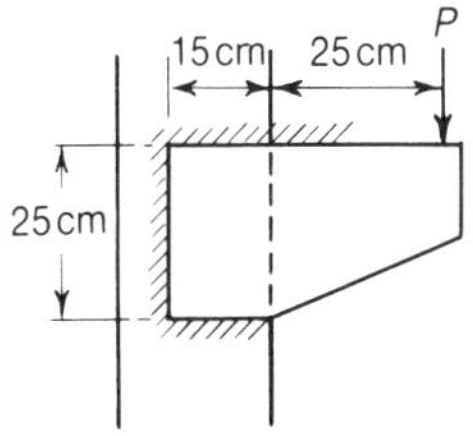

5 Analysis of stress and strain

5.1 Introduction

Up to the present we have confined our attention to considerations of simple direct and shearing stresses. But in most practical problems we have to deal with combinations of these stresses.

The strengths and elastic properties of materials are determined usually by simple tensile and compressive tests. How are we to make use of the results of such tests when we know that stress in a given practical problem is compounded from a tensile stress in one direction, a compressive stress in some other direction, and a shearing stress in a third direction? Clearly we cannot make tests of a material under all possible combinations of stress to determine its strength. It is essential, in fact, to study stresses and strains in more general terms; the analysis which follows should be regarded as having a direct and important bearing on practical strength problems, and is not merely a display of mathematical ingenuity.

5.2 Shearing stresses in a tensile test specimen

A long uniform bar, Figure 5.1, has a rectangular cross-section of area A. The edges of the bar are parallel to perpendicular axes Ox, Oy, Oz. The bar is uniformly stressed in tension in the x-direction, the tensile stress on a cross-section of the bar parallel to Ox being σ_x. Consider the stresses acting on an inclined cross-section of the bar; an inclined plane is taken at an angle θ to the yz-plane. The resultant force at the end cross-section of the bar is acting parallel to Ox.

$$P = A\sigma_x$$

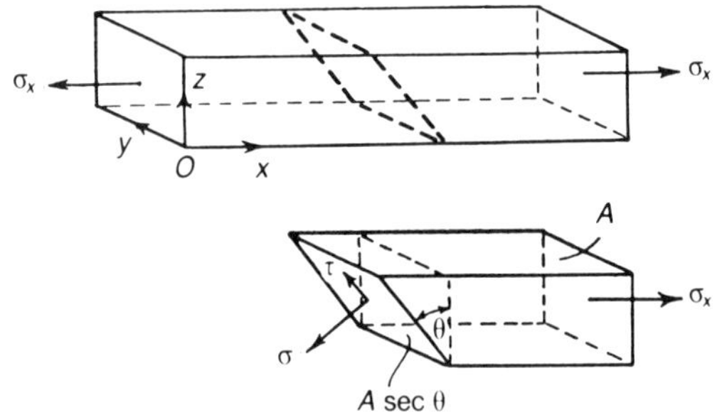

Figure 5.1 Stresses on an inclined plane through a tensile test piece.

For equilibrium the resultant force parallel to Ox on an inclined cross-section is also $P = A\sigma_x$. At the inclined cross-section in Figure 5.1, resolve the force $A\sigma_x$ into two components—one perpendicular, and the other tangential, to the inclined cross-section, the latter component acting parallel to the *xz*-plane. These two components have values, respectively, of

$$A\sigma_x \cos\theta \text{ and } A\sigma_x \sin\theta$$

The area of the inclined cross-section is

$$A \sec\theta$$

so that the normal and tangential stresses acting on the inclined cross-section are

$$\sigma = \frac{A\sigma_x \cos\theta}{A \sec\theta} = \sigma_x \cos^2\theta \tag{5.1}$$

$$\tau = \frac{A\sigma_x \sin\theta}{A \sec\theta} = \sigma_x \cos\theta \sin\theta \tag{5.2}$$

σ is the *direct stress* and τ the *shearing stress* on the inclined plane. It should be noted that the stresses on an inclined plane are not simply the resolutions of σ_x perpendicular and tangential to that plane; the important point in Figure 5.1 is that the area of an inclined cross-section of the bar is different from that of a normal cross-section. The shearing stress τ may be written in the form

$$\tau = \sigma_x \cos\theta \sin\theta = \tfrac{1}{2}\sigma_x \sin 2\theta$$

At $\theta = 0°$ the cross-section is perpendicular to the axis of the bar, and $\tau = 0$; τ increases as θ increases until it attains a maximum of ½ σ_x at $\theta = 45°$; τ then diminishes as θ increases further until it is again zero at $\theta = 90°$. Thus on any inclined cross-section of a tensile test-piece, shearing stresses are always present; the shearing stresses are greatest on planes at 45° to the longitudinal axis of the bar.

Problem 5.1 A bar of cross-section 2.25 cm by 2.25 cm is subjected to an axial pull of 20 kN. Calculate the normal stress and shearing stress on a plane the normal to which makes an angle of 60° with the axis of the bar, the plane being perpendicular to one face of the bar.

Solution

We have $\theta = 60°$, $P = 20$ kN and $A = 0.507 \times 10^{-3}$ m^2. Then

$$\sigma_x = \frac{20 \times 10^3}{0.507 \times 10^{-3}} = 39.4 \text{ MN/m}^2$$

The normal stress on the oblique plane is

$$\sigma = \sigma_x \cos^2 60° = \left(39.4 \times 10^6\right)\tfrac{1}{4} = 9.85 \text{ MN/m}^2$$

The shearing stress on the oblique plane is

$$\frac{1}{2}\sigma_x \sin 120^\circ = \frac{1}{2}\left(39.4 \times 10^6\right)\sqrt{\frac{3}{2}} = 17.05 \text{ MN / m}^2$$

5.3 Strain figures in mild steel; Lüder's lines

If a tensile specimen of mild steel is well polished and then stressed, it will be found that, when the specimen yields, a pattern of fine lines appears on the polished surface; these lines intersect roughly at right-angles to each other, and at 45° approximately to the longitudinal axis of the bar; these lines were first observed by Lüder in 1854. Lüder's lines on a tensile specimen of mild steel are shown in Figure 5.2. These strain figures suggest that yielding of the material consists of slip along the planes of greatest shearing stress; a single line represents a slip band, containing a large number of metal crystals.

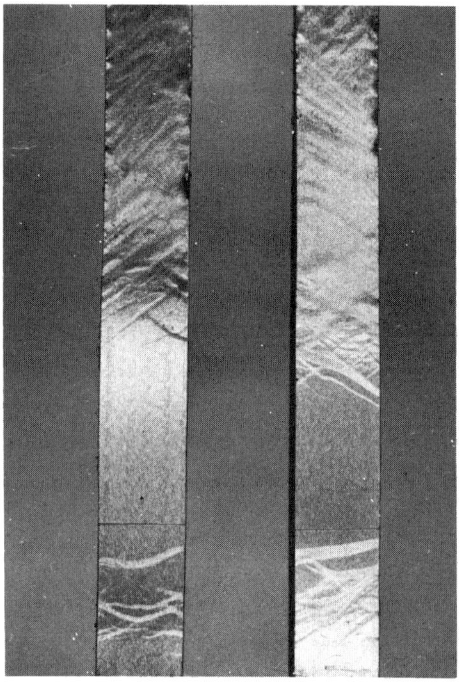

Figure 5.2 Lüder's lines in the yielding of a steel bar in tension.

5.4 Failure of materials in compression

Shearing stresses are also developed in a bar under uniform compression. The failure of some materials in compression is due to the development of critical shearing stresses on planes inclined to the direction of compression. Figure 5.3 shows two failures of compressed timbers; failure is due primarily to breakdown in shear on planes inclined to the direction of compression.

Figure 5.3 Failures of compressed specimens of timber, showing breakdown of the material in shear.

5.5 General two-dimensional stress system

A *two-dimensional* stress system is one in which the stresses at any point in a body act in the same plane. Consider a thin rectangular block of material, *abcd*, two faces of which are parallel to the *xy*-plane, Figure 5.4. A two-dimensional state of stress exists if the stresses on the remaining four faces are parallel to the *xy*-plane. In general, suppose the *forces* acting on the faces are *P*, *Q*, *R*, *S*, parallel to the *xy*-plane, Figure 5.4. Each of these forces can be resolved into components P_x, P_y etc., Figure 5.5. The perpendicular components give rise to direct stresses, and the tangential components to shearing stresses.

The system of *forces* in Figure 5.5 is now replaced by its equivalent system of *stresses*; the rectangular block of Figure 5.6 is in uniform state of two-dimensional stress; over the two faces parallel to *Ox* are direct and shearing stresses σ_y and τ_{yx}, respectively. The thickness is assumed to be 1 unit of length, for convenience, the other sides having lengths *a* and *b*. Equilibrium of the block in the *x*- and *y*-directions is already ensured; for rotational equilibrium of the block in the *xy*-plane we must have

$$[\tau_{xy}(a \times 1)] \times b = [\tau_{yx}(b \times 1)] \times a$$

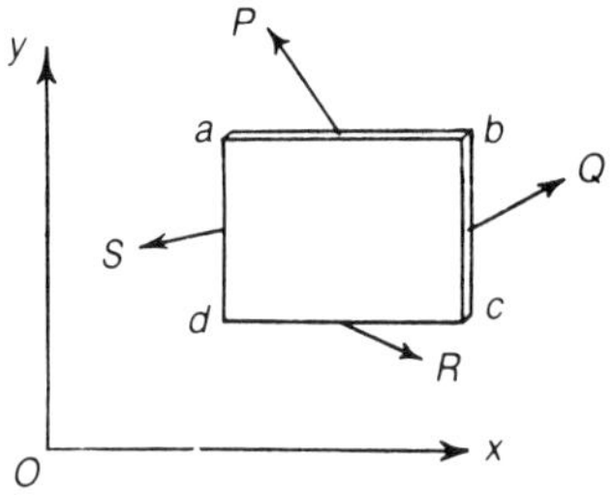

Figure 5.4 Resultant force acting on the faces of a 'two-dimensional' rectangular block.

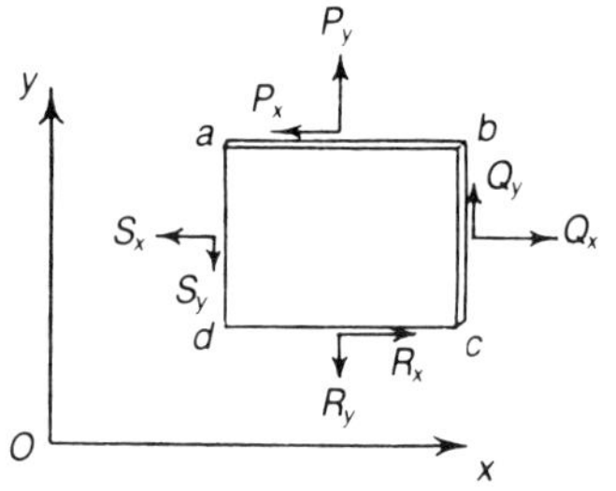

Figure 5.5 Components of resultant forces parallel to O_x and O_y.

Thus $\quad (ab)\, \tau_{xy} = (ab)\, \tau_{yx}$

or

$$\tau_{xy} = \tau_{yx} \tag{5.3}$$

Then the shearing stresses on perpendicular planes are equal and *complementary* as we found in the simpler case of pure shear in Section 3.3.

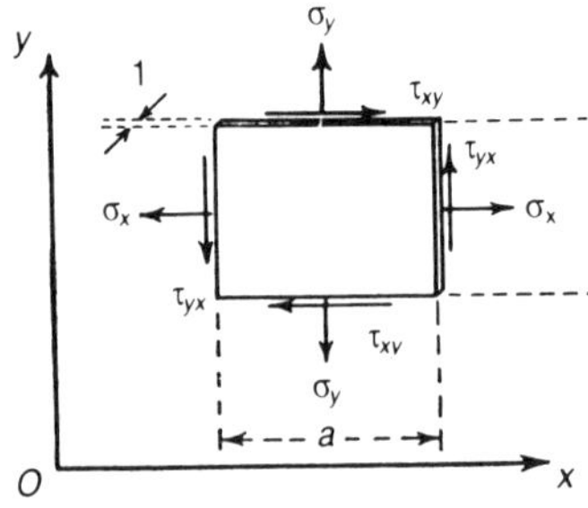

Figure 5.6 General two-dimensional state of stress.

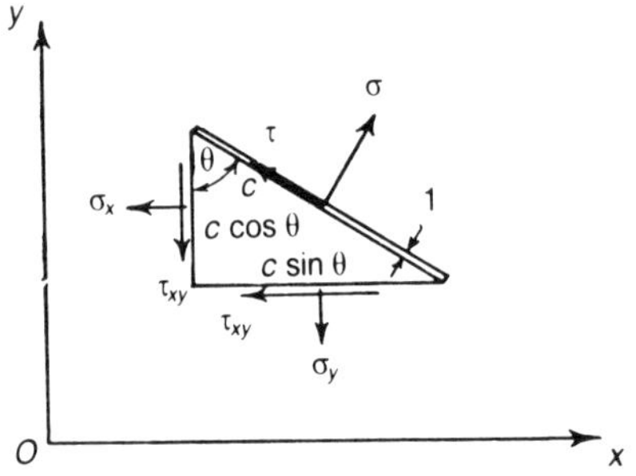

Figure 5.7 Stresses on an inclined plane in a two-dimensional stress system.

5.6 Stresses on an inclined plane

Consider the stresses acting on an inclined plane of the uniformly stressed rectangular block of Figure 5.6; the inclined plane makes an angle θ with O_y, and cuts off a 'triangular' block, Figure 5.7. The length of the hypotenuse is c, and the thickness of the block is taken again as one unit of length, for convenience. The values of direct stress, σ, and shearing stress, τ, on the inclined plane are found by considering equilibrium of the triangular block. The direct stress acts along the normal to the inclined plane. Resolve the forces on the three sides of the block parallel to this normal: then

$$\sigma\,(c.1) = \sigma_x\,(c\cos\theta\cos\theta) + \sigma_y\,(c\sin\theta\sin\theta) + \tau_{xy}\,(c\cos\theta\sin\theta) + \tau_{xy}\,(c\sin\theta\cos\theta)$$

This gives

$$\sigma = \sigma_x\cos^2\theta + \sigma_y\sin^2\theta + 2\tau_{xy}\sin\theta\cos\theta \tag{5.4}$$

Now resolve forces in a direction parallel to the inclined plane:

$$\tau.(c\ 1) = -\sigma_x\,(c\cos\theta\sin\theta) + \sigma_y\,(c\sin\theta\cos\theta) + \tau_{xy}\,(c\cos\theta\cos\theta) - \tau_{xy}\,(c\sin\theta\sin\theta)$$

This gives

$$\tau = -\sigma_x\cos\theta\sin\theta + \sigma_y\sin\theta\cos\theta + \tau_{xy}(\cos^2\theta - \sin^2\theta) \tag{5.5}$$

The expressions for σ and τ are written more conveniently in the forms:

$$\sigma = \tfrac{1}{2}(\sigma_x + \sigma_y) + \tfrac{1}{2}(\sigma_x - \sigma_y)\cos2\theta + \tau_{xy}\sin2\theta \tag{5.6}$$

$$\tau = -\tfrac{1}{2}(\sigma_x - \sigma_y)\sin2\theta + \tau_{xy}\cos2\theta \tag{5.7}$$

The shearing stress τ vanishes when

$$\tfrac{1}{2}(\sigma_x - \sigma_y)\sin2\theta = \tau_{xy}\cos2\theta$$

that is, when

$$\tan2\theta = \frac{2\tau_{xy}}{\sigma_x - \sigma_y} \tag{5.8}$$

or when

$$2\theta = \tan^{-1}\frac{2\tau_{xy}}{\sigma_x - \sigma_y} \quad \text{or} \quad \tan^{-1}\frac{2\tau_{xy}}{\sigma_x - \sigma_y} + 180°$$

These may be written

$$\theta = \frac{1}{2}\tan^{-1}\frac{2\tau_{xy}}{\sigma_x - \sigma_y} \quad \text{or} \quad \frac{1}{2}\tan^{-1}\frac{2\tau_{xy}}{\sigma_x - \sigma_y} + 90° \tag{5.9}$$

In a two-dimensional stress system there are thus two planes, separated by 90°, on which the shearing stress is zero. These planes are called the *principal planes*, and the corresponding values of σ are called the *principal stresses*. The direct stress σ is a maximum when

$$\frac{d\sigma}{d\theta} = -(\sigma_x - \sigma_y)\sin 2\theta + 2\tau_{xy}\cos 2\theta = 0$$

that is, when

$$\tan 2\theta = \frac{2\tau_{xy}}{\sigma_x - \sigma_y}$$

which is identical with equation (5.8), defining the directions of the principal stresses; thus the *principal stresses* are also the *maximum* and *minimum direct stresses* in the material.

5.7 Values of the principal stresses

The directions of the principal planes are given by equation (5.8). For any two-dimensional stress system, in which the values of σ_x, σ_y and τ_{xy} are known, $\tan 2\theta$ is calculable; two values of θ, separated by 90°, can then be found. The principal stresses are then calculated by substituting these vales of θ into equation (5.6).

Alternatively, the principal stresses can be calculated more directly without finding the principal planes. Earlier we defined a principal plane as one on which there is no shearing stress; in Figure 5.8 it is assumed that no shearing stress acts on a plane at an angle θ to *Oy*.

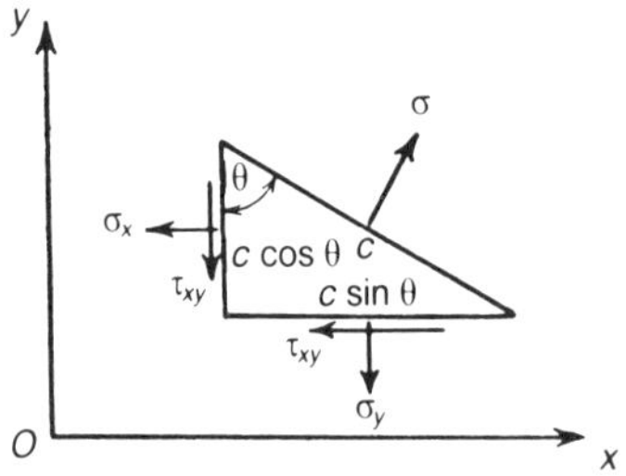

Figure 5.8 A principal stress acting on an inclined plane; there is no shearing stress τ associated with a principal stress σ.

For equilibrium of the triangular block in the *x*-direction,

$$\sigma(c\cos\theta) - \sigma_x(c\cos\theta) = \tau_{xy}(c\sin\theta)$$

and so

$$\sigma - \sigma_x = \tau_{xy}\tan\theta \qquad (5.10)$$

For equilibrium of the block in the y-direction

$$\sigma\,(c\sin\theta) - \sigma_y\,(c\sin\theta) = \tau_{xy}\,(c\cos\theta)$$

and thus

$$\sigma - \sigma_y = \tau_{xy}\cot\theta \qquad (5.11)$$

On eliminating θ between equations (5.10) and (5.11); by multiplying these equations together, we get

$$(\sigma - \sigma_x)(\sigma - \sigma_y) = \tau^2_{xy}$$

This equation is quadratic in σ; the solutions are

$$\sigma_1 = \frac{1}{2}(\sigma_x + \sigma_y) + \frac{1}{2}\sqrt{(\sigma_x - \sigma_y)^2 + 4\tau_{xy}^2} = \text{maximum principal stress}$$

$$(5.12)$$

$$\sigma_2 = \frac{1}{2}(\sigma_x + \sigma_y) - \frac{1}{2}\sqrt{(\sigma_x - \sigma_y)^2 + 4\tau_{xy}^2} = \text{minimum principal stress}$$

which are the values of the principal stresses; these stresses occur on mutually perpendicular planes.

5.8 Maximum shearing stress

The principal planes define directions of zero shearing stress; on some intermediate plane the shearing stress attains a maximum value. The shearing stress is given by equation (5.7); τ attains a maximum value with respect to θ when

$$\frac{d\tau}{d\theta} = -(\sigma_x - \sigma_y)\cos 2\theta - 2\tau_{xy}\sin 2\theta = 0$$

i.e., when

$$\cot 2\theta = -\frac{2\tau_{xy}}{\sigma_x - \sigma_y}$$

The planes of maximum shearing stress are inclined then at 45° to the principal planes. On substituting this value of cot 2θ into equation (5.7), the maximum numerical value of τ is

$$\tau_{max} = \sqrt{\left[\frac{1}{2}\left(\sigma_x - \sigma_y\right)\right]^2 + \left[\tau_{xy}\right]^2} \tag{5.13}$$

But from equations (5.12),

$$\sqrt{\left[\frac{1}{2}\left(\sigma_x - \sigma_y\right)\right]^2 + \left[\tau_{xy}\right]^2} = \sigma_1 - \frac{1}{2}\left(\sigma_x + \sigma_y\right) = \frac{1}{2}\left(\sigma_x + \sigma_y\right) - \sigma_2$$

where σ_1 and σ_2 are the principal stresses of the stress system. Then by adding together the two equations on the right hand side, we get

$$2\sqrt{\left[\frac{1}{2}\left(\sigma_x - \sigma_y\right)\right]^2 + \left[\tau_{xy}\right]^2} = \sigma_1 - \sigma_2$$

and equation (5.13) becomes

$$\tau_{max} = \frac{1}{2}\left(\sigma_1 - \sigma_2\right) \tag{5.14}$$

The maximum shearing stress is therefore half the difference between the principal stresses of the system.

Problem 5.2 At a point of a material the two-dimensional stress system is defined by

σ_x = 60.0 MN/m^2, tensile

σ_y = 45.0 MN/m^2, compressive

τ_{xy} = 37.5 MN/m^2, shearing

where σ_x, σ_y, τ_{xy} refer to Figure 5.7. Evaluate the values and directions of the principal stresses. What is the greatest shearing stress?

Solution

Now, we have

$$\frac{1}{2}\left(\sigma_x + \sigma_y\right) = \frac{1}{2}\ (60.0 - 45.0) = 7.5 \text{ MN/m}^2$$

$$\frac{1}{2}\left(\sigma_x - \sigma_y\right) = \frac{1}{2}\ (60.0 + 45.0) = 52.5 \text{ MN/m}^2$$

Then, from equations (5.12),

$$\sigma_1 = 7.5 + \left[(52.5)^2 + (37.5)^2\right]^{\frac{1}{2}} = 7.5 + 64.4 = 71.9 \text{ MN/m}^2$$

$$\sigma_2 = 7.5 - \left[(52.5)^2 + (37.5)^2\right]^{\frac{1}{2}} = 7.5 - 64.4 = -56.9 \text{ MN/m}^2$$

From equation (5.8)

$$\tan 2\theta = \frac{2\tau_{xy}}{\sigma_x - \sigma_y} = \frac{37.5}{52.5} = 0.714$$

Thus,

$$2\theta = \tan^{-1}(0.714) = 35.5° \text{ or } 215.5°$$

Then

$$\theta = 17.8° \text{ or } 107.8°$$

From equation (5.14)

$$\tau_{max} = \frac{1}{2}(\sigma_1 - \sigma_2) = \frac{1}{2}(71.9 + 56.9) = 64.4 \text{ MN/m}^2$$

This maximum shearing stress occurs on planes at 45° to those of the principal stresses.

5.9 Mohr's circle of stress

A geometrical interpretation of equations (5.6) and (5.7) leads to a simple method of stress analysis. Now, we have found already that

$$\sigma = \frac{1}{2}(\sigma_x + \sigma_y) + \frac{1}{2}(\sigma_x - \sigma_y)\cos 2\theta + \tau_{xy}\sin 2\theta$$

$$\tau = -\frac{1}{2}(\sigma_x - \sigma_y)\sin 2\theta + \tau_{xy}\cos 2\theta$$

Take two perpendicular axes $O\sigma$, $O\tau$, Figure 5.9; on this co-ordinate system set off the point having co-ordinates (σ_x, τ_{xy}) and $(\sigma_y, -\tau_{xy})$, corresponding to the known stresses in the x- and y-directions. The line PQ joining these two points is bisected by the $O\sigma$ axis at a point O'. With a centre at O', construct a circle passing through P and Q. The stresses σ and τ on a plane at an angle θ to Oy are found by setting off a radius of the circle at an angle 2θ to PQ, Figure 5.9; 2θ is measured in a clockwise direction from $O'P$.

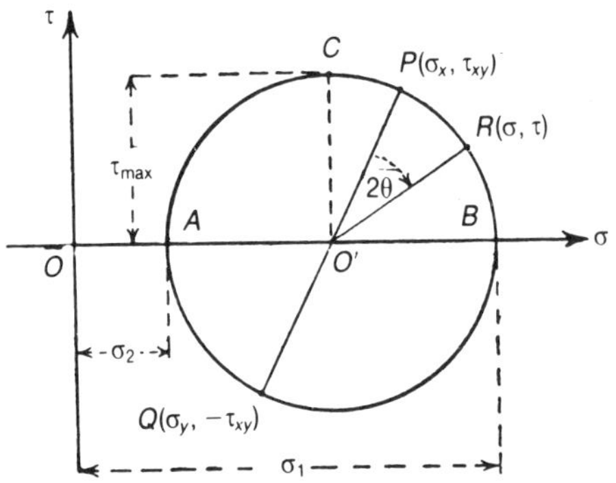

Figure 5.9 Mohr's circle of stress. The points *P* and *Q* correspond to the stress states (σ_x, τ_{xy}) and (σ_y, $-\tau_{xy}$) respectively, and are diametrically opposite; the state of stress (σ, τ) on a plane inclined at an angle θ to *Oy* is given by the point *R*.

The co-ordinates of the point $R(\sigma, \tau)$ give the direct and shearing stresses on the plane. We may write the above equations in the forms

$$\sigma - \frac{1}{2}\left(\sigma_x + \sigma_y\right) = \frac{1}{2}\left(\sigma_x - \sigma_y\right)\cos 2\theta + \tau_{xy}\sin 2\theta$$

$$-\tau = \frac{1}{2}\left(\sigma_x - \sigma_y\right)\sin 2\theta - \tau_{xy}\cos 2\theta$$

Square each equation and add; then we have

$$\left[\sigma - \frac{1}{2}\left(\sigma_x + \sigma_y\right)\right]^2 + \tau^2 = \left[\frac{1}{2}\left(\sigma_x - \sigma_y\right)\right]^2 + \left[\tau_{xy}\right]^2 \quad (5.15)$$

Thus all corresponding values of σ and τ lie on a circle of radius

$$\sqrt{\left[\frac{1}{2}\left(\sigma_x - \sigma_y\right)\right]^2 + \tau_{xy}^2}$$

with its centre at the point (½[$\sigma_x + \sigma_y$], 0), Figure 5.9.

This circle defining all possible states of stress is known as *Mohr's Circle of Stress*; the principal stresses are defined by the points *A* and *B*, at which $\tau = 0$. The maximum shearing stress, which is given by the point *C*, is clearly the radius of the circle.

Problem 5.3 At a point of a material the stresses forming a two-dimensional system are shown in Figure 5.10. Using Mohr's circle of stress, determine the magnitudes and directions of the principal stresses. Determine also the value of the maximum shearing stress.

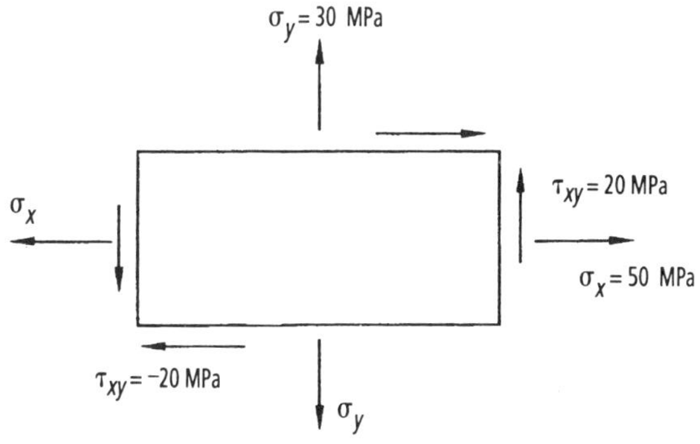

Figure 5.10. Stress at a point.

<u>*Solution*</u>

From Figure 5.10, the shearing stresses acting in conjunction with σ_x are counter-clockwise, hence, τ_{xy} is said to be positive on the vertical planes. Similarly, the shearing stresses acting in conjunction with τ_y are clockwise, hence, τ_{xy} is said to be negative on the horizontal planes.

On the $\sigma - \tau$ diagram of Figure 5.11, construct a circle with the line joining the point (σ_x, τ_{xy}) or (50, 20) and the point $(\sigma_y, -\tau_{xy})$ or (30, −20) as the diameter, as shown by *A* and *B*, respectively

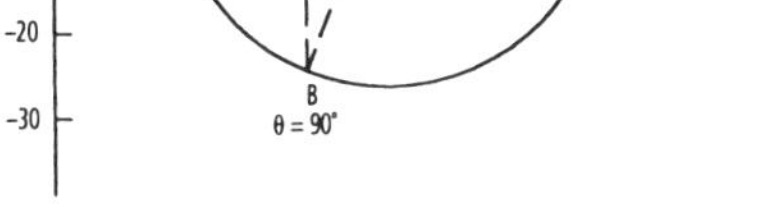

Figure 5.11 Problem 5.3.

The principal stresses and their directions can be obtained from a scaled drawing, but we shall calculate σ_1, σ_2 etc.

$$DA = 20 \text{ MPa}$$

$$OD = \sigma_x = 50 \text{ MPa}$$

$$OG = \sigma_y = 30 \text{ MPa}$$

$$OC = \frac{(OD + OG)}{2} = \frac{(50 + 30)}{2} = 40 \text{ MPa}$$

$$CD = OD - OC = 50 - 40 = 10 \text{ MPa}$$

$$AC^2 = CD^2 + DA^2$$

$$= 10^2 + 20^2$$

or $AC = 22.36$ MPa

$$\sigma_1 = OE = OC + AC = 40 + 22.36$$

$$\sigma_1 = 62.36 \text{ MPa}$$

$$\sigma_2 = OF = OC - AC$$

$$= 40 - 22.36$$

or $\sigma_2 = 17.64$ MPa

$$2\theta = \tan^{-1}\left(\frac{AD}{CD}\right)$$

$$= \tan^{-1}\left(\frac{20}{10}\right) = 63.43°$$

$\therefore \theta = 31.7°$ see below

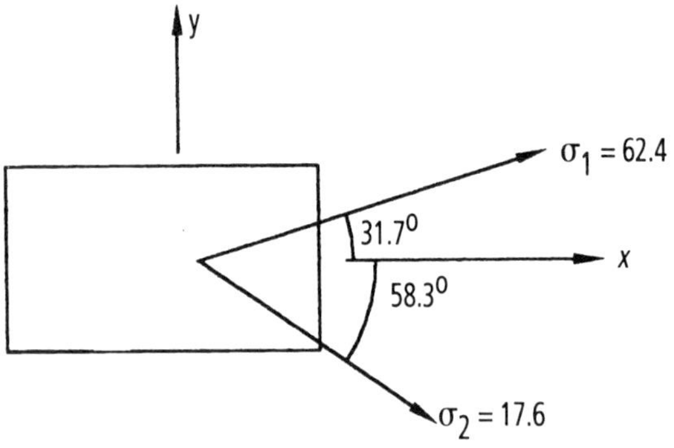

Maximum shear stress $= \tau_{max} = AC = 22.36$ MPa which occurs on planes at 45° to those of the principal stresses.

Problem 5.4 At a point of a material the two-dimensional state of stress is shown in Figure 5.12. Determine σ_1, σ_2, θ and τ_{max}

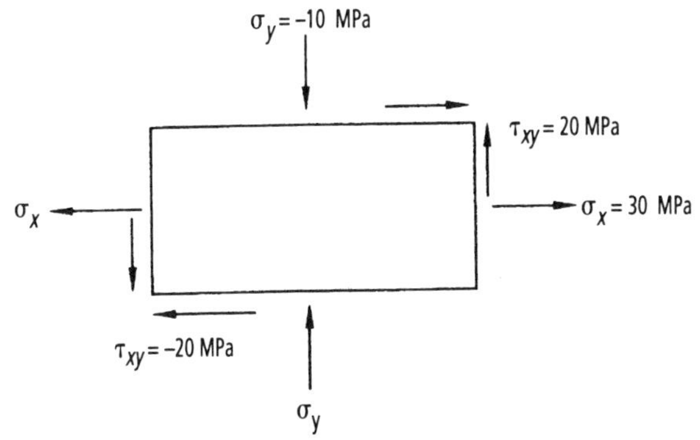

Figure 5.12 Stress at a point.

<u>*Solution*</u>

On the σ–τ diagram of Figure 5.13, construct a circle with the line joining the point (σ_x, τ_{xy}) or (30, 20) to the point (σ_y, $-\tau_{xy}$) or (−10, −20), as the diameter, as shown by the points *A* and *B* respectively. It should be noted that τ_{xy} is positive on the vertical planes of Figure 5.12, as these shearing stresses are causing a counter-clockwise rotation; vice-versa for the shearing stresses on the horizontal planes.

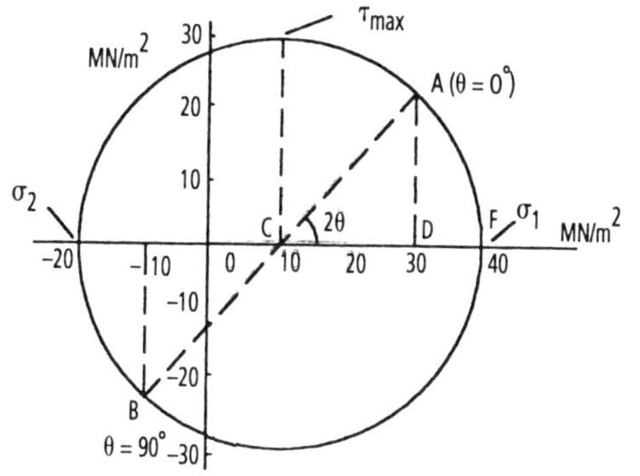

Figure 5.13 Problem 5.4.

From Figure 5.13,

$$AD = \tau_{xy} = 20$$

$$OD = \sigma_x = 30$$

$$OE = \sigma_y = -10$$

$$OC = \frac{(OD + OE)}{2} = \frac{(30 - 10)}{2}$$

or $OC = 10$

$$CD = OD - OC = 30 - 10 = 20$$

$$AC^2 = CD^2 + AD^2$$

$$= 20^2 + 20^2 = 800$$

or $AC = 28.28$

$$\sigma_1 = OF = OC + AC = 10 + 28.28$$

or $\sigma_1 = 38.3$ MPa

$$\sigma_2 = OG = OC - AC$$

$$= 10 - 28.3$$

or $\sigma_2 = -18.3$ MPa

$$2\theta = \tan^{-1}\left(\frac{AD}{CD}\right) = \left(\frac{20}{20}\right) = 45°$$

$\therefore \theta = 22.5$ (see below)

τ_{max} = Maximum shearing stress = AC

or $\underline{\tau_{max}}$ = 28.3 MPa acting on planes at 45° to σ_1 and σ_2.

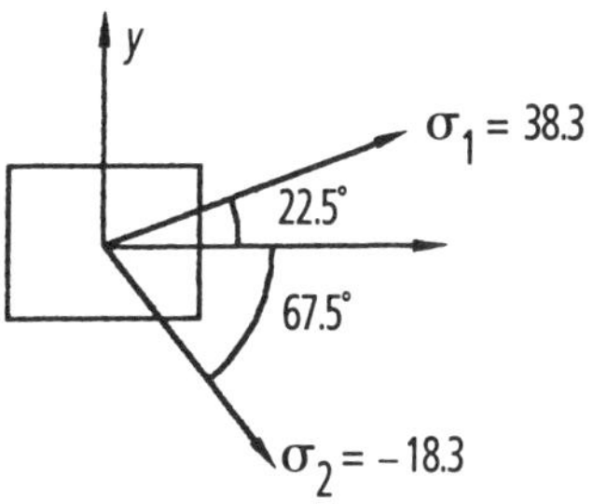

5.10 Strains in an inclined direction

For two-dimensional system of strains the direct and shearing strains in any direction are known if the direct and shearing strains in two mutually perpendicular directions are given. Consider a rectangular element of material, *OABC*, in the *xy*-plane, Figure 5.14, it is required to find the direct and shearing strains in the direction of the diagonal *OB*, when the direct and shearing strains in the directions *Ox, Oy* are given. Suppose ε_x is the strain in the direction *Ox*, ε_y the strain in the direction *Oy*, and γ_{xy} the shearing strain relative to *Ox* and *Oy*.

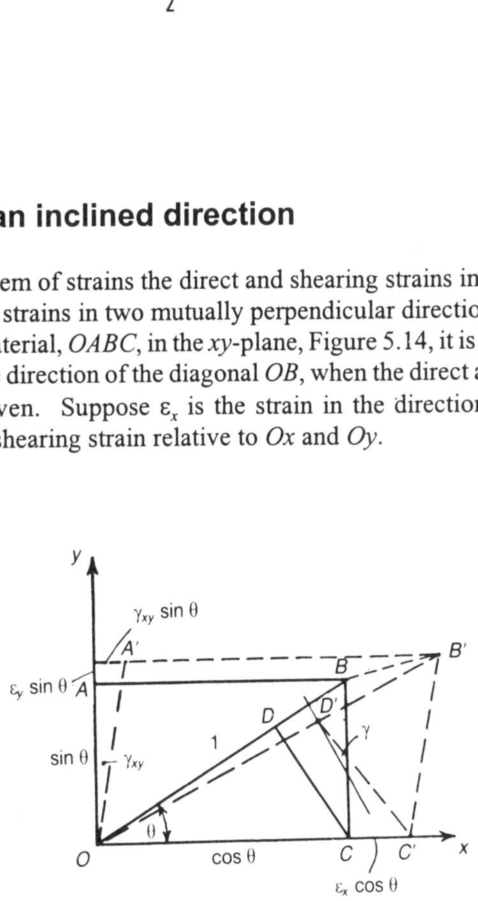

Figure 5.14 Strains in an inclined direction; strains in the directions O_x and O_y and defined by ε_x, ε_y and γ_{xy}, lead to strains ε, γ along the inclined direction *OB*.

All the strains are considered to be small; in Figure 5.14, if the diagonal *OB* of the rectangle is taken to be of unit length, the sides *OA, OB* are of lengths $\sin\theta$, $\cos\theta$, respectively, in which θ is the angle *OB* makes with *Ox*. In the strained condition *OA* extends a small amount $\varepsilon_y \sin\theta$, *OC* extends a small amount $\varepsilon_x \cos\theta$, and due to shearing strain *OA* rotates through a small angle γ_{xy}.

If the point B moves to point B', the movement of B parallel to Ox is

$$\varepsilon_x \cos\theta + \gamma_{xy} \sin\theta$$

and the movement parallel to Oy is

$$\varepsilon_y \sin\theta$$

Then the movement of B parallel to OB is

$$\left(\varepsilon_x \cos\theta + \gamma_{xy} \sin\theta\right) \cos\theta + \left(\varepsilon_y \sin\theta\right) \sin\theta$$

Since the strains are small, this is equal to the extension of the OB in the strained condition; but OB is of unit length, so that the extension is also the direct strain in the direction OB. If the direct strain in the direction OB is denoted by ε, then

$$\varepsilon = \left(\varepsilon_x \cos\theta + \gamma_{xy} \sin\theta\right) \cos\theta + \left(\varepsilon_y \sin\theta\right) \sin\theta$$

This may be written in the form

$$\varepsilon = \varepsilon_x \cos^2\theta + \varepsilon_y \sin^2\theta + \gamma_{xy} \sin\theta \cos\theta$$

and also in the form

$$\varepsilon = \frac{1}{2}\left(\varepsilon_x + \varepsilon_y\right) + \frac{1}{2}\left(\varepsilon_x - \varepsilon_y\right) \cos 2\theta + \frac{1}{2}\gamma_{xy} \sin 2\theta \qquad (5.16)$$

This is similar in form to equation (5.6), defining the direct stress on an inclined plane; ε_x and ε_y replace σ_x and σ_y, respectively, and $\frac{1}{2}\gamma_{xy}$ replaces τ_{xy}.

To evaluate the shearing strain in the direction OB we consider the displacements of the point D, the foot of the perpendicular from C to OB, in the strained condition, Figure 5.10. The point D, is displaced to a point D'; we have seen that OB extends an amount ε, so that OD extends an amount

$$\varepsilon\, OD = \varepsilon \cos^2\theta$$

During straining the line CD rotates anti-clockwise through a small angle

$$\frac{\varepsilon_x \cos^2\theta - \varepsilon \cos^2\theta}{\cos\theta \sin\theta} = \left(\varepsilon_x - \varepsilon\right) \cot\theta$$

At the same time OB rotates in a clockwise direction through a small angle

$$\left(\varepsilon_x \cos\theta + \gamma_{xy} \sin\theta\right) \sin\theta - \left(\varepsilon_y \sin\theta\right) \cos\theta$$

The amount by which the angle *ODC* diminishes during straining is the shearing strain γ in the direction *OB*. Thus

$$\gamma = -(\varepsilon_x - \varepsilon)\cot\theta - (\varepsilon_x \cos\theta + \gamma_{xy}\sin\theta)\sin\theta + (\varepsilon_y \sin\theta)\cos\theta$$

On substituting for ε from equation (5.16) we have

$$\gamma = -2(\varepsilon_x - \varepsilon_y)\cos\theta\sin\theta + \gamma_{xy}(\cos^2\theta - \sin^2\theta)$$

which may be written

$$\frac{1}{2}\gamma = -\frac{1}{2}(\varepsilon_x - \varepsilon_y)\sin 2\theta + \frac{1}{2}\gamma_{xy}\cos 2\theta \qquad (5.17)$$

This is similar in form to equation (5.7) defining the shearing stress on an inclined plane; σ_x and σ_y in that equation are replaced by ε_x and ε_y, respectively, and τ_{xy} by $\frac{1}{2}\gamma_{xy}$.

5.11 Mohr's circle of strain

The direct and shearing strains in an inclined direction are given by relations which are similar to equations (5.6) and (5.7) for the direct and shearing stresses on an inclined plane. This suggests that the strains in any direction can be represented graphically in a similar way to the stress system. We may write equations (5.16) and (5.17) in the forms

$$\varepsilon - \frac{1}{2}(\varepsilon_x + \varepsilon_y) = \frac{1}{2}(\varepsilon_x - \varepsilon_y)\cos 2\theta + \frac{1}{2}\gamma_{xy}\sin 2\theta$$

$$\frac{1}{2}\gamma = -\frac{1}{2}(\varepsilon_x - \varepsilon_y)\sin 2\theta + \frac{1}{2}\gamma_{xy}\cos 2\theta$$

Square each equation, and then add; we have

$$\left[\varepsilon - \frac{1}{2}\left(\varepsilon_x + \varepsilon_y\right)\right]^2 + \left[\frac{1}{2}\gamma\right]^2 = \left[\frac{1}{2}\left(\varepsilon_x - \varepsilon_y\right)\right]^2 + \left[\frac{1}{2}\gamma_{xy}\right]^2$$

Thus all values of ε and $\frac{1}{2}\gamma$ lie on a circle of radius

$$\sqrt{\left[\frac{1}{2}\left(\varepsilon_x - \varepsilon_y\right)\right]^2 + \left[\frac{1}{2}\gamma_{xy}\right]^2}$$

with its centre at the point

$$\left[\frac{1}{2}(\varepsilon_x + \varepsilon_y)\,,\ 0\right]$$

This circle defining all possible states of strain is usually called *Mohr's circle of strain*. For given values of ε_x, ε_y and γ_{xy} it is constructed in the following way: two mutually perpendicular axes, ε and $\frac{1}{2}\gamma$, are set up, Figure 5.15; the points $(\varepsilon_x, \frac{1}{2}\gamma_{xy})$ and $(\varepsilon_y, -\frac{1}{2}\gamma_{xy})$ are located; the line joining these points is a diameter of the circle of strain. The values of ε and $\frac{1}{2}\gamma$ in an inclined direction making an angle θ with Ox (Figure 5.10) are given by the points on the circle at the ends of a diameter making an angle 2θ with PQ; the angle 2θ is measured clockwise.

We note that the maximum and minimum values of ε, given by ε_1 and ε_2 in Figure 5.15, occur when $\frac{1}{2}\gamma$ is zero; ε_1, ε_2 are called *principal strains*, and occur for directions in which there is no shearing strain.

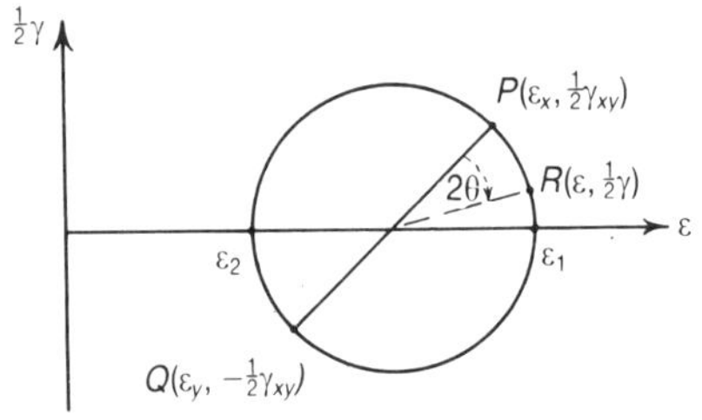

Figure 5.15 Mohr's circle of strain; the diagram is similar to the circle of stress, except that $\frac{1}{2}\gamma$ is plotted along the ordinates and not γ.

An important feature of this strain analysis is that we have *not* assumed that the strains are elastic; we have taken them to be small, however, with this limitation Mohr's circle of strain is applicable to both elastic and inelastic problems.

5.12 Elastic stress–strain relations

When a point of a body is acted upon by stresses σ_x and σ_y in mutually perpendicular directions the strains are found by superposing the strains due to σ_x and σ_y acting separately.

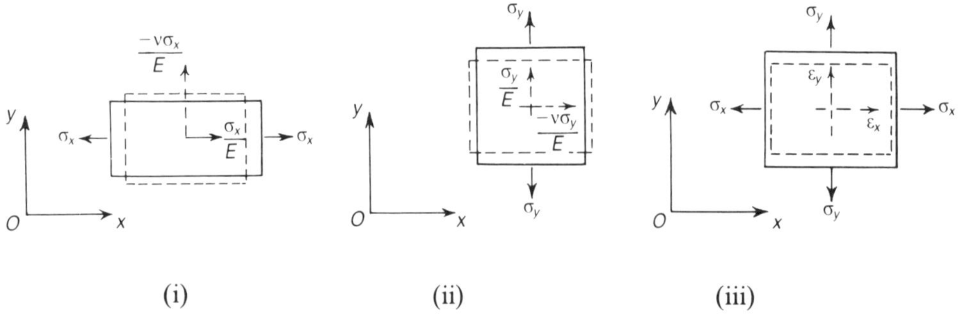

Figure 5.16 Strains in a two-dimensional linear-elastic stress system; the strains can be regarded as compounded of two systems corresponding to uni-axial tension in the x- and y- directions.

The rectangular element of material in Figure 5.16(i) is subjected to a tensile stress σ_x in the x direction; the tensile strain in the x-direction is

$$\frac{\sigma_x}{E}$$

and the compressive strain in the y-direction is

$$-\frac{\nu\sigma_x}{E}$$

in which E is Young's modulus, and ν is Poisson's ratio (see section 1.10). If the element is subjected to a tensile stress σ_y in the y-direction as in Figure 5.12(ii), the compressive strain in the x-direction is

$$-\frac{\nu\sigma_y}{E}$$

and the tensile strain in the y-direction is

$$\frac{\sigma_y}{E}$$

These elastic strains are small, and the state of strain due to both stresses σ_x and σ_y, acting simultaneously, as in Figure 5.16(iii), is found by superposing the strains of Figures 5.16(i) and (ii); taking tensile strain as positive and compressive strain as negative, the strains in the x- and y-directions are given, respectively, by

$$\varepsilon_x = \frac{\sigma_x}{E} - \frac{\nu\sigma_y}{E} \tag{5.18}$$

$$\varepsilon_y = \frac{\sigma_y}{E} - \frac{\nu\sigma_x}{E}$$

On multiplying each equation by E, we have

$$\begin{aligned} E\varepsilon_x &= \sigma_x - \nu\sigma_y \\ E\varepsilon_y &= \sigma_y - \nu\sigma_x \end{aligned} \tag{5.19}$$

These are the elastic stress-strain relations for two-dimensional system of direct stresses. When a shearing stress τ_{xy} is present in addition to the direct stresses σ_x and σ_y, as in Figure 5.17, the shearing stress τ_{xy} is assumed to have no effect on the direct strains ε_x and ε_y caused by σ_x and σ_y.

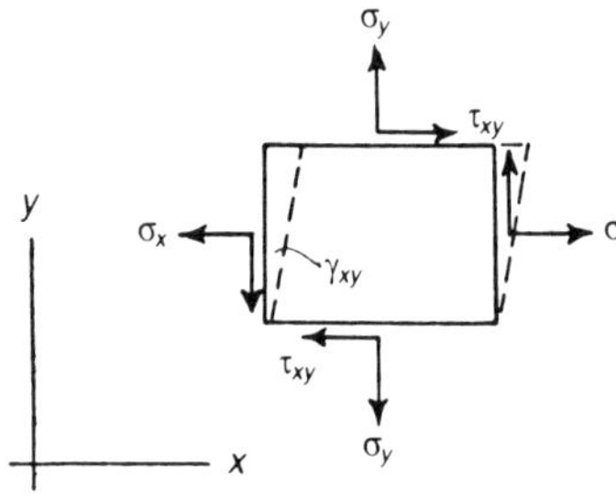

Figure 5.17 Shearing strain in a two-dimensional system.

Similarly, the direct stresses σ_x and σ_y are assumed to have no effect on the shearing strain γ_{xy} due to τ_{xy}. When shearing stresses are present, as well as direct stresses, there is therefore an additional stress-strain relation having the form in which G is the shearing modulus.

$$\frac{\tau_{xy}}{\gamma_{xy}} = G$$

Then, in addition to equations (5.19) we have the relation

$$\tau_{xy} = G\gamma_{xy} \tag{5.20}$$

5.13 Principal stresses and strains

We have seen that in a two-dimensional system of stresses there are always two mutually perpendicular directions in which there are no shearing stresses; the direct stresses on these planes were referred to as principal stresses, σ_1 and σ_2. As there are no shearing stresses in these two mutually perpendicular directions, there are also no shearing strains; for the principal directions the corresponding direct strains are given by

$$\begin{aligned} E\varepsilon_1 &= \sigma_1 - \nu\sigma_2 \\ E\varepsilon_2 &= \sigma_2 - \nu\sigma_1 \end{aligned} \tag{5.21}$$

The direct strains, ε_1, ε_2, are the principal strains already discussed in Mohr's circle of strain. It follows that the principal strains occur in directions parallel to the principal stresses.

5.14 Relation between *E, G* and *ν*

Consider an element of material subjected to a tensile stress σ_0 in one direction together with a compressive stress σ_0 in a mutually perpendicular direction, Figure 5.18(i). The Mohr's circle for this state of stress has the form shown in Figure 5.18(ii); the circle of stress has a centre at the origin and a radius of σ_0. The direct and shearing stresses on an inclined plane are given by the co-ordinates of a point on the circle; in particular we note that there is no direct stress when $2\theta = 90°$, that is, when $\theta = 45°$ in Figure 5.18(i).

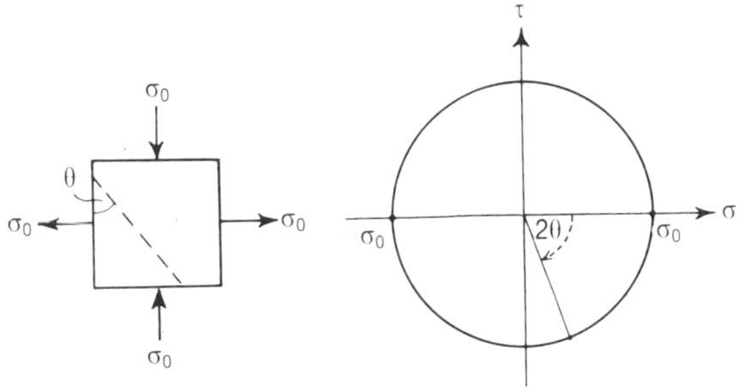

Figure 5.18 (i) A stress system consisting of tensile and compressive stresses of equal magnitude, but acting in mutually perpendicular directions. (ii) Mohr's circle of stress for this system.

Moreover when $\theta = 45°$, the shearing stress on this plane is of magnitude σ_0. We conclude then that a state of equal and opposite tension and compression, as indicated in Figure 5.18(i), is equivalent, from the stress standpoint, to a condition of simple shearing in directions at 45°, the shearing stresses having the same magnitudes as the direct stresses σ_0 (Figure 5.19). This system of stresses is called *pure shear*.

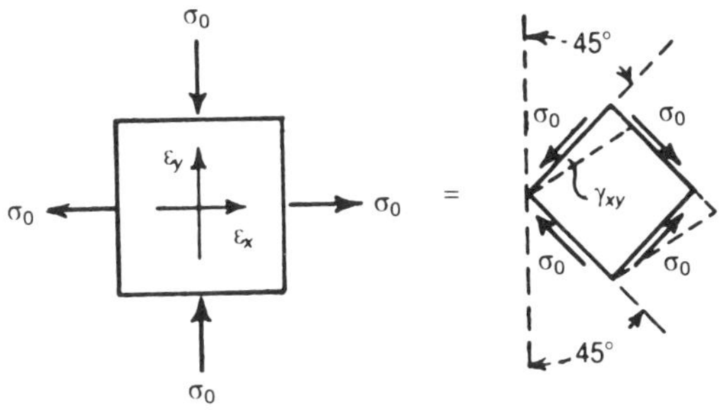

Figure 5.19 Pure Shear. Equality of (i) equal and opposite tensile and compressive stresses and (ii) pure shearing stress.

If the material is elastic, the strains ε_x and ε_y caused by the direct stresses σ_0 are, from equations (5.18),

$$\varepsilon_x = \frac{1}{E}\left(\sigma_0 + \nu\sigma_0\right) = \frac{\sigma_0}{E}(1 + \nu)$$

$$\varepsilon_y = \frac{1}{E}\left(-\sigma_0 - \nu\sigma_0\right) = -\frac{\sigma_0}{E}(1 + \nu)$$

If the sides of the element are of unit length, the work done in distorting the element is

$$W = \frac{1}{2}\sigma_0\,\varepsilon_x - \frac{1}{2}\sigma_0\,\varepsilon_y = \frac{\sigma_0^2}{E}(1 + \nu) \tag{5.22}$$

per unit volume of the material.

In the state of pure shearing under stresses σ_0, the shearing strain is given by equation (5.20),

$$\gamma_{xy} = \frac{\sigma_0}{G}$$

The work done in distorting an element of sides unit length is

$$W = \frac{1}{2}\sigma_0\gamma_{xy} = \frac{\sigma_0^2}{2G} \tag{5.23}$$

per unit volume of the material. As the one state of stress is equivalent to the other, the values of work done per unit volume of the material are equal. Then

$$\frac{\sigma_0^2}{E}(1 + \nu) = \frac{\sigma_0^2}{2G}$$

and hence

$$E = 2G(1 + \nu) \tag{5.24}$$

Thus ν can be calculated from measured values E and G.

The shearing stress–strain relation is given by equation (5.20), which may now be written in the form

$$E\gamma_{xy} = 2(1 + \nu)\tau_{xy} \tag{5.25}$$

For most metals v is approximately 0.3; then, approximately,

$$E = 2(1 + \nu)G = 2.6G \tag{5.26}$$

Problem 5.5 From tests on a magnesium alloy it is found that E is 45 GN/m^2 and G is 17 GN/m^2. Estimate the value of Poisson's ratio.

Solution

From equation (5.24),

$$\nu = \frac{E}{2G} - 1 = \frac{45}{34} - 1 = 1.32 - 1$$

Then

$$\nu = 0.32$$

Problem 5.6 A thin sheet of material is subjected to a tensile stress of 80 MN/m^2, in a certain direction. One surface of the sheet is polished, and on this surface fine lines are ruled to form a square of side 5 cm, one diagonal of the square being parallel to the direction of the tensile stresses. If $E = 200$ GN/m^2, and $\nu = 0.3$, estimate the alteration in the lengths of the sides of the square, and the changes in the angles at the corners of the square.

Solution

The diagonal parallel to the tensile stresses increases in length by an amount

$$\frac{(80 \times 10^6)(0.05\sqrt{2})}{200 \times 10^9} = 28.3 \times 10^{-6} \text{ m}$$

The diagonal perpendicular to the tensile stresses diminishes in length by an amount

$$0.3\,(28.3 \times 10^{-6}) = 8.50 \times 10^{-6} \text{ m}$$

The change in the corner angles is then

$$\frac{1}{0.05}\left[(28.3 + 8.50)10^{-6}\right]\frac{1}{\sqrt{2}} = 52.0 \times 10^{-3} \text{ radians} = 0.0405°$$

The angles in the line of pull are diminished by this amount, and the others increased by the same amount. The increase in length of each side is

$$\frac{1}{2\sqrt{2}}\left[(28.3 - 8.50)10^{-6}\right] = 7.00 \times 10^{-6} \text{ m}$$

5.15 Strain 'rosettes'

To determine the stresses in a material under practical loading conditions, the strains are measured by means of small gauges; many types of gauges have been devised, but perhaps the most convenient is the electrical resistance strain gauge, consisting of a short length of fine wire which is glued to the surface of the material. The resistance of the wire changes by small amounts as the wire is stretched, so that as the surface of the material is strained the gauge indicates a change of resistance which is measurable on a Wheatstone bridge. The lengths of wire resistance strain gauges can be as small as 0.4 mm, and they are therefore extremely useful in measuring local strains.

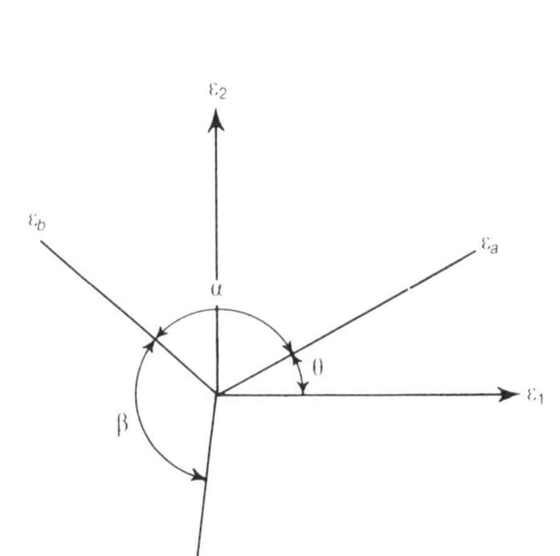

Figure 5.20 Finding the principal strains in a two-dimensional system by recording three linear strains, ε_a, ε_b and ε_c in the vicinity of a point.

The state of strain at a point of a material is defined in the two-dimensional case if the direct strains, ε_x and ε_y, and the shearing strain, γ_{xy}, are known. Unfortunately, the shearing strain γ_{xy} is not readily measured; it is possible, however, to measure the direct strains in three different directions by means of strain gauges. Suppose ε_1, ε_2 are the unknown principal strains in a two-dimensional system, Figure 5.20. Then from equation (5.16) we have that the measured direct strains ε_a, ε_b and ε_c in directions inclined at θ, $(\theta + \alpha)$, $(\theta + \alpha + \beta)$ to ε_1 are

$$\varepsilon_a = \frac{1}{2}\left(\varepsilon_1 + \varepsilon_2\right) + \frac{1}{2}\left(\varepsilon_1 - \varepsilon_2\right)\cos 2\theta$$

$$\varepsilon_b = \frac{1}{2}\left(\varepsilon_1 + \varepsilon_2\right) + \frac{1}{2}\left(\varepsilon_2 - \varepsilon_2\right)\cos 2\left(\theta + \alpha\right) \tag{5.27}$$

$$\varepsilon_c = \frac{1}{2}\left(\varepsilon_1 + \varepsilon_2\right) + \frac{1}{2}\left(\varepsilon_1 - \varepsilon_2\right)\cos 2\left(\theta + \alpha + \beta\right)$$

In practice the directions of the principal strains are not known usually; but if the three direct strains ε_a, ε_b and ε_c are measured in known directions, then the three unknowns in equations (5.27) are

ε_1, ε_2 and θ

Three strain gauges arranged so that $\alpha = \beta = 45°$ form a 45° rosette, Figure 5.22. Equations (5.27) become

$$\varepsilon_a = \frac{1}{2}\left(\varepsilon_1 + \varepsilon_2\right) + \frac{1}{2}\left(\varepsilon_1 - \varepsilon_2\right)\cos 2\theta \tag{5.28a}$$

$$\varepsilon_b = \frac{1}{2}\left(\varepsilon_1 + \varepsilon_2\right) - \frac{1}{2}\left(\varepsilon_1 - \varepsilon_2\right)\sin 2\theta \tag{5.28b}$$

$$\varepsilon_c = \frac{1}{2}\left(\varepsilon_1 + \varepsilon_2\right) - \frac{1}{2}\left(\varepsilon_1 - \varepsilon_2\right)\cos 2\theta \tag{5.28c}$$

Adding together equations (5.28a) and (5.28c), we get

$$\varepsilon_a + \varepsilon_c = \varepsilon_1 + \varepsilon_2 \tag{5.29}$$

Equation (5.29) is known as the *first invariant of strain*, which states that the sum of two mutually perpendicular normal strains is a constant.

From equations (5.28a) and (5.28b).

$$-\frac{1}{2}\left(\varepsilon_1 - \varepsilon_2\right)\sin 2\theta = \varepsilon_b - \frac{1}{2}\left(\varepsilon_1 - \varepsilon_2\right) \tag{5.30a}$$

$$-\frac{1}{2}\left(\varepsilon_1 - \varepsilon_2\right)\cos 2\theta = -\varepsilon_a + \frac{1}{2}\left(\varepsilon_1 + \varepsilon_2\right) \tag{5.30b}$$

Dividing equation (5.30a) by (5.30b), we obtain

$$\tan 2\theta = \frac{\varepsilon_b - \frac{1}{2}(\varepsilon_1 + \varepsilon_2)}{-\varepsilon_a + \frac{1}{2}(\varepsilon_1 + \varepsilon_2)} \tag{5.31}$$

Substituting equation (5.29) into (5.31)

$$\tan 2\theta = \frac{(\varepsilon_a - 2\varepsilon_b + \varepsilon_c)}{(\varepsilon_a - \varepsilon_c)} \tag{5.32}$$

To determine ε_1 and ε_2 in terms of the known strains, namely ε_a, ε_b and ε_c, put equation (5.32) in the form of the mathematical triangle of Figure 5.21.

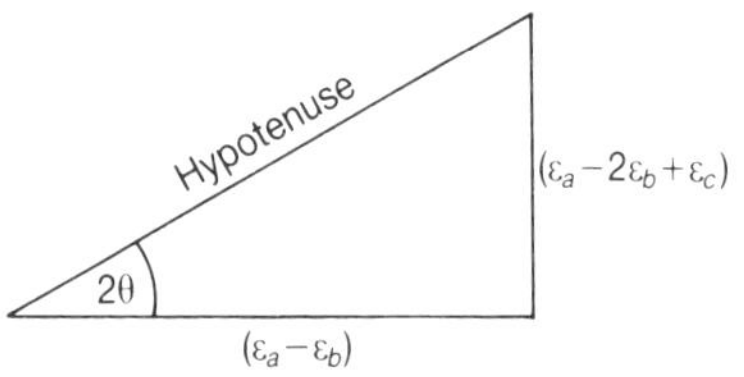

Figure 5.21 Mathematical triangle from equation (5.32).

$$\text{hypotenuse} = \sqrt{\varepsilon_a^2 + 4\varepsilon_b^2 + \varepsilon_c^2 - 4\varepsilon_a\varepsilon_b - 4\varepsilon_b\varepsilon_c + 2\varepsilon_a\varepsilon_c + \varepsilon_a^2 + \varepsilon_b^2 - 2\varepsilon_a\varepsilon_b}$$

$$= \sqrt{2}\sqrt{(\varepsilon_a - \varepsilon_b)^2 + (\varepsilon_c - \varepsilon_b)^2}$$

$$\therefore \cos 2\theta = \frac{\varepsilon_a - \varepsilon_c}{\sqrt{2}\sqrt{(\varepsilon_a - \varepsilon_b)^2 + (\varepsilon_c - \varepsilon_b)^2}} \tag{5.33}$$

and

$$\sin 2\theta = \frac{\varepsilon_a - 2\varepsilon_b + \varepsilon_c}{\sqrt{2}\sqrt{(\varepsilon_a - \varepsilon_b)^2 + (\varepsilon_c - \varepsilon_b)^2}} \tag{5.34}$$

Substituting equations (5.33) and (5.34) into equations (5.30a) and (5.30b) and solving,

$$\varepsilon_1 = \frac{1}{2}(\varepsilon_a + \varepsilon_c) + \frac{\sqrt{2}}{2}\sqrt{(\varepsilon_a - \varepsilon_b)^2 + (\varepsilon_c - \varepsilon_b)^2} \tag{5.35}$$

$$\varepsilon_2 = \frac{1}{2}(\varepsilon_a + \varepsilon_c) - \frac{\sqrt{2}}{2}\sqrt{(\varepsilon_a - \varepsilon_b)^2 + (\varepsilon_c - \varepsilon_b)^2} \tag{5.36}$$

θ is the angle between the directions of ε_1 and ε_a, and is measured clockwise from the direction of ε_1.

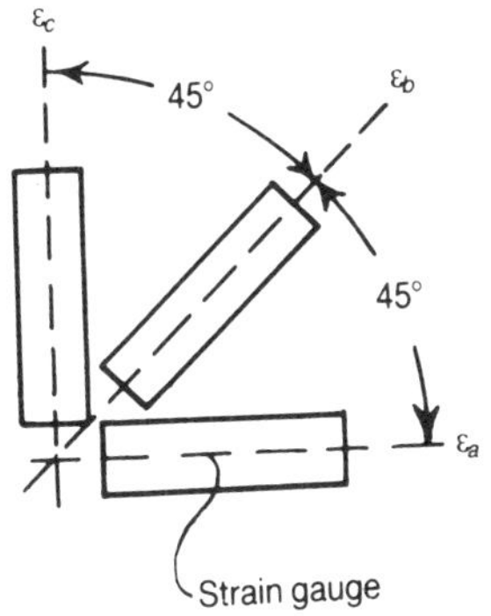

Figure 5.22 A 45° strain rosette.

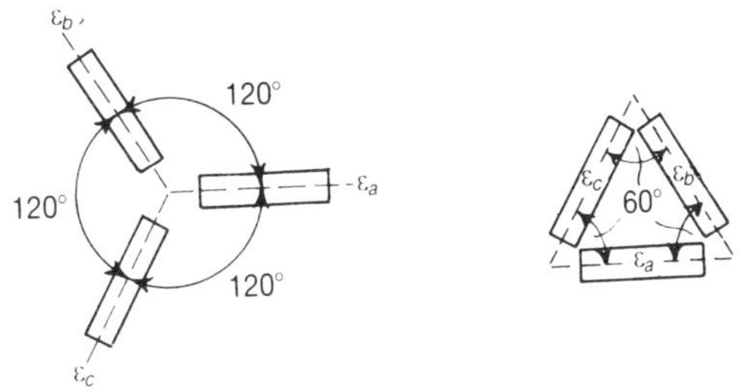

Figure 5.23 Alternative arrangements of 120° rosettes.

The alternative arrangements of gauges in Figure 5.23 correspond to 120° rosettes. On putting $\alpha = \beta = 120°$ in equations (5.27), we have

$$\varepsilon_a = \frac{1}{2}(\varepsilon_1 + \varepsilon_2) + \frac{1}{2}(\varepsilon_1 - \varepsilon_2)\cos 2\theta \tag{5.37a}$$

$$\varepsilon_b = \frac{1}{2}(\varepsilon_1 + \varepsilon_2) - \frac{1}{2}(\varepsilon_1 - \varepsilon_2)\left(\frac{1}{2}\cos 2\theta - \frac{\sqrt{3}}{2}\sin 2\theta\right) \tag{5.37b}$$

$$\varepsilon_c = \frac{1}{2}(\varepsilon_1 + \varepsilon_2) - \frac{1}{2}(\varepsilon_1 - \varepsilon_2)\left(\frac{1}{2}\cos 2\theta + \frac{\sqrt{3}}{2}\sin 2\theta\right) \tag{5.37c}$$

Equations (5.37b) and (5.37c) can be written in the forms

$$\varepsilon_b = \frac{1}{2}(\varepsilon_1 + \varepsilon_2) - \frac{1}{2}(\varepsilon_1 - \varepsilon_2)\left(\frac{1}{2}\cos 2\theta - \frac{\sqrt{3}}{2}\sin 2\theta\right) \tag{5.38a}$$

$$\varepsilon_c = \frac{1}{2}(\varepsilon_1 + \varepsilon_2) - \frac{1}{2}(\varepsilon_1 - \varepsilon_2)\left(\frac{1}{2}\cos 2\theta + \frac{\sqrt{3}}{2}\sin 2\theta\right) \tag{5.38b}$$

Adding together equations (5.37a), (5.38a) and (5.38b), we get:

$$\varepsilon_a + \varepsilon_b + \varepsilon_c = \frac{3}{2}(\varepsilon_1 + \varepsilon_2)$$

or

$$\varepsilon_1 + \varepsilon_2 = \frac{2}{3}(\varepsilon_a + \varepsilon_b + \varepsilon_c) \tag{5.39}$$

Taking away equation (5.38b) from (5.38a),

$$\varepsilon_b - \varepsilon_c = \frac{\sqrt{3}}{2}(\varepsilon_1 - \varepsilon_2)\sin 2\theta \tag{5.40}$$

Taking away equation (5.38b) from (5.37a)

$$\varepsilon_a - \varepsilon_c = \frac{1}{2}(\varepsilon_1 - \varepsilon_2)\left(\frac{3}{2}\cos 2\theta + \frac{\sqrt{3}}{2}\sin 2\theta\right) \tag{5.41}$$

Dividing equation (5.41) by (5.40)

$$\frac{\varepsilon_a - \varepsilon_c}{\varepsilon_b - \varepsilon_c} = \frac{1}{2}\left(\frac{3}{2}\,\frac{\cot 2\theta}{\sqrt{3/2}} + 1\right)$$

or

$$\sqrt{3}\cot 2\theta = \frac{2(\varepsilon_a - \varepsilon_c)}{(\varepsilon_b - \varepsilon_c)} - \frac{(\varepsilon_b - \varepsilon_c)}{(\varepsilon_b - \varepsilon_c)}$$

or

$$\tan 2\theta = \frac{\sqrt{3}(\varepsilon_b - \varepsilon_c)}{(2\varepsilon_a - \varepsilon_b - \varepsilon_c)} \tag{5.42}$$

To determine ε_1 and ε_2 in terms of the measured strains, namely ε_a, ε_b and ε_c, put equation (5.42) in the form of the mathematical triangle of Figure 5.24.

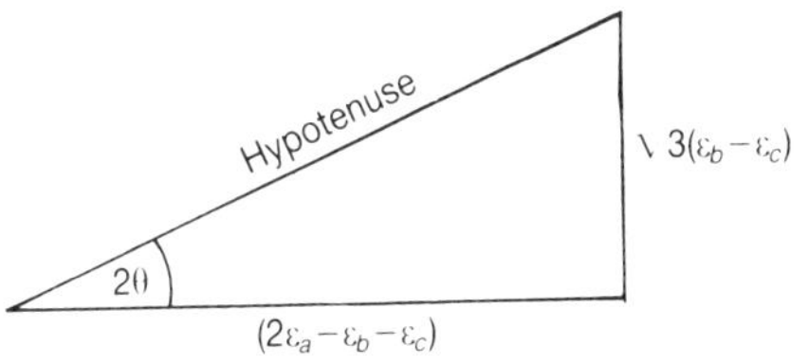

Figure 5.24 Mathematical triangle from Pythagoras' theorem.

Then,

$$\text{hypotenuse} = \sqrt{\left[(\varepsilon_a-\varepsilon_b)^2+(\varepsilon_b-\varepsilon_c)^2+(\varepsilon_a-\varepsilon_b)^2\right]}$$

Hence,

$$\cos 2\theta = \frac{(2\varepsilon_a-\varepsilon_b-\varepsilon_c)}{\text{hypotenuse}} \tag{5.43}$$

and

$$\sin 2\theta \quad \frac{\sqrt{3}(\varepsilon_b-\varepsilon_c)}{\text{hypotenuse}} \tag{5.44}$$

Substituting equation (5.43) and (5.44) into equations (5.38a) and (5.38b), and solving the two simultaneous equations, we get

$$\varepsilon_1 = \frac{1}{3}(\varepsilon_a+\varepsilon_b+\varepsilon_c)+\frac{\sqrt{2}}{3}\sqrt{\left[(\varepsilon_a-\varepsilon_b)^2+(\varepsilon_b-\varepsilon_c)^2+(\varepsilon_a-\varepsilon_c)^2\right]} \tag{5.45}$$

and

$$\varepsilon_2 = \frac{1}{3}(\varepsilon_a+\varepsilon_b+\varepsilon_c)-\frac{\sqrt{2}}{3}\sqrt{\left[(\varepsilon_a-\varepsilon_b)^2+(\varepsilon_b-\varepsilon_c)^2+(\varepsilon_a-\varepsilon_c)^2\right]} \tag{5.46}$$

When the principal strains ε_1 and ε_2 have been estimated, the corresponding principal stresses are deduced from the relations

$$E\varepsilon_1 = \sigma_1 - \nu\sigma_2$$

$$E\varepsilon_2 = \sigma_2 - \nu\sigma_1$$

These give

$$\sigma_1 = \frac{E}{1 - \nu^2}(\varepsilon_1 + \nu\varepsilon_2)$$

$$\sigma_2 = \frac{E}{1 - \nu^2}(\varepsilon_2 + \nu\varepsilon_1) \qquad (5.47)$$

Equations (5.18) and (5.47) are for the *plane stress* condition, which is a two-dimensional system of stress, as discussed in Section 5.12.

Another two-dimensional system is known as a *plane strain* condition, which is a two-dimensional system of strain and a three-dimensional system of stress, as in Figure 5.25, where

$$\varepsilon_z = 0 = \frac{\sigma_z}{E} - \frac{\nu\sigma_x}{E} - \frac{\nu\sigma_y}{E} \qquad (5.48a)$$

$$\varepsilon_y = \frac{\sigma_y}{E} - \frac{\nu\sigma_x}{E} - \frac{\nu\sigma_z}{E} \qquad (5.48b)$$

$$\varepsilon_x = \frac{\sigma_x}{E} - \frac{\nu\sigma_y}{E} - \frac{\nu\sigma_z}{E} \qquad (5.48c)$$

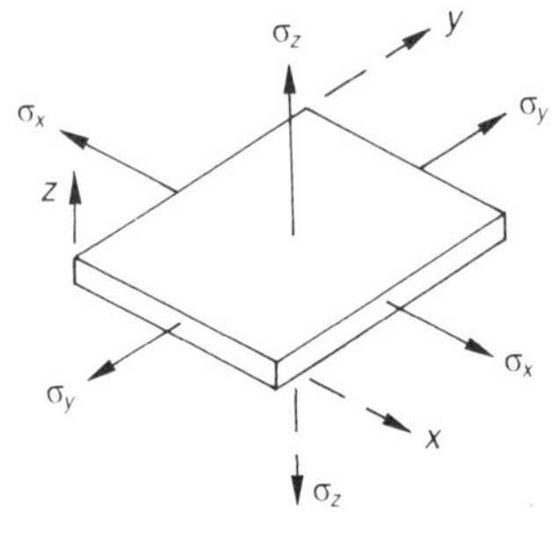

Figure 5.25 Plane strain condition.

From equation (5.48a)

$$\sigma_z = \nu(\sigma_x + \sigma_y) \qquad (5.49)$$

Substituting equation (5.49) into equations (5.48b) and (5.48c), we get,

$$\varepsilon_y = \frac{\sigma_y}{E} - \frac{\nu\sigma_x}{E} - \frac{\nu^2}{E}(\sigma_x + \sigma_y)$$

$$= \frac{\sigma_y}{E}(1 - \nu^2) - \frac{\nu\sigma_x}{E}(1 + \nu) \qquad (5.50a)$$

and

$$\varepsilon_x = \frac{\sigma_x}{E} - \frac{\nu\sigma_y}{E} - \frac{\nu^2(\sigma_x + \sigma_y)}{E}$$

$$= \frac{\sigma_x}{E}(1 - \nu^2) - \frac{\nu\sigma_y}{E}(1 + \nu) \qquad (5.50b)$$

Multiplying equation (5.50a) by $(1 - \nu^2)/(1 + \nu)\,\nu$ we get

$$\frac{(1 - \nu^2)}{(1 + \nu)\nu}\varepsilon_y = \frac{\sigma_y(1 - \nu^2)^2}{E(1 + \nu)\nu} - \frac{\sigma_x}{E}(1 - \nu^2) \qquad (5.51)$$

Adding equation (5.50b) to (5.51), we get

$$\frac{(1 - \nu^2)}{(1 + \nu)\nu}\varepsilon_y + \varepsilon_x = \frac{-\nu\sigma_y}{E}(1 + \nu) + \frac{\sigma_y(1 - \nu^2)^2}{E(1 + \nu)\nu}$$

or

$$(1 - \nu)\,\varepsilon_y + \nu\varepsilon_x = \frac{-\nu^2(1 + \nu)\sigma_y}{E} + \frac{\sigma_y(1 - \nu^2)^2}{E(1 + \nu)}$$

or

$$E\left[(1 - \nu)\,\varepsilon_y + \nu\varepsilon_x\right] = \frac{\sigma_y}{(1 + \nu)}\left[-\nu^2(1 + \nu)^2 + (1 - \nu^2)^2\right]$$

or

$$E\left[(1 - \nu)\varepsilon_y + \nu\varepsilon_x\right] = \sigma_y\left[-\nu^2(1 + \nu) + (1 - \nu)\left(1 - \nu^2\right)\right]$$

$$= \sigma_y\left[-\nu^2 - \nu^3 + 1 - \nu - \nu^2 + \nu^3\right]$$

$$= \sigma_y\left(1 - \nu - 2\nu^2\right) \qquad (5.52a)$$

$$= \sigma_y (1 + \nu)(1 - 2\nu)$$

$$\therefore \sigma_y = \frac{E\left[(1 - \nu)\varepsilon_y + \nu\varepsilon_x\right]}{(1 + \nu)(1 - 2\nu)}$$

Similarly

$$\sigma_x = \frac{E\left[(1 - \nu)\varepsilon_x + \nu\varepsilon_y\right]}{(1 + \nu)(1 - 2\nu)} \tag{5.52b}$$

Obviously the values of E and ν must be known before the stresses can be estimated from either equations (5.19), (5.47) or (5.52).

5.16 Strain energy for a two-dimensional stress system

If σ_1 and σ_2 are the principal stresses in a two-dimensional stress system, the corresponding principal strains for an elastic material are, from equations (5.21),

$$\varepsilon_1 = \frac{1}{E}(\sigma_1 - \nu\sigma_2)$$

$$\varepsilon_2 = \frac{1}{E}(\sigma_2 - \nu\sigma_1)$$

Consider a cube of material having sides of unit length, and therefore having also unit volume. The edges parallel to the direction of σ_1 extend amounts ε_1, and those parallel to the direction of σ_2 by amounts ε_2. The work done by the stresses σ_1 and σ_2 during straining is then

$$W = \frac{1}{2}\sigma_1\varepsilon_1 + \frac{1}{2}\sigma_2\varepsilon_2$$

per unit volume of material. On substituting for ε_1 and ε_2 we have

$$W = \frac{1}{2}\sigma_1\left[\frac{1}{E}(\sigma_1 - \nu\sigma_2)\right] + \frac{1}{2}\sigma_2\left[\frac{1}{E}(\sigma_2 - \nu\sigma_1)\right]$$

This is equal to the strain energy U *per unit volume*; thus

$$U = \frac{1}{2E}\left[\sigma_1^2 + \sigma_2^2 - 2\nu\sigma_1\sigma_2\right] \tag{5.53}$$

5.17 Three-dimensional stress systems

In any two-dimensional stress system we found there were two mutually perpendicular directions in which only direct stresses, σ_1 and σ_2, acted; these were called the principal stresses. In any three-dimensional stress system we can always find three mutually perpendicular directions in which only direct stresses, σ_1, σ_2 and σ_3 in Figure 5.26, are acting. No shearing stresses act on the faces of a rectangular block having its edges parallel to the axes 1, 2 and 3 in Figure 5.26. These direct stresses are again called *principal stresses.*

If $\sigma_1 > \sigma_2 > \sigma_3$, then the three-dimensional stress system can be represented in the form of Mohr's circles, as shown in Figure 5.27. Circle *a* passes through the points σ_1 and σ_2 on the σ-axis, and defines all states of stress on planes parallel to the axis 3, Figure 5.26, but inclined to axis 1 and axis 2, respectively.

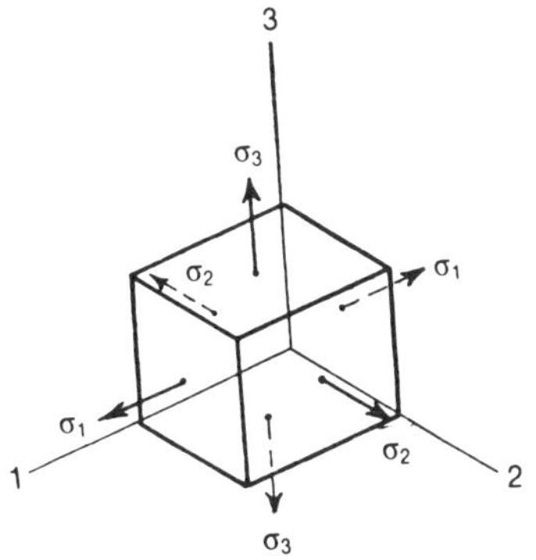

Figure 5.26 Principal stresses in a three-dimensional system.

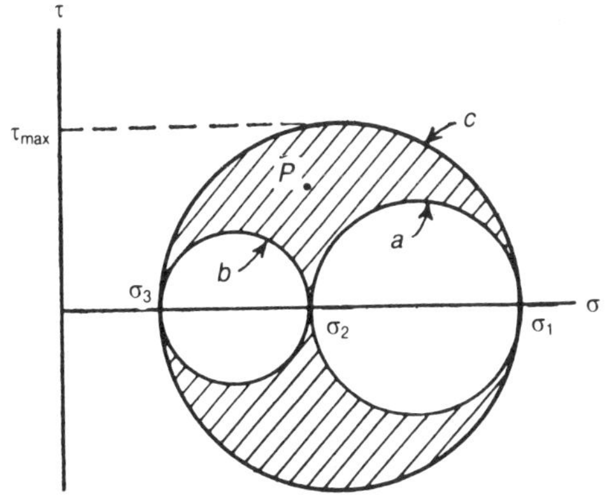

Figure 5.27 Mohr's circle of stress for a three-dimensional system; circle *a* is the Mohr's circle of the two-dimensional system σ_1, σ_2; *b* corresponds to σ_2, σ_3 and *c* to σ_3, σ_1. The resultant direct and tangential stress on any plane through the point must correspond to a point *P* lying on or between the three circles.

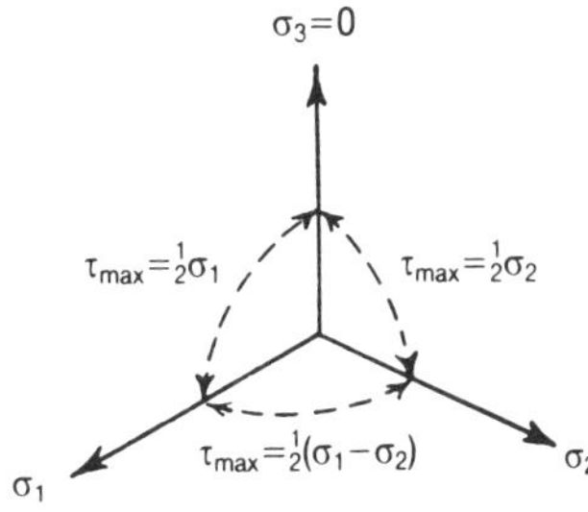

Figure 5.28 Two-dimensional stress system as a particular case of a three-dimensional system with one of the three principal stresses equal to zero.

Circle *c*, having a diameter $(\sigma_1 - \sigma_3)$, embraces the two smaller circles. For a plane inclined to all three axes the stresses are defined by a point such as *P* within the shaded area in Figure 5.27. The maximum shearing stress is

$$\tau_{max} = \frac{1}{2}(\sigma_1 - \sigma_3)$$

and occurs on a plane parallel to the axis 2.

From our discussion of three-dimensional stress systems we note that when one of the principal stresses, σ_3 say, is zero, Figure 5.28, we have a two-dimensional system of stresses σ_1, σ_2; the maximum shearing stresses in the planes 1–2, 2–3, 3–1 are, respectively,

$$\frac{1}{2}(\sigma_1 - \sigma_2), \quad \frac{1}{2}\sigma_1, \quad \frac{1}{2}\sigma_2$$

Suppose, initially, that σ_1 and σ_2 are both tensile and that $\sigma_1 > \sigma_2$; then the greatest of the three maximum shearing stresses is $\frac{1}{2}\sigma_1$ which occurs in the 2–3 plane. If, on the other hand, σ_1 is tensile and σ_2 is compressive, the greatest of the maximum shearing stresses is $\frac{1}{2}(\sigma_1 - \sigma_2)$ and occurs in the 1–2 plane.

We conclude from this that the presence of a zero stress in a direction perpendicular to a two-dimensional stress system may have an important effect on the maximum shearing stresses in the material and cannot be disregarded therefore. The direct strains corresponding to σ_1, σ_2 and σ_3 for an elastic material are found by taking account of the Poisson ratio effects in the three directions; the principal strains in the directions 1, 2 and 3 are, respectively,

$$\varepsilon_1 = \frac{1}{E}(\sigma_1 - \nu\sigma_2 - \nu\sigma_3)$$

$$\varepsilon_2 = \frac{1}{E}(\sigma_2 - \nu\sigma_3 - \nu\sigma_1)$$

$$\varepsilon_3 = \frac{1}{E}(\sigma_3 - \nu\sigma_1 - \nu\sigma_2)$$

The strain energy stored per unit volume of the material is

$$U = \frac{1}{2}\sigma_1\varepsilon_1 + \frac{1}{2}\sigma_2\varepsilon_2 + \frac{1}{2}\sigma_3\varepsilon_3$$

In terms of σ_1, σ_2 and σ_3, this becomes

$$U = \frac{1}{2E}\left(\sigma_1^2 + \sigma_2^2 + \sigma_3^2 - 2\nu\sigma_1\sigma_2 - 2\nu\sigma\ \sigma_3 - 2\nu\sigma_3\sigma_1\right) \tag{5.54}$$

5.18 Volumetric strain in a material under hydrostatic pressure

A material under the action of equal compressive stresses σ in three mutually perpendicular directions, Figure 5.29, is subjected to a *hydrostatic pressure*, σ. The term hydrostatic is used because the material is subjected to the same stresses as would occur if it were immersed in a fluid at a considerable depth.

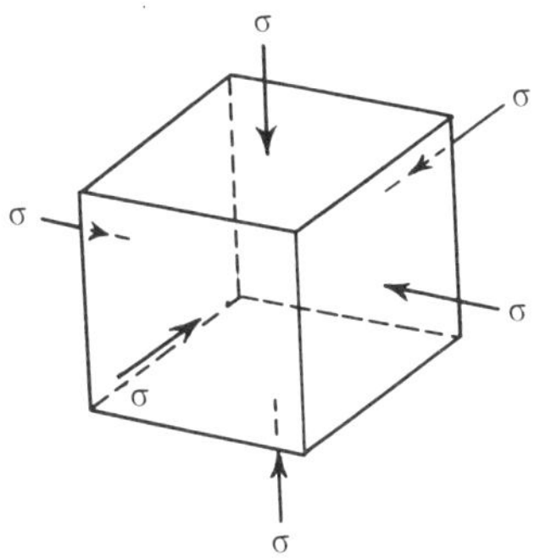

Figure 5.29 Region of a material under a hydrostatic pressure.

If the initial volume of the material is V_0, and if this diminishes an amount δV due to the hydrostatic pressure, the volumetric strain is

$$\frac{\delta V}{V_0}$$

The ratio of the hydrostatic pressure, σ, to the volumetric strain, $\delta V/V_0$, is called the *bulk modulus* of the material, and is denoted by K. Then

$$K = \frac{\sigma}{\left(\dfrac{\delta V}{V_0}\right)} \tag{5.55}$$

If the material remains elastic under hydrostatic pressure, the strain in each of the three mutually perpendicular directions is

$$\varepsilon = -\frac{\sigma}{E} + \frac{\nu\sigma}{E} + \frac{\nu\sigma}{E}$$

$$= -\frac{\sigma}{E}(1 - 2\nu)$$

because there are two Poisson ratio effects on the strain in any of the three directions. If we consider a cube of material having sides of unit length in the unstrained condition, the volume of the strained cube is

$$(1 - \varepsilon)^3$$

Now ε is small, so that this may be written approximately

$$1 - 3\varepsilon$$

The change in volume of a unit volume is then

$$3\varepsilon$$

which is therefore the volumetric strain. Then equation (5.55) gives the relationship

$$K = \frac{\sigma}{\left(\dfrac{\delta V}{V_0}\right)} = \frac{\sigma}{-3\varepsilon} = \frac{E}{3(1 - 2\nu)}$$

We should expect the volume of a material to diminish under a hydrostatic pressure. In general, if K is always positive, we must have

$$1 - 2\nu > 0$$

or

$$\nu < \frac{1}{2}$$

Then Poisson's ratio is *always less than ½*. For plastic strains of a metallic material there is a negligible change of volume, the Poisson's ratio is equal to ½, approximately.

5.19 Strain energy of distortion

In the three-dimensional stress system of Figure 5.22 we may consider the principal stress σ_1 to be the resultant of stresses

$$\frac{1}{3}(\sigma_1 + \sigma_2 + \sigma_3)$$

and stresses

$$\frac{1}{3}(2\sigma_1 - \sigma_2 - \sigma_3)$$

since

$$\frac{1}{3}(\sigma_1 + \sigma_2 + \sigma_3) + \frac{1}{3}(2\sigma_1 - \sigma_2 - \sigma_3) = \sigma_1$$

Similarly, we write

$$\sigma_2 = \frac{1}{3}(\sigma_1 + \sigma_2 + \sigma_3) + \frac{1}{3}(2\sigma_2 - \sigma_3 - \sigma_1)$$

$$\sigma_3 = \frac{1}{3}(\sigma_1 + \sigma_2 + \sigma_3) + \frac{1}{3}(2\sigma_3 - \sigma_1 - \sigma_2)$$

Now, the component ⅓ $(\sigma_1 + \sigma_2 + \sigma_2)$ which occurs in σ_1, σ_2 and σ_3, represents a *hydrostatic* tensile stress; the strains associated with this stress give rise to no distortion, i.e., a cube of material under stress ⅓ $(\sigma_1 + \sigma_2 + \sigma_3)$ in three mutually perpendicular directions is strained into a cube. The remaining components of σ_1, σ_2 and σ_3, are

$$\frac{1}{3}(2\sigma_1 - \sigma_2 - \sigma_3), \quad \frac{1}{3}(2\sigma_2 - \sigma_3 - \sigma_1), \quad \frac{1}{3}(2\sigma_3 - \sigma_1 - \sigma_2)$$

The strain energy associated with these stresses, which are the only stresses giving rise to distortion, is called the *strain energy of distortion*. The strains due to these distorting stresses are

$$\varepsilon_1 = \frac{1}{3E}(1 + \nu)(2\sigma_1 - \sigma_2 - \sigma_3) = \frac{1}{6G}\left[(\sigma_1 - \sigma_2) + (\sigma_1 - \sigma_3)\right]$$

$$\varepsilon_2 = \frac{1}{3E}(1 + \nu)(2\sigma_2 - \sigma_3 - \sigma_1) = \frac{1}{6G}\left[(\sigma_2 - \sigma_3) + (\sigma_2 - \sigma_1)\right]$$

$$\varepsilon_3 = \frac{1}{3E}(1 + \nu)(2\sigma_3 - \sigma_1 - \sigma_2) = \frac{1}{6G}\left[(\sigma_3 - \sigma_1) + (\sigma_3 - \sigma_2)\right]$$

The strain energy of distortion is therefore

$$U_D = \frac{1}{36G}\left[(2\sigma_1 - \sigma_2 - \sigma_3)^2 + (2\sigma_2 - \sigma_3 - \sigma_1)^2 + (2\sigma_3 - \sigma_1 - \sigma_2)^2\right]$$

per unit volume. Then

$$U_D = \frac{1}{12G}\left[(\sigma_1 - \sigma_2)^2 + (\sigma_2 - \sigma_3)^2 + (\sigma_3 - \sigma_1)^2\right] \tag{5.56}$$

For a two-dimensional stress system, σ_3 (say) = 0, and U_D reduces to

$$U_D = \frac{1}{12G}\left[(\sigma_1 - \sigma_2)^2 + \sigma_2^2 + \sigma_1^2\right]$$

We shall see later that the strain energy of distortion plays an important part in the yielding of ductile materials under combined stresses.

5.20 Isotropic, orthotropic and anisotropic

A material is said to be *isotropic* when its material properties are the same in all directions. An *orthotropic* material is said to exhibit symmetric material properties about three mutually perpendicular planes. In two dimensions, typical orthotropic materials are in the form of many composites. An *anisotropic* material is a material that exhibits different material properties in all directions.

5.21 Fibre composites

Fibre composites are very important for structures which require a large strength:weight ratio, especially when the weight of the structure is at a premium. They are likely to become even more important in the 21st century and will probably revolutionise the design and construction of aircraft, rockets, submarines and warships.

To represent the elasticity of a composite, tensile modulus is used in preference to Young's modulus of elasticity. Additionally, as most composites are usually assumed to be of orthotropic form, their material properties in one direction, (say) 'x' are likely to be different to a direction perpendicular to the 'x' direction, (say) 'y'. Composites usually consist of several layers of fibre matting, set in a resin, as shown by Figure 5.30. To gain maximum strength the layers of fibre matting are laid in different directions. In this Chapter, the term *lamina* or ply will be used to describe a single layer of the composite structure and the term *laminate* or composite will be used to define the entire mixture of plies and resin.

If the material properties of the fibre composite are orthogonal, the following relationship applies:

$$\nu_x E_y = \nu_y E_x \tag{5.57}$$

Figure 5.30 Five layers of fibre reinforcement.

where

E_x	=	tensile modulus in the x-direction.
E_y	=	tensile modulus in the y-direction.
ν_x	=	Poisson's ratio due to the effects of σ_x
ν_y	=	Poisson's ratio due to the effects of σ_y
σ_x	=	direct stress in the local x-direction.
σ_y	=	direct stress in the local y-direction.

} see Figure 5.31

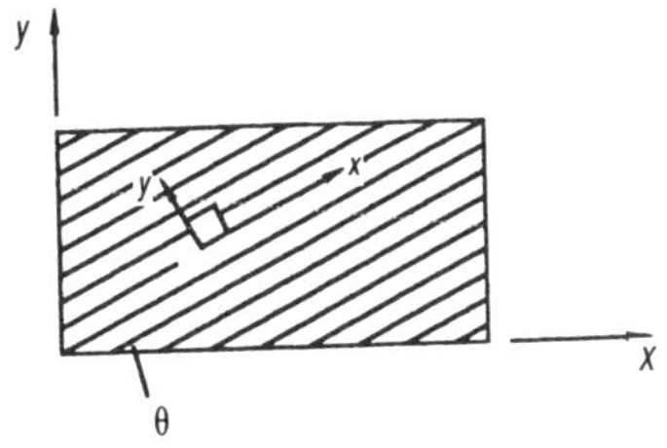

Figure 5.31 A lamina from a composite.

It is evident from the theory of Section 5.12 that the following relationships between stress and strain apply for orthotropic materials:

$$\varepsilon_x = \frac{\sigma_x}{E_x} - \frac{\nu_y \sigma_y}{E_y}$$

$$\varepsilon_y = \frac{\sigma_y}{E_y} - \frac{\nu_x \sigma_x}{E_x} \qquad (5.58)$$

$$\gamma_{xy} = \frac{\tau_{xy}}{G_{xv}} \qquad (5.59)$$

where

ε_x	=	direct strain in the x-direction
ε_y	=	direct strain in the y-direction
γ_{xy}	=	shear strain in the x–y plane

Solving equations (5.58), the following alternative relationship is obtained:

$$\sigma_x = \frac{E_x}{(1 - \nu_x \nu_y)} (\varepsilon_x + \nu_y \varepsilon_y)$$

$$\sigma_y = \frac{E_y}{(1 - \nu_x \nu_y)} (\varepsilon_y + \nu_x \varepsilon_x) \tag{5.60}$$

In matrix form, equations (5.58) and (5.59) can be written as

$$\begin{Bmatrix} \varepsilon_x \\ \varepsilon_y \\ \gamma_{xy} \end{Bmatrix} = \begin{bmatrix} S_{11} & S_{12} & S_{16} \\ S_{21} & S_{22} & S_{26} \\ S_{61} & S_{62} & S_{66} \end{bmatrix} \begin{Bmatrix} \sigma_x \\ \sigma_y \\ \tau_{xy} \end{Bmatrix} \tag{5.61}$$

or $$\{\varepsilon_{xy}\} = [\boldsymbol{S}] \{\sigma_{xy}\} \tag{5.62}$$

where $[\boldsymbol{S}]$ is the *compliance* matrix.

From equations (5.59) and (5.60)

$$\{\sigma_{xy}\} = \begin{Bmatrix} \sigma_x \\ \sigma_y \\ \tau_{xy} \end{Bmatrix} = \begin{bmatrix} Q_{11} & Q_{12} & Q_{16} \\ Q_{21} & Q_{22} & Q_{26} \\ Q_{61} & Q_{62} & Q_{66} \end{bmatrix} \begin{Bmatrix} \varepsilon_x \\ \varepsilon_y \\ \gamma_{xy} \end{Bmatrix} \tag{5.63}$$

where

$$Q_{11} = \frac{E_x}{(1 - \nu_x \nu_y)}$$

$$Q_{22} = \frac{E_y}{(1 - \nu_x \nu_y)}$$

$$Q_{66} = G_{xy} = \text{shear modulus}$$

$$Q_{12} = Q_{21} = \frac{\nu_x E_y}{(1 - \nu_x \nu_y)} = \frac{\nu_y E_x}{(1 - \nu_x \nu_y)} \tag{5.64}$$

$$Q_{16} = Q_{61} = Q_{26} = Q_{62} = 0$$

$$Q_{66} = G_{xy}$$

or $\{\sigma_{xy}\} = [Q]\{\varepsilon_{xy}\} = [S^{-1}]\{\varepsilon_{xy}\}$

[Q] = the *stiffness* matrix

= the inverse of [S]

The problem with the above relationships are that they are all in the local co-ordinate system of the lamina, namely x and y. However, as each layer of fibres may have a different direction for its local co-ordinate system, it will be necessary to refer all relationships to a fixed global system, namely, X and Y, as shown by Figure 5.31.

Now from equations (5.4) and (5.5)

$$\begin{aligned} \sigma_x &= \sigma_X \cos^2\theta + \sigma_Y \sin^2\theta + 2\tau_{XY} \sin\theta \cos\theta \\ \sigma_y &= \sigma_{X+90^o} = \sigma_X \sin^2\theta + \sigma_Y \cos^2\theta - 2\tau_{XY} \sin\theta \cos\theta \\ \tau_{xy} &= -\sigma_X \sin\theta \cos\theta + \sigma_Y \sin\theta \cos\theta + \tau_{XY} \left(\cos^2\theta - \sin^2\theta\right) \end{aligned} \tag{5.65}$$

where σ_x, σ_y and τ_{xy} are local stresses and σ_X, σ_Y and τ_{XY} are global or reference stresses; in matrix form equations (5.65) appear as:

$$\begin{Bmatrix} \sigma_x \\ \sigma_y \\ \tau_{xy} \end{Bmatrix} = \begin{bmatrix} C^2 & S^2 & 2SC \\ S^2 & C^2 & -2SC \\ -SC & SC & (C^2 - S^2) \end{bmatrix} \begin{Bmatrix} \sigma_X \\ \sigma_Y \\ \tau_{XY} \end{Bmatrix} \tag{5.66}$$

where S = sin θ and C = cos θ

$$\{\sigma_{xy}\} = [DC]\{\sigma_{XY}\} \tag{5.67}$$

and

$$\{\sigma_{XY}\} = [DC]^{-1}\{\sigma_{xy}\} \tag{5.68}$$

where

$$[DC] = \begin{bmatrix} C^2 & S^2 & 2SC \\ S^2 & C^2 & -2SC \\ -SC & SC & \left(C^2 - S^2\right) \end{bmatrix} \; and \left[DC^{-1}\right] = \begin{bmatrix} C^2 & S^2 & -2SC \\ S^2 & C^2 & 2SC \\ SC & -SC & \left(C^2 - S^2\right) \end{bmatrix} \tag{5.69}$$

Similarly from Section (5.10)

$$\varepsilon_x = \varepsilon_X \cos^2\theta + \varepsilon_Y \sin^2\theta + \gamma_{XY} \sin\theta \cos\theta$$

$$\varepsilon_y = \varepsilon_X \sin^2\theta + \varepsilon_Y \cos^2\theta - \gamma_{XY} \sin\theta \cos\theta \tag{5.70}$$

$$\gamma_{xy} = -2\varepsilon_X \sin\theta \cos\theta + 2\varepsilon_Y \sin\theta \cos\theta + \gamma_{XY}\left(\cos^2\theta - \sin^2\theta\right)$$

or, in matrix form,

$$\begin{Bmatrix} \varepsilon_x \\ \varepsilon_y \\ \gamma_{xy} \end{Bmatrix} = \begin{bmatrix} C^2 & S^2 & SC \\ S^2 & C^2 & -SC \\ -2SC & 2SC & (C^2 - S^2) \end{bmatrix} \begin{Bmatrix} \varepsilon_X \\ \varepsilon_Y \\ \gamma_{XY} \end{Bmatrix} \tag{5.71}$$

or $$\{\varepsilon_{xy}\} = [DC_1]\{\varepsilon_{XY}\} \tag{5.72}$$

Now from equation (5.63),

$$\{\sigma_{xy}\} = [Q]\{\varepsilon_{xy}\}$$

but from equation (5.72),

$$\{\sigma_{xy}\} = [Q][DC_1]\{\varepsilon_{XY}\}$$

but from equation (5.67),

$$\{\sigma_{xy}\} = [DC]\{\sigma_{XY}\}$$

$$\therefore [DC]\{\sigma_{XY}\} = [Q][DC_1]\{\varepsilon_{XY}\}$$

or

$$\{\sigma_{XY}\} = [DC]^{-1}[Q][DC_1]\{\varepsilon_{XY}\} \tag{5.73}$$

or

$$\{\sigma_{XY}\} = [Q^1]\{\varepsilon_{XY}\} \tag{5.74}$$

where

$$[Q^1] = \begin{bmatrix} q^1_{11} & q^1_{12} & q^1_{16} \\ q^1_{21} & q^1_{22} & q^1_{26} \\ q^1_{61} & q^1_{62} & q^1_{66} \end{bmatrix}$$

$$= [DC]^{-1}[Q][DC_1]$$

$$q^1_{11} = \frac{1}{\gamma}\left[E_x \cos^4\theta + E_y \sin^4\theta + (2\nu_x E_y + 4\gamma G)\cos^2\theta\sin^2\theta\right]$$

$$q^1_{12} = q^1_{21} = \frac{1}{\gamma}\left[\nu_x E_y(\cos^4\theta + \sin^4\theta) + (E_x + E_y - 4\gamma G)\cos^2\theta\sin^2\theta\right]$$

$$q_{16}^1 = q_{61}^1 = \frac{1}{\gamma}\left[\cos^3\theta \sin\theta \left(E_x - \nu_x E_y - 2\gamma G\right) - \cos\theta \sin^3\theta \left(E_y - \nu_x E_y - 2\gamma G\right)\right]$$

$$q_{22}^1 = \frac{1}{\gamma}\left[E_y \cos^4\theta + E_x \sin^4\theta + \sin^2\theta \cos^2\theta \left(2\nu_x E_y + 4\gamma G\right)\right]$$

$$q_{26}^1 = q_{62}^1 = \frac{1}{\gamma}\left[\cos\theta \sin^3\theta \left(E_x - \nu_x E_y - 2\gamma G\right) - \cos^3\theta \sin\theta \left(E_y - \nu_x E_y - 2\gamma G\right)\right]$$

$$q_{66}^1 = \frac{1}{\gamma}\left[\sin^2\theta \cos^2\theta \left(E_x + E_y - 2\nu_x E_y - 2\gamma G\right) + \gamma G \left(\cos^4\theta + \sin^4\theta\right)\right]$$

where

$$\gamma = \left(1 - \nu_x \nu_y\right)$$

Similarly, to obtain the global strains of the lamina or ply of Figure 5.32 in terms of the global stresses, consider equation (5.61), as follows.

Now

$$\{\varepsilon_{xy}\} = [S]\{\sigma_{xy}\}$$

so that from equation (5.67)

$$\{\varepsilon_{xy}\} = [S][DC]\{\sigma_{XY}\}$$

and from equation (5.72)

$$[DC_1]\{\varepsilon_{XY}\} = [S][DC]\{\sigma_{XY}\}$$

$$\text{or } \{\varepsilon_{XY}\} = [DC_1]^{-1}[S][DC]\{\sigma_{XY}\} \tag{5.75}$$

$$\text{or } \{\varepsilon_{XY}\} = [S^1]\{\sigma_{XY}\}$$

where

$$[S^1] = \begin{bmatrix} S_{11}^1 & S_{12}^1 & S_{16}^1 \\ S_{21}^1 & S_{22}^1 & S_{26}^1 \\ S_{61}^1 & S_{62}^1 & S_{66}^1 \end{bmatrix} = [DC_1]^{-1}\,[S]\,[DC]$$

$$S_{11}^1 = S_{11}\cos^4\theta + S_{22}\sin^4\theta + (2S_{12} + S_{66})\cos^2\theta\sin^2\theta$$

$$S_{12}^1 = S_{21}^1 = (S_{11} + S_{22} - S_{66})\cos^2\theta\sin^2\theta + S_{12}(\cos^4\theta\sin^4\theta)$$

$$S_{16}^1 = S_{61}^1 = (2S_{22} - 2S_{12} - S_{66})\cos^3\theta\sin\theta - (2S_{22} - 2S_{12} - S_{66})\sin^3\theta\cos\theta$$

$$S_{22}^1 = S_{11}\sin^4\theta + S_{22}\cos^4\theta + (2S_{12} + S_{66})\cos^2\theta\sin^2\theta$$

$$S_{26}^1 = S_{62}^1 = (2S_{11} - 2S_{12} - S_{66})\cos\theta\sin^3\theta - (2S_{22} - 2S_{12} - S_{66})\sin\theta\cos^3\theta$$

$$S_{66}^1 = 4\,(S_{11} - 2S_{12} + S_{66})\cos^2\theta\sin^2\theta + S_{66}\,(\cos^2\theta - \sin^2\theta)^2$$

5.22 In-plane equations for a symmetric laminate or composite

Consider a section of the symmetric laminate of Figure 5.33(a), which is under in-plane loading.

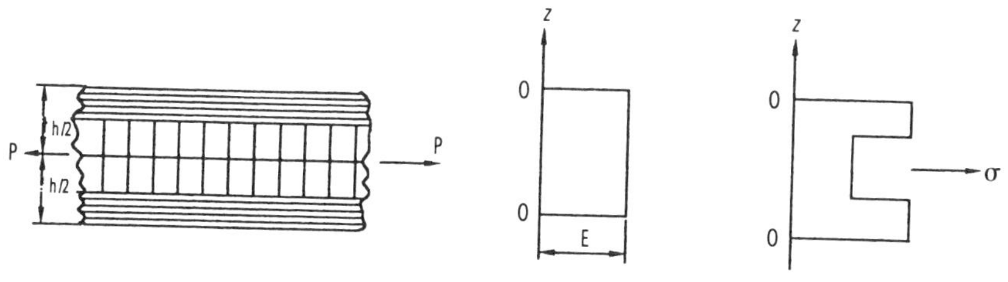

(i) Section through the laminate (ii) strain distribution (iii) stress distribution

Figure 5.33 In-plane stresses and strains in a laminate.

As the load P is in-plane and symmetrical, the strain distribution across the laminate will be constant, as shown by Figure 5.33(ii). However, as the stiffness of each layer is different the stresses in each layer will be different, as shown by Figure 5.33(iii). Now, in order to define the overall equivalent stress-strain behaviour of a laminate, it will be necessary to adopt the equivalent average stresses or in matrix form σ_X^{1}, σ_Y^{1} and τ_{XY}^{1}; these are obtained as follows:

$$\sigma_X^1 = \frac{1}{h}\int_{-\frac{h}{2}}^{\frac{h}{2}} \sigma_X dz\ , \qquad \sigma_Y^1 = \frac{1}{h}\int_{-\frac{h}{2}}^{\frac{h}{2}} \sigma_Y dz\ , \qquad \tau_{XY}^1 = \frac{1}{h}\int_{-\frac{h}{2}}^{\frac{h}{2}} \tau_{XY} dz$$

or in matrix form

$$\begin{Bmatrix} \sigma_X^1 \\ \sigma_Y^1 \\ \tau_{XY}^1 \end{Bmatrix} = \frac{1}{h}\int_{-\frac{h}{2}}^{\frac{h}{2}} \begin{Bmatrix} \sigma_X \\ \sigma_Y \\ \tau_{XY} \end{Bmatrix} dz$$

but from equation (5.74)

$$\begin{Bmatrix} \sigma_X^1 \\ \sigma_Y^1 \\ \tau_{XY}^1 \end{Bmatrix} = \left[Q^1\right] \begin{Bmatrix} \varepsilon_X \\ \varepsilon_Y \\ \gamma_{XY} \end{Bmatrix}$$

$$\therefore \begin{Bmatrix} \sigma_X^1 \\ \sigma_Y^1 \\ \tau_{XY}^1 \end{Bmatrix} = \frac{1}{h}\int_{-\frac{h}{2}}^{\frac{h}{2}} \begin{bmatrix} q_{11}^1 & q_{12}^1 & q_{16}^1 \\ q_{21}^1 & q_{22}^1 & q_{26}^1 \\ q_{62}^1 & q_{61}^1 & q_{66}^1 \end{bmatrix} \begin{Bmatrix} \varepsilon_X \\ \varepsilon_Y \\ \gamma_{XY} \end{Bmatrix} dz \qquad (5.76)$$

However, as $[\varepsilon_X \, \varepsilon_Y \, \gamma_{XY}]^T$ is not a function of 'z', equation (5.76) can be written as follows:

$$\begin{Bmatrix} \sigma_X^1 \\ \sigma_Y^1 \\ \tau_{XY}^1 \end{Bmatrix} = \frac{1}{h} \int_{-\frac{h}{2}}^{\frac{h}{2}} \left[Q^1\right] dz \begin{Bmatrix} \varepsilon_X \\ \varepsilon_Y \\ \gamma_{XY} \end{Bmatrix}$$

$$= [A] \begin{Bmatrix} \varepsilon_X \\ \varepsilon_Y \\ \gamma_{XY} \end{Bmatrix} \tag{5.77}$$

where

$$A_{11} = \frac{1}{h} \int_{-\frac{h}{2}}^{\frac{h}{2}} q_{11}^1 \, dz$$

$$= \frac{2}{h} \int_{o}^{\frac{h}{2}} q_{11}^1 \, dx \tag{5.78}$$

$$A_{12} = \frac{2}{h} \int_{o}^{\frac{h}{2}} q_{12}^1 \, dz$$

or, in general,

$$A_{ij} = \frac{2}{h} \int_{o}^{\frac{h}{2}} q_{ij}^1 \, dz$$

For the kth lamina of the laminate, the q^1 terms are constant, hence the integrals for the A terms can be replaced by summations:

$$A_{11} = \frac{2}{h} \sum q^1_{11(k)} \; h_k = \sum q^1_{11(k)} \left(\frac{2h_k}{h} \right) \tag{5.79}$$

and similarly for the other values of A_{ij},

where

h_k = thickness of the kth lamina or ply

$q_{11(k)}$ = kth value of $q_{11}{}^1$

v_k = $(2h_k/h)$ = the volume fraction in the kth lamina

Once the *stiffness* matrix $[A]$ is obtained, it can be inverted to obtain the *compliance* matrix $[a]$ and hence, the equivalent material for the laminate properties are as follows:

$$E_X = \frac{1}{a_{11}}, \quad E_Y = \frac{1}{a_{22}}, \quad G_{XY} = \frac{1}{a_{66}}, \quad \nu_X = \frac{-a_{12}}{a_{11}} \text{ and } \nu_Y = \frac{-a_{12}}{a_{22}}$$

Experience has shown that the diagonal terms in the laminate's stiffness matrix are considerably larger than the off-diagonal terms, so that E_x etc. can be approximated by

$$E_X \doteqdot \sum v_k \; E_{X(k)} \; \cos^4 \theta_k$$

where k refers to the kth lamina of the laminate.

5.23 Equivalent elastic constants for problems involving bending and twisting

For problems in this category, the equivalent stress resultants for the laminate are $\sigma_X{}^1$, $\sigma_Y{}^1$, $\tau_{XY}{}^1$, $M_X{}^1$, $M_Y{}^1$ and $M_{XY}{}^1$, where the former three symbols are in-plane and the latter three are out-of-plane bending and twisting terms.

The equivalent stress–strain relationships for the laminate are:

$$\begin{Bmatrix} \sigma_X^1 \\ \sigma_Y^1 \\ \tau_{XY}^1 \\ M_X^1 \\ M_Y^1 \\ M_{XY}^1 \end{Bmatrix} = \begin{bmatrix} A & B \\ B & D \end{bmatrix} \begin{Bmatrix} \varepsilon_X \\ \varepsilon_Y \\ \gamma_{XY} \\ \dfrac{-\partial^2 w}{\partial X^2} \\ \dfrac{-\partial^2 w}{\partial Y^2} \\ \dfrac{-2\partial^2 w}{\partial X \partial Y} \end{Bmatrix} \tag{5.80}$$

or

$$\begin{Bmatrix} \sigma \\ M \end{Bmatrix} = \begin{bmatrix} A & B \\ B & D \end{bmatrix} \begin{Bmatrix} \varepsilon \\ \chi \end{Bmatrix}$$

where

$$[\varepsilon]^T = [\varepsilon_X \;\; \varepsilon_Y \;\; \gamma_{XY}]^T$$

$$[\chi]^T = \left[\frac{-\partial^2 w}{\partial X^2} \;\; \frac{-\partial^2 w}{\partial Y^2} \;\; \frac{-2\partial^2 w}{\partial \times \partial Y}\right]^T$$

A_{ij} are as described in Section 5.21.

$$B_{ij} = \frac{1}{h^2} \int_{\frac{-h}{2}}^{\frac{h}{2}} q_{ij}^1 \, z \, . \, dz = \sum_{k=1}^{n} q_{ij(k)}^1 h_k \, z_k$$

$$D_{ij} = \frac{1}{h^3} \int_{\frac{-h}{2}}^{\frac{h}{2}} q_{ij}^1 \, z^2 \, dz$$

$$D_{ij} = \frac{1}{h^3} \sum_{k=1}^{n} q_{ij(k)}^1 \left(h_k \, z_k^2 + h_k^3/12\right) \tag{5.81}$$

where

w = out-of-plane deflection

n = number of laminates or plies

k = the kth ply or lamina

z_k = distance of the centre plane of the kth ply

For symmetrical laminates, $B_{ij} = 0$, however, for design purposes, the following relationship is often used:

$$\begin{Bmatrix} \varepsilon \\ M \end{Bmatrix} = \begin{bmatrix} a & b_1 \\ b_2 & d \end{bmatrix} \begin{Bmatrix} \sigma \\ \chi \end{Bmatrix}$$

where

$[a] = [A]^{-1}$ (see Section 5.21)

$[b_1] = -[A]^{-1}\,[B]$

$[b_2] = [B]\,[A]^{-1}$

$[d] = [D] - [B]\,[A]^{-1}\,[B]$

Another way of looking at the components of [D] are as follows:

$$D_{ij} = \sum_{k=1}^{n} q^{1}_{ij(k)} \times \left(\frac{I_k}{I_{comp}} \right) \tag{5.82}$$

where k = the kth ply

I_k = the second moment of area of the kth ply or lamina about the neutral axis of the laminate or composite

I_{comp} = the second moment of area of the entire laminate or composite about the neutral axis

5.24 Yielding of ductile materials under combined stresses

It was noted in Section 5.3 that when a polished bar of mild steel is loaded in tension, strain figures are observable on the surface of the bar after the yield point has been exceeded. The figures take the form of 'lines' inclined at about 45° to the axis of the bar; this direction corresponds to the planes of maximum shearing stress in the bar; the 'lines' are, in fact, bands of metal crystals shearing over similar bands. That yielding takes place in this way suggests that the crystal structure of the metal is relatively weak in shear; yielding takes the form of sliding of one crystal plane over another, and not the tearing apart of two crystal planes.

This form of behaviour—yielding by a shearing action—is typical of ductile materials. We note firstly that if a material is subjected to a hydrostatic pressure σ, the three principal stresses σ_1, σ_2 and σ_3 in a three-dimensional system are each equal to σ. A state of stress of this sort exists in a solid sphere of material subjected to an external pressure σ, Figure 5.34. As the three principal stresses are equal in magnitude, there are no shearing stresses in the material; if yielding is governed by the presence of shearing on some planes in a material, then no yielding is theoretically possible when the material is under hydrostatic pressure.

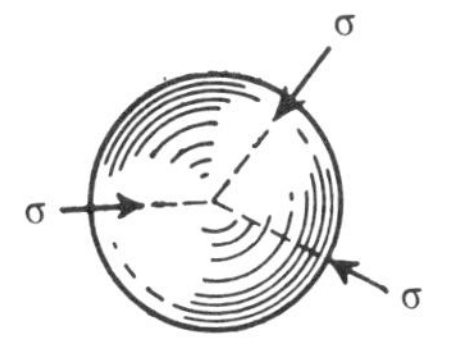

Figure 5.34 A solid sphere of material under hydrostatic pressure.

For a two-dimensional stress system one of the three principal stresses of a three-dimensional system is zero. We consider now the yielding of a mild steel under different combinations of the principal stresses, σ_1 and σ_2, of a two-dimensional system; in discussing the problem we keep in mind the presence of a zero stress perpendicular to the plane of σ_1 and σ_2, Figure 5.27.

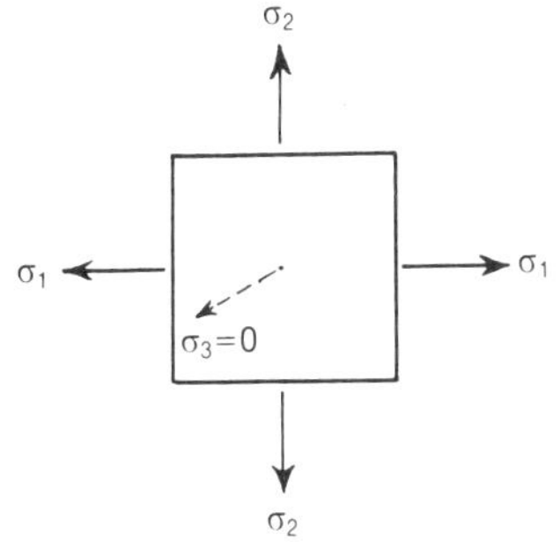

Figure 5.35 Yield envelope of a two dimensional stress system when the material yields according to the maximum shearing stress criterion.

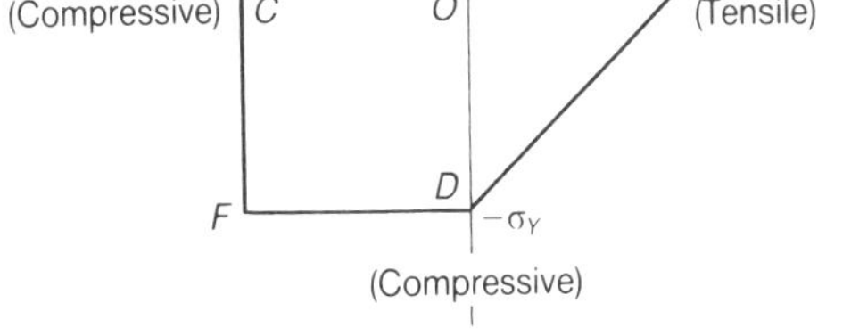

Figure 5.36 In a two-dimensional stress system, one of the three principal stresses - (σ_3 say) is zero.

Suppose we conduct a simple tension test on the material; we may put $\sigma_2 = 0$, and yielding occurs when $\sigma_1 = \sigma_Y$, (say)

This yielding condition corresponds to the point *A* in Figure 5.35. If the material has similar properties in tension and compression, yielding under a compressive stress σ_1 occurs when $\sigma_1 = -\sigma_Y$; this condition corresponds to the point *C* in Figure 5.35. We could, however, perform the tension and compression tests in the direction of σ_2, Figure 5.35; if the material is isotropic —that is, it has the same properties in all directions—yielding occurs at the yield stress σ_Y; we can thus derive points *B* and *D* in the yield diagram, Figure 5.35.

We consider now yielding of the material when both σ_1 and σ_2, Figure 5.36, are present; we shall assume that yielding of the mild steel occurs when the maximum shearing stress attains a critical value; from the simple tensile test, the maximum shearing stress at yielding is

$$\tau_{max} = \frac{1}{2}\sigma_Y$$

which we shall take as the critical value. Suppose that $\sigma_1 > \sigma_2$, and that both principal stresses are tensile; the maximum shearing stress is

$$\tau_{max} = \frac{1}{2}(\sigma_1 - 0) = \frac{1}{2}\sigma_1$$

and occurs in the 3–1 plane of Figure 5.36; τ_{max} attains the critical value when

$$\frac{1}{2}\sigma_1 = \frac{1}{2}\sigma_Y, \quad \text{or} \quad \sigma_1 = \sigma_Y$$

Thus, yielding for these stress conditions is unaffected by σ_2. In Figure 5.35, these stress conditions are given by the line *AH*. If we consider similarly the case when σ_1 and σ_2 are both tensile, but $\sigma_2 > \sigma_1$, yielding occurs when $\sigma_2 = \sigma_Y$, giving the line *BH* in Figure 5.35.

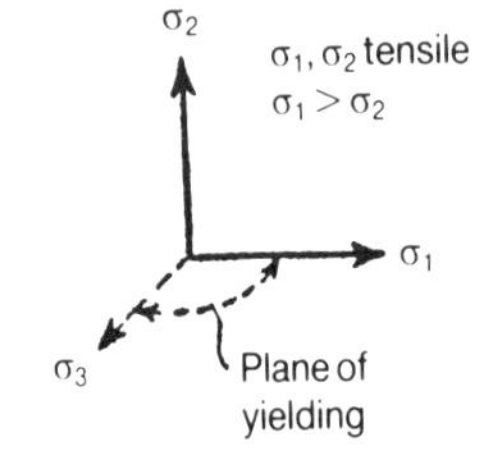

Figure 5.37 Plane of yielding when both principal stresses tensile and $\sigma_1 > \sigma_2$.

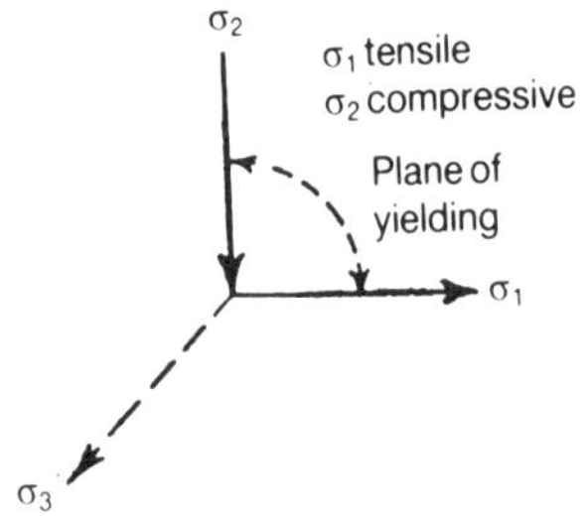

Figure 5.38 Plane of yielding when the principal stresses are of opposite sign.

By making the stresses both compressive, we can derive in a similar fashion the lines *CF* and *DF* of Figure 5.36.

But when σ_1 is tensile and σ_2 is compressive, Figure 5.36, the maximum shearing stress occurs in the 1–2 plane, and has the value

$$\tau_{max} = \frac{1}{2}(\sigma_1 - \sigma_2)$$

Yielding occurs when

$$\frac{1}{2}(\sigma_1 - \sigma_2) = \frac{1}{2}\sigma_Y, \quad \text{or} \quad \sigma_1 - \sigma_2 = \sigma_Y$$

This corresponds to the line *AD* of Figure 5.36. Similarly, when σ_1 is compressive and σ_2 is tensile, yielding occurs when corresponding to the line *BC* of Figure 5.36.

$$\sigma_2 - \sigma_1 = \sigma_Y$$

The hexagon *AHBCFD* of Figure 5.36 is called a *yield locus*, because it defines all combinations of σ_1 and σ_2 giving yielding of mild steel; for any state of stress within the hexagon the material remains elastic; for this reason the hexagon is also sometimes called a *yield envelope*. The *criterion of yielding* used in the derivation of the hexagon of Figure 5.36 was that of maximum shearing stress; the use of this criterion was first suggested by Tresca in 1878.

Not all ductile metals obey the maximum shearing stress criterion; the yielding of some metals, including certain steels and alloys of aluminium, is governed by a critical value of the strain energy of distortion. For a two-dimensional stress system the strain energy of distortion per unit of volume of the material is given by equation (5.83). In the simple tension test for which $\sigma_2 = 0$, say, yielding occurs when $\sigma_1 = \sigma_Y$. The critical value of U_D is therefore

$$U_D = \frac{1}{6G}\left[\sigma_1^2 - \sigma_1\sigma_2 + \sigma_2^2\right] = \frac{1}{6G}\left[\sigma_Y^2 - \sigma_Y(0) + 0^2\right] = \frac{\sigma_Y^2}{6G}$$

Then for other combinations of σ_1 and σ_2, yielding occurs when

$$\sigma_1^2 - \sigma_1\sigma_2 + \sigma_2^2 = \sigma_Y^2 \tag{5.83}$$

The yield locus given by this equation is an ellipse with major and minor axes inclined at 45° to the directions of σ_1 and σ_2, Figure 5.39. This locus was first suggested by von Mises in 1913.

For a three-dimensional system the yield locus corresponding to the strain energy of distortion is of the form

$$(\sigma_1 - \sigma_2)^2 + (\sigma_2 - \sigma_3)^2 + (\sigma_3 - \sigma_1)^2 = \text{constant}$$

This relation defines the surface of a cylinder of circular cross-section, with its central axis on the line $\sigma_1 = \sigma_2 = \sigma_3$; the axis of the cylinder passes through the origin of the σ_1, σ_2, σ_3 co-ordinate system, and is inclined at equal angles to the axes σ_1, σ_2 and σ_3, Figure 5.40. When σ_3 is zero, critical values of σ_1 and σ_2 lie on an ellipse in the σ_1–σ_2 plane, corresponding to the ellipse of Figure 5.39.

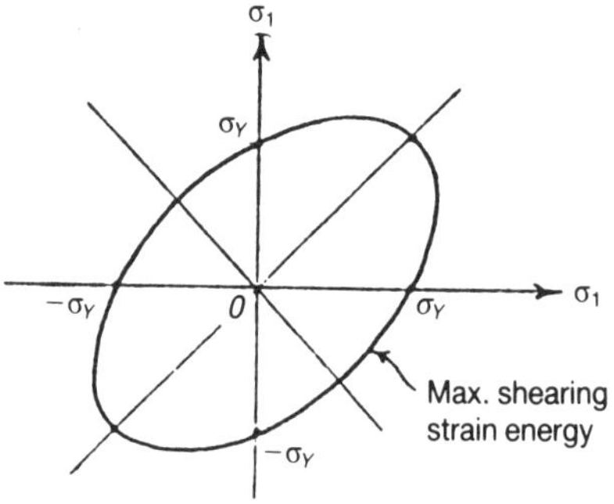

Figure 5.39 The von Mises yield locus for a two-dimensional system of stresses.

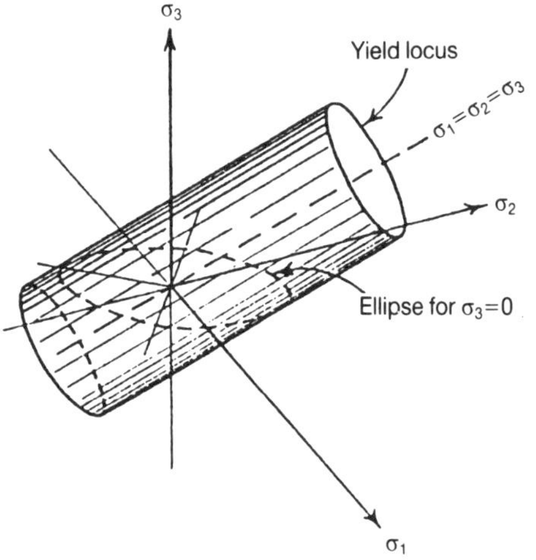

Figure 5.40 The von Mises yield locus for a three-dimensional stress system.

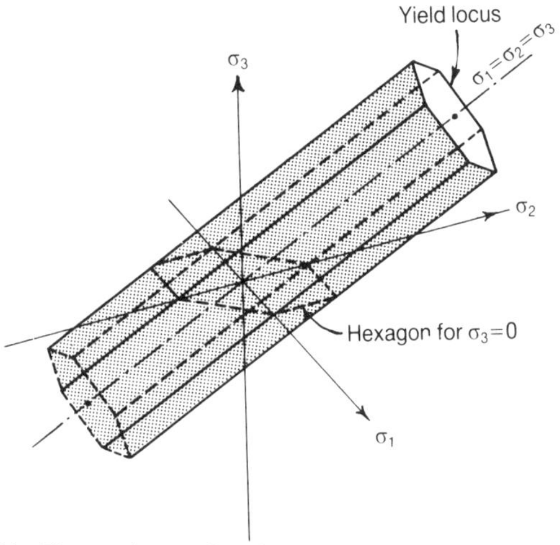

Figure 5.41 The maximum shearing stress (or Tresca) yield locus for a three-dimensional stress system.

When a material obeys the maximum shearing stress criterion, the three-dimensional yield locus is a regular hexagonal cylinder with its central axis on the line $\sigma_1 = \sigma_2 = \sigma_3 = 0$, Figure 5.40. When σ_3 is zero, the locus is an irregular hexagon, of the form already discussed in Figure 5.36.

The surfaces of the yield loci in Figures 5.40 and 5.41 extend indefinitely parallel to the line $\sigma_1 = \sigma_2 = \sigma_3$, which we call the hydrostatic stress line. Hydrostatic stress itself cannot cause yielding, and no yielding occurs at other stresses provided these fall within the cylinders of Figures 5.40 and 5.41.

The problem with the maximum principal stress and maximum principal strain theories is that they break down in the hydrostatic stress case; this is because under hydrostatic stress, failure does not occur as there is no shear stress. It must be pointed out that under uniaxial tensile stress, all the major theories give the same predictions for elastic failure, hence, all apply in the uniaxial case. However, in the case of a ductile specimen under pure torsion, the maximum shear stress theory predicts that yield occurs when the maximum shear reaches 0.5 σ_y, but in practice, yield occurs when the maximum shear stress reaches 0.577 of the yield stress. This last condition is only satisfied by the von Mises or distortion energy theory and for this reason, this theory is currently very much in favour for ductile materials.

Another interpretation of the von Mises or distortion energy theory is that yield occurs when the von Mises stress, namely σ_{vm}, reaches yield.

In *three dimensions*, σ_{vm} is calculated as follows:

$$\underline{\sigma_{vm}} = \underline{\sqrt{\left[\left(\sigma_1 - \sigma_2\right)^2 + \left(\sigma_1 - \sigma_3\right)^2 + \left(\sigma_2 - \sigma_3^2\right)\right]}/\sqrt{2}} \tag{5.84}$$

In *two-dimensions*, $\sigma_3 = 0$, therefore equation (5.84) becomes:

$$\underline{\sigma_{vm}} = \underline{\sqrt{\left(\sigma_1^2 + \sigma_2^2 - \sigma_1 \sigma_2\right)}} \tag{5.85}$$

5.25 Elastic breakdown and failure of brittle material

Unlike ductile materials the failure of brittle materials occurs at relatively low strains, and there is little, or no, permanent yielding on the planes of maximum shearing stress.

Some brittle materials, such as cast iron and concrete, contain large numbers of holes and microscopic cracks in their structures. These are believed to give rise to high stress concentrations, thereby causing local failure of the material. These stress concentrations are likely to have a greater effect in reducing tensile strength than compressive strength; a general characteristic of brittle materials is that they are relatively weak in tension. For this reason elastic breakdown and failure in a brittle material are governed largely by the maximum principal tensile stress; as an example of the application of this criterion consider a concrete: in simple tension the breaking stress is about 1.5 MN/m^2, whereas in compression it is found to be about 30 MN/m^2, or 20 times as great; in pure shear the breaking stress would be of the order of 1.5 MN/m^2, because the principal stresses are of the same magnitude, and one of these stresses is tensile, Figure 5.42. Cracking in the concrete would occur on planes inclined at 45° to the directions of the applied shearing stresses.

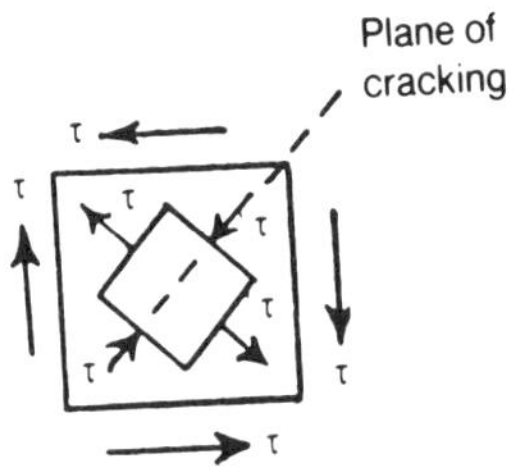

Figure 5.42 Elastic breakdown of a brittle metal under shearing stresses (pure shear).

5.26 Failure of composites

Accurate prediction of the failure of laminates is a much more difficult task than it is for steels and aluminium alloys. The failure load of the laminate is also dependent on whether the laminate is under in-plane loading, or bending or shear. Additionally, under compression, individual plies can buckle through a microscopic form of beam-column buckling (see Chapter 18). In general, it is better to depend on experimental data than purely on theories of elastic failure. Theories, however, exist and Hill, Azzi and Tsai produced theories based on the von Mises theory of yield. One such popular two-dimensional theory is the Azzi–Tsai theory, as follows:

$$\frac{\sigma_x^2}{X^2} + \frac{\sigma_y^2}{Y^2} - \frac{\sigma_x \sigma_y}{X^2} + \frac{\tau_{xy}^2}{S^2} = 1 \qquad (5.86)$$

where X and Y are the uniaxial strengths related to σ_x and σ_y respectively and S is the shear strength in the x–y directions, which are not principal planes.

For the isotropic case, where $X = Y = \sigma_Y$ and $S = \sigma_Y / \sqrt{3}$, equation (5.86) reduces to the von Mises form:

$$\sigma_x^2 + \sigma_y^2 - \sigma_x \sigma_y + 3\tau_{xy}^2 = \sigma_Y^2$$

and when $\sigma_x = \sigma_1$ and $\sigma_y = \sigma_2$ so that $\tau_{1-2} = 0$, we get

$$\sigma_1^2 + \sigma_2^2 - \sigma_1 \sigma_2 = \sigma_Y^2 \quad \text{[See equation (5.85)]}$$

Further problems *(answers on page 692)*

5.7 A tie-bar of steel has a cross-section 15 cm by 2 cm, and carries a tensile load of 200 kN. Find the stress normal to a plane making an angle of 30° with the cross-section and the shearing stress on this plane. *(Cambridge)*

5.8 A rivet is under the action of shearing stress of 60 MN/m^2 and a tensile stress, due to contraction, of 45 MN/m^2. Determine the magnitude and direction of the greatest tensile and shearing stresses in the rivet. *(RNEC)*

5.9 A propeller shaft is subjected to an end thrust producing a stress of 90 MN/m^2, and the maximum shearing stress arising from torsion is 60 MN/m^2. Calculate the magnitudes of the principal stresses. *(Cambridge)*

5.10 At a point in a vertical cross-section of a beam there is a resultant stress of 75 MN/m^2, which is inclined upwards at 35° to the horizontal. On the horizontal plane through the point there is only shearing stress. Find in magnitude and direction, the resultant stress on the plane which is inclined at 40° to the vertical and 95° to the resultant stress. *(Cambridge)*

5.11 A plate is subjected to two mutually perpendicular stresses, one compressive of 45 MN/m^2, the other tensile of 75 MN/m^2, and a shearing stress, parallel to these directions, of 45 MN/m^2. Find the principal stresses and strains, taking Poisson's ratio as 0.3 and E = 200 GN/m^2. *(Cambridge)*

5.12 At a point in a material the three principal stresses acting in directions O_x, O_y, O_z, have the values 75, 0 and −45 MN/m^2, respectively. Determine the normal and shearing stresses for a plane perpendicular to the xz-plane inclined at 30° to the xy-plane. *(Cambridge)*

6 Thin shells under internal pressure

6.1 Thin cylindrical shell of circular cross-section

A problem in which combined stresses are present is that of a cylindrical shell under internal pressure. Suppose a long circular shell is subjected to an internal pressure p, which may be due to a fluid or gas enclosed within the cylinder, Figure 6.1. The internal pressure acting on the long sides of the cylinder gives rise to a circumferential stress in the wall of the cylinder; if the ends of the cylinder are closed, the pressure acting on these ends is transmitted to the walls of the cylinder, thus producing a longitudinal stress in the walls.

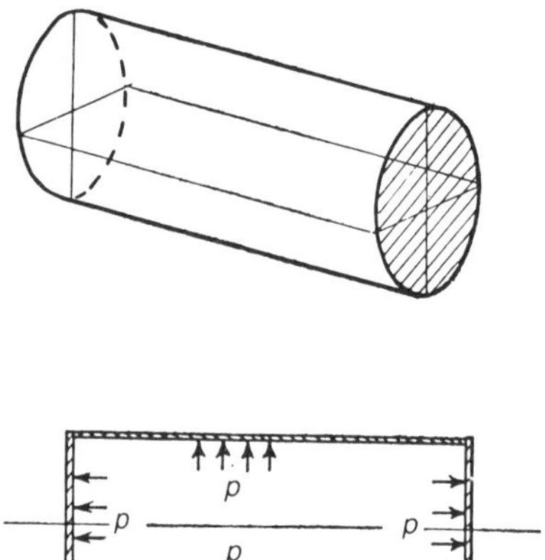

Figure 6.1 Long thin cylindrical shell with closed ends under internal pressure.

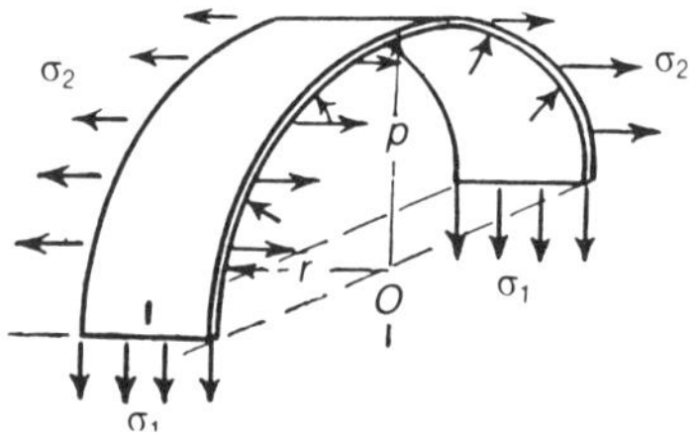

Figure 6.2 Circumferential and longitudinal stresses in a thin cylinder with closed ends under internal pressure.

Suppose r is the mean radius of the cylinder, and that its thickness t is small compared with r. Consider a unit length of the cylinder remote from the closed ends, Figure 6.2; suppose we cut this unit length with a diametral plane, as in Figure 6.2. The tensile stresses acting on the cut sections are σ_1, acting circumferentially, and σ_2, acting longitudinally. There is an internal pressure p on the inside of the half-shell. Consider equilibrium of the half-shell in a plane perpendicular to the axis of the cylinder, as in Figure 6.3; the total force due to the internal pressure p in the direction OA is

$$p \times (2r \times 1)$$

because we are dealing with a unit length of the cylinder. This force is opposed by the stresses σ_1; for equilibrium we must have

$$p \times (2r \times 1) = \sigma_1 \times 2(t \times 1)$$

Then

$$\sigma_1 = \frac{pr}{t} \tag{6.1}$$

We shall call this the *circumferential (or hoop) stress.*

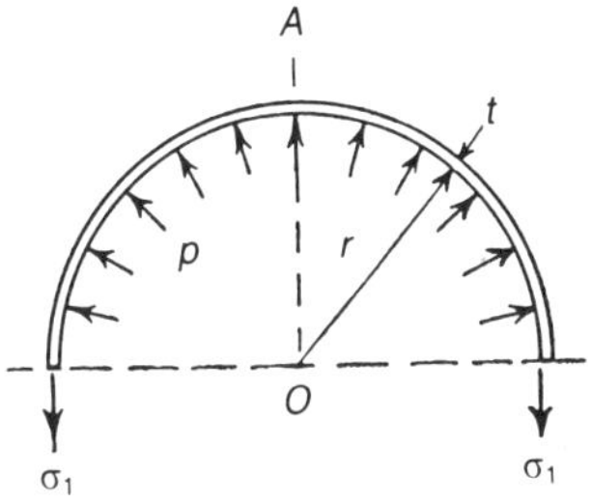

Figure 6.3 Derivation of circumferential stress.

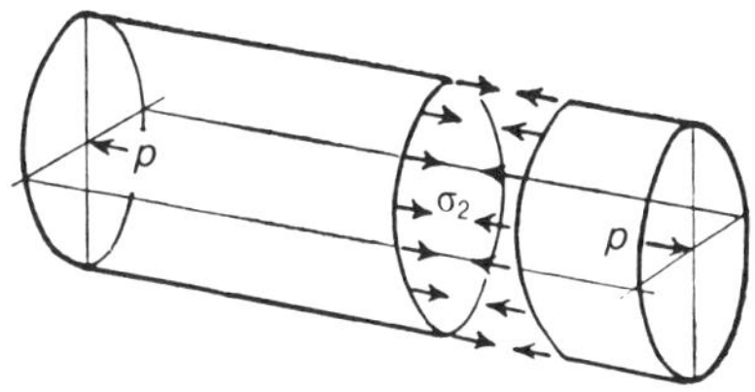

Figure 6.4 Derivation of longitudinal stress.

Now consider any transverse cross-section of the cylinder remote from the ends, Figure 6.4; the total longitudinal force on each closed end due to internal pressure is

$$p \times \pi r^2$$

At any section this is resisted by the internal stresses σ_2, Figure 6.4. For equilibrium we must have

$$p \times \pi r^2 = \sigma_2 \times 2\,\pi r t$$

which gives

$$\sigma_2 = \frac{pr}{2t} \tag{6.2}$$

We shall call this the *longitudinal stress*. Thus the longitudinal stress, σ_2, is only half the circumferential stress, σ_1.

The stresses acting on an element of the wall of the cylinder consist of a circumferential stress σ_1, a longitudinal stress σ_2, and a radial stress p on the internal face of the element, Figure 6.5. As (r/t) is very much greater than unity, p is small compared with σ_1 and σ_2. The state of stress in the wall of the cylinder approximates then to a simple two-dimensional system with principal stresses σ_1 and σ_2.

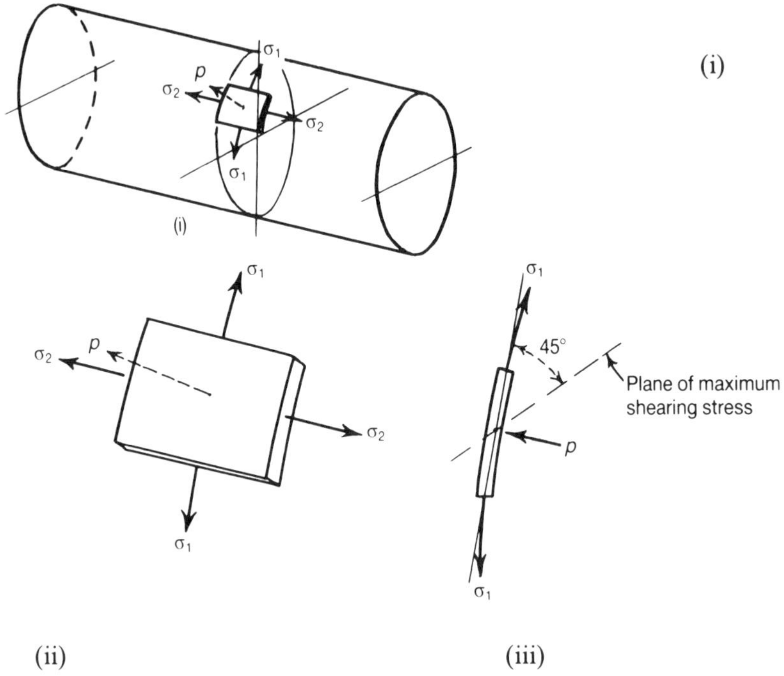

Figure 6.5 Stresses acting on an element of the wall of a circular cylindrical shell with closed ends under internal pressure.

The maximum shearing stress in the plane of σ_1 and σ_2 is therefore

$$\tau_{max} = \frac{1}{2}\left(\sigma_1 - \sigma_2\right) = \frac{pr}{4t}$$

This is not, however, the maximum shearing stress in the wall of the cylinder, for, in the plane of σ_1 and p, the maximum shearing stress is

$$\tau_{max} = \frac{1}{2}(\sigma_1) = \frac{pr}{2t} \tag{6.3}$$

since p is negligible compared with σ_1; again, in the plane of σ_2 and p, the maximum shearing stress is

$$\tau_{max} = \frac{1}{2}(\sigma_2) = \frac{pr}{4t}$$

The greatest of these maximum shearing stresses is given by equation (6.3); it occurs on a plane at 45° to the tangent and parallel to the longitudinal axis of the cylinder, Figure 6.5(iii).

The circumferential and longitudinal stresses are accompanied by direct strains. If the material of the cylinder is elastic, the corresponding strains are given by

$$\varepsilon_1 = \frac{1}{E}(\sigma_1 - \nu\sigma_2) = \frac{pr}{Et}\left(1 - \frac{1}{2}\nu\right)$$

$$\varepsilon_1 = \frac{1}{E}(\sigma_2 - \nu\sigma_1) = \frac{pr}{Et}\left(\frac{1}{2} - \nu\right) \tag{6.4}$$

The circumference of the cylinder increases therefore by a small amount $2\pi r\varepsilon_1$; the increase in mean radius is therefore $r\varepsilon_1$ The increase in length of a unit length of the cylinder is ε_2, so the change in internal volume of a unit length of the cylinder is

$$\delta V = \pi(r + r\varepsilon_1)^2(1 + \varepsilon_2) - \pi r^2$$

The volumetric strain is therefore

$$\frac{\delta V}{\pi r^2} = (1 + \varepsilon_1)^2(1 + \varepsilon_2) - 1$$

But ε_1 and ε_2 are small quantities, so the volumetric strain is

$$(1 + \varepsilon_1)^2(1 + \varepsilon_2) - 1 \doteq (1 + 2\varepsilon_1)(1 + \varepsilon_2) - 1$$

$$\doteq 2\varepsilon_1 + \varepsilon_2$$

In terms of σ_1 and σ_2 this becomes

$$2\varepsilon_1 + \varepsilon_2 = \frac{pr}{Et}\left[2\left(1 - \frac{1}{2}\nu\right) + \left(\frac{1}{2} - \nu\right)\right] = \frac{pr}{Et}\left(\frac{5}{2} - 2\nu\right) \tag{6.5}$$

Problem 6.1 A thin cylindrical shell has an internal diameter of 20 cm, and is 0.5 cm thick. It is subjected to an internal pressure of 3.5 MN/m^2. Estimate the circumferential and longitudinal stresses if the ends of the cylinders are closed.

Solution

From equations (6.1) and (6.2),

$$\sigma_1 = \frac{pr}{t} = \left(3.5 \times 10^6\right)(0.1025)/(0.005) = 71.8 \text{ MN/m}^2$$

and

$$\sigma_2 = \frac{pr}{2t} = \left(3.5 \times 10^6\right)(0.1025)/(0.010) = 35.9 \text{ MN/m}^2$$

Problem 6.2 If the ends of the cylinder in Problem 6.1 are closed by pistons sliding in the cylinder, estimate the circumferential and longitudinal stresses.

Solution

The effect of taking the end pressure on sliding pistons is to remove the force on the cylinder causing longitudinal stress. As in Problem 6.1, the circumferential stress is

$$\sigma_1 = 71.8 \text{ MN/m}^2$$

but the longitudinal stress is zero.

Problem 6.3 A pipe of internal diameter 10 cm, and 0.3 cm thick is made of mild-steel having a tensile yield stress of 375 MN/m^2. What is the maximum permissible internal pressure if the stress factor on the maximum shearing stress is to be 4?

Solution

The greatest allowable maximum shearing stress is

$$\frac{1}{4}\left(\frac{1}{2} \times 375 \times 10^6\right) = 46.9 \text{ MN/m}^2$$

The greatest shearing stress in the cylinder is

$$\tau_{max} = \frac{pr}{2t}$$

Then $$p = \frac{2t}{r}\left(\tau_{max}\right) = \frac{2 \times 0.003}{0.0515} \times \left(46.9 \times 10^6\right) = 5.46 \text{ MN / m}^2$$

Problem 6.4 Two boiler plates, each 1 cm thick, are connected by a double-riveted butt joint with two cover plates, each 0.6 cm thick. The rivets are 2 cm diameter and their pitch is 0.90 cm. The internal diameter of the boiler is 1.25 m, and the pressure is 0.8 MN/m^2. Estimate the shearing stress in the rivets, and the tensile stresses in the boiler plates and cover plates.

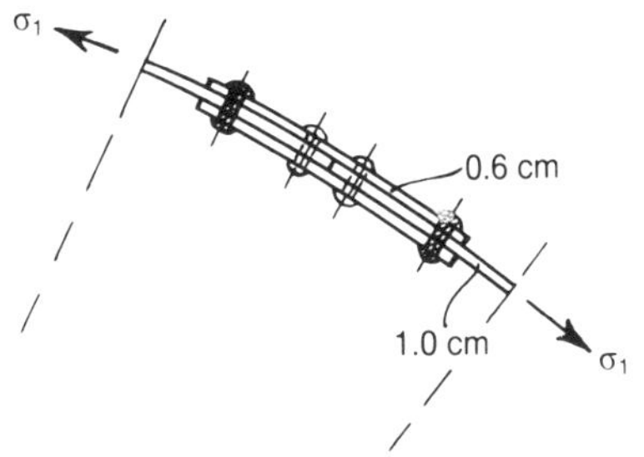

Solution

Suppose the rivets are staggered on each side of the joint. Then a single rivet takes the circumferential load associated with a ½ (0.090) = 0.045 m length of boiler. The load on a rivet is

$$\left[\frac{1}{2}(1.25)\right](0.045)\left(0.8 \times 10^6\right) = 22.5 \text{ kN}$$

Area of a rivet is

$$\frac{\pi}{4}(0.02)^2 = 0.314 \times 10^{-3} \text{ m}^2$$

The load of 22.5 kN is taken in double shear, and the shearing stress in the rivet is then

$$\frac{1}{2}\ (22.5 \times 10^3) / (0.314 \times 10^{-3}) = 35.8 \text{ MN/m}^2$$

The rivet holes in the plates give rise to a loss in plate width of 2 cm in each 9 cm of rivet line. The effective area of boiler plate in a 9 cm length is then

$$(0.010)\ (0.090 - 0.020) = (0.010)\ (0.070) = 0.7 \times 10^{-3} \text{ m}^2$$

The tensile load taken by this area is

$$\frac{1}{2}\ (1.25)\ (0.090)\ (0.8 \times 10^6) = 45.0 \text{ kN}$$

The average circumferential stress in the boiler plates is therefore

$$\sigma_1 = \frac{45.0 \times 10^3}{0.7 \times 10^{-3}} = 64.2 \text{ MN/m}^2$$

This occurs in the region of the riveted connection. Remote from the connection, the circumferential tensile stress is

$$\sigma_1 = \frac{pr}{t} = \frac{\left(0.8 \times 10^6\right)(0.625)}{(0.010)} = 50.0 \text{ MN/m}^2$$

In the cover plates, the circumferential tensile stress is

$$\frac{45.0 \times 10^3}{2(0.006)\,(0.070)} = 53.6 \text{ MN/m}^2$$

The longitudinal tensile stresses in the plates in the region of the connection are difficult to estimate; except very near to the rivet holes, the stress will be

$$\sigma_2 = \frac{pr}{2t} = 25.0 \text{ MN/m}^2$$

Problem 6.5 A long steel tube, 7.5 cm internal diameter and 0.15 cm thick, has closed ends, and is subjected to an internal fluid pressure of 3 MN/m^2. If $E = 200 \text{ GN/m}^2$, and $\nu = 0.3$, estimate the percentage increase in internal volume of the tube.

<u>Solution</u>

The circumferential tensile stress is

$$\sigma_1 = \frac{pr}{t} = \frac{\left(3 \times 10^6\right)(0.0383)}{(0.0015)} = 76.6 \text{ MN/m}^2$$

The longitudinal tensile stress is

$$\sigma_2 = \frac{pr}{2t} = 38.3 \text{ MN/m}^2$$

The circumferential strain is

$$\varepsilon_1 = \frac{1}{E}\left(\sigma_1 - \nu\sigma_2\right)$$

and the longitudinal strain is

$$\varepsilon_2 = \frac{1}{E}\left(\sigma_2 - \nu\sigma_1\right)$$

The volumetric strain is then

$$2\varepsilon_1 + \varepsilon_2 = \frac{1}{E}\left[2\sigma_1 - 2\nu\sigma_2 + \sigma_2 - \nu\sigma_1\right]$$

$$= \frac{1}{E}\left[\sigma_1 (2 - \nu) + \sigma_2 (1 - 2\nu)\right]$$

Thus

$$2\varepsilon_1 + \varepsilon_2 = \frac{(76.6 \times 10^6)\left[(2 - 0.3) + (1 - 0.6)\right]}{200 \times 10^9}$$

$$= \frac{(76.6 \times 10^6)(1.9)}{(200 \times 10^9)} = 0.727 \times 10^{-3}$$

The percentage increase in volume is therefore 0.0727%

Problem 6.6 An air vessel, which is made of steel, is 2 m long; it has an external diameter of 45 cm and is 1 cm thick. Find the increase of external diameter and the increase of length when charged to an internal air pressure of 1 MN/m^2.

Solution

For steel, we take

$$E = 200 \text{ GN/m}^2, \quad \nu = 0.3$$

The mean radius of the vessel is $r = 0.225$ m; the circumferential stress is then

$$\sigma_1 = \frac{pr}{t} = \frac{(1 \times 10^6)(0.225)}{0.010} = 22.5 \text{ MN/m}^2$$

The longitudinal stress is

$$\sigma_2 = \frac{pr}{2t} = 11.25 \text{ MN/m}^2$$

The circumferential strain is therefore

$$\varepsilon_1 = \frac{1}{E}(\sigma_1 - \nu\sigma_2) = \frac{\sigma_1}{E}\left(1 - \frac{1}{2}\nu\right) = \frac{(22.5 \times 10^6)(0.85)}{200 \times 10^9}$$

$$= 0.957 \times 10^{-4}$$

The longitudinal strain is

$$\varepsilon_2 = \frac{1}{E}(\sigma_2 - \nu\sigma_1) = \frac{\sigma_1}{E}\left(\frac{1}{2} - \nu\right) = \frac{(22.5 \times 10^6)(0.2)}{200 \times 10^9}$$

$$= 0.225 \times 10^{-4}$$

The increase in external diameter is then

$$0.450\,(0.957 \times 10^{-4}) = 0.430 \times 10^{-4} \text{ m}$$

$$= 0.0043 \text{ cm}$$

The increase in length is

$$2\,(0.225 \times 10^{-4}) = 0.450 \times 10^{-4} \text{ m}$$

$$= 0.0045 \text{ cm}$$

Problem 6.7 A thin cylindrical shell is subjected to internal fluid pressure, the ends being closed by:

(a) two watertight pistons attached to a common piston rod;

(b) flanged ends.

Find the increase in internal diameter in each case, given that the internal diameter is 20 cm, thickness is 0.5 cm, Poisson's ratio is 0.3, Young's modulus is 200 GN/m^2, and the internal pressure is 3.5 MN/m^2. *(RNC)*

Solution

We have

$$p = 3.5 \text{ MN/m}^2, \qquad r = 0.1 \text{ m}, \qquad t = 0.005 \text{ m}$$

In both cases the circumferential stress is

$$\sigma_1 = \frac{pr}{t} = \frac{\left(3.5 \times 10^6\right)(0.1)}{(0.005)} = 70 \text{ MN/m}^2$$

(a) In this case there is no longitudinal stress. The circumferential strain is then

$$\varepsilon_1 = \frac{\sigma_1}{E} = \frac{70 \times 10^6}{200 \times 10^9} = 0.35 \times 10^{-3}$$

The increase of internal diameter is

$$0.2\left(0.35 \times 10^{-3}\right) = 0.07 \times 10^{-3} \text{ m} = 0.007 \text{ cm}$$

(b) In this case the longitudinal stress is

$$\sigma_2 = \frac{pr}{2t} = 35 \text{ MN/m}^2$$

The circumferential strain is therefore

$$\varepsilon_1 = \frac{1}{E}\left(\sigma_1 - \nu\sigma_2\right) = \frac{\sigma_1}{E}\left(1 - \frac{1}{2}\nu\right) = 0.85\,\frac{\sigma_1}{E}$$

$$= 0.85\left(0.35 \times 10^{-3}\right) = 0.298 \times 10^{-3}$$

The increase of internal diameter is therefore

$$0.2\left(0.298 \times 10^{-3}\right) = 0.0596 \times 10^{-3} \text{ m} = 0.00596 \text{ cm}$$

Equations (6.1) and (6.2) are for determining stress in perfect thin-walled circular cylindrical shells. If, however, the circular cylinder is fabricated, so that its joints are weaker than the rest of the vessel, then equations (6.1) and (6.2) take on the following modified forms:

$$\sigma_1 = \text{hoop or circumferential stress} = \frac{pr}{\eta_L t} \tag{6.6}$$

$$\sigma_2 = \text{longitudinal stress} = \frac{pr}{2\eta_c t} \tag{6.7}$$

where

$$\eta_c = \text{circumferential joint efficiency} \leq 1$$

$$\eta_L = \text{longitudinal joint efficiency} \leq 1$$

NB The *circumferential stress* is associated with the *longitudinal joint* efficiency, and the *longitudinal stress* is associated with the *circumferential joint* efficiency.

6.2 Thin spherical shell

We consider next a thin spherical shell of means radius r, and thickness t, which is subjected to an internal pressure p. Consider any diameter plane through the shell, Figure 6.6; the total force normal to this plane due to p acting on a hemisphere is

$$p \times \pi r^2$$

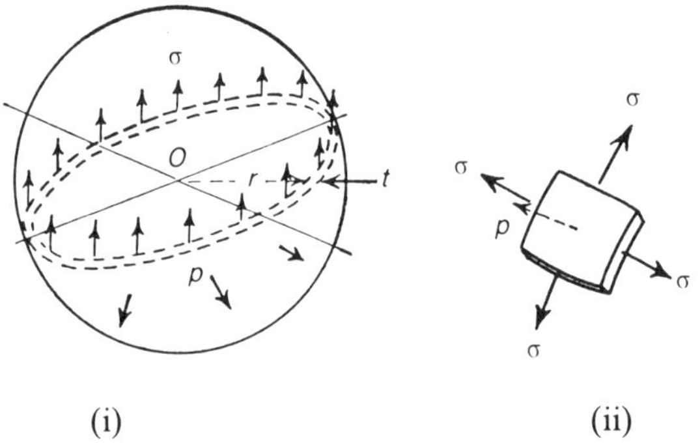

Figure 6.6 Membrane stresses in a thin spherical shell under internal pressure.

This is opposed by a tensile stress σ in the walls of the shell. By symmetry σ is the same at all points of the shell; for equilibrium of the hemisphere we must have

$$p \times \pi r^2 = \sigma \times 2\pi r t$$

This gives

$$\sigma = \frac{pr}{2t} \tag{6.8}$$

At any point of the shell the direct stress σ has the same magnitude in all directions in the plane of the surface of the shell; the state of stress is shown in Figure 6.6(ii). As p is small compared with σ, the maximum shearing stress occurs on planes at 45° to the tangent plane at any point.

If the shell remains elastic, the circumference of the sphere in any diametral plane is strained an amount

$$\varepsilon = \frac{1}{E}(\sigma - \nu\sigma) = (1 - \nu)\frac{\sigma}{E} \tag{6.9}$$

The volumetric strain of the enclosed volume of the sphere is therefore

$$3\varepsilon = 3(1 - \nu)\frac{\sigma}{E} = 3(1 - \nu)\frac{pr}{2Et} \tag{6.10}$$

Equation (6.8) is intended for determining membrane stresses in a perfect thin-walled spherical shell. If, however, the spherical shell is fabricated, so that its joint is weaker than the remainder of the shell, then equation (6.8) takes on the following modified form:

$$\sigma = \text{stress} = \frac{pr}{2\eta t} \tag{6.11}$$

where

$$\eta = \text{joint efficiency} \leq 1$$

6.3 Cylindrical shell with hemispherical ends

Some pressure vessels are fabricated with hemispherical ends; this has the advantage of reducing the bending stresses in the cylinder when the ends are flat. Suppose the thicknesses t_1 and t_2 of the cylindrical section and the hemispherical end, respectively (Figure 6.7), are proportioned so that the radial expansion is the same for both cylinder and hemisphere; in this way we eliminate bending stresses at the junction of the two parts.

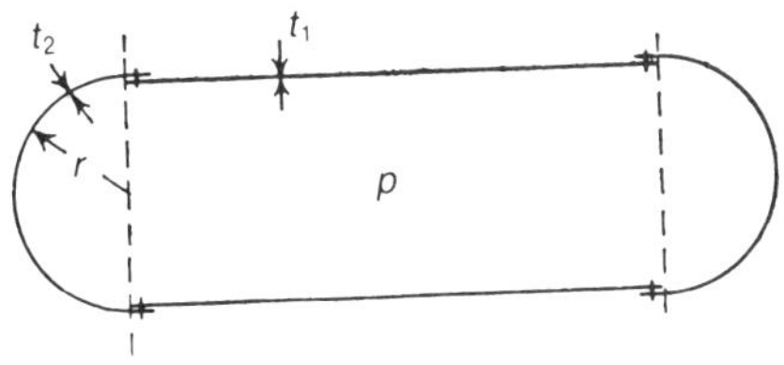

Figure 6.7 Cylindrical shell with hemispherical ends, so designed as to minimise the effects of bending stresses.

From equations (6.4), the circumferential strain in the cylinder is

$$\frac{pr}{Et_1}\left(1 - \frac{1}{2}v\right)$$

and from equation (6.7) the circumferential strain in the hemisphere is

$$\left(1 - \nu\right)\frac{pr}{2Et_2}$$

If these strains are equal, then

$$\frac{pr}{Et_1}\left(1 - \frac{1}{2}\nu\right) = \frac{pr}{2Et_2}\left(1 - \nu\right)$$

This gives

$$\frac{t_1}{t_2} = \frac{2 - \nu}{1 - \nu} \tag{6.12}$$

For most metals ν is approximately 0.3, so an average value of (t_1/t_2) is 1.7/0.7 ∔ 2.4. The hemispherical end is therefore thinner than the cylindrical section.

6.4 Bending stresses in thin-walled circular cylinders

The theory presented in Section 6.1 is based on membrane theory and neglects bending stresses due to end effects and ring stiffness. To demonstrate these effects, Figures 6.9 to 6.13 show plots of the theoretical predictions for a ring stiffened circular cylinder[3] together with experimental values, shown by crosses. This ring stiffened cylinder, which was known as Model No. 2, was firmly fixed at its ends, and subjected to an external pressure of 0.6895 MPa (100 psi), as shown by Figure 6.8.

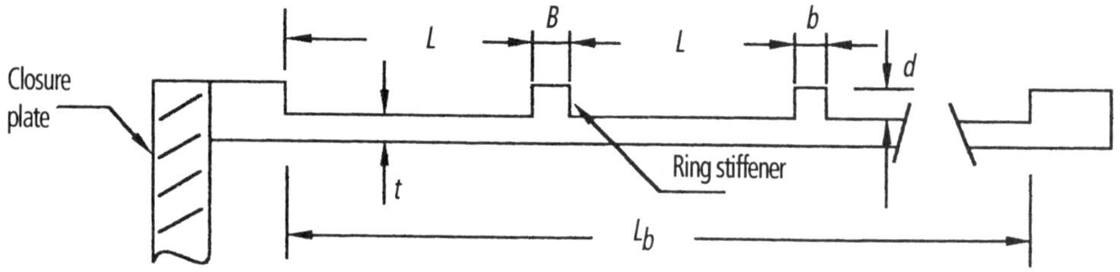

L	L_b	B	b	d	N
95.25	616.6	10.16	8.26	15.75	5

t = 0.08 N = number of ring stiffeners
E = Young's modulus = 71 GPa υ = Poisson's ratio = 0.3

Figure 6.8 Details of model No. 2 (mm).

The theoretical analysis was based on *beam on elastic foundations*, and is described by Ross[3].

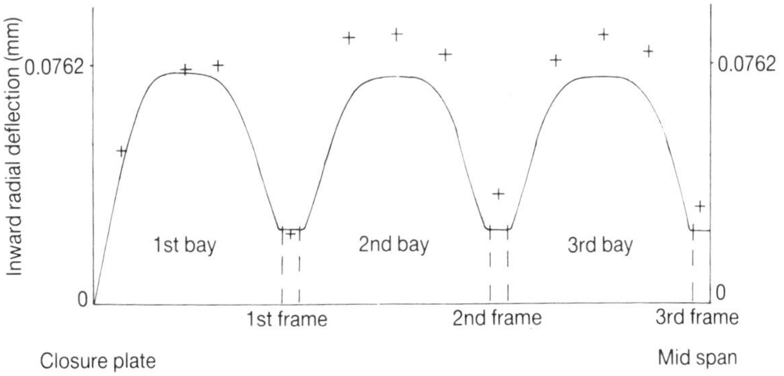

Figure 6.9 Deflection of longitudinal generator at 0.6895 MPa (100 psi), Model No. 2.

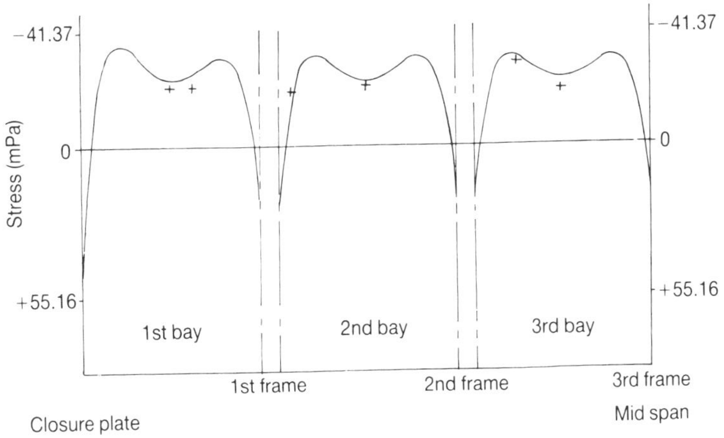

Figure 6.10 Longitudinal stress of the outermost fibre at 0.6895 MPa (100 psi), Model No. 2.

[3]Ross, C T F, *Pressure vessels under external pressure,* Elsevier Applied Science 1990.

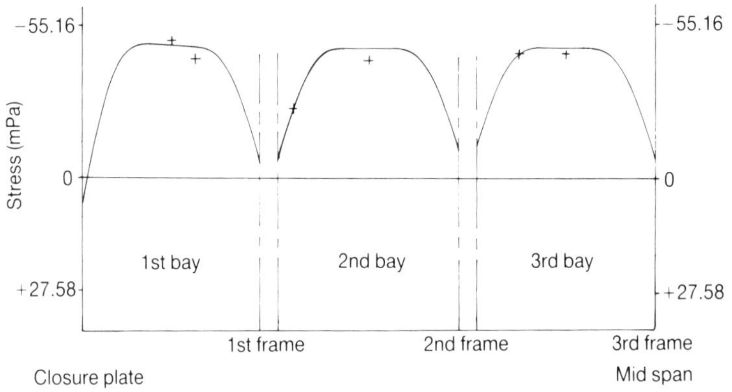

Figure 6.11 Circumferential stress of the outermost fibre at 0.6895 MPa (100 psi), Model No. 2.

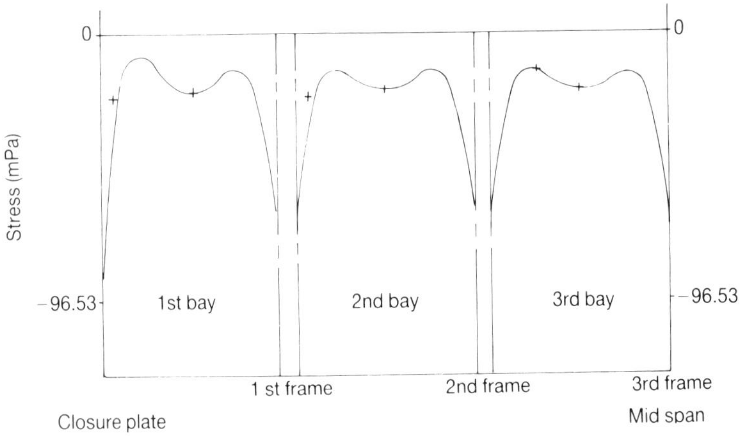

Figure 6.12 Longitudinal stress of the innermost fibre at 0.6895 MPa (100 psi), Model No. 2.

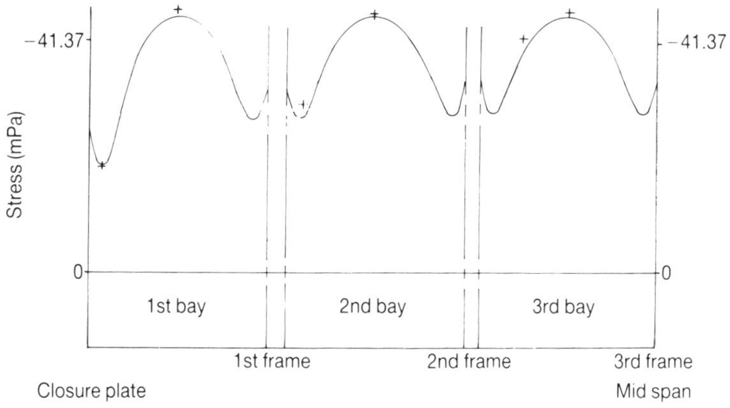

Figure 6.13 Circumferential stress of the innermost fibre at 0.6895 MPa (100 psi), Model No.2.

From Figures 6.9 to 6.13, it can be seen that bending stresses in thin-walled circular cylinders are very localised.

Further problems *(answers on page 692)*

6.8 A pipe has an internal diameter of 10 cm and is 0.5 cm thick. What is the maximum allowable internal pressure if the maximum shearing stress does not exceed 55 MN/m^2? Assume a uniform distribution of stress over the cross-section. *(Cambridge)*

6.9 A long boiler tube has to withstand an internal test pressure of 4 MN/m^2, when the mean circumferential stress must not exceed 120 MN/m^2. The internal diameter of the tube is 5 cm and the density is 7840 kg/m^3. Find the mass of the tube per metre run. *(RNEC)*

6.10 A long, steel tube, 7.5 cm internal diameter and 0.15 cm thick, is plugged at the ends and subjected to internal fluid pressure such that the maximum direct stress in the tube is 120 MN/m^2. Assuming $\nu = 0.3$ and $E = 200$ GN/m^2, find the percentage increase in the capacity of the tube. *(RNC)*

6.11 A copper pipe 15 cm internal diameter and 0.3 cm thick is closely wound with a single layer of steel wire of diameter 0.18 cm, the initial tension of the wire being 10 N. If the pipe is subjected to an internal pressure of 3 MN/m^2 find the stress in the copper and in the wire (a) when the temperature is the same as when the tube was wound, (b) when the temperature throughout is raised 200°C. E for steel = 200 GN/m^2, E for copper = 100 GN/m^2, coefficient of linear expansion for steel = 11×10^{-6}, for copper 18×10^{-6} per 1°C. *(Cambridge)*

6.12 A thin spherical copper shell of internal diameter 30 cm and thickness 0.16 cm is just full of water at atmospheric pressure. Find how much the internal pressure will be increased if 25 cc of water are pumped in. Take $\nu = 0.3$ for copper and $K = 2$ GN/m^2 for water. *(Cambridge)*

6.13 A spherical shell of 60 cm diameter is made of steel 0.6 cm thick. It is closed when just full of water at 15°C, and the temperature is raised to 35°C. For this range of temperature, water at atmospheric pressure increases 0.0059 per unit volume. Find the stress induced in the steel. The bulk modulus of water is 2 GN/m^2, E for steel is 200 GN/m^2, and the coefficient of linear expansion of steel is 12×10^{-6} per 1°C, and Poisson's ratio = 0.3. *(Cambridge)*

7 Bending moments and shearing forces

7.1 Introduction

In Chapter 1 we discussed the stresses set up in a bar due to axial forces of tension and compression. When a bar carries lateral forces, two important types of loading action are set up at any section: these are a bending moment and a shearing force.

Consider first the simple case of a beam which is fixed rigidly at one end B and is quite free at its remote end D, Figure 7.1; such a beam is called a *cantilever*, a familiar example of which is a fishing rod held at one end. Imagine that the cantilever is horizontal, with one end B embedded in a wall, and that a lateral force W is applied at the remote end D. Suppose the cantilever is divided into two lengths by an imaginary section C; the lengths BC and CD must individually be in a state of statical equilibrium. If we neglect the mass of the cantilever itself, the loading actions over the section C of CD balance the actions of the force W at C. The length CD of the cantilever is in equilibrium if we apply an upwards vertical force F and an anti-clockwise couple M at C; F is equal in magnitude to W, and M is equal to $W(L - z)$, where z is measured from B. The force F at C is called a *shearing force*, and the couple M is a *bending moment*.

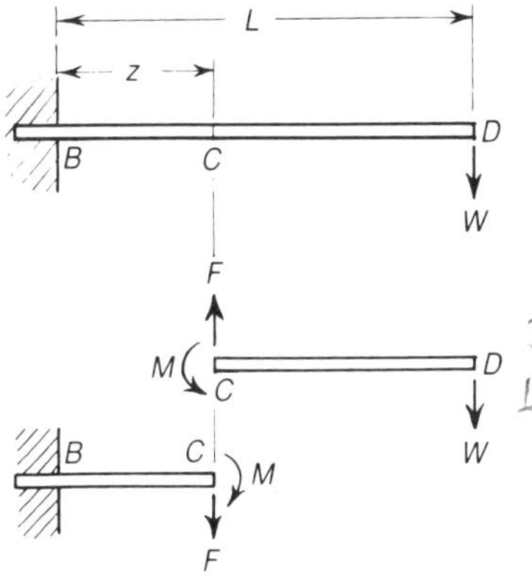

Figure 7.1 Bending moment and shearing force in a simple cantilever beam.

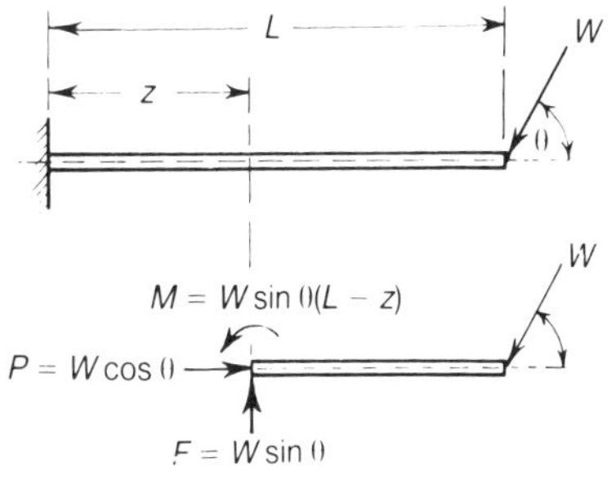

Figure 7.2 Cantilever with and inclined end load.

But at the imaginary section C of the cantilever, the actions F and M on CD are provided by the length BC of the cantilever. In fact, equal and opposite actions F and M are applied by CD to BC. For the length BC, the actions at C are a downwards shearing force F, and a clockwise couple M.

When the cantilever carries external loads which are not applied normally to the axis of the beam, Figure 7.2, axial forces are set up in the beam. If W is inclined at an angle θ to the axis of the beam, Figure 7.2, the axial thrust in the beam at any section is

$$P = W \cos \theta \tag{7.1}$$

The bending moment and shearing force at a section a distance z from the built-in end are

$$M = W(L - z) \sin \theta \qquad F = W \sin \theta \tag{7.2}$$

7.2 Concentrated and distributed loads

A concentrated load on a beam is one which can be regarded as acting wholly at one point of the beam. For the purposes of calculation such a load is localised at a point of the beam; in reality this would imply an infinitely large bearing pressure on the beam at the point of application of a concentrated load. All loads must be distributed in practice over perhaps only a small length of beam, thereby giving a finite bearing pressure. Concentrated loads arise frequently on a beam where the beam is connected to other transverse beams.

In practice there are many examples of distributed loads: they arise when a wall is built on a girder; they occur also in many problems of fluid pressure, such as wind pressure on a tall building, and aerodynamic forces on an aircraft wing.

7.3 Relation between the intensity of loading, the shearing force, and bending moment in a straight beam

Consider a straight beam under any system of lateral loads and external couples, Figure 7.3; an element length δz of the beam at a distance z from one end is acted upon by an external lateral load, and internal bending moments and shearing forces. Suppose external lateral loads are distributed so that the intensity of loading on the elemental length δz is w.

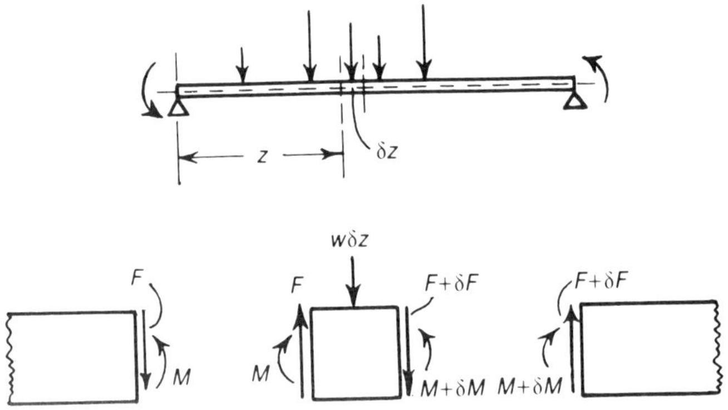

Figure 7.3 Shearing and bending actions on an elemental length of a straight beam.

Then the external vertical force on the element is $w\delta z$, Figure 7.3; this is reacted by an internal bending moment M and shearing force F on one face of the element, and $M + \delta M$ and $F + \delta F$ on the other face of the element. For vertical equilibrium of the element we have

$$(F + \delta F) - F + w\delta z = 0$$

If δz is infinitesimally small,

$$\frac{dF}{dz} = -w \tag{7.3}$$

Suppose this relation is integrated between the limits z_1 and z_2, then

$$\int_{z=z_1}^{z=z_2} dF = -\int_{z_1}^{z_2} w dz$$

If F_1 and F_2 are the shearing forces at $z = z_1$ and $z = z_2$ respectively, then

$$(F_2 - F_1) = -\int_{z_1}^{z_2} w dz$$

or

$$F_1 - F_2 = \int_{z_1}^{z_2} w dz \tag{7.4}$$

Then, the decrease of shearing force from z_1 to z_2 is equal to the area below the load distribution curve over this length of the beam, or the difference between F_1 and F_2 is the net lateral load over this length of the beam.

Furthermore, for rotational equilibrium of the elemental length δz,

$$(F + \delta F)\,\delta z - (M + \delta M) + M + wdz\left(\frac{1}{2}\,\delta z\right) = 0$$

Then, to the first order of small quantities,

$$F\delta z - \delta M = 0$$

Then, in the limit as δz approaches zero,

$$\frac{dM}{dz} = F \tag{7.5}$$

On integrating between the limits $z = z_1$ and $z = z_2$, we have

$$\int_{z=z_1}^{z=z_2} dM = \int_{z_1}^{z_2} F dz$$

Thus

$$M_2 - M_1 = \int_{z_1}^{z_2} F dz \tag{7.6}$$

where M_1 and M_2 are the values M at $z = z_1$ and $z = z_2$, respectively. Then the increase of bending moment from z_1 to z_2 is the area below the shearing force curve for that length of the beam.

Equations (7.4) and (7.6) are extremely useful for finding the bending moments and shearing forces in beams with irregularly distributed loads. From equation (7.4) the shearing force F at a section distance z from one end of the beam is

$$F = F_1 - \int_{z_1}^{z} w dz \tag{7.7}$$

On substituting this value of F into equation (7.6),

$$M_2 - M_1 = \int_{z_1}^{z_2} \left\{ F_1 - \int_{z_1}^{z} w dz \right\} dz$$

Thus

$$M_2 = M_1 + F_1 (z_2 - z_1) - \int_{z_1}^{z_2} \left\{ \int_{z_1}^{z} w dz \right\} dz \tag{7.8}$$

From equation (7.5) we have that the bending moment M has a stationary value when the shearing force F is zero. Equations (7.3) and (7.5) give

$$\frac{d^2 M}{dz^2} = \frac{dF}{dz} = -w \tag{7.9}$$

For the directions of M, F and w considered in Figure 7.3, M is *mathematically* a maximum, since d^2M/dz^2 is negative; the significance of the word mathematically will be made clearer in Section 7.8.

All the relations developed in this section are merely statements of statical equilibrium, and are therefore true independently of the state of the material of the beam.

7.4 Sign conventions for bending moments and shearing forces

The bending moments on the elemental length δz of Figure 7.3 tend to make the beam concave on its upper surface and convex on its lower surface; such bending moments are sometimes called *sagging* bending moments. The shearing forces on the elemental length tend to rotate the element in a *clockwise* sense. In deriving the equations in this section it is assumed implicitly, therefore, that

(i) *downwards* vertical loads are positive;

(ii) *sagging* bending moments are positive; and

(iii) *clockwise* shearing forces are positive.

These sign conventions are shown in Figure 7.4. Any other system of sign conventions can be used, provided the signs of the loads, bending moments and shearing forces are considered when equations (7.3) and (7.5) are applied to any particular problem.

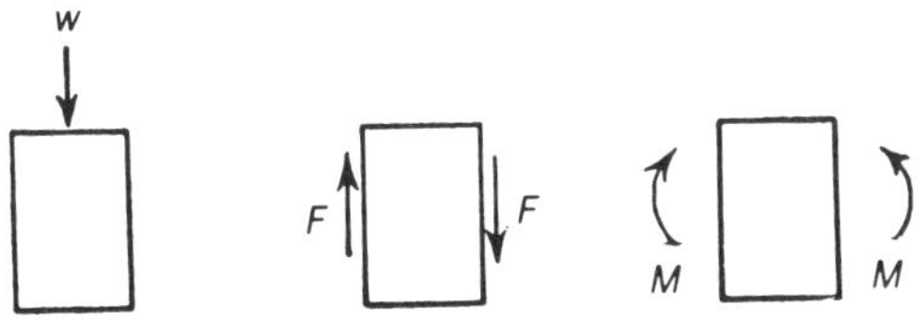

Figure 7.4 Positive values of w, F and M, (i) downward vertical loading, (ii) clockwise shearing forces, (iii) sagging bending-moment.

Figures that show graphically the variations of bending moment and shearing force along the length of a beam are called *bending moment diagrams* and *shearing force diagrams*. Sagging bending moments are considered positive, and clockwise shearing forces taken as positive. The two quantities are plotted above the centre line of the beam when positive, and below when negative. Before we can calculate the stresses and deformations of beams, we must be able to find the bending moment and shearing force at any section.

7.5 Cantilevers

A cantilever is a beam supported at one end only; for example, the beam already discussed in Section 7.1, and shown in Figure 7.1, is held rigidly at B. Consider first the cantilever shown in Figure 7.5(a), which carries a concentrated lateral load W at the free end. The bending moment at a section a distance z from B is

$$M = -W(L - z)$$

the negative sign occurring since the moment is hogging, as shown in Figure 7.5(b). The variation of bending moment is linear, as shown in Figure 7.5(c). The shearing force at any section is

$$F = +W$$

the shearing force being positive as it is clockwise, as shown in Figure 7.5(d). The shearing force is constant throughout the length of the cantilever. We note that

$$\frac{dM}{dz} = W = F$$

Further $dF/dz = 0$, as there are no lateral loads between B and D.

The bending moment diagram is shown in Figure 7.5(c) and the shearing force diagram is shown in Figure 7.5(e)

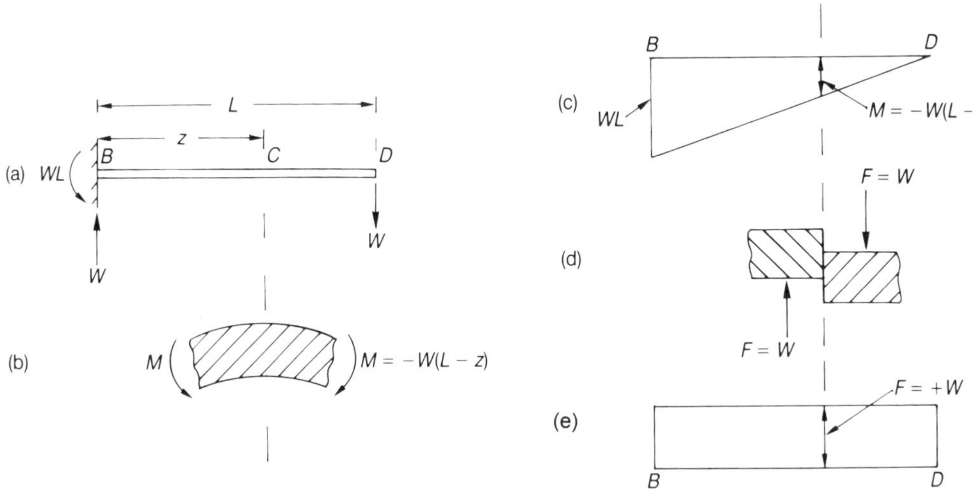

Figure 7.5 Bending-moment and shearing-force diagrams for a cantilever with a concentrated load at the free end.

Now consider a cantilever carrying a uniformly distributed downwards vertical load of intensity w, Figure 7.6(a). The shearing force at a distance z from B is

$$F = +w(L - z)$$

as shown in Figure 7.6 (d). The bending moment at a distance z from B is

$$M = -\tfrac{1}{2}w(L - z)^2$$

as shown in Figure 7.6(b). The shearing force varies linearly and the bending moment parabolically along the length of the beam, as shown in Figure 7.6(e) and 7.6(c), respectively. We see that

$$\frac{dM}{dz} = w(L - z) = +F$$

7.6 Cantilever with non-uniformly distributed load

Where a cantilever carries a distributed lateral load of variable intensity, we can find the bending moments and shearing forces from equations (7.4) and (7.6). When the loading intensity w cannot be expressed as a simple analytic function of z, equations (7.4) and (7.6) can be integrated numerically.

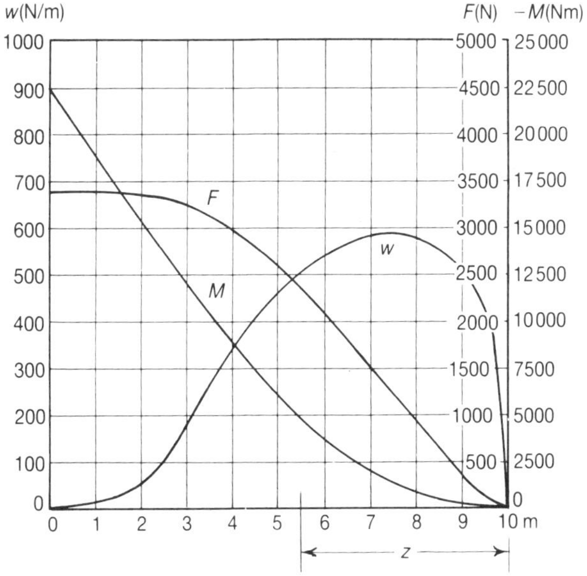

Problem 7.2 A cantilever of length 10 m, built in at its left end, carries a distributed lateral load of varying intensity w N per metre length. Construct curves of shearing force and bending moment in the cantilever.

<u>*Solution*</u>

If z is the distance from the free end of cantilever, the shearing force at a distance z from the free end is

$$F = \int_0^z w dz$$

We find first the shearing force F by numerical integration of the w-curve. The greatest force occurs at the built-in end, and has the value

$$F_{max} \doteqdot 3400 \text{ N}$$

The bending moment at a section a distance z from the free end is

$$M = - \int_0^z F dz$$

and is found therefore by numerical integration of the F-curve. The greatest bending moment occurs at the built-in end, and has the value

$$M_{max} = 22500 \text{ Nm}$$

NB It should be noted that by inspection the bending moment and the shearing force at the free end of the cantilever are zero; these are boundary conditions.

7.7 Simply-supported beams

By *simply-supported* we mean that the supports are of such a nature that they do not apply any resistance to bending of a beam; for instance, knife-edges or frictionless pins perpendicular to the plane of bending cannot transmit couples to a beam. The remarks concerning bending moments and shearing forces, which were made in Section 7.5 in relation to cantilevers, apply equally to beams simply-supported at each end, or with any conditions of end support.

As an example, consider the beam shown in Figure 7.7(a), which is simply-supported at B and C, and carries a vertical load W a distance a from B. If the ends are simply-supported no bending moments are applied to the beam at B and C. By taking moments about B and C we find that the reactions at these supports are

$$\frac{W}{L}(L - a) \text{ and } \frac{Wa}{L}$$

respectively. Now consider a section of the beam a distance z from B; if $z < a$, the bending moment and shearing force are

$$M = +\frac{Wz}{L}(L - a), \quad F = +\frac{W}{L}(L - a), \text{ as shown by Figures 7.7(b) and 7.7(d)}$$

If $z > a$,

$$M = +\frac{Wz}{L}(L - a) - W(z - a) = +\frac{Wa}{L}(L - z)$$

$$F = -\frac{Wa}{L}$$

The bending moment and shearing force diagrams show discontinuities at $z = a$; the maximum bending moment occurs under the load W, and has the value

$$M_{max} = \frac{Wa}{L}(L - a) \tag{7.10}$$

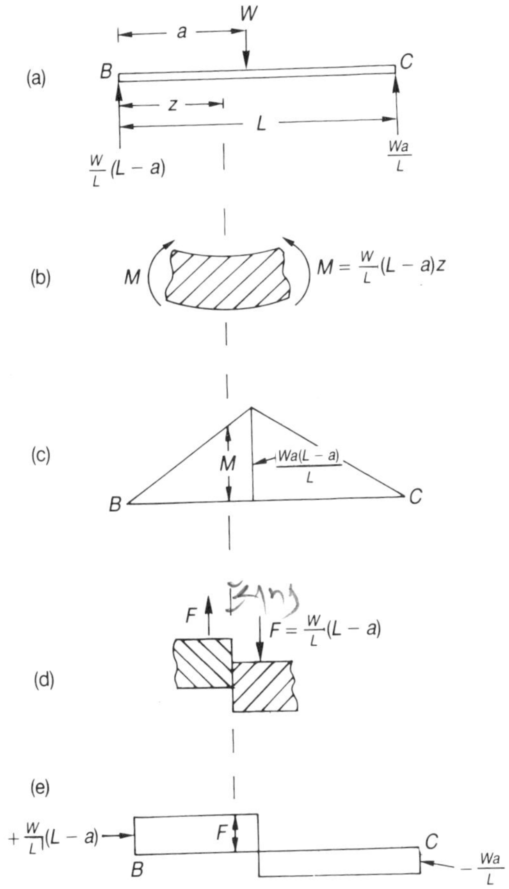

Figure 7.7 Bending-moment and shearing-force diagrams for a simply-supported beam with a single concentrated lateral load.

The simply-supported beam of Figure 7.8(a) carries a uniformly-distributed load of intensity w. The vertical reactions at B and C are $\frac{1}{2}wL$. Consider a section at a distance z from B. The bending moment at this section is

$$M = \frac{1}{2}wLz - \frac{1}{2}wz^2$$

$$= \frac{1}{2}wz(L - z)$$

as shown in Figure 7.8(b) and the shearing force is

$$F = +\frac{1}{2}wL - wz$$

$$= w\left(\frac{1}{2}L - z\right)$$

as shown in Figure 7.8(d).

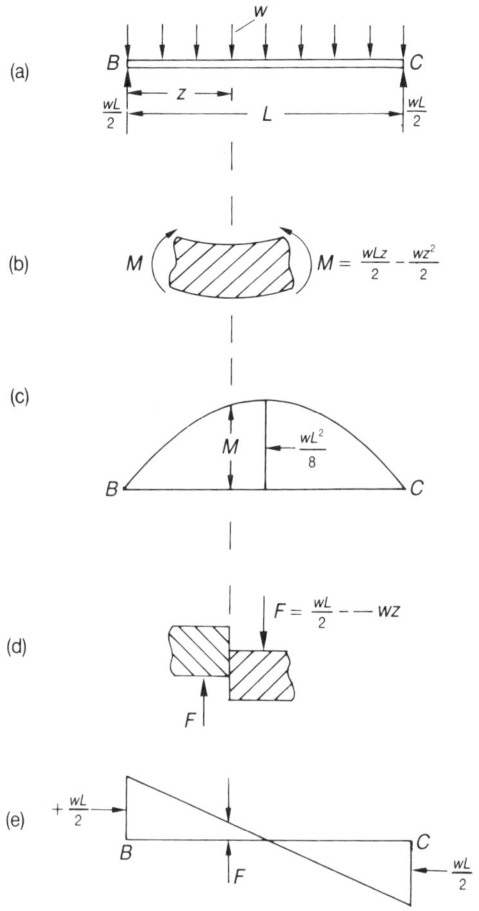

Figure 7.8 Bending-moment and shearing-force diagrams for a simply-supported beam with a uniformly distributed lateral load.

The bending moment is a maximum at $z = \frac{1}{2}L$, where

$$M_{max} = \frac{wL^2}{8} \tag{7.11}$$

At $z = \frac{1}{2}L$, we note that

$$\frac{dM}{dz} = +F = 0$$

The bending moment diagram is shown in Figure 7.8(c) and the shearing force diagram is shown in Figure 7.8(e).

Problem 7.3 A simply-supported beam carries concentrated lateral loads at C and D, and a uniformly distributed lateral load over the length DF. Construct the bending moment and shearing force diagrams.

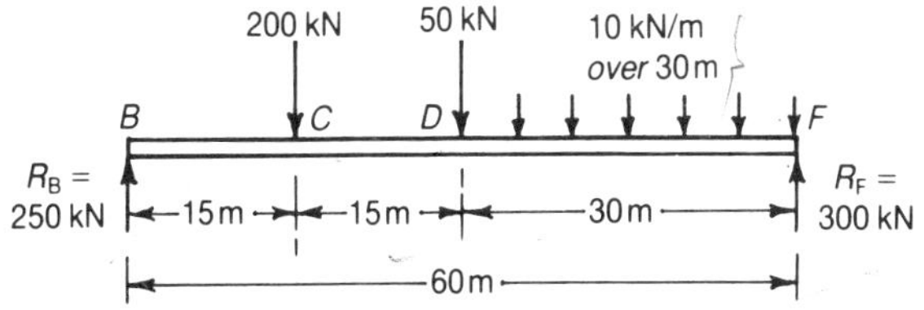

Solution

First we calculate the vertical reactions at B and F. On taking moments about F,

$$60\ R_B = (200 \times 10^3)(45) + (50 \times 10^3)(30) + (300 \times 10^3)(15) = 15\ 000 \times 10^3$$

Then

$$R_B = 250 \text{ kN}$$

and

$$R_F = (200 \times 10^3) + (50 \times 10^3) + (300 \times 10^3) - R_B = 300 \text{ kN}$$

The bending moment varies linearly between B and C, and between C and D, and parabolically from D to F. The maximum bending moment is 4.5 MNm, and occurs at D. The maximum shearing force is 300 kN, and occurs at F.

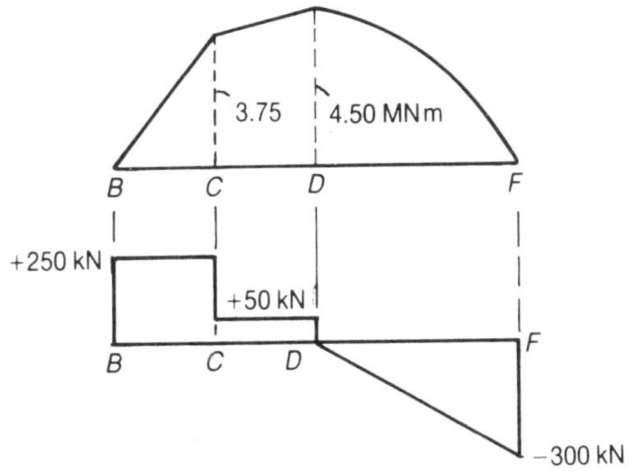

Problem 7.4 A beam rests on knife-edges at each end, and carries a clockwise moment M_B at B, and an anticlockwise moment M_C at C. Construct bending moment and shearing force diagrams for the beam.

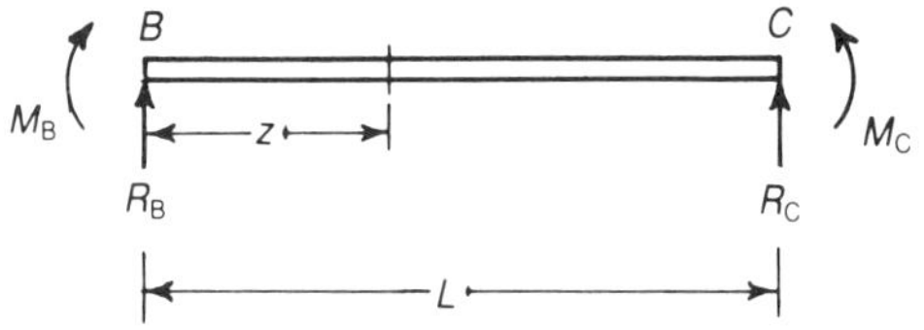

Solution

Suppose R_B and R_C are vertical reactions at B and C; then for statical equilibrium of the beam

$$R_B = -R_C = \frac{1}{L}\left(M_C - M_B\right)$$

The shearing force at all sections is then

$$F = R_B = \frac{1}{L}\left(M_C - M_B\right)$$

The bending moment a distance z from B is

$$M = M_B + R_B z = \frac{M_B}{L}\left(L - z\right) + \frac{M_C z}{L}$$

so M varies linearly between B and C.

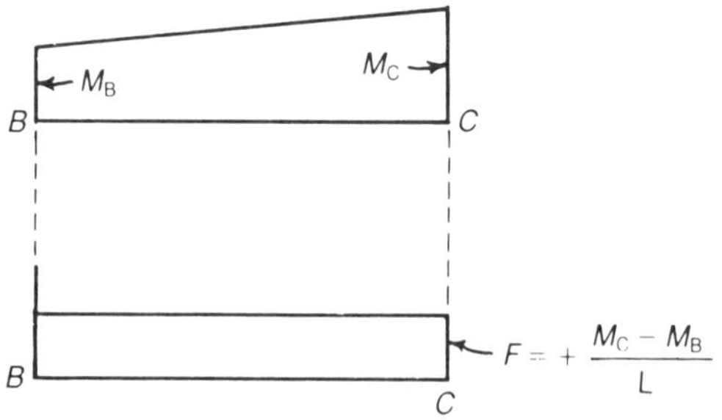

Problem 7.5 A simply-supported beam carries a couple M_0 applied at a point distant a from B. Construct bending moment and shearing force diagrams for the beam.

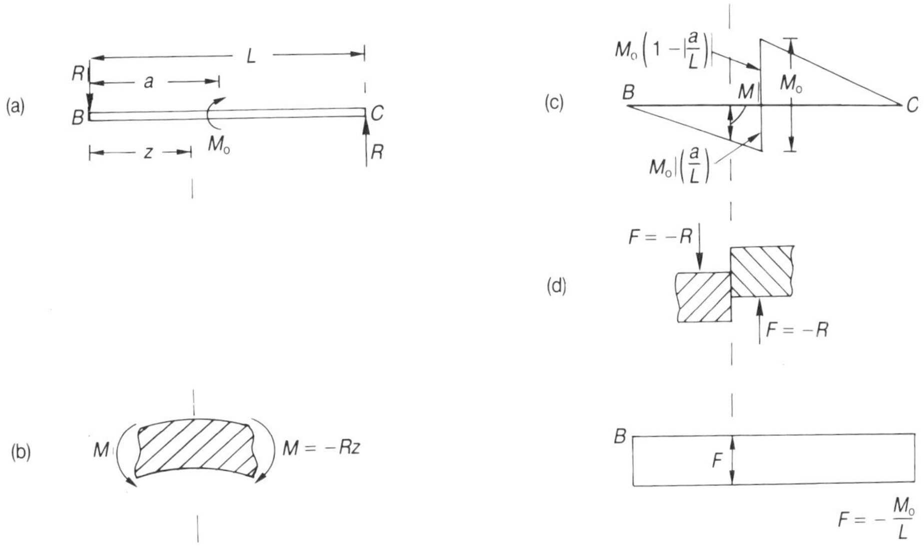

<u>Solution</u>

The vertical reactions R at B and C are equal and opposite. For statical equilibrium of BC,

$$M_0 = RL, \text{ or } R = \frac{M_0}{L}$$

The shearing force at all sections is

$$F = -R = -\frac{M_0}{L}$$

as shown in Figure (d), above. The bending moment at $z < a$ is

$$M = -Rz = -\frac{M_0 z}{L}$$

as shown in Figure (c), above, and for $z > a$

$$M = -Rz + M_0 = M_0\left(1 - \frac{z}{L}\right)$$

as shown in Figure (c), above.

7.8 Simply-supported beam carrying a uniformly distributed load and end couples

Consider a simply-supported beam BC, carrying a uniformly distributed load w per unit length, and couples M_B and M_C applied to ends, Figure 7.9(i). The reactions R_B and R_C can be found directly by taking moments about B and C in turn; we have

$$R_B = \frac{wL}{2} - \frac{1}{L}(M_B - M_C) \tag{7.12}$$

$$R_C = \frac{wL}{2} + \frac{1}{L}(M_B - M_C)$$

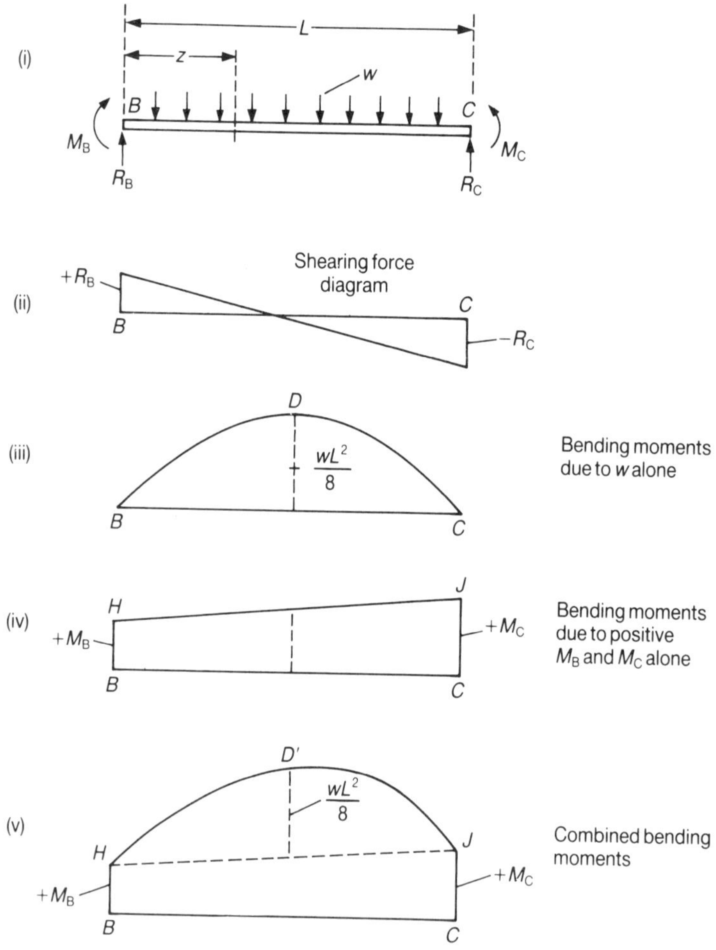

Figure 7.9 Simply-supported beam with uniformly distributed lateral load and end couples.

These give the shearing forces at the end of the beam, and the shearing force at any point of the beam can be deduced, Figure 7.9(ii). In discussing bending moments we consider the total loading actions on the beam as the superposition of a uniformly distributed load and end couples; the distributed load gives rise to a parabolic bending moment curve, *BDC* in Figure 7.9(iii), whereas the end couples M_B and M_C give the straight line *HJ*, Figure 7.9(iv). The combined effects of the lateral load and the end couples give the curve *BHD'JC*, Figure 7.9(v). The bending moment at a distance z from B is

$$M = \frac{1}{2}wz(L - z) + \frac{M_B}{L}(L - z) + \frac{M_C z}{L} \tag{7.13}$$

The 'maximum' bending moment occurs when

$$\frac{dM}{dz} = \frac{1}{2}w(L - 2z) - \frac{M_B}{L} + \frac{M_C}{L} = 0$$

that is, when

$$z = \frac{1}{2}L - \frac{1}{wL}(M_B - M_C)$$

The value of M for this value of z is

$$M_{\max} = \frac{1}{8}wL^2 + \frac{1}{2}(M_B + M_C) + \frac{1}{2wL^2}(M_B - M_C)^2 \tag{7.14}$$

This, however, is only a mathematical 'maximum'; if M_B or M_C is negative, the numerically greatest bending moment may occur at B or C. Care should therefore be taken to find the truly greatest bending moment in the beam.

7.9 Points of inflection

When either, or both, of the end couples in Figure 7.9 is reversed in direction, there is at least one section of the beam where the bending moment is zero.

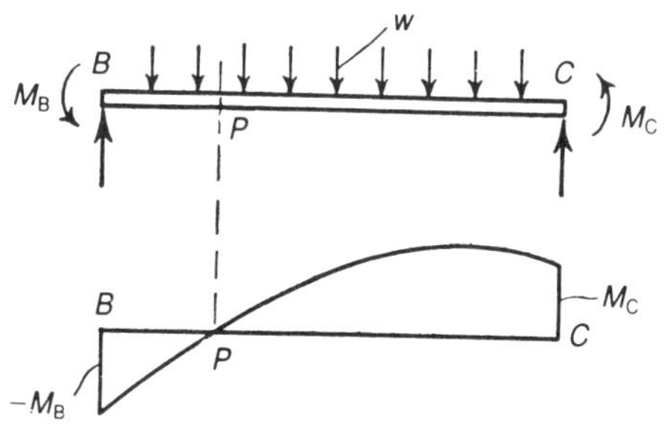

Figure 7.10 Single point of inflection in a beam.

In Figure 7.10 the end couple M_B is applied in an anticlockwise direction; the bending moment at a distance z from B is

$$M = \frac{1}{2}wz(L - z) - \frac{M_B}{L}(L - z) + \frac{M_C z}{L} \tag{7.15}$$

and this is zero when

$$z^2 - zL\left(1 + \frac{2}{wL^2}\left[M_B + M_C\right]\right) + \frac{2M_B}{w} = 0 \tag{7.16}$$

The distance PB is the relevant root of this quadratic equation.

When the end couple M_C is also reversed in direction, Figure 7.11, there are two points, P and Q, in the beam at which the bending moment is zero. The distances P and Q from B are given by the roots of the equation

$$z^2 - zL\left[1 + \frac{2}{wL^2}\left(M_B - M_C\right)\right] + \frac{2M_B}{w} = 0 \tag{7.17}$$

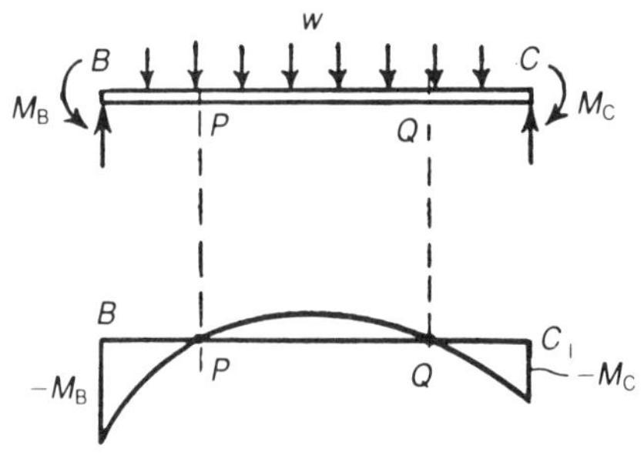

Figure 7.11 Beam with two points of inflection.

The distance PQ is

$$2\sqrt{\frac{L^2}{4} - \left(\frac{M_B - M_C}{w}\right) + \left(\frac{M_B - M_C}{wL}\right)^2} \tag{7.18}$$

The points P and Q are called *points of inflection*, or *points of contraflexure*; as we shall see later, the curvature of the deformed beam changes sign at these points.

7.10 Simply-supported beam with a uniformly distributed load over part of a span

The beam *BCDF*, shown in Figure 7.12, carries a uniformly distributed vertical load *w* per unit length over the portion *CD*. On taking moments about *B* and *F*,

$$V_B = \frac{bw}{2L}(b + 2c), \quad V_F = \frac{bw}{2L}(b + 2a) \tag{7.19}$$

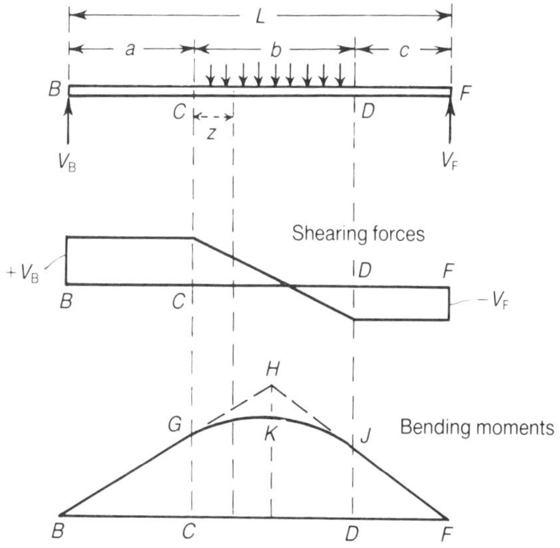

Figure 7.12 Shearing-force and bending-moment diagrams for simply-supported beam with distributed load over part of the span.

The bending moments at *C* and *D* are

$$M_C = aV_B = \frac{baw}{2L}(b + 2c)$$

$$M_D = cV_F = \frac{bcw}{2L}(b + 2a) \tag{7.20}$$

The bending moments in *BC* and *FD* vary linearly. The bending moment in *CD*, at a distance *z* from *C*, is

$$M = \left(1 - \frac{z}{b}\right) M_C + \frac{z}{b} M_D + \frac{1}{2} wz(b - z) \tag{7.21}$$

Then

$$\frac{dM}{dz} = \frac{1}{b}\left(M_D - M_C\right) + \frac{1}{2}w\,(b - 2z)$$

On substituting for M_C and M_D from equations (7.20)

$$\frac{dM}{dz} = \frac{bw}{2L}\,(c - a) + \frac{1}{2}w\,(b - 2z)$$

At C, $z = 0$, and

$$\frac{dM}{dz} = \frac{bw}{2L}\,(b + 2c) = V_B$$

But V_B is the slope of the line BG in the bending moment diagram, so the curve of equation (7.21) is tangential to BG at G. Similarly, the curve of equation (7.21) is tangential to FJ at J. Between C and D the bending moment varies parabolically; the simplest method of constructing the bending moment diagram for CD is to produce BG and FJ to meet at H, and then to draw a parabola between G and J, having tangents BG and FJ.

7.11 Simply-supported beam with non-uniformly distributed load

Suppose a simply-supported beam of span L, Figure 7.13, carries a lateral distributed load of variable intensity w. Then, from equation (7.4), if F is the shearing force a distance z from B,

$$F_0 - F = \int_0^z w\,dz$$

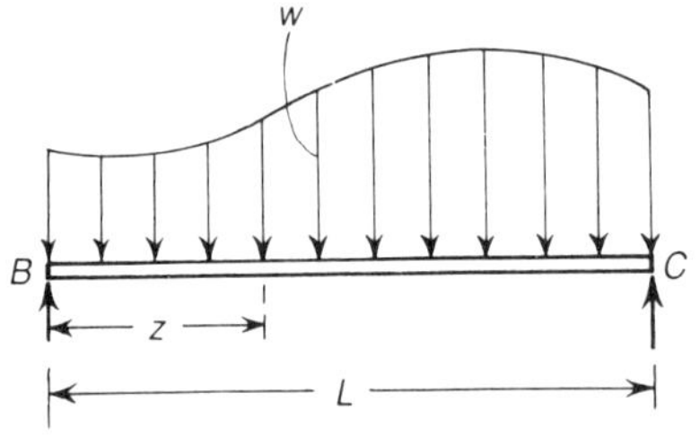

Figure 7.13 Simply-supported beam with lateral load of varying intensity.

where F_0 is the shearing force at $z = 0$. Then

$$F = F_0 - \int_0^z w\,dz \tag{7.22}$$

Furthermore, from equation (7.6), the bending moment a distance z from B is

$$M = M_0 + F_0 z - \int_0^z \int_0^z w\,dz\,dz \tag{7.23}$$

where M_0 is the bending moment at $z = 0$. However, as the beam is simply-supported at $z = 0$, we have $M_0 = 0$, and so

$$M = F_0 z - \int_0^z \int_0^z w\,dz\,dz$$

The end $z = L$ is also simply-supported, so for this end $M = 0$; then

$$F_0 L - \int_0^L \int_0^z w\,dz\,dz = 0$$

This gives

$$F_0 = \frac{1}{L} \int_0^L \int_0^z w\,dz\,dz \tag{7.24}$$

Equations (7.22), (7.23) and (7.24) may be used in the graphical solution of problems in which w is not an analytic function of z. The value of F_0 is found firstly from equation (7.24); numerical integrations then give the values of F and M, from equations (7.22) and (7.23), respectively.

7.12 Plane curved beams

Consider a beam BCD, Figure 7.14, which is curved in the plane of the figure. The beam is loaded so that no twisting occurs, and bending is confined to the plane of Figure 7.14. Suppose an imaginary cross-section of the beam is taken at C; statical equilibrium of the length CD of the beam is ensured if, in general, a force and a couple act at C; it is convenient to consider the resultant force at C as consisting of two components—an axial force P, acting along the centre line of the beam, and a lateral force F, acting along the normal to the centre line of the beam. The couple M at C acts about an axis perpendicular to the plane of bending and passing through the centre line of the beam. The actions at C on the length BC of the beam, are equal and opposite to those at C on the length CD.

As before the couple M is the *bending moment* in the beam at C, and the lateral force F is the *shearing force*.

As an example, consider the beam of Figure 7.15, which has a centre line of constant radius R. The beam carries a radial load W at its free end. Consider a section of the beam at some angular position θ: for statical equilibrium of the length of the bar shown in Figure 7.15(ii),

$$M = WR\sin\theta$$

$$F = W\cos\theta$$

$$P = W\sin\theta \qquad (7.25)$$

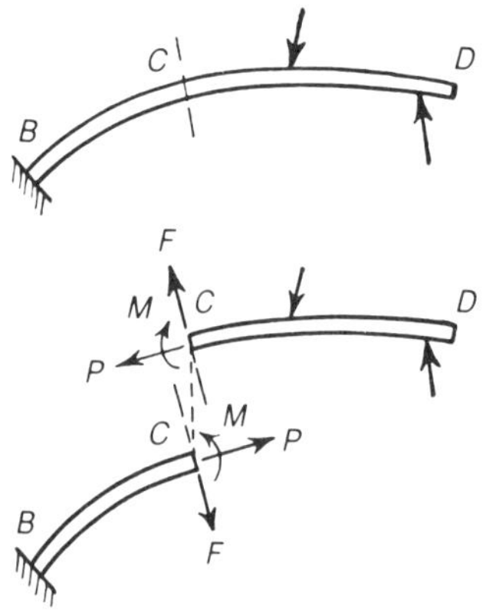

Figure 7.14 Bending and shearing actions in a plane curved beam.

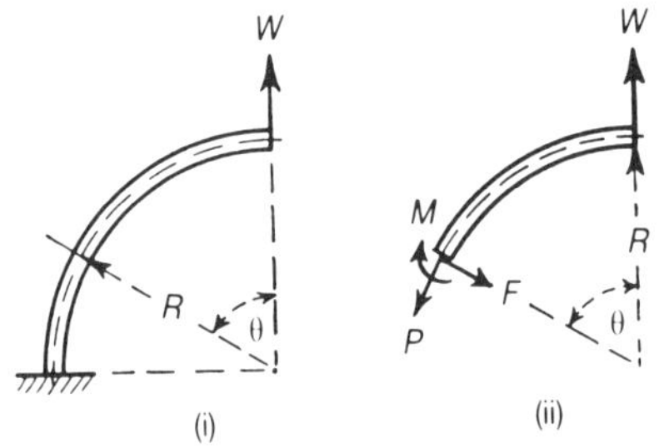

Figure 7.15 Plane curved beam of circular form carrying an end load.

Consider again, the beam shown in Figure 7.16, consisting of two straight limbs, BC and CD, connected at C. In CD the bending moment varies linearly, from zero at D to 70 000 Nm at C. In BC the bending moment is constant and equal to 70 000 Nm. In Figure 7.17 the bending moments are plotted on the concave sides of the bent limbs; this is equivalent to following the sign convention of Section 7.4, that sagging bending moments are positive.

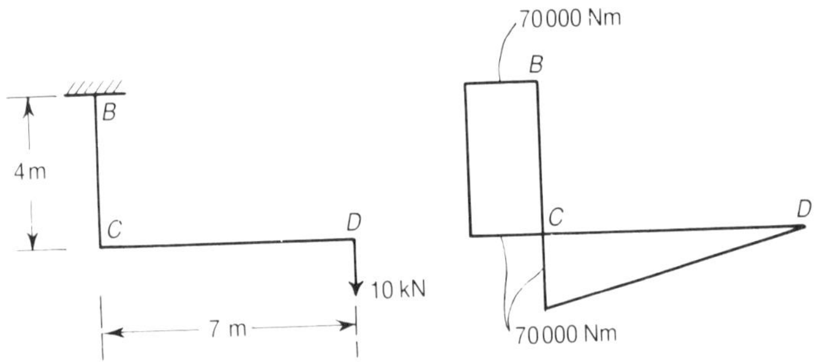

Figure 7.16 Bending moments in a bracket.

Problem 7.6 *AB* is a vertical post of a crane; the sockets at *A* and *B* offer no constraint against flexure. The horizontal arm *CD* is hinged to *AB* at *C* and supported by the strut *FE* which is freely hinged at its two extremities to *AB* and *CD*. Construct the bending moment diagrams for *AB* and *CD*. *(Cambridge)*

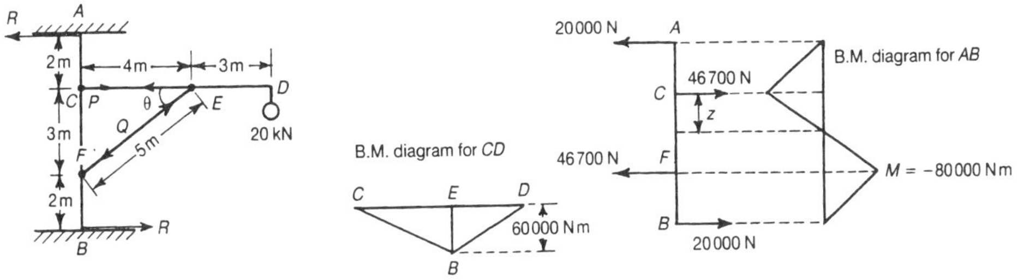

<u>*Solution*</u>

It is clear from considering the equilibrium of the whole crane that the horizontal reactions at *A* and *B* must be equal and opposite, and that the couple due to them must equal the moment of the 20 kN force. Let *R* be the magnitude of the horizontal reactions at *A* and *B*, then

$$7R = 7(20000)$$

and therefore

$$R = 20000\ N$$

Let P = the pull in *CE*, and Q = the thrust in *FE*. Then taking moments about *C* for the rod *CD* we have

$$4Q\ \sin\theta = 7(20000)$$

and therefore

$$Q = 58300\ N$$

Resolving horizontally for *AB* we have

$$P = Q\ \cos\theta = \frac{1}{2}(70000)\ \cot\theta = 46700\ N$$

The vertical reaction at $E = Q\sin\theta = 35\,000$ N.

We can now draw the bending moment diagrams for *AB* and *CD*, considering only the forces at right-angles to each beam; let us take *CD* first. *CD* is a beam freely supported at *C* and *E* and loaded at *D*. The bending moment at $E = 3 \times 20\,000 = 60\,000$ Nm, to which value it rises uniformly from zero at *D*; from *E* to *C* the bending moment decreases uniformly to zero.

AB is supported at A and B and loaded with equal and opposite loads at C and F.

The bending moment at C is

$$(2)\,(20\,000) \;=\; 40\,000 \text{ Nm.}$$

The bending moment at F is

$$(2)\,(-20\,000) \;=\; -40\,000 \text{ Nm.}$$

At any point z between C and F, the bending moment is

$$M \;=\; 20\,000\,(z + 2) - 46\,700z \;=\; 40\,000 - 26\,700z$$

In the bending moment diagram positive bending moments are those which make the beam concave to the left, and are plotted to the left in the figure.

7.13 More general case of bending of a curved bar

In Figure 7.17, OBC represents the centre line of a beam of any shape; the line OBC is curved in space in general. Suppose the beam carries any system of external loads; consider the actions over a section of the beam at B. For statical equilibrium of BC we require at B a force and a couple.

The force is resolved into two components—an axial force P along the centre line of the beam, and a shearing force F normal to the centre line; the couple is resolved into two components—a torque T about the centre line of the beam, and a bending moment M about an axis perpendicular to the centre line. The axis of M is not necessarily coincident with the axis of F.

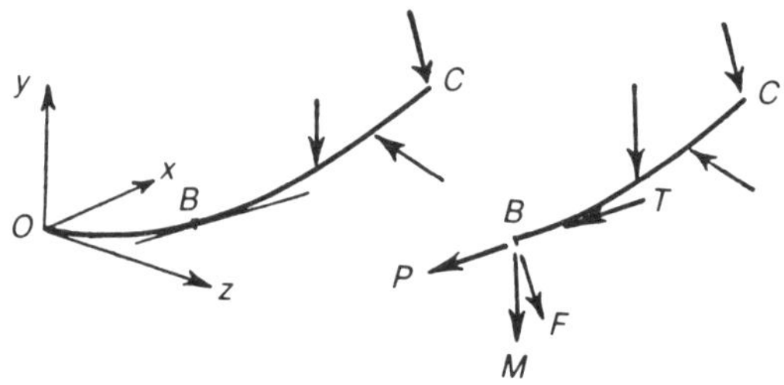

Fig. 7.17 Lateral loading of a curved beam.

Problem 7.7 The centre line of a beam is curved in the plane xz with a radius a. Find the loading actions at any section of the beam when a concentrated load W is applied at C in a direction parallel to yO.

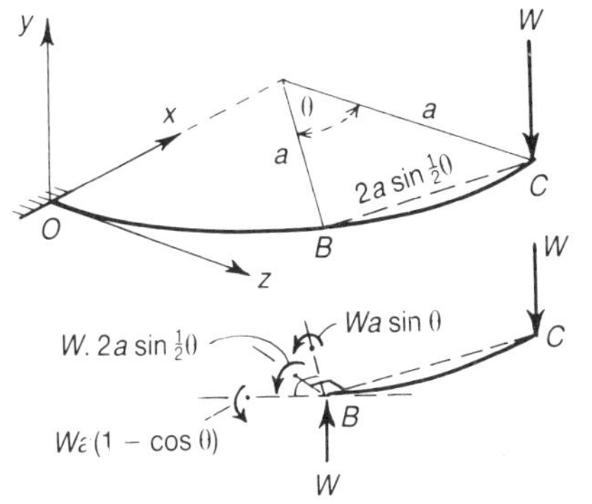

Solution

Consider any section at an angular position θ in the *xz*-plane; there is no axial force on the centre line, and the shearing force at any section is *W*. The torque about he centre line is

$$W(a - a\cos\theta) = Wa(1 - \cos\theta)$$

The bending moment acts about the radius, and has the value

$$Wa\sin\theta$$

Problem 7.8 The axis of a beam consists of two lines *BC* and *CD* in a horizontal plane and at right angles to each other. Estimate the greatest bending moment and torque when the beam carries a vertical load of 10 kN at *D*.

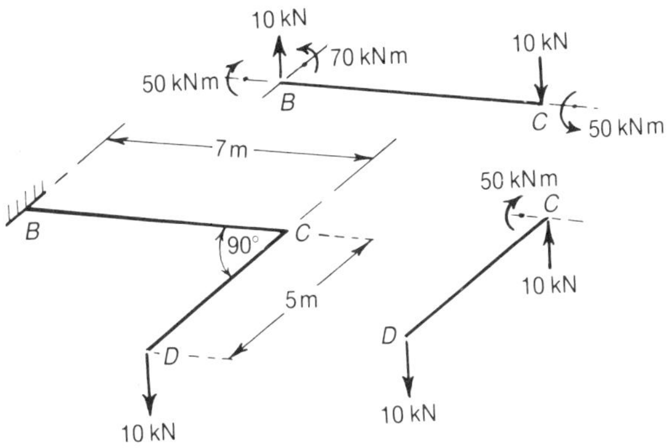

Solution

Consider the statical equilibrium of *DC* alone; there is no torque in *DC*, and the only internal actions at *C* in *DC* are a shearing force of 10 kN and a bending moment of 50 kNm. Now reverse the actions at C on DC and consider these reversed actions at C on BC. Equilibrium of BC is ensured if there is a shearing force of 10 kN at B, a bending moment of 70 kNm, and a torque of 50 kNm.

7.14 Rolling loads and influence lines

In the design of bridge girders it is frequently necessary to know the maximum bending moment and shearing force which each section will have to bear when a travelling load, such as a train, passes from one end of the bridge to the other. The diagrams which we have considered so far show the simultaneous values of the bending moment, or shearing force, for all sections of the beam with the loads in one fixed position; we shall now see how to construct a diagram which shows the greatest value of these quantities for all positions of the loads. These diagrams are called *maximum bending moment* or *maximum shearing force*, diagrams.

We assume that the loads on a beam are moving slowly; then there are negligible inertia effects from the mass of the beam and any moving masses.

7.15 A single concentrated load traversing a beam

Suppose a single concentrated vertical load W travels slowly along a beam BC, which is simply-supported at each end, Figure 7.18(i). If a is the distance of the load from B, the reactions at B and C are

$$R_B = \frac{W}{L}(L - a) \qquad R_C = \frac{Wa}{L}$$

The bending moment at a distance z from B, is

$$M = \frac{Wz}{L}(L - a) \text{ for } z < a \tag{7.26}$$

$$M = \frac{Wa}{L}(L - z) \text{ for } z > a \tag{7.27}$$

Consider the load rolling slowly from C to B: initially $z < a$, and the bending moment, given by equation 7.26, increases as a decreases; when $a = z$,

$$M = \frac{Wz}{L}(L - z) \tag{7.28}$$

As W proceeds further, we have $z > a$, and the bending moment, given by equation (7.27), decreases as a decreases further.

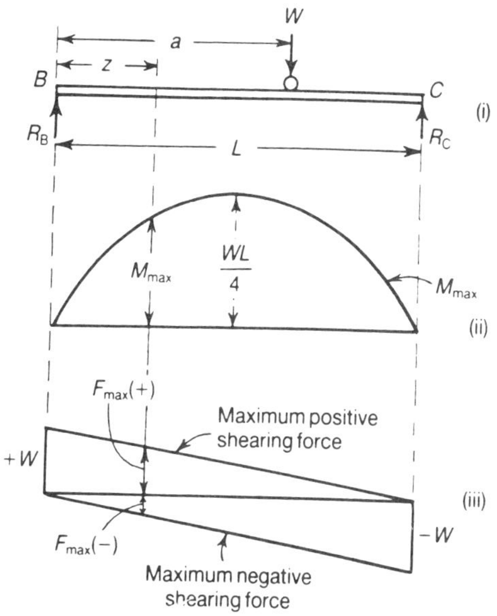

Figure 7.18 Bending moments and shearing forces due to a rolling load traversing a simply-supported beam.

Clearly, equation 7.28 is the greatest bending moment which can occur at the section; thus, for any section a distance z from B, the maximum bending moment that can be induced is

$$M_{max} = \frac{Wz}{L}(L - z) \tag{7.29}$$

and this occurs when the load W is at that section of the beam. The variation of M_{max} for different values of z is shown in Figure 7.18(ii); the curve of M_{max} is a parabola, attaining a peak value when $z = \frac{1}{2}L$, for which

$$M_{max} = \frac{WL}{4}$$

The shearing force a distance z from B is

$$F = R_B = \frac{W}{L}(L - a) \quad \text{for} \quad z < a \tag{7.30}$$

$$F = -R_C = -\frac{Wa}{L} \quad \text{for} \quad z > a \tag{7.31}$$

Consider again a load rolling slowly from C to B; initially $z < a$, and the shearing force, given by equation (7.30), is positive and increases as a diminishes. The greatest positive shearing force occurs just before the load W passes the section under consideration; it has the value

$$F_{max}(+) = \frac{W}{L}(L - z) \tag{7.32}$$

After the load has passed the section being considered, that is, when $z > a$, the shearing force, given by equation (7.31) is negative and decreases as a diminishes further. The greatest negative shearing force occurs when the load W has just passed the section at a distance z; it has the value

$$F_{max}(-) = -\frac{Wz}{L} \tag{7.33}$$

The variations of maximum positive and negative shearing forces are shown in Figure 7.18(iii).

7.16 Influence lines of bending moment and shearing force

A curve that shows the value of the bending moment at a given section of a beam, for all positions of a travelling load, is called the bending-moment *influence line* for that section; similarly, a curve that shows the shearing force at the section for all positions of the load is called the shearing force *influence line* for the section. The distinction between influence lines and maximum bending-moment (or shearing force) diagrams must be carefully noted: for a given load there will be only one maximum bending-moment diagram for the beam, but an infinite number of bending-moment influence lines, one for each section of the beam.

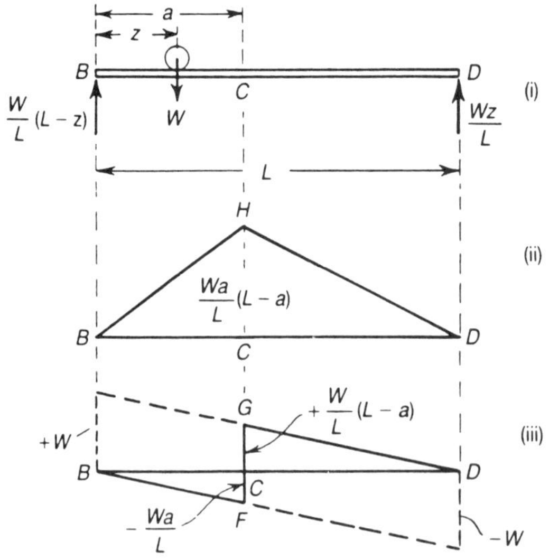

Figure 7.19 (i) Single rolling load on a simply-supported beam. (ii) Bending-moment influence line for section C. (iii) Shearing force influence line for Section C.

Consider a simply-supported beam, Figure 7.19, carrying a single concentrated load, W. As the load rolls across the beam, the bending moments at a section C of the beam vary with the position of the load. Suppose W is a distance z from B; then the bending moment at a section C is given by

$$M = \frac{Wz}{L}(L - a) \quad \text{for } z < a$$

and

$$M = \frac{Wa}{L}(L - z) \quad \text{for } z > a$$

The first of these equations gives the straight line BH in Figure 7.19(ii), and the second the line HD. The influence line for bending moments at C is then BHD; the bending moment is greatest when the load acts at the section.

Again, the shearing force at C is

$$F = -\frac{Wz}{L} \quad \text{for } z < a$$

and

$$F = +\frac{W}{L}(L - z) \quad \text{for } z > a$$

These relationships give the lines $BFCGD$ for the shearing force influence line for C. There is an abrupt change of shearing force as the load W crosses the section C.

Further problems *(answers on page 692)*

7.9 Draw the shearing-force and bending-moment diagrams for the following beams:

(i) A cantilever of length 20 m carrying a load of 10 kN at a distance of 15 m from the supported end.

(ii) A cantilever of length 20 m carrying a load of 10 kN uniformly distributed over the inner 15 m of its length.

(iii) A cantilever of length 12 m carrying a load of 8 kN, applied 5 m from the supported end, and a load of 2kN/m over its whole length.

(iv) A beam, 20 m span, simply-supported at each end and carrying a vertical load of 20 kN at a distance 5 m from one support.

(v) A beam, 16 m span, simply-supported at each end and carrying a vertical load of 2.5 kN at a distance of 4 m from one support and the beam itself weighing 500 N per metre.

7.10 A pair of lock gates are strengthened by two girders AC and BC. If the load on each girder amounts to 15 kN per metre run, find the bending moment at the middle of each girder. *(Cambridge)*

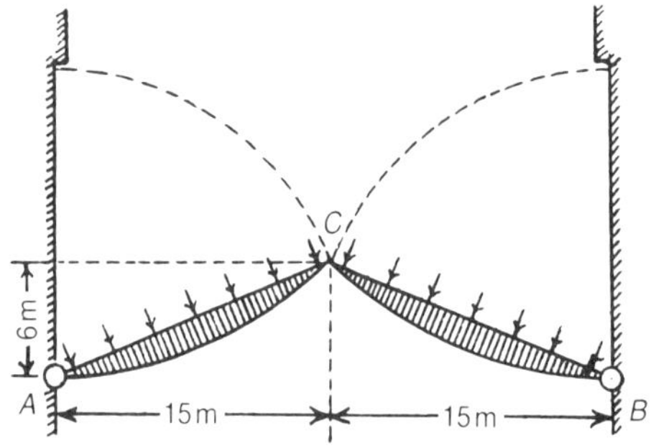

7.11 A girder $ABCDE$ bears on a wall for a length BC and is prevented from overturning by a holding-down bolt at A. The packing under BC is so arranged that the pressure over the bearing is uniformly distributed and the 30 kN load may also be taken as a uniformly distributed load. Neglecting the mass of the beam, draw its bending moment and shearing force diagrams. *(Cambridge)*

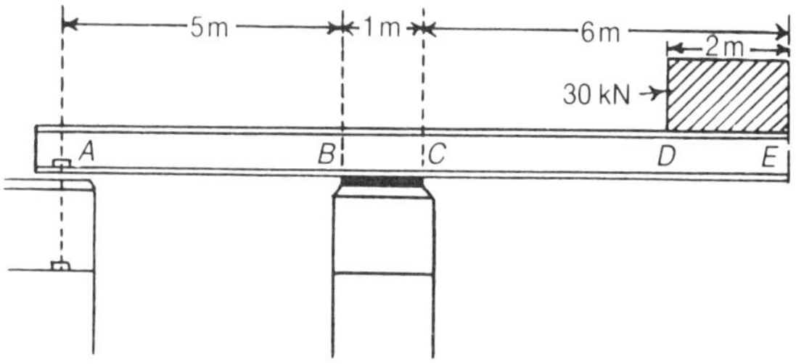

7.12 Draw the bending moment and shearing force diagrams for the beam shown. The beam is supported horizontally by the strut DE, hinged at one end to a wall, and at the other end to the projection CD which is firmly fixed at right angles to AB. The beam is freely hinged to the wall at B. The masses of the beam and strut can be neglected. *(Cambridge)*

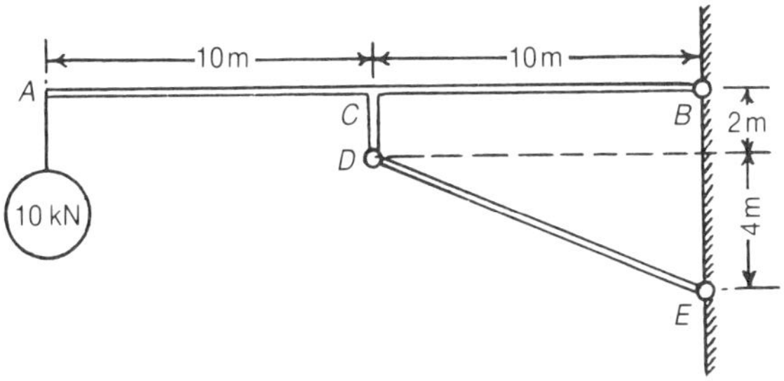

7.13 A timber dam is made of planking backed by vertical piles. The piles are built-in at the section A where they enter the ground and they are supported by horizontal struts whose centre lines are 10 m above A. The piles are spaced 1 m apart between centres and the depth of water against the dam is 10 m.

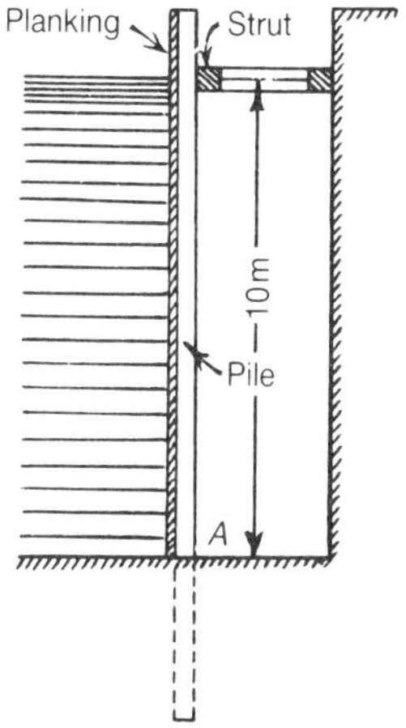

Assuming that the thrust in the strut is two-sevenths the total water pressure resisted by each pile, sketch the form of the bending moment and shearing force diagrams for a pile. Determine the magnitude of the bending moment at A and the position of the section which is free from bending moment. *(Cambridge)*

7.14 The load distribution (full lines) and upward water thrust (dotted lines) for a ship are given, the numbers indicating kN per metre run. Draw the bending moment diagram for the ship. *(Cambridge)*

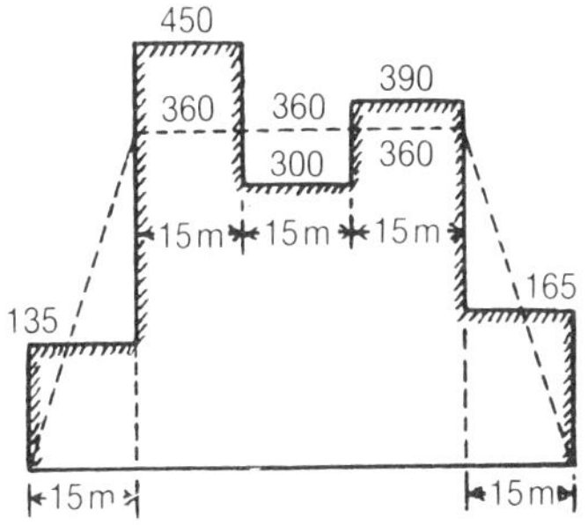

8 Geometrical properties of cross-sections

8.1 Introduction

The strength of a component of a structure is dependent on the geometrical properties of its cross-section in addition to its material and other properties. For example, a beam with a large cross-section will, in general, be able to resist a bending moment more readily than a beam with a smaller cross-section. Typical cross-section of structural members are shown in Figure 8.1.

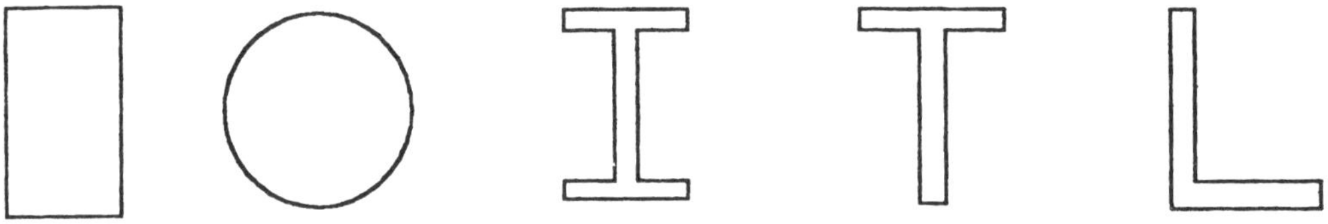

(a) Rectangle (b) Circle (c) 'I' beam (d) 'Tee' beam (e) Angle bar

Figure 8.1 Some typical cross-sections of structural components.

The cross-section of Figure 8.1(c) is also called a *rolled steel joist* (RSJ); it is used extensively in structural engineering. It is quite common to make cross-sections of metal structural members in the form of the cross-sections of Figure 8.1(c) to (e), as such cross-sections are structurally more efficient in bending than cross-sections such as Figures 8.1(a) and (b). Wooden beams are usually of rectangular cross-section and not of the forms shown in Figures 8.1(c) to (e). This is because wooden beams have grain and will have lines of weakness along their grain if constructed as in Figures 8.1(c) to (e).

8.2 Centroid

The position of the centroid of a cross-section is the centre of the moment of area of the cross-section. If the cross-section is constructed from a homogeneous material, its centroid will lie at the same position as its centre of gravity.

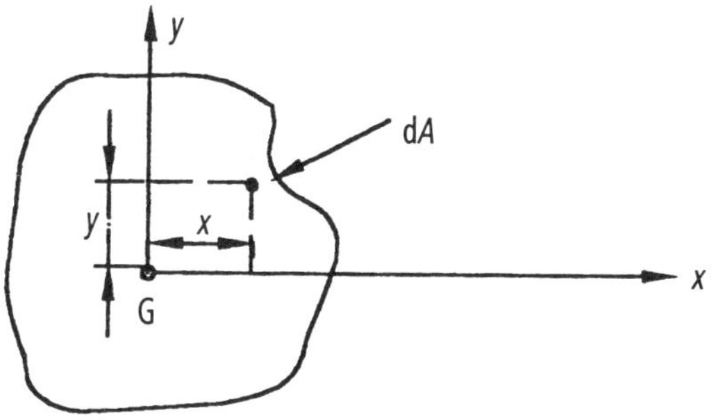

Figure 8.2 Cross-section.

Let G denote the position of the centroid of the plane lamina of Figure 8.2. At the centroid the moment of area is zero, so that the following equations apply

$$\Sigma\, x \;\; dA \;=\; \Sigma\, y \;\; dA \;=\; 0 \tag{8.1}$$

where dA = elemental area of the lamina

x = horizontal distance of dA from G

y = vertical distance of dA from G

8.3 Centroidal axes

These are the axes that pass through the centroid.

8.4 Second moment of area (I)

The second moments of area of the lamina about the $x - x$ and $y - y$ axes, respectively, are given by

$$I_{xx} = \sum y^2 \;\; dA = \text{second moment of area about } x - x \tag{8.2}$$

$$I_{yy} = \sum x^2 \;\; dA = \text{second moment of area about } y - y \tag{8.3}$$

Now from Pythagoras' theorem

$$x^2 + y^2 = r^2$$

$$\therefore \;\; \sum x^2 \;\; dA + \sum y^2 \;\; dA = \sum r^2 \;\; dA$$

or

$$I_{yy} + I_{xx} = J \tag{8.4}$$

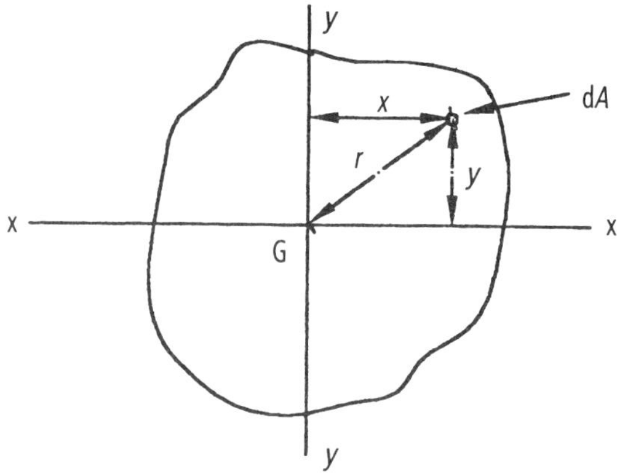

Figure 8.3 Cross-section.

where

$$J = \text{polar second moment of area} = \sum r^2 \; dA \tag{8.5}$$

Equation (8.4) is known as the *perpendicular axes theorem* which states that the sum of the second moments of area of two mutually perpendicular axes of a lamina is equal to the polar second moment of area about a point where these two axes cross.

8.5 Parallel axes theorem

Consider the lamina of Figure 8.4, where the x–x axis passes through its centroid. Suppose that I_{xx} is known and that I_{XX} is required, where the X–X axis lies parallel to the x–x axis and at a perpendicular distance h from it.

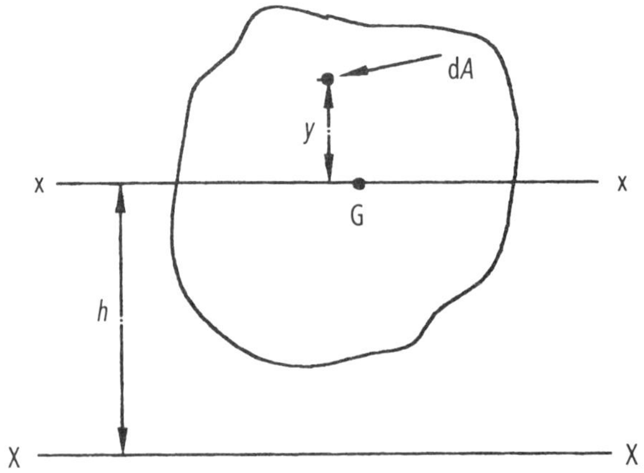

Figure 8.4 Parallel axes.

Now from equation (8.2)

$$I_{xx} = \Sigma y^2 \, dA$$

and

$$I_{XX} = \Sigma (y + h)^2 \, dA \quad (8.6)$$

$$= \Sigma (y^2 + h^2 + 2\,hy) \, dA, \quad (8.7)$$

but $\Sigma\, 2\,hy\, dA = 0$, as 'y' is measured from the centroid.

$$\therefore I_{XX} = \Sigma (y^2 + h^2) \, dA \quad (8.8)$$

but

$$I_{xx} = \Sigma y^2 \, dA$$

$$\therefore I_{XX} = I_{xx} + h^2 \, \Sigma\, dA$$

$$= I_{xx} + h^2 \, A \quad (8.9)$$

where

$$A = \text{area of lamina} = \Sigma\, dA$$

Equation (8.9) is known as the *parallel axes theorem*, which states that the second moment of area about the X–X axis is equal to the second moment of area about the x–x axis + $h^2 \times A$, where x–x and X–X are parallel.

h = the perpendicular distance between the x–x and X–X axes.

I_{xx} = the second moment of area about x–x

I_{XX} = the second moment of area about X–X

The importance of the parallel axes theorem is that it is useful for calculating second moments of area of sections of RSJs, tees, angle bars etc. The geometrical properties of several cross-sections will now be determined.

Problem 8.1 Determine the second moment of area of the rectangular section about its centroid (x–x) axis and its base (X–X) axis; see Figure 8.5. Hence or otherwise, verify the parallel axes theorem.

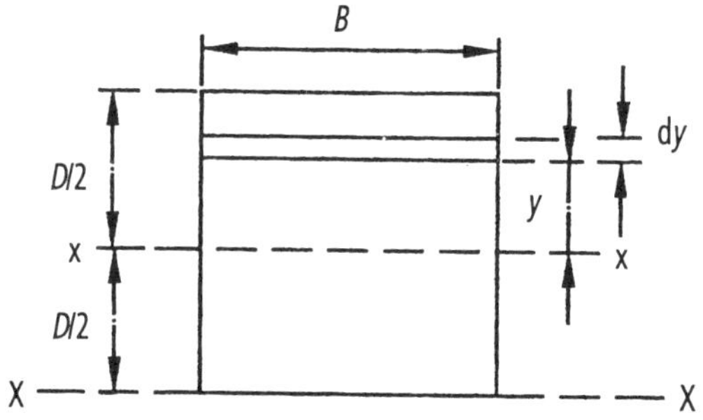

Figure 8.5 Rectangular section.

Solution

From equation (8.2)

$$I_{xx} = \int y^2 \, dA = \int_{-D/2}^{D/2} y^2 \, (B \, dy)$$

$$= B\left[\frac{y^3}{3}\right]_{-D/2}^{D/2} = \frac{2B}{3}\left[y^3\right]_o^{D/2} \tag{8.10}$$

$$I_{xx} = BD^3/12 \text{ (about centroid)}$$

$$I_{XX} = \int_{-D/2}^{D/2} (y + D/2)^2 \; B \, dy$$

$$= B\int_{-D/2}^{D/2} \left(y^2 + D^2/4 + Dy\right) dy$$

$$= B\left[\frac{y^3}{3} + \frac{D^2 y}{4} + \frac{Dy^2}{2}\right]_{-D/2}^{D/2} \tag{8.11}$$

$$I_{XX} = BD^3/3 \text{ (about base)}$$

To verify the parallel axes theorem,

from equation (8.9)

$$I_{XX} = I_{xx} + h^2 \times A$$

$$= \frac{BD^3}{12} + \left(\frac{D}{2}\right)^2 \times BD$$

$$= BD^3\left(\frac{1}{12} + \frac{1}{4}\right)$$

$$I_{XX} = BD^3/3 \quad \text{QED}$$

Problem 8.2 Determine the second moment of area about x–x, of the circular cross-section of Figure 8.6. Using the perpendicular axes theorem, determine the polar second moment of area, namely 'J'.

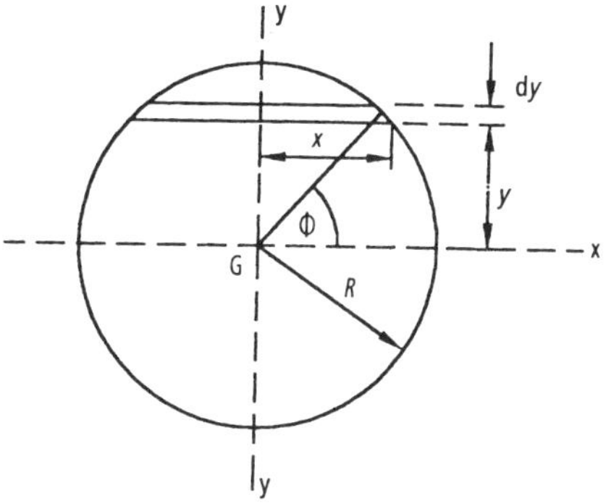

Figure 8.6 Circular section.

Solution

From the theory of a circle,

$$x^2 + y^2 = R^2$$

or $$y^2 = R^2 - x^2 \tag{8.12}$$

Let $$x = R\cos\varphi \quad \text{(see Figure 8.6)}$$

$$\therefore y^2 = R^2 - R^2\cos^2\varphi \tag{8.13}$$

$$= R^2\sin^2\varphi \tag{8.14}$$

or $\quad y = R \sin \varphi$

and $$\frac{dy}{d\phi} = R \cos \varphi \tag{8.15}$$

or $$dy = R \cos \varphi \; d\varphi \tag{8.16}$$

Now $$A = \text{area of circle}$$

$$= 4 \int_0^R x \, dy$$

$$= 4 \int_0^{\pi/2} R \cos \varphi \;\; R \cos \varphi \; d\varphi$$

$$= 4R^2 \int_0^{\pi/2} \cos^2 \varphi \; d\varphi$$

but $$\cos^2 \varphi = \frac{1 + \cos 2\phi}{2}$$

$$\therefore A = 2R^2 \int_0^{\pi/2} \left(1 + \cos 2\phi\right) d\varphi$$

$$= 2R^2 \left[\phi + \frac{\sin 2\phi}{2}\right]_0^{\pi/2}$$

$$= 2R^2 \left[\left(\frac{\pi}{2} + 0\right) - \left(0 + 0\right)\right]$$

or $$A = \pi R^2 \text{ QED} \tag{8.17}$$

Now $$I_{xx} = 4 \int_0^R y^2 \; x \; dy \tag{8.18}$$

Substituting equations (8.14), (8.13) and (8.16) into equation (8.18), we get

$$I_{xx} = 4 \int_0^{\pi/2} R^2 \sin^2 \varphi \;\; R \cos \varphi \; R \cos \varphi \; d\varphi$$

$$= 4R^4 \int_0^{\pi/2} \sin^2 \varphi \; \cos^2 \varphi \; d\varphi$$

but $$\sin^2 \varphi = (1 - \cos 2\varphi)/2$$

and $\cos^2 \varphi = (1 + \cos 2\varphi)/2$

$$\therefore \; I_{xx} = R^4 \int_0^{\pi/2} (1 - \cos 2\varphi)\,(1 + \cos 2\varphi)\; d\,\varphi$$

$$= R^4 \int_0^{\pi/2} (1 - \cos^2 2\varphi)\; d\,\varphi$$

but $\cos^2 2\,\varphi = \dfrac{1 + \cos 4\phi}{2}$

$$I_{xx} = R^4 \int_0^{\pi/2} \left[1 - \frac{1 + \cos 4\phi}{2}\right] d\,\varphi$$

$$= R^4 \left[\phi - \phi/2 - \frac{\sin 4\phi}{8}\right]_0^{\pi/2}$$

$$= R^4\,[(\pi/2 - \pi/4 - 0) - (0 - 0 - 0)]$$

or $$I_{xx} = \pi R^4/4 = \pi D^4/64 \tag{8.19}$$

where

D = diameter = $2R$

As the circle is symmetrical about x–x and y–y

$$I_{yy} = I_{xx} = \pi D^4/64$$

From the perpendicular axes theorem of equation (8.4),

J = polar second moment of area

$$= I_{xx} + I_{yy} = \pi\, D^4/64 + \pi\; D^4/64 \tag{8.20}$$

or $$J = \pi\, D^4/32 = \pi R^4/2$$

Problem 8.3 Determine the second moment of area about its centroid of the RSJ of Figure 8.7.

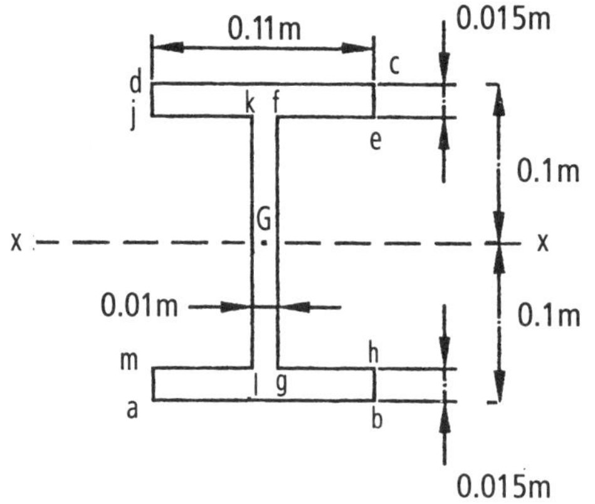

Figure 8.7 RSJ.

Solution

I_{xx} = *'I'* of outer rectangle (*abcd*) about *x–x* minus the sum of the I's of the two inner rectangles (efgh and jklm) about *x–x*.

$$= \frac{0.11 \times 0.2^3}{12} - \frac{2 \times 0.05 \times 0.17^3}{12}$$

$$= 7.333 \times 10^{-5} - 4.094 \times 10^{-5}$$

or $I_{xx} = 3.739 \times 10^{-5}\ \text{m}^4$

Problem 8.4 Determine I_{xx} for the cross-section of the RSJ as shown in Figure 8.8.

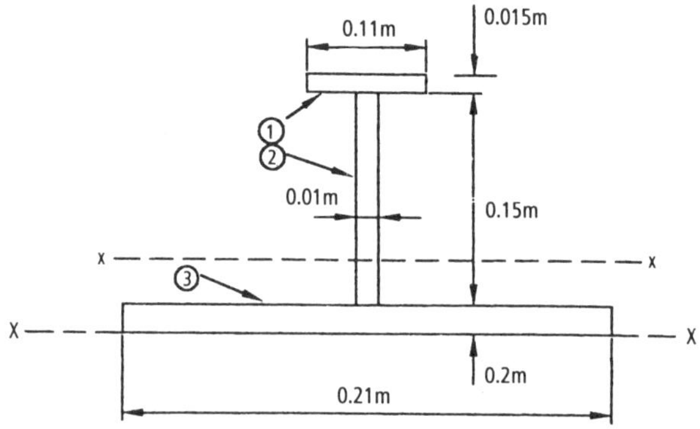

Figure 8.8 RSJ (dimensions in metres).

<u>*Solution*</u>

The calculation will be carried out with the aid of Table 8.1. It should be emphasised that this method is suitable for almost any *computer spreadsheet*. To aid this calculation, the RSJ will be subdivided into three rectangular elements, as shown in Figure 8.8.

Table 8.1

Col. 1	**Col. 2**	**Col. 3**	**Col. 4**	**Col. 5**	**Col. 6**
Element	$a = bd$	y	ay	ay^2	$i = bd^3/_{12}$
1	$0.11 \times 0.015 = 0.00165$	0.1775	2.929×10^{-4}	5.199×10^{-5}	$0.11 \times 0.015^3/12 = 3 \times 10^{-8}$
2	$0.01 \times 0.15 = 0.0015$	0.095	1.425×10^{-4}	1.354×10^{-5}	$0.01 \times 0.15^3/12 = 2.812 \times 10^{-6}$
3	$0.02 \times 0.21 = 0.0042$	0.01	4.2×10^{-5}	4.2×10^{-7}	$0.21 \times 0.02^3/12 = 1.4 \times 10^{-7}$
Σ	$\Sigma a = 0.00735$	—	$\Sigma ay = 4.77 \times 10^{-4}$	$\Sigma ay^2 = 6.595 \times 10^{-5}$	$\Sigma i = 2.982 \times 10^{-6}$

a = area of an element (column 2)

y = vertical distance of the local centroid of an element from XX (column 3)

ay = the product $a \times y$ (column 4 = column 2 × column 3)

ay^2 = the product $a \times y \times y$ (column 5 = column 3 × column 4)

i = the second moment of area of an element about its own local centroid = $bd^3/12$

b = 'width' of element (horizontal dimension)

d = 'depth' of element (vertical dimension)

Σ = summation of the column

$\bar{y}$ = distance of centroid of the cross-section about XX

$$= \Sigma ay / \Sigma a \tag{8.21}$$

$$= 4.774 \times 10^{-4}/0.00735 = 0.065 \text{ m} \tag{8.22}$$

Now from equation (8.9)

$$I_{XX} = \sum ay^2 + \sum i$$

$$= 6.595 \times 10^{-5} + 2.982 \times 10^{-6} \qquad (8.23)$$

$$I_{XX} = 6.893 \times 10^{-5}\ \text{m}^4$$

From the parallel axes theorem (8.9),

$$I_{xx} = I_{XX} - \overline{y^2} \sum a$$

$$= 6.893 \times 10^{-5} - 0.065^2 \times 0.00735 \qquad (8.24)$$

$$\text{or } I_{xx} = 3.788 \times 10^{-5}\ \text{m}^4$$

Further problems *(for answers, see page 692)*

8.5 Determine I_{xx} for the thin-walled sections shown in Figures 8.9(a) to 8.9(c), where the wall thicknesses are 0.01 m.

NB Dimensions are in metres. I_{xx} = second moment of area about a horizontal axis passing through the centroid.

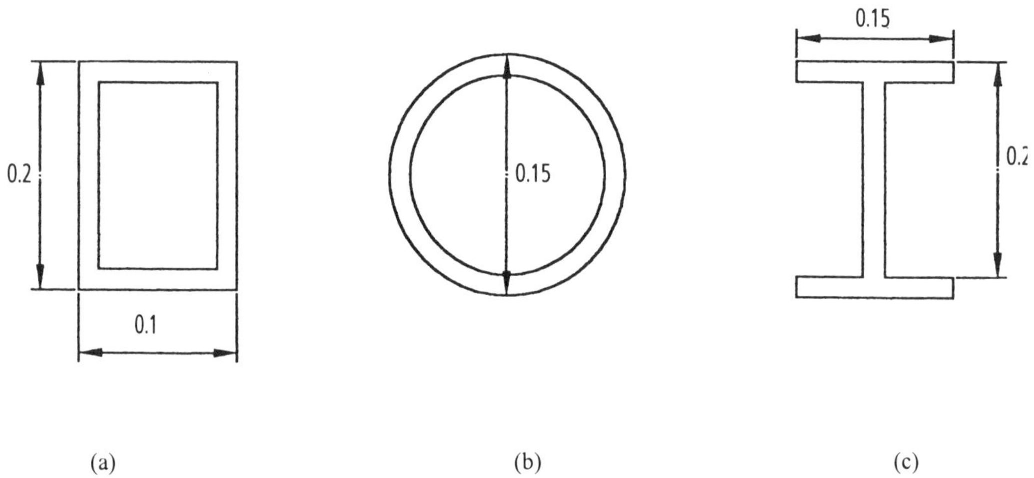

Figure 8.9 Thin-walled sections.

8.6 Determine I_{xx} for the thin-walled sections shown in Figure 8.10, which have wall thicknesses of 0.01 m.

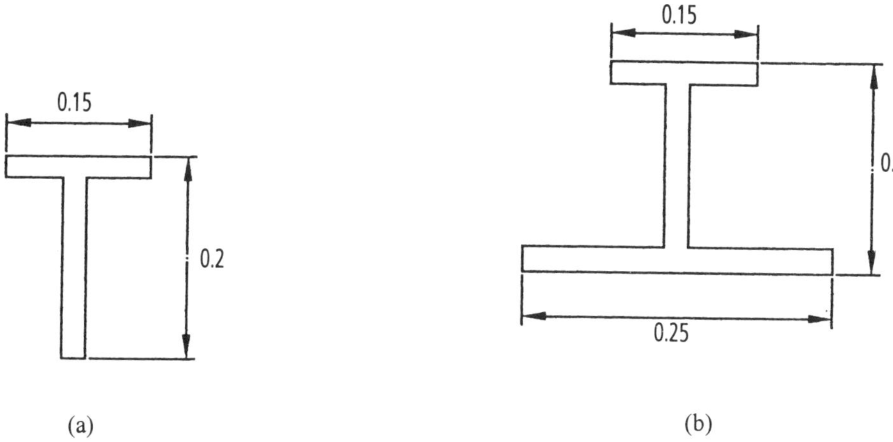

Figure 8.10

8.7 Determine the position of the centroid of the section shown in Figure 8.11, namely $\bar{y}$. Determine also I_{xx} for this section.

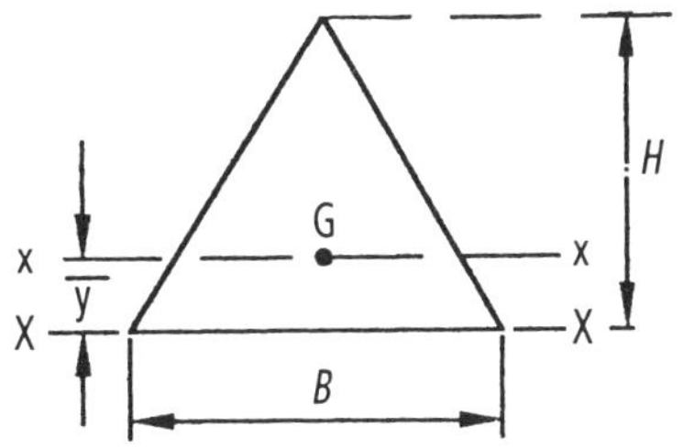

Figure 8.11 Isosceles triangular section.

9 Longitudinal stresses in beams

9.1 Introduction

We have seen that when a straight beam carries lateral loads the actions over any cross-section of the beam comprise a bending moment and shearing force; we have also seen how to estimate the magnitudes of these actions. The next step in discussing the strength of beams is to consider the stresses caused by these actions.

As a simple instance consider a cantilever carrying a concentrated load W at its free end, Figure 9.1. At sections of the beam remote from the free end the upper longitudinal fibres of the beam are stretched, i.e. tensile stresses are induced; the lower fibres are compressed. There is thus a variation of direct stress throughout the depth of any section of the beam. In any cross-section of the beam, as in Figure 9.2, the upper fibres which are stretched longitudinally contract laterally owing to the Poisson ratio effect, while the lower fibres extend laterally; thus the whole cross-section of the beam is distorted.

In addition to longitudinal direct stresses in the beam, there are also shearing stresses over any cross-section of the beam. In most engineering problems shearing *distortions* in beams are relatively unimportant; this is not true, however, of shearing *stresses*.

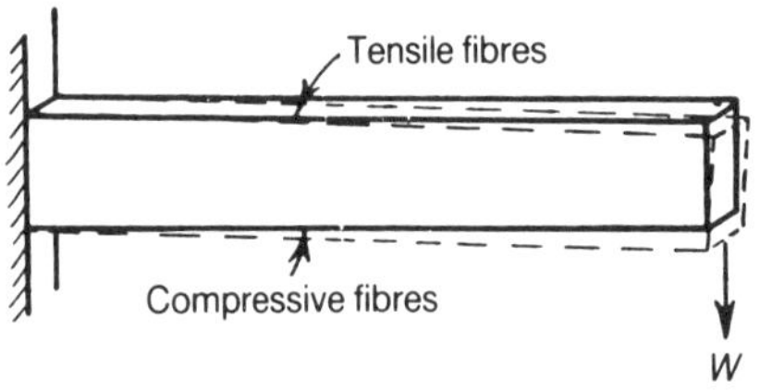

Figure 9.1 Bending strains in a loaded cantilever.

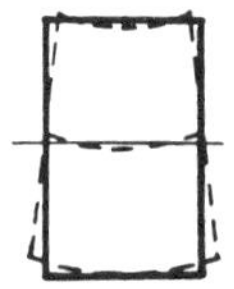

Figure 9.2 Cross-sectional distortion of a bent beam.

9.2 Pure bending of a rectangular beam

An elementary bending problem is that of a rectangular beam under end couples. Consider a straight uniform beam having a rectangular cross-section of breadth b and depth h, Figure 9.3; the axes of symmetry of the cross-section are Cx, Cy.

A long length of the beam is bent in the yz-plane, Figure 9.4, in such a way that the longitudinal centroidal axis, Cz, remains unstretched and takes up a curve of uniform radius of curvature, R.

We consider an elemental length δz of the beam, remote from the ends; in the unloaded condition, AB and FD are transverse sections at the ends of the elemental length, and these sections are initially parallel. In the bent form we assume that planes such as AB and FD remain flat planes; $A'B'$ and $F'D'$ in Figure 9.4 are therefore cross-sections of the bent beam, but are no longer parallel to each other.

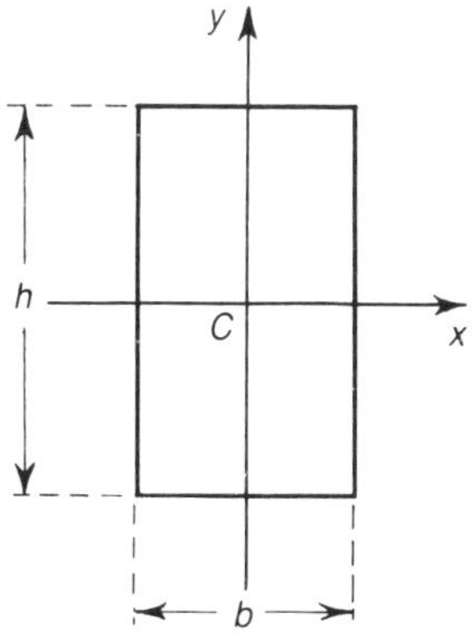

Figure 9.3 Cross-section of a rectangular beam.

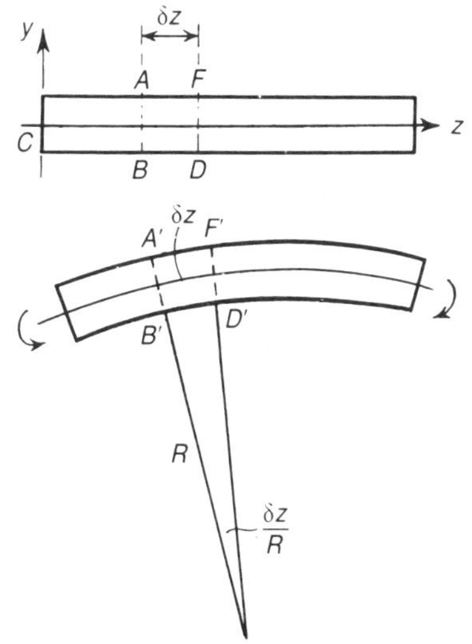

Figure 9.4 Beam bent to a uniform radius of curvature R in the yz-plane.

In the bent form, some of the longitudinal fibres, such as $A'F'$, are stretched, whereas others, such as $B'D'$ are compressed. The unstrained middle surface of the beam is known as the *neutral axis*. Now consider an elemental fibre HJ of the beam, parallel to the longitudinal axis Cz, Figure 9.5; this fibre is at a distance y from the neutral surface and on the tension side of the beam. The original length of the fibre HJ in the unstrained beam is δz; the strained length is

$$H'J' = (R + y)\frac{\delta z}{R}$$

because the angle between $A'B'$ and $F'D'$ in Figure 9.4 and 9.5 is $(\delta z/R)$. Then during bending HJ stretches an amount

$$H'J' - HJ = (R + y)\frac{\delta z}{R} - \delta z = \frac{y}{R}\,\delta z$$

The longitudinal strain of the fibre HJ is therefore

$$\varepsilon = \left(\frac{y}{R}\,\delta z\right) / \delta z = \frac{y}{R}$$

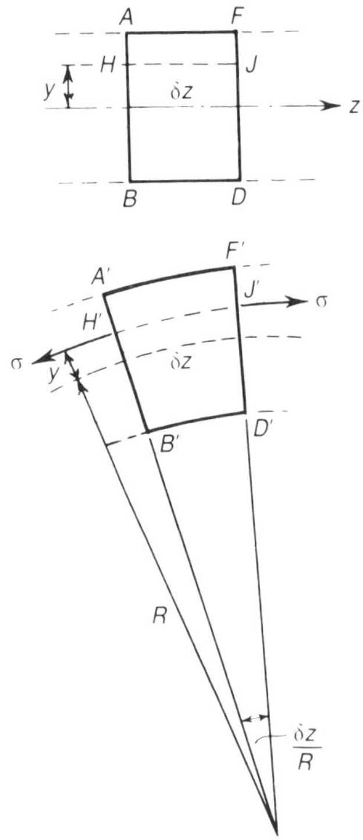

Figure 9.5 Stresses on a bent element of the beam.

Then the longitudinal strain at any fibre is proportional to the distance of that fibre from the neutral surface; over the compressed fibres, on the lower side of the beam, the strains are of course negative.

If the material of the beam remains elastic during bending then the longitudinal stress on the fibre *HJ* is

$$\sigma = E\varepsilon = \frac{Ey}{R} \tag{9.1}$$

The distribution of longitudinal stresses over the cross-section takes the form shown in Figure 9.6; because of the symmetrical distribution of these stresses about *Cx*, there is no resultant longitudinal thrust on the cross-section of the beam. The resultant hogging moment is

$$M = \int_{-\frac{1}{2}h}^{+\frac{1}{2}h} \sigma b y \, dy \tag{9.2}$$

On substituting for σ from equation (9.1), we have

$$M = \frac{E}{R}\int_{-\frac{1}{2}h}^{+\frac{1}{2}h} by^2\,dy = \frac{EI_x}{R} \tag{9.3}$$

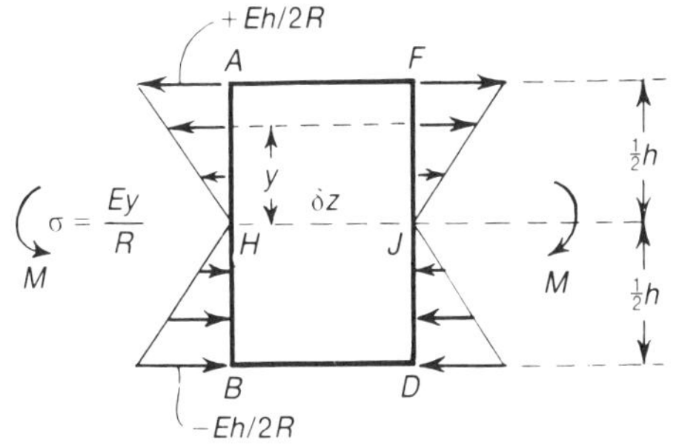

Figure 9.6 Distribution of bending stresses giving zero resultant longitudinal force and a resultant couple M.

where I_x is the second moment of area of the cross-section about Cx. From equations (9.1) and (9.3), we have

$$\frac{\sigma}{y} = \frac{E}{R} = \frac{M}{I_x} \tag{9.4}$$

We deduce that a uniform radius of curvature, R, of the centroidal axis Cz can be sustained by end couples M, applied about the axes Cx at the ends of the beam.

Equation (9.3) implies a linear relationship between M, the applied moment, and (1/R), the curvature of the beam. The constant EI_x in this linear relationship is called the *bending stiffness* or sometimes the *flexural stiffness* of the beam; this stiffness is the product of Young's modulus, E, and the second moment of area, I_x, of the cross-section about the axis of bending.

Problem 9.1 A steel bar of rectangular cross-section, 10 cm deep and 5 cm wide, is bent in the planes of the longer sides. Estimate the greatest allowable bending moment if the bending stresses are not to exceed 150 MN/m² in tension and compression.

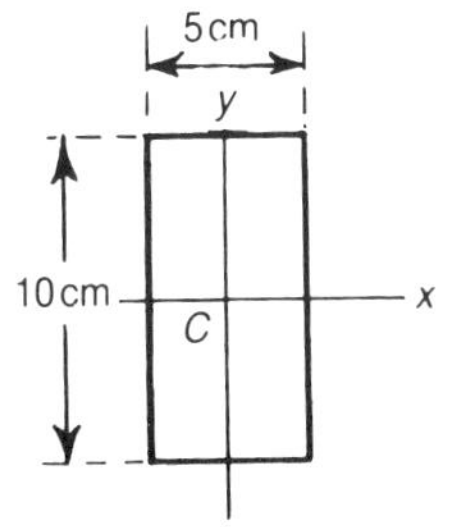

Solution

The bending moment is applied about *Cx*. The second moment of area about this axis is

$$I_x = \frac{1}{12}(0.05)(0.10)^3 = 4.16 \times 10^{-6}\ \text{m}^2$$

The bending stress, σ, at a fibre a distance *y* from *Cx* is, by equation (9.4)

$$\sigma = \frac{My}{I_x}$$

where *M* is the applied moment. If the greatest stresses are not to exceed 150 MN/m², we must have

$$\frac{My}{I_x} \leq 150\ \text{MN/m}^2$$

The greatest bending stresses occur in the extreme fibres where $y = 5$ cm. Then

$$M \leq \frac{(150 \times 10^6)\, I_x}{(0.05)} = \frac{(150 \times 10^6)(4.16 \times 10^{-6})}{(0.05)}$$

$$= 12\,500\ \text{Nm}$$

The greatest allowable bending moment is therefore 12 500 Nm. (The second moment of area about *Cy* is

$$I_y = \frac{1}{12}(0.10)(0.05)^3 = 1.04 \times 10^{-6}\ \text{m}^2$$

The greatest allowable bending moment about *Cy* is

$$M = \frac{(150 \times 10^6)\, I_y}{(0.025)} = \frac{(150 \times 10^6)(1.04 \times 10^{-6})}{(0.025)}$$

$$= 6250\ \text{Nm}$$

which is only half that about *Cx*.

9.3 Bending of a beam about a principal axis

In section 9.2 we considered the bending of a straight beam of rectangular cross-section; this form of cross-section has two axes of symmetry. More generally we are concerned with sections having only one, or no, axis of symmetry.

Consider a long straight uniform beam having any cross-sectional form, Figure 9.7; the axes Cx and Cy are *principal axes* of the cross-section. The principal axes of a cross-section are those centroid axes for which the product second moments of area are zero. In Figure 9.7, C is the centroidal of the cross-section; Cz is the longitudinal centroidal axis.

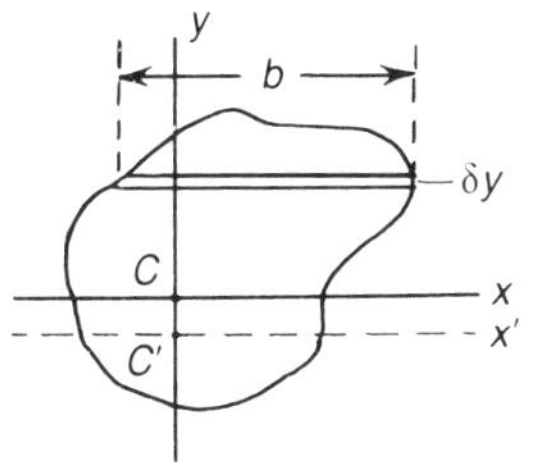

Figure 9.7 General cross-sectional form of a beam.

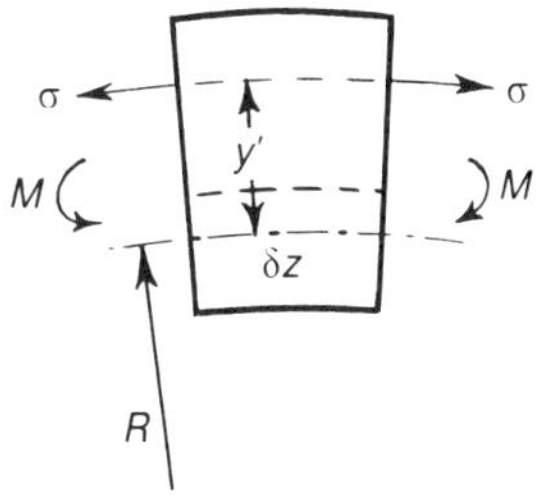

Figure 9.8 Elemental length of a beam.

When end couples M are applied to the beam, we assume as before that transverse sections of the beam remain plane during bending. Suppose further that, if the beam is bent in the yz-plane only, there is a neutral axis $C'x'$, Figure 9.7, which is parallel to Cx and is unstrained; radius of curvature of this neutral surface is R, Figure 9.8. As before, the strain in a longitudinal fibre at a distance y' from $C'x'$ is

$$\varepsilon = \frac{y'}{R}$$

If the material of the beam remains elastic during bending the longitudinal stress on this fibre is

$$\sigma = \frac{Ey'}{R}$$

If there is to be no resultant longitudinal thrust on the beam at any transverse section we must have

$$\int_A \sigma b \, dy' = 0$$

Where b is the breadth of an elemental strip of the cross-section parallel to Cx, and the integration is performed over the whole cross-sectional area, A. But

$$\int_A \sigma b \, dy' = \frac{E}{R} \int_A y' b \, dy'$$

This can be zero only if $C'x'$ is a centroidal axis; now, Cx is a principal axis, and is therefore a centroidal axis, so that $C'x'$ and Cx are coincident, and the neutral axis is Cx in any cross-section of the beam. The total moment about Cx of the internal stresses is

$$M = \int_A \sigma b y \, dy = \frac{E}{R} \int_A b y^2 \, dy$$

But $\int_A by^2 \, dy$ is the second moment of area of the cross-section about Cx; if this is denoted by I_x, then

$$M = \frac{EI_x}{R} \tag{9.5}$$

The stress in any fibre a distance y from Cx is

$$\sigma = \frac{Ey}{R} = \frac{My}{I_x} \tag{9.6}$$

No moment about Cy is implied by this stress system, for

$$\int_A \sigma x \, dA = \frac{E}{R} \int_A xy \, dA = 0$$

because Cx and Cy are principal axes for which $\int_A xy \, dA$, or the product second moment of area, is zero; δA is an element of area of the cross-section.

9.4 Beams having two axes of symmetry in the cross-section

Many cross-sectional forms used in practice have two axes of symmetry; examples are the I-section and circular sections, Figure 9.9, besides the rectangular beam already discussed.

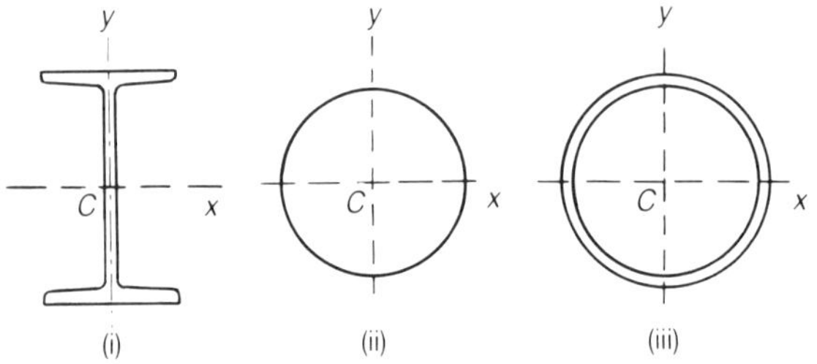

Figure 9.9 (i) I-section beam. (ii) Solid circular cross-section. (iii) Hollow circular cross-section.

An axis of symmetry of a cross-section is also a principal axis; then for bending about the axis Cx we have, from equation (9.6),

$$\sigma = \frac{Ey}{R_x} = \frac{M_x y}{I_x} \tag{9.7}$$

where R_x is the radius of curvature in the yz-plane, M_x is the moment about Cx, and I_x is the second moment of area about Cx. Similarly for bending by a couple M_y about Cy,

$$\sigma = \frac{Ex}{R_y} = \frac{M_y x}{I_y} \tag{9.8}$$

where R_y is the radius of curvature in the xz-plane, and I_y is the second moment of area about Cy. The longitudinal centroid axis is Cz. From equations (9.7) and (9.8) we see that the greatest bending stresses occur in the extreme longitudinal fibres of the beams.

Problem 9.2 A light-alloy I-beam of 10 cm overall depth has flanges of overall breadth 5 cm and thickness 0.625 cm, the thickness of the web is 0.475 cm. If the bending stresses are not to exceed 150 MN/m² in tension and compression estimate the greatest moments which may be applied about the principal axes of the cross-section.

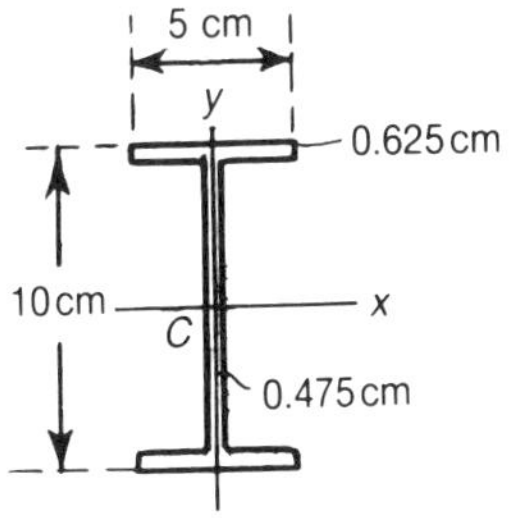

Solution

Consider, first, bending about Cx. From equation (8.10), the second moment of area about Cx is

$$I_x = 0.05 \times 0.1^3/12 - (0.05 - 0.00475) \times (0.1 - 2 \times 0.00625)^3/12$$

$$= 4.167 \times 10^{-6} - 0.04525 \times 0.0875^3/12$$

$$= 4.167 \times 10^{-6} - 2.526 \times 10^{-6}$$

$$I_x = 1.641 \times 10^{-6} \text{ m}^4$$

The above calculation has been obtained by taking away the second moments of area of the two inner rectangles from the second moment of area of the outer rectangle, as previously demonstrated in Chapter 8. The allowable moment M_x is

$$M_x = \frac{\sigma I_x}{y} = \frac{(150 \times 10^6)\,(1.64 \times 10^{-6})}{0.05} = 4926 \text{ Nm}$$

Second, for bending about *Cy*.

$$I_y = (0.1 - 2 \times 0.00625) \times 0.00475^3 / 12 + 2 \times 0.00625 \times 0.05^3 / 12$$

The first term, which is the contribution of the web, is negligible compared with the second. With sufficient accuracy

$$I_y = 2\left(\frac{1}{12}\right)(0.00625)\,(0.05)^3 = 0.130 \times 10^{-6} \text{ m}^4$$

The allowable moment about *Cy* is

$$M_y = \frac{\sigma I_y}{x} = \frac{(150 \times 10^6)\,(0.130 \times 10^{-6})}{0.025} = 780 \text{ Nm}$$

Problem 9.3 A steel scaffold tube has an external diameter of 5 cm, and a thickness of 0.5 cm. Estimate the allowable bending moment on the tube if the bending stresses are limited to 100 MN/m^2.

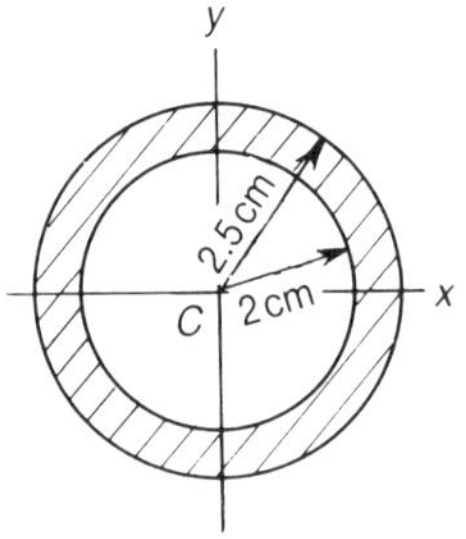

Solution

From equation (8.19), the second moment of area about a centroid axis *Cx* is

$$I_x = \frac{\pi}{4}\left[(0.025)^4 - (0.020)^4\right] = 0.181 \times 10^{-6} \text{ m}^4$$

The allowable bending moment about *Cx* is

$$M_x = \frac{(100 \times 10^6)\,(0.181 \times 10^{-6})}{0.025} = 724 \text{ Nm}$$

9.5 Beams having only one axis of symmetry

Other common sections in use, as shown in Figure 9.10, have only one axis of symmetry Cx. In each of these, Cx is the axis of symmetry, and Cx and Cy are both principal axes. When bending moments M_x and M_y are applied about Cx and Cy, respectively, the bending stresses are again given by equations (9.7) and (9.8). However, an important feature of beams of this type is that their behaviour in bending when shearing forces are also present is not as simple as that of beams having two axes of symmetry. This problem is discussed in Chapter 10.

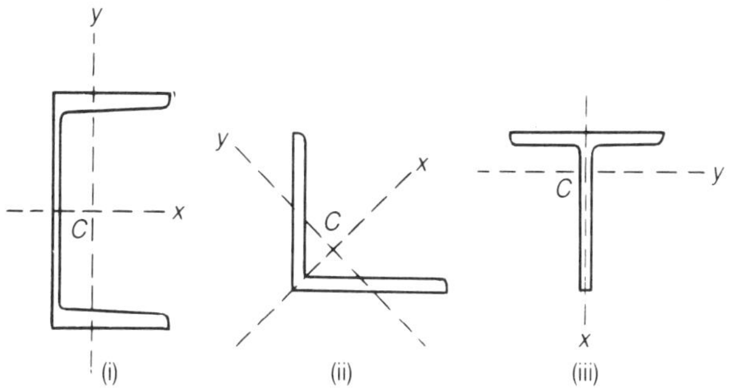

Figure 9.10 (i) Channel section. (ii) Equal angle section. (iii) T–section.

Problem 9.4 A T-section of uniform thickness 1 cm has a flange breadth of 10 cm and an overall depth of 10 cm. Estimate the allowable bending moments about the principal axes if the bending stresses are limited to 150 MN/m^2.

<u>*Solution*</u>

Suppose $\bar{y}$ is the distance of the principal axis Cx from the remote edge of the flange. The total area of the section is

$$A = (0.10)\,(0.01) + (0.09)\,(0.01) = 1.90 \times 10^{-3}\ \mathrm{m}^2$$

On taking first moments of areas about the upper edge of the flange,

$$A\bar{y} = (0.10)(0.01)(0.005) + (0.09)(0.01)(0.055) = 0.0545 \times 10^{-3}\mathrm{m}^3$$

Then

$$\bar{y} = \frac{0.0545 \times 10^{-3}}{1.9 \times 10^{-3}} = 0.0287 \text{ m}$$

The second moment of area of the flange about Cx is

$$\frac{1}{12}(0.10)(0.01)^3 + (0.10)(0.01)(0.0237)^2 = 0.570 \times 10^{-6} \text{ m}^4$$

The second moment of area of the web about Cx is

$$\frac{1}{12}(0.01)(0.09)^3 + (0.09)(0.01)(0.0263)^2 = 1.230 \times 10^{-6} \text{ m}^4$$

Then

$$I_x = (0.570 + 1.230)\,10^{-6} = 1.800 \times 10^{-6} \text{ m}^4$$

For bending about Cx, the greatest bending stress occurs at the toe of the web, as shown in the figure. The maximum allowable moment is

$$M_x = \frac{(150 \times 10^6)(1.800 \times 10^{-6})}{0.0713} = 3790 \text{ Nm}$$

The bending stress in the extreme fibres of the flange is only 60.4 MN/m^2 at this bending moment. The second moment of area about Cy is

$$I_y = \frac{1}{12}(0.01)(0.10)^3 + \frac{1}{12}(0.09)(0.01)^3 = 0.841 \times 10^{-6} \text{ m}^4$$

The T-section is symmetrical about Cy, and for bending about this axis equal tensile and compressive stresses are induced in the extreme fibres of the flange; the greatest allowable moment is

$$M_y = \frac{(150 \times 10^6)(0.841 \times 10^{-6})}{0.05} = 2520 \text{ Nm}$$

9.6 More general case of pure bending

In the analysis of the preceding sections we have assumed either that the cross-section has two axes of symmetry, or that bending takes place about a principal axis. In the more general case we are interested in bending stress in the beam when moments are applied about any axis of the cross-section. Consider a long uniform beam, Figure 9.11, having any cross-section; the centroid of a cross-section is C, and Cz is the longitudinal axis of the beam; Cx and Cy are any two mutually perpendicular axes in the cross-section. The axes Cx, Cy and Cz are therefore centroidal axes of the beam.

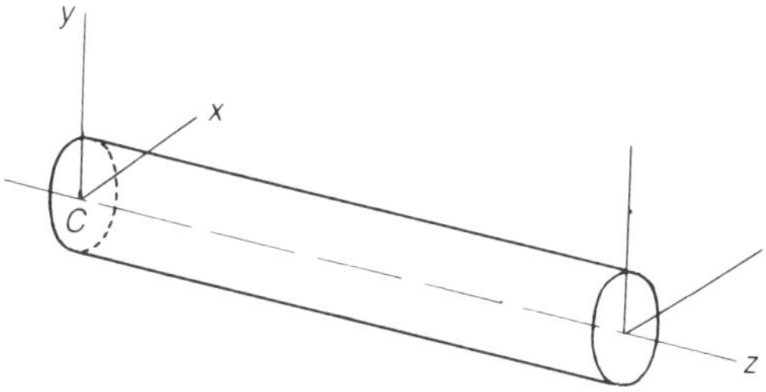

Figure 9.11 Co-ordinate system for a beam of any cross-sectional form.

We suppose first that the beam is bent in the yz-plane only, in such a way that the axis Cz takes up the form of a circular arc of radius R_x, Figure 9.12. Suppose further there is no longitudinal strain of Cx; this axis is then a neutral axis. The strain at a distance y from the neutral axis is

$$\varepsilon = \frac{y}{R_x}$$

If the material of a beam is elastic, the longitudinal stress in this fibre is

$$\sigma = \frac{Ey}{R_x}$$

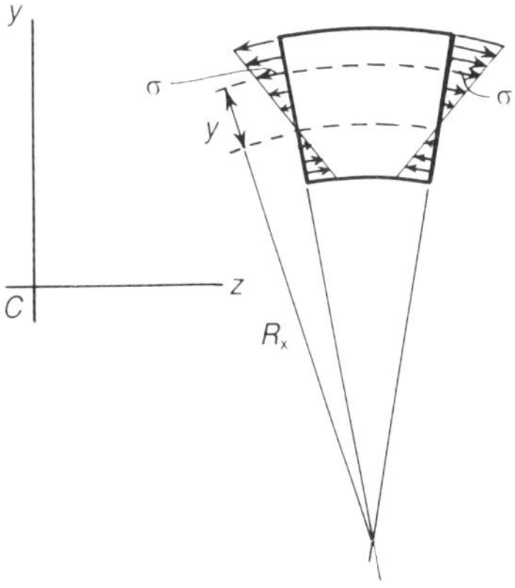

Figure 9.12 Bending in the yz-plane.

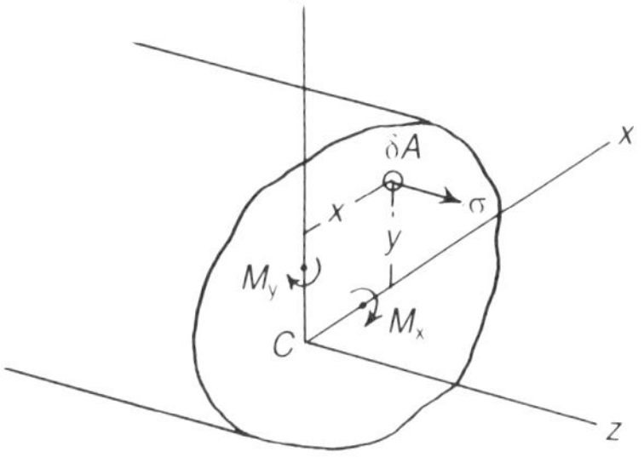

Figure 9.13 Bending moments about the axes C_x and C_y.

Suppose δA is a small element of area of the cross-section of the beam acted upon by the direct stress σ, Figures 9.12 and 9.13. Then the total thrust on any cross-section in the direction Cz is

where the integration is performed over the whole area A of the beam. But, as Cx is a centriodal axis, we have

$$\int_A y dA = 0$$

and no resultant longitudinal thrust is implied by the stresses σ. The moment about Cx due to the stresses σ is

$$M_x = \int_A \sigma y dA = \frac{E}{R_x} \int_A y^2 dA = \frac{EI_x}{R_x} \tag{9.9}$$

where I_x is the second moment of area of the cross-section about Cx. For the resultant moment about Cy we have

$$M_y = \int_A \sigma x dA = \frac{E}{R_x} \int_A xy dA = \frac{EI_{xy}}{R_x} \tag{9.10}$$

where I_{xy} is the product second moment of area of the cross-section about Cx and Cy. Unless I_{xy} is zero, in which case Cx and Cy are the principal axes, bending in the yz-plane implies not only a couple M_x about the Cx axis, but also a couple M_y about Cy.

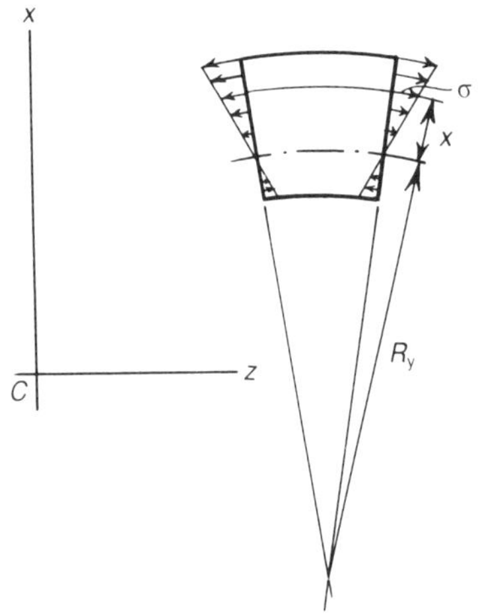

Figure 9.14 Bending in the xz-plane.

When the beam is bent in the xz-plane only, Figure 9.14, so that Cz again lies in the neutral surface, and takes up a curve of radius R_y, the longitudinal stress in a fibre a distance x from the neutral axis is

$$\sigma = \frac{Ex}{R_y}$$

The thrust implied by these stresses is again zero as

$$\int_A \sigma dA = \frac{E}{R_y}\int_A x dA = 0$$

because Cy is a centroidal axis of the cross-section. The bending moment about Cy due to stresses σ is

$$M_y = \int_A \sigma x dA = \frac{E}{R_y}\int_A x^2\, dA = \frac{EI_y}{R_y} \tag{9.11}$$

where I_y is the second moment of area of the cross-section about Cy. Furthermore,

$$M_x = \int_A \sigma y dA = \frac{E}{R_y}\int_A xy dA = \frac{EI_{xy}}{R_y} \tag{9.12}$$

where I_{xy} is again the product second moment of area.

If we now superimpose the two loading conditions, the total moments about the axes Cx and Cy, respectively, are

$$M_x = \frac{EI_x}{R_x} + \frac{EI_{xy}}{R_y} \tag{9.13}$$

$$M_y = \frac{EI_y}{R_y} + \frac{EI_{xy}}{R_x} \tag{9.14}$$

These equations may be rearranged in the forms

$$\frac{1}{R_x} = \frac{M_x I_y - M_y I_{xy}}{E\left(I_x I_y - I_{xy}^2\right)} \tag{9.15}$$

$$\frac{1}{R_y} = \frac{M_y I_x - M_x I_{xy}}{E\left(I_x I_y - I_{xy}^2\right)} \tag{9.16}$$

where $(1/R_x)$ and $(1/R_y)$ are the curvatures in the yz-and xz-planes caused by any set of moments

M_x and M_y. If C_x and C_y are the principal centroid axes then $I_{xy} = 0$, and equations (9.15) and (9.16) reduce to

$$\frac{1}{R_x} = \frac{M_x}{EI_x}, \qquad \frac{1}{R_y} = \frac{M_y}{EI_y} \tag{9.17}$$

In general we require a knowledge of three geometrical properties of the cross-section, namely I_x, I_y and I_{xy}. The resultant longitudinal stress at any point (x, y) of the cross-section of the beam is

$$\sigma = \frac{Ex}{R_y} + \frac{Ey}{R_x} = \frac{x\left(M_y I_x - M_x I_{xy}\right) + y\left(M_x I_y - M_y I_{xy}\right)}{\left(I_x I_y - I_{xy}^2\right)} \tag{9.18}$$

This stress is zero for points of the cross-section on the line

$$x\left(M_y I_x - M_x I_{xy}\right) + y\left(M_x I_y - M_y I_{xy}\right) = 0 \tag{9.19}$$

which is the equation of the unstressed fibre, or neutral axis, of the beam.

Problem 9.5 The I-section of Problem 9.2 is bent by couples of 2500 Nm about *Cx* and 500 Nm about *Cy*. Estimate the maximum bending stress in the cross-section, and find the equation of the neutral axis of the beam.

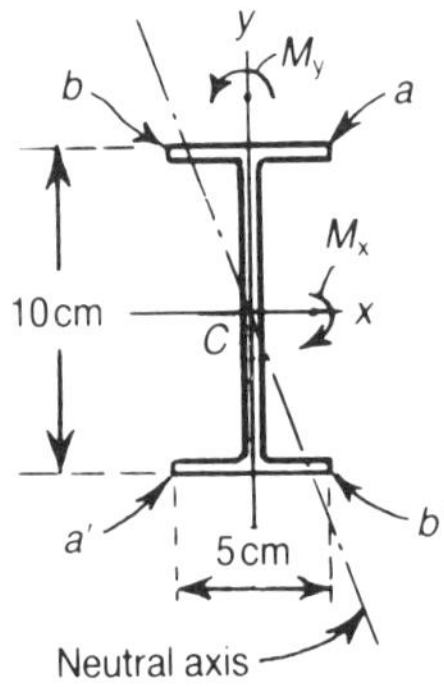

Solution

From Problem 9.2

$$I_x = 1.641 \times 10^{-6} \text{ m}^4, \quad I_y = 0.130 \times 10^{-6} \text{ m}^4$$

For bending about *Cx* the bending stresses in the extreme fibres of the flanges are

$$\sigma = \frac{M_x y}{I_x} = \frac{(2500)(0.05)}{1.641 \times 10^{-6}} = 76.1 \text{ MN/m}^2$$

For bending about *Cy* the bending stresses at the extreme ends of the flanges are

$$\sigma = \frac{M_y x}{I_y} = \frac{(500)(0.025)}{0.130 \times 10^{-6}} = 96.1 \text{ MN/m}^2$$

On superposing the stresses due to the separate moments, the stress at the corner *a* is tensile, and of magnitude

$$\sigma_a = (76.1 + 96.1) = 172.2 \text{ MN/m}^2$$

The total stress at the corner a' is also 172.2 MN/m^2, but compressive. The total stress at the corner *b* is compressive, and of magnitude

$$\sigma_b = (96.1 - 76.1) = 20.2 \text{ MN/m}^2$$

The total stress at the corner b' is also 20.0 MN/m^2, but tensile. The equation of the neutral axis is given by

$$xM_y I_x + yM_x I_y = 0$$

Then

$$\frac{y}{x} = -\frac{M_y I_x}{M_x I_y} = -\frac{(500)(1.641 \times 10^{-6})}{(2500)(0.130 \times 10^{-6})} = -2.53$$

The greatest bending stresses occur at points most remote from the neutral axis; these are the points *a* and a'. the greatest bending stresses are therefore ±172.2 MN/m^2.

9.7 Elastic section modulus

For bending of a section about a principal axis Cx, the longitudinal bending stress at a fibre a distance y from Cx, due to a moment M_x, is from equation (9.18) (in which we put $I_{xy} = 0$ and $M_y = 0$),

$$\sigma = \frac{M_x y}{I_x}$$

where I_x is the second moment of area about Cx. The greatest bending stress occurs at the fibre most remote from Cx. If the distance to the extreme fibre is y_{max}, the maximum bending stress is

$$\sigma_{max} = \frac{M_x y_{max}}{I_x}$$

The allowable moment for a given value of σ_{max} is therefore

$$M_x = \frac{I_x \sigma_{max}}{y_{max}} \tag{9.20}$$

The geometrical quantity (I_x/y_{max}) is the *elastic section modulus*, and is denoted by Z_e.

Then

$$M_x = Z_e \, \sigma_{max} \tag{9.21}$$

The allowable bending moment is therefore the product of a geometrical quantity, Z_e, and the maximum allowable stress, σ_{max}. The quantity $Z_e \, \sigma_{max}$ is frequently called the *elastic moment of resistance*.

Problem 9.6 A steel I-beam is to be designed to carry a bending moment of 10^5 Nm, and the maximum bending stress is not to exceed 150 MN/m^2. Estimate the required elastic section modulus, and find a suitable beam.

Solution

The required elastic section modulus is

$$Z_e = \frac{M}{\sigma} = \frac{10^5}{150 \times 10^6} = 0.667 \times 10^{-3} \ m^3$$

The elastic section modulus of a 22.8 cm by 17.8 cm standard steel I-beam about its axis of greatest bending stiffness is 0.759×10^{-3} m^3, which is a suitable beam.

9.8 Longitudinal stresses while shearing forces are present

The analysis of the proceeding paragraphs deals with longitudinal stresses in beams under uniform bending moment. No shearing forces are present at cross-sections of the beam in this case.

When a beam carries lateral forces, bending moments may vary along the length of the beam. Under these conditions we may assume with sufficient accuracy in most engineering problems that the longitudinal stresses at any section are dependant only on the bending moment at that section, and are unaffected by the shearing force at that section.

Where a shearing force is present at the section of a beam, an elemental length of the beam undergoes a slight shearing distortion; these shearing distortions make a negligible contribution to the total deflection of the beam in most engineering problems.

Problem 9.7 A 4 m length of the I-beam of Problem 9.2 is simply-supported at each end. What maximum central lateral load may be applied if the bending stresses are not to exceed 150 MN/m^2?

<u>*Solution*</u>

Suppose W is the central load. If this is applied in the plane of the web, then bending takes place about Cx. The maximum bending moment is

$$M_x = \frac{1}{2}W(2) = W \text{ Nm}$$

From Problem 9.2,

$$I_x = 1.641 \times 10^{-6} \text{ m}^4$$

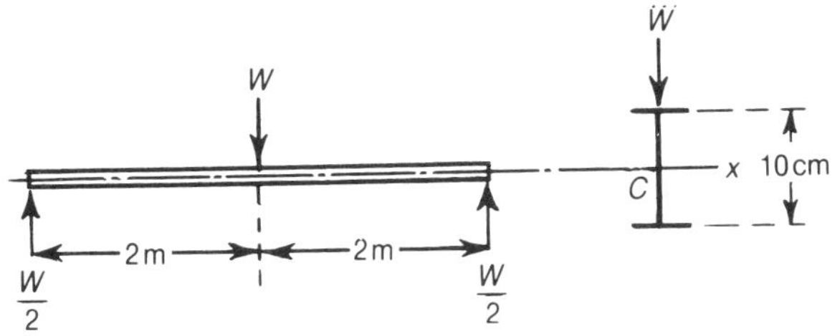

Then, the greatest bending stress is

$$\sigma = \frac{M_x \, y_{max}}{I_x} = \frac{(W)\,(0.05)}{1.641 \times 10^{-6}}$$

If this is equal to 150 MN/m^2, then

$$W = \frac{(150 \times 10^6)\,(1.641 \times 10^{-6})}{0.05} = 4920 \text{ N}$$

Problem 9.8 If the bending stresses are again limited to 150 MN/m^2, what total uniformly-distributed load may be applied to the beam of Problem 9.7?

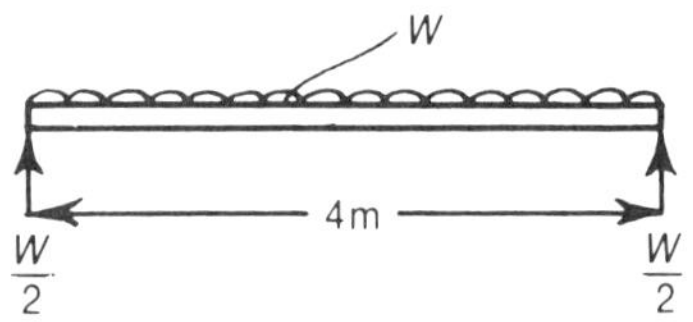

Solution

The maximum bending moment occurs at mid-span, and has the value

$$M_x = \frac{WL}{8} = \frac{1}{2} W \text{ Nm}$$

Then

$$\frac{1}{2} W = \frac{(150 \times 10^6)\,(1.641 \times 10^{-6})}{0.05} = 4920 \text{ N}$$

and

$$W = 9840 \text{ N}$$

9.9 Calculation of the principal second moments of area

In problems of bending involving beams of unsymmetrical cross-section we have frequently to find the principal axes of the cross-section.

Suppose Cx and Cy are any two centroidal axes of the cross-section of the beam, Figure 9.15.

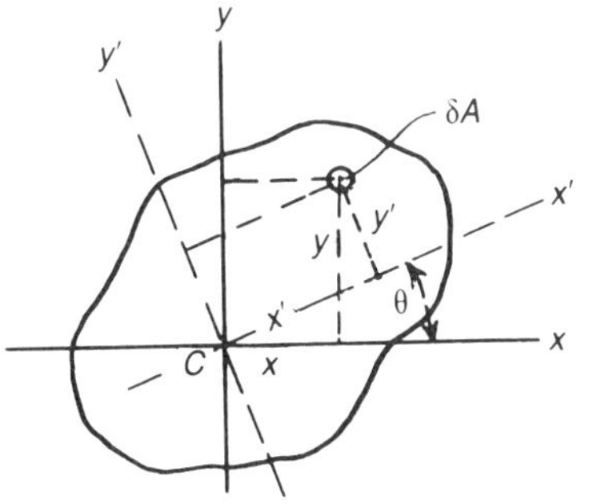

Figure 9.15 Derivation of the principal axes of a section.

If δA is an elemental area of the cross-section at the point (x, y), then the property of the axes Cx and Cy is that

$$\int_A x\,dA = \int_A y\,dA = 0$$

The second moments of area about the axes Cx and Cy, respectively, are

$$I_x = \int_A y^2 dA, \quad I_y = \int_A x^2 dA \tag{9.22}$$

The product second moment of area is

$$I_{xy} = \int_A xy\,dA \tag{9.23}$$

Now consider two mutually perpendicular axes Cx' and Cy', which are the principal axes of bending, inclined at an angle θ to the axes Cx and Cy. A point having co-ordinates (x, y) in the xy-system, now has co-ordinates (x', y') in the $x'y'$-system. Further, we have

$$x' = x\cos\theta + y\sin\theta$$

$$y' = y\cos\theta - x\sin\theta$$

The second moment of area of the cross-section about Cx' is

$$I_{x'} = \int_A y'1^2 dA$$

which becomes

$$I_{x'} = \int_A (y \cos\theta - x \sin\theta)^2 \, dA$$

This may be written

$$I_{x'} = \cos^2\theta \int_A y^2 dA - 2 \cos\theta \sin\theta \int_A xy dA + \sin^2\theta \int_A x^2 dA$$

But

$$\int_A y^2 \, dA = I_x, \int_A x^2 \, dA = I_y, \text{ and } \int_A xy dA = I_{xy}$$

Then

$$I_{x'} = I_x \cos^2\theta - 2I_{xy} \cos\theta \sin\theta + I_y \sin^2\theta \tag{9.24}$$

Similarly, the second moment of area about Cy' is

$$I_{y'} = \int_A x'^2 dA = \int_A (x \cos\theta + y \sin\theta)^2 \, dA$$

Then

$$I_{y'} = I_y \cos^2\theta + 2I_{xy} \cos\theta \sin\theta + I_x \sin^2\theta \tag{9.25}$$

Finally, the product second moment of area about Cx' and Cy' is

$$I_{x'y'} = \int_A x'y' dA \int_A (x \cos\theta + y \sin\theta) (y \cos\theta - x \sin\theta) dA$$

Then

$$I_{x'y'} = I_x \sin\theta \cos\theta + I_{xy}\left(\cos^2\theta - \sin^2\theta\right) - I_y \cos\theta \sin\theta \tag{9.26}$$

We note from equations (9.24) and (9.25), that

$$I_{x'} + I_{y'} = I_x + I_y \tag{9.27}$$

that is, the sum of the second moments of area about any perpendicular axes is independent of θ. The sum is in fact the polar second moment of area, or the second moment of area about an axis through C, perpendicular to the xy-plane.

We may write equation (9.26) in the form

$$I_{x'y'} = \frac{1}{2}\left(I_x - I_y\right) \sin 2\theta + I_{xy} \cos 2\theta \tag{9.28}$$

The principal axes Cx' and Cy' are defined as those for which $I_{x'y'} = 0$; then for the principal axes

$$\frac{1}{2}\left(I_x - I_y\right) \sin 2\theta + I_{xy} \cos 2\theta = 0$$

or

$$\tan 2\theta = \frac{2I_{xy}}{I_y - I_x} \tag{9.29}$$

This relationship gives two values of θ differing by 90°. On making use of equation (9.27), we may write equations (9.24) and (9.25) in the forms

$$I_{x'} = \frac{1}{2}\left(I_x + I_y\right) + \frac{1}{2}\left(I_x - I_y\right)\cos 2\theta + I_{xy} \sin 2\theta$$

$$I_{y'} = \frac{1}{2}\left(I_x + I_y\right) - \frac{1}{2}\left(I_x - I_y\right)\cos 2\theta + I_{xy} \sin 2\theta \tag{9.30}$$

Now

$$I_{x'} I_{y'} = \left[\frac{1}{2}\left(I_x + I_y\right) + \frac{1}{2}\left(I_x - I_y\right) \cos 2\theta - I_{xy} \sin 2\theta\right]$$

$$\times \left[\frac{1}{2}\left(I_x + I_y\right) - \frac{1}{2}\left(I_x - I_y\right) \cos 2\theta + I_{xy} \sin 2\theta\right]$$

$$= \frac{1}{4}\left(I_x + I_y\right)^2 - \frac{1}{4}\left(I_x + I_y\right)\left(I_x - I_y\right) \cos 2\theta$$

$$+ \frac{1}{2}\left(I_x + I_y\right) . I_{xy} \sin 2\theta$$

$$+ \frac{1}{4}\left(I_x + I_y\right) \cos 2\theta . \left(I_x - I_y\right)$$

$$-\frac{1}{4}\left(I_x - I_y\right)^2 \cos^2 2\theta + \frac{1}{2}\left(I_x - I_y\right) \cos 2\theta \, . \, I_{xy} \, . \, \sin 2\theta$$

$$-\frac{1}{2} I_{xy} \left(I_x + I_y\right) \sin 2\theta + \frac{1}{2} I_{xy} \left(I_x - I_y\right) \sin 2\theta \cos 2\theta - I_{xy}^2 \sin^2 2\theta$$

$$= \frac{1}{4}\left(I_x + I_y\right)^2 - \frac{1}{4}\left(I_x - I_y\right)^2 \cos^2 2\theta - I_{xy}^2 \sin^2 2\theta$$

$$+ I_{xy} \left(I_x - I_y\right) \sin 2\theta \cos 2\theta$$

or

$$I_{x'} I_{y'} = \left[\frac{1}{2}\left(I_x + I_y\right)\right]^2 - \left\{\frac{1}{2}\left[\left(I_x - I_y\right)\cos 2\theta - 2 I_{xy} \sin 2\theta\right]\right\}^2 \tag{9.31}$$

From equation (9.29), the mathematical triangle of the figure below is obtained:

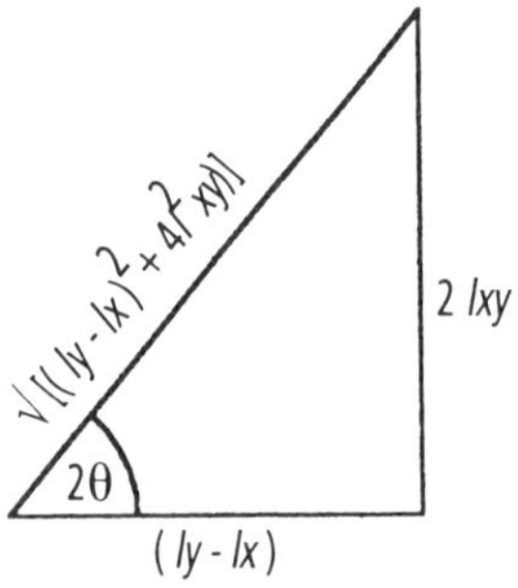

From the mathematical triangle

$$\cos 2\theta = \frac{\left(I_y - I_x\right)}{\sqrt{\left(I_y - I_x\right)^2 + 4 I_{xy}^2}}$$

and $$\sin 2\theta = \frac{2 \, I_{xy}}{\sqrt{\left(I_y - I_x\right)^2 + 4 \, I_{xy}^2}}$$

$$\therefore I_{x'} I_{y'} = \left[\frac{1}{2}(I_x + I_y)\right]^2 - \left\{\frac{\frac{1}{2}(I_x - I_y) \cdot (I_y - I_x)}{\sqrt{(I_y - I_x)^2 + 4 I_{xy}^2}} - \frac{2 I_{xy} \cdot 2 I_{xy}}{\sqrt{(I_y - I_x)^2 + 4 I_{xy}^2}}\right\}^2$$

$$= \left[\frac{1}{2}(I_x + I_y)\right]^2 - \left\{\frac{1}{2} \cdot \frac{-[I_y - I_x]^2 - 4 I_{xy}^2}{\sqrt{(I_y - I_x)^2 + 4 I_{xy}^2}}\right\}^2$$

$$= \left[\frac{1}{2}(I_x + I_y)\right]^2 - \left\{\frac{1}{2}\sqrt{(I_y - I_x)^2 + 4 I_{xy}^2}\right\}^2$$

$$= \frac{1}{4}\left(I_x^2 + I_y^2 + 2 I_x I_y\right) - \frac{1}{4}\left[(I_y - I_x)^2 + 4 I_{xy}^2\right]$$

$$= \frac{1}{4}\left(I_x^2 + I_y^2 + 2 I_x I_y\right) - \frac{1}{4}\left(I_y^2 + I_x^2 - 2 I_x I_y + 4 I_{xy}^2\right)$$

or $$I_{x'} I_{y'} = I_x I_y - I_{xy}^2 \tag{9.32}$$

Substituting equation (9.27) into equation (9.32) we get

$$(I_x + I_y - I_{y'}) I_{y'} = I_x I_y - I_{xy}^2$$

or $$I_{y'}^2 - (I_x + I_y) I_{y'} + (I_x I_y - I_{xy}^2) \tag{9.33a}$$

Similarly,

$$I_{x'}^2 - (I_x + I_y) I_{x'} + (I_x I_y - I_{xy}^2), \tag{9.33b}$$

which are both quadratic equations.

In general, equations (9.33a) and (9.33b) can be written as the following quadratic equation, where I = a *principal* second moment of area

$$I^2 - (I_x + I_y) I + (I_x I_y - I_{xy}^2) = 0 \tag{9.34}$$

Then

$$I = \frac{1}{2}\left(I_x + I_y\right) \pm \sqrt{\frac{1}{4}\left(I_x + I_y\right)^2 - \left(I_x I_y - I_{xy}^2\right)} \tag{9.35}$$

which may be written

$$I = \frac{1}{2}\left(I_x - I_y\right) \pm \sqrt{\frac{1}{4}\left(I_x - I_y\right)^2 + I_{xy}^2} \tag{9.36}$$

Equations (9.30) and (9.26) may be written in the forms

$$I_{x'} - \frac{1}{2}\left(I_x + I_y\right) = \frac{1}{2}\left(I_x - I_y\right)\cos 2\theta - I_{xy}\sin 2\theta$$

$$I_{x'y'} = \frac{1}{2}\left(I_x - I_y\right)\sin 2\theta + I_{xy}\cos 2\theta \tag{9.37}$$

Square each equation, and then add; we have

$$\left[I_{x'} - \frac{1}{2}\left(I_x - I_y\right)\right]^2 + \left[I_{x'y'}\right]^2 = \left[\frac{1}{2}\left(I_x - I_y\right)\right]^2 + \left[I_{xy}\right]^2 \tag{9.38}$$

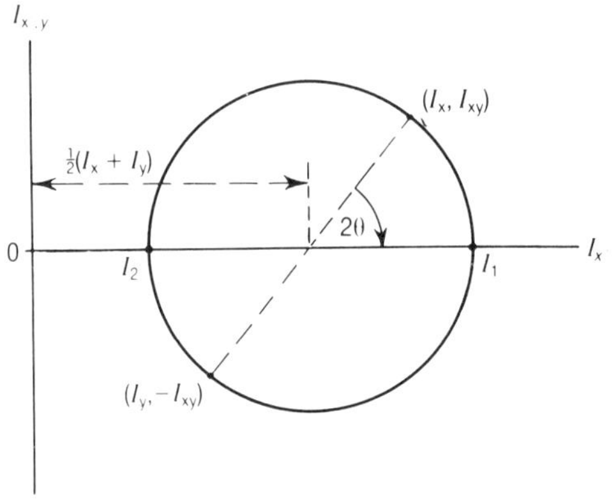

Figure 9.16 Graphical representation of the second moments of area.

Then I_x, $I_{xy'}$ lie on a circle of radius

$$\left\{\left[\frac{1}{2}\left(I_x - I_y\right)\right]^2 + \left[I_{xy}^2\right]\right\}^{\frac{1}{2}} \tag{9.39}$$

and centre

$$\left[\frac{1}{2}\left(I_x + I_y\right),\ 0\right] \tag{9.40}$$

in the $I_{x'}$, $I_{x'y'}$ diagram.

Suppose $OI_{x'}$ and $OI_{x'y'}$ are mutually perpendicular axes; then equation (9.38) has the graphical representation shown in Figure 9.16. To find the principal second moments of area, locate the points (I_x, I_{xy}) and $(I_y, -I_{xy})$ in the $(I_{x'}, I_{x'y'})$ plane. With the line joining these points as a diameter construct a circle. The principal second moments of area, I_1 and I_2, are given by the points where the circle cuts the axis $OI_{x'}$. Figure 9.16 might be referred to as the *circle of second moments of area*.

Problem 9.9 An unequal angle section of uniform thickness 0.5 cm has legs of lengths 6 cm and 4 cm. Estimate the positions of the principal axes, and the principal second moments of area.

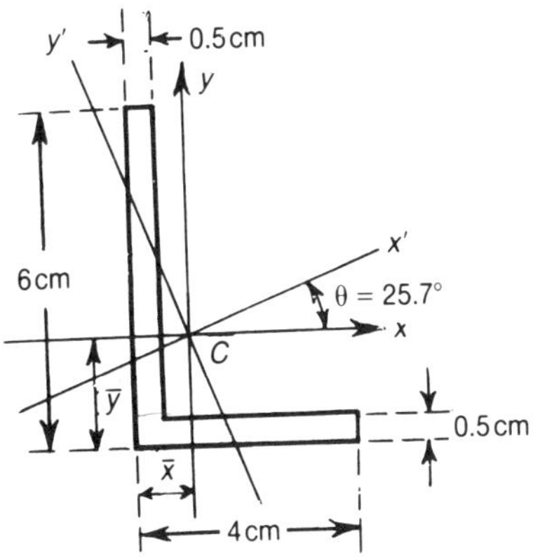

Solution

Firstly, find the position of the centroid of the cross-section. Total area is

$$A = (0.06)\,(0.005) + (0.035)\,(0.005)$$

$$= 0.475 \times 10^{-3}\ \text{m}^2$$

Now

$$A\bar{x} = (0.055)\,(0.005)\,(0.0025) + (0.04)\,(0.005)\,(0.02)$$

$$= 4.69 \times 10^{-6}\ \text{m}^3$$

Then

$$\bar{x} = \frac{4.69 \times 10^{-6}}{0.475 \times 10^{-3}} = 9.86 \times 10^{-3} \text{ m}$$

Again

$$A\bar{y} = (0.035)\ (0.005)\ (0.0025) + (0.06)\ (0.005)\ (0.03) = 9.44 \times 10^{-6} \text{ m}^3$$

Then

$$\bar{y} = \frac{9.44 \times 10^{-6}}{0.475 \times 10^{-3}} = 19.85 \times 10^{-3} \text{ m}$$

Now

$$I_x = \frac{1}{3}(0.005)\ (0.06)^3 + \frac{1}{3}(0.035)\ (0.005)^3 - (0.475 \times 10^{-3})\ (0.01985)^2$$

$$= 0.174 \times 10^{-6} \text{ m}^4$$

and

$$I_y = \frac{1}{3}(0.005)(0.04)^3 + \frac{1}{3}(0.055)(0.005)^3 - (0.475 \times 10^{-3})(0.00986)^2$$

$$= 0.063 \times 10^{-6} \text{ m}^4$$

With the axes Cx and Cy having the positive directions shown,

$$I_{xy} = \int \int xy\, dx\, dy$$

$$= \int_{-\bar{x}}^{0.04 - \bar{x}} x dx \int_{-\bar{y}}^{0.005 - \bar{y}} y dy + \int_{\bar{x}}^{0.005 - \bar{x}} x dx \int_{-\bar{y} + 0.005}^{0.06 - \bar{y}} y dy$$

$$= \frac{1}{4}\{\left[(0.04 - \bar{x})^2 - (-\bar{x})^2\right]\left[(0.005 - \bar{y})^2 - (-\bar{y})^2\right]$$

$$+ \left[(0.005 - \bar{x})^2 - (-\bar{x})^2\right]\left[(0.06 - \bar{y})^2 - (-\bar{y} + 0.005)^2\right]\}$$

$$= -0.06 \times 10^{-6} \text{ m}^4$$

From equation (9.29),

$$\tan 2\theta = \frac{2 \times (-0.06)}{0.063 - 0.174} = 1.08$$

Then

$$2\theta = 47.2°$$

and

$$\theta = 23.6°$$

From equations (9.36) the principal second moments of area are

$$\frac{1}{2}\left(I_x + I_y\right) \pm \left[\left\{\frac{1}{2}\left(I_x - I_y\right)\right\}^2 + I_{xy}^2\right]^{\frac{1}{2}} = (0.1185 \pm 0.0988)\ 10^{-6}$$

$$= 0.2173 \text{ or } 0.0197 \times 10^{-6}\ \text{m}^4$$

9.10 Elastic strain energy of bending

As couples are applied to a beam, strain energy is stored in the fibres. Consider an elemental length δz of a beam, which is bent about a principal axis Cx by a moment M_x, Figure 9.17. During bending, the moments M_x at each end of the element are displaced with respect to each other an angular amount

$$\theta = \frac{\delta z}{R_x} \tag{9.41}$$

where R_x is the radius of curvature in the yz-plane. But from equation (9.6)

$$M_x = \frac{EI_x}{R_x}$$

and thus

$$\theta = \frac{M_x \delta z}{EI_x} \qquad (9.42)$$

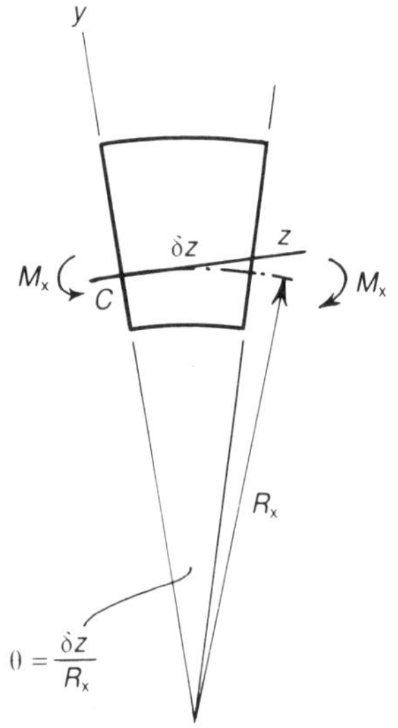

Figure 9.17 Bent form of an elemental length of beam.

As there is a linear relation between θ and M_x, the total work done by the moments M_x during bending of the element is

$$\frac{1}{2} M_x \theta = \frac{M_x^2 \delta z}{2EI_x} \qquad (9.43)$$

which is equal to the strain energy of bending of the element. For a uniform beam of length L under a moment M_x, constant throughout its length, the bending strain energy is then

$$U = \frac{M_x^2 L}{2EI_x} \qquad (9.44)$$

When the bending moment varies along the length, the total bending strain energy is

$$U = \int_L \frac{M_x^2 \, dz}{2EI_x} \qquad (9.45)$$

where the integration is carried out over the whole length L of the beam.

9.11 Change of cross-section in pure bending

In Section 9.1 we pointed out the change which takes place in the shape of the cross-section when a beam is bent. This change involves infinitesimal lateral strains in the beam. The upper and lower edges of a cross-section which was originally rectangular, are strained into concentric circular arcs with their centre on the opposite side of the beam to the axis of bending. The upper and lower surfaces of the beam then have *anticlastic curvature*, the general nature of the strain being as shown in Figure 9.18. The anticlastic curvature effect can be readily observed by bending a flat piece of india-rubber. If the beam is bent to a mean radius R, we find that cross-sections are bent to a mean radius (R/v).

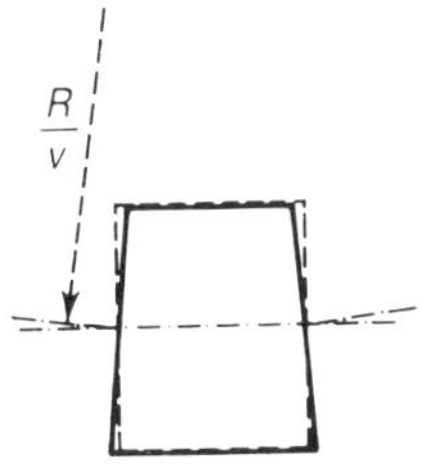

Figure 9.18 Anticlastic curvature in the cross-section of a bent rectangular beam.

Problem 9.10 What load can a beam 4 m long carry at its centre, if the cross-section is a hollow square 30 cm by 30 cm outside and 4 cm thick, the permissible longitudinal stress being 75 MN/m^2?

Solution

We must find the second moment of area of cross-section about its neutral axis. The inside is a square 22 cm by 22 cm. Then

$$\frac{1}{12}\left(0.3^4 - 0.22^4\right) = 0.47 \times 10^{-3} \text{ m}^4$$

The length of the beam is 4 m; therefore if W N be a concentrated load at the middle, the maximum bending moment is

$$M_x = \frac{WL}{4} = W \text{ Nm}$$

Hence the maximum stress is

$$\sigma = \frac{M_x y}{I_x} = \frac{W(0.15)}{0.47 \times 10^{-3}}$$

If $\sigma = 75$ MN/m^2 we must therefore have

$$W = \frac{(75 \times 10^6)(0.47 \times 10^{-3})}{0.15} = 235 \text{ kN}$$

Problem 9.11 Estimate the elastic section modulus and the maximum longitudinal stress in a built-up I-girder, with equal flanges carrying a load of 50 kN per metre run, with a clear span of 20 m. The web is of thickness 1.25 cm and the depth between flanges 2 m. Each flange consists of four 1 cm plates 65 cm wide, and is attached to the web by angle iron sections 10 cm by 10 cm by 1.25 cm thick. *(Cambridge)*

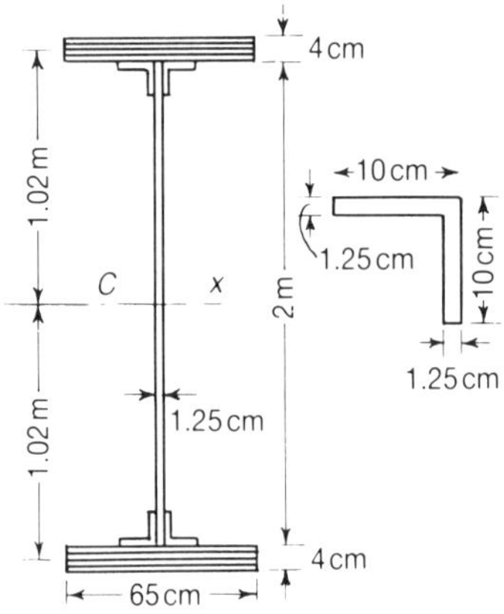

Solution

The second moment of area of each flange about Cx is

$$(0.04)(0.65)(1.02)^2 = 0.0270 \text{ m}^4$$

The second moment of area of the web about Cx is

$$\frac{1}{12}(0.0125)(2)^3 = 0.0083 \text{ m}^4$$

The horizontal part of each angle section has an area 0.00125 m^2, and its centroid is 0.944 m from the neutral axis. Therefore the corresponding second moment of area is approximately

$$(0.00125)(0.994)^2 = 0.0012 \text{ m}^4$$

The area of the vertical part of each angle section is 0.001093 m^2, and its centroid is 0.944 m from the neutral axis. Therefore the corresponding second moment of area is approximately

$$(0.001093)(0.944)^2 = 0.00097 \text{ m}^4$$

The second moment of area of the whole section of the angle section about Cx is then

$$0.0012 + 0.00097 = 0.0022 \text{ m}^4$$

The second moment of area of the whole cross-section of the beam is then

$$I_x = 2(0.0270) + (0.0083) + 4(0.0022)$$

$$= 0.0711 \text{ m}^4$$

The elastic section modulus is therefore

$$Z_e = \frac{0.0711}{1.04} = 0.0684 \text{ m}^3$$

The bending moment at the mid-span is

$$M_x = \frac{wL^2}{8} = \frac{(50)(20)^2}{8} = 2.50 \text{ MN.m}$$

The greatest longitudinal stress is then

$$\sigma = \frac{M_x}{Z_e} = \frac{2.50 \times 10^6}{0.0684} = 36.6 \text{ MN/m}^2$$

Further problems *(answers on page 692)*

9.12 A beam of I-section is 25 cm deep and has equal flanges 10 cm broad. The web is 0.75 cm thick and the flanges 1.25 cm thick. If the beam may be stressed in bending to 120 MN/m^2, what bending moment will it carry? *(Cambridge)*

9.13 The front-axle beam of a motor vehicle carries the loads shown. The axle is of I-section: flanges 7.5 cm by 2.5 cm, web 5 cm by 2.5 cm. Calculate the tensile stress at the bottom of the axle beam. *(Cambridge)*

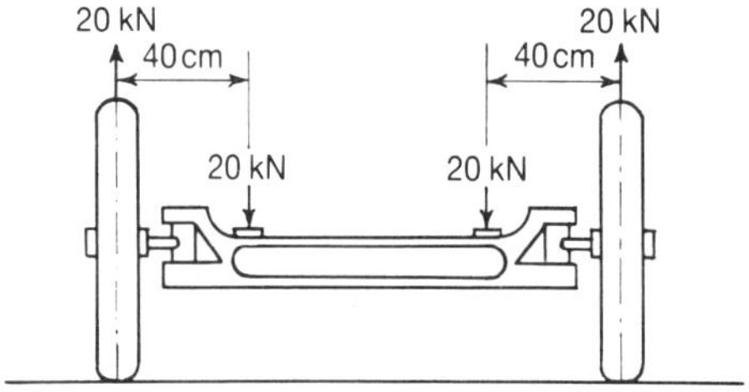

9.14 A water trough 8 m long, is simply-supported at the ends. It is supported at its extremities and is filled with water. If the metal has a density 7840 kg/m^3, and the water a density 1000 kg/m^3, calculate the greatest longitudinal stress for the middle cross-section of the trough. *(Cambridge)*

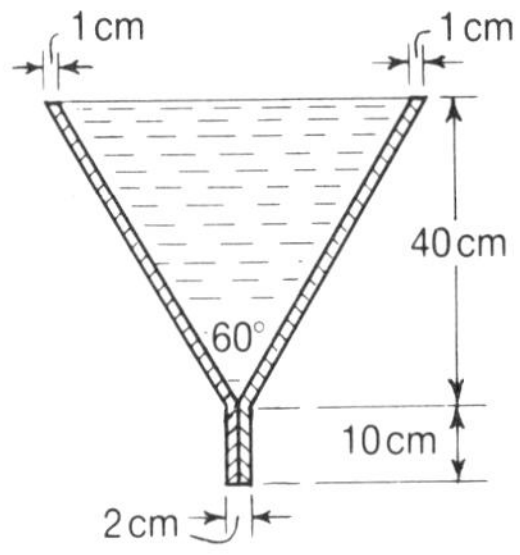

9.15 A built-up steel I-girder is 2 m deep over the flanges, each of which consists of four 1 cm plates, 1 m wide, riveted together. The web is 1 cm thick and is attached to the flanges by four 9 cm by 9 cm by 1 cm angle sections. The girder has a clear run of 30 m between the supports and carries a superimposed load of 60 kN per metre. Find the maximum longitudinal stress. *(Cambridge)*

9.16 A beam rests on supports 3 m apart carries a load of 10 kN uniformly distributed. The beam is rectangular in section 7.5 cm deep. How wide should it be if the skin-stress must not exceed 60 MN/m^2? *(RNEC)*

10 Shearing stresses in beams

10.1 Introduction

We referred earlier to the existence of longitudinal direct stresses in a cantilever with a lateral load at the free end; on a closer study we found that these stresses are distributed linearly over the cross-section of a beam carrying a uniform bending moment. In general we are dealing with bending problems in which there are shearing forces present at any cross-section, as well as bending moments. In practice we find that the longitudinal direct stresses in the beam are almost unaffected by the shearing force at any section, and are governed largely by the magnitude of the bending moment at that section. Consider again the bending of a cantilever with a concentrated lateral load F, at the free end, Figure 10.1; Suppose the beam is of rectangular cross-section. If we cut the beam at any transverse cross-section, we must apply bending moments M and shearing forces F at the section to maintain equilibrium. The bending moment M is distributed over the cross-section in the form of longitudinal direct stresses, as already discussed.

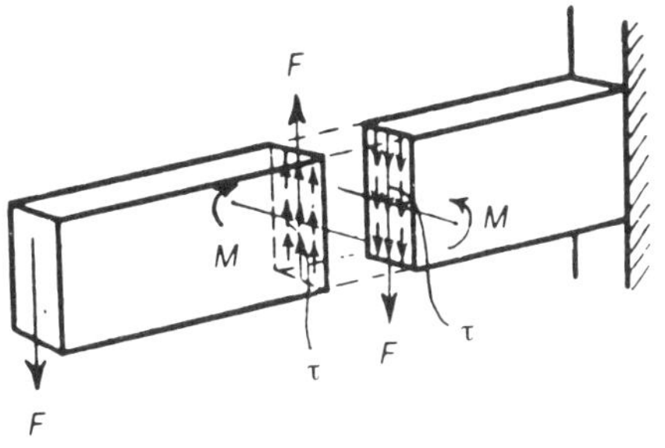

Figure 10.1 Shearing actions in a cantilever carrying an end load.

The shearing force F is distributed in the form of shearing stresses τ, acting tangentially to the cross-section of the beam; the form of the distribution of τ is dependent on the shape of the cross-section of the beam, and on the direction of application of the shearing force F. An interesting feature of these shearing stresses is that, as they give rise to complementary shearing stresses, we find that shearing stresses are also set up in longitudinal planes parallel to the axis of the beam.

10.2 Shearing stresses in a beam of narrow rectangular cross-section

We consider first the simple problem of a cantilever of *narrow* rectangular cross-section, carrying a concentrated lateral load F at the free end, Figure 10.2; h is the depth of the cross-section, and t is the thickness, Figure 10.3; the depth is assumed to be large compared with the thickness. The load is applied in a direction parallel to the longer side h.

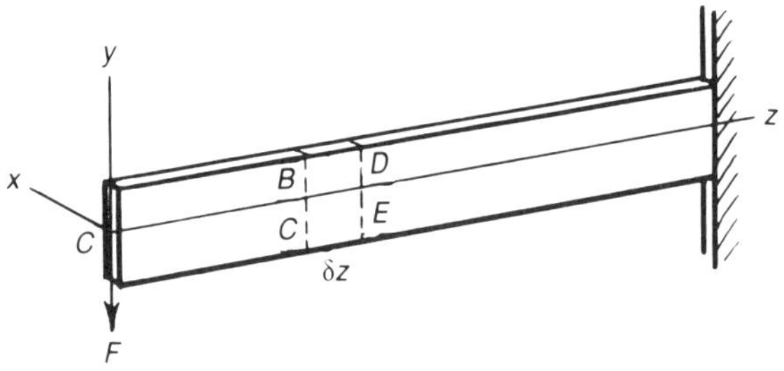

Figure 10.2 Shearing stresses in a cantilever of narrow rectangular cross-section under end load.

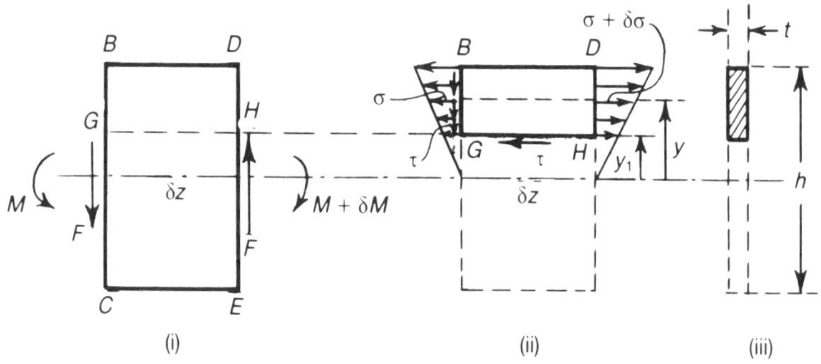

Figure 10.3 Shearing actions on an elemental length of a beam of narrow rectangular cross-section.

Consider an elemental length δz of the beam at a distance z from the loaded end. On the face BC of the element the hogging bending moment is

$$M = Fz$$

We suppose the longitudinal stress σ at a distance y from the centroidal axis Cx is the same as that for uniform bending of the element. Then

$$\sigma = \frac{My}{I_x} = \frac{Fyz}{I_x}$$

Where I_x is the second moment of area about the centroidal axis of bending, Cx, which is also a neutral axis . On the face DE of the element the bending moment has increased to

$$M + \delta M = F(z + \delta z)$$

The longitudinal bending stress at a distance y from the neutral axis has increased correspondingly to

$$\sigma + \delta\sigma = \frac{F(z + \delta z)y}{I_x}$$

Now consider a depth of the beam contained between the upper extreme fibre *BD*, given by $y = \frac{1}{2}h$, and the fibre *GH*, given by $y = y_1$, Figure 10.3(ii). The total longitudinal force on the face *BG* due to bending stresses σ is

$$\int_{y_1}^{h/2} \sigma t dy = \frac{Fzt}{I_x}\int_{y_1}^{\frac{1}{2}h} y dy + \frac{Fzt}{2I_x}\left[\frac{h^2}{4} - y_1^2\right]$$

By a similar argument we have that the total force on the face *DH* due to bending stresses σ + δσ is

$$\frac{Ft}{2I_x}\left(\frac{h^2}{4} - y_1^2\right)(z + \delta z)$$

These longitudinal force, which act in opposite directions, are not quite in balance; they differ by a small amount

$$\frac{Ft}{2I_x}\left(\frac{h^2}{4} - y_1^2\right)\delta z$$

Now the upper surface *BD* is completely free of shearing stress, and this out-of-balance force can only be equilibrated by a shearing force on the face *GH*. We suppose this shearing force is distributed uniformly over the face *GH*; the shearing stress on this face is then

$$\tau = \frac{Ft}{2I_x}\left(\frac{h^2}{4} - y_1^2\right)\delta z \,/\, t\delta z$$

$$= \frac{F}{2I_x}\left(\frac{h^2}{4} - y_1^2\right) \qquad (10.1)$$

This shearing stress acts on a plane parallel to the neutral surface of the beam; it gives rise therefore to a complementary shearing stress τ at a point of the cross-section a distance y_1 from the neutral axis, and acting tangentially to the cross-section. Our analysis gives then the variation of shearing stress over the depth of the cross-section. For this simple type of cross-section

$$I_x = \frac{1}{12}h^3 t$$

and so

$$\tau = \frac{6F}{h^3 t}\left(\frac{h^2}{4} - y_1^2\right) = \frac{6F}{ht}\left[\frac{1}{4} - \left(\frac{y_1}{h}\right)^2\right] \tag{10.2}$$

We note firstly that τ is independent of z; this is so because the resultant shearing force is the same for all cross-sections, and is equal to F. The resultant shearing force implied by the variation of τ is

$$\int_{-h/2}^{+h/2} \tau t dy_1 = \frac{6F}{h}\int_{-h/2}^{+h/2}\left[\frac{1}{4} - \left(\frac{y_1}{h}\right)^2\right] dy_1 = F$$

The shearing stresses τ are sufficient then to balance the force F applied to every cross-section of the beam.

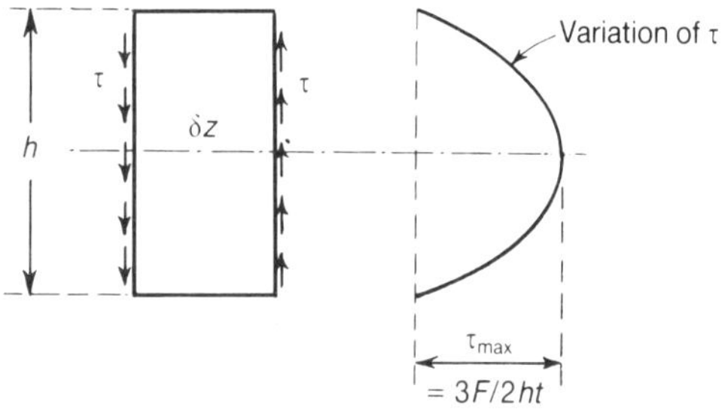

Figure 10.4 Variation of shearing stresses over the depth of a beam of rectangular cross-section.

The variation of τ over the cross-section of the beam is parabolic, Figure 10.4; τ attains a maximum value on the neutral axis of the beam, where $y_1 = 0$, and

$$\tau_{max} = \frac{3F}{2ht} \tag{10.3}$$

The shearing stresses must necessarily be zero at the extreme fibres as there can be no complementary shearing stresses in the longitudinal direction on the upper and lower surfaces of the beam.

In the case of a cantilever with a single concentrated load F at the free end the shearing force is the same for all cross-sections, and the distribution of shearing stresses is also the same for all cross-sections. In a more general case the shearing force is variable from one cross-section to another: in this case the value of F to be used is the shearing force at the section being considered.

10.3 Beam of any cross-section having one axis of symmetry

We are concerned generally with more complex cross-sectional forms than narrow rectangles. Consider a beam having a uniform cross-section which is symmetrical about Cy, Figure 10.5. Suppose, as before, that the beam is a cantilever carrying an end load F acting parallel to Cy and passing through the centroid C of the cross-section. Then Cx is the axis of bending.

Consider an elemental length δz of the beam; on the near face of this element, which is at a distance z from the free end of the cantilever, the bending moment is

$$M = Fz$$

This gives rise to bending stresses in the cross-section; the longitudinal bending stress at a point of the cross-section a distance y from the neutral axis Cx is

$$\sigma = \frac{My}{I_x} = \frac{Fyz}{I_x}$$

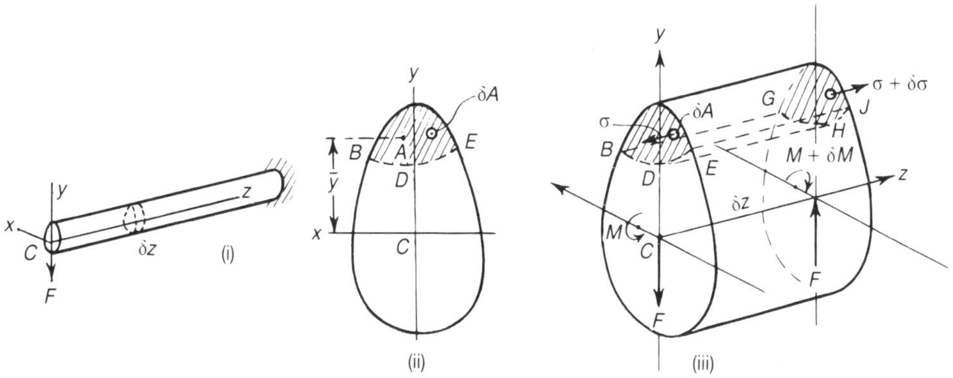

Figure 10.5 Shearing stresses in a bent beam having one axis of symmetry.

Now consider a section of the element cut off by the cylindrical surface $BDEGHJ$, Figure 10.5(iii), which is parallel to Cz. Suppose A is the area of each end of this cylindrical element; then the total longitudinal force on the end BDE due to bending stresses is

$$\int_A \sigma dA = \frac{Fz}{I_x} \int_A y dA$$

where δA is an element of the area A, and y is the distance of this element from the neutral axis Cx. The total longitudinal force on the remote end GHJ due to bending stresses is

$$\int_A (\sigma + \delta\sigma)\, dA = \frac{F}{I_x}(z + \delta z) \int_A y dA$$

as the bending moment at this section is

$$M + \delta M = F(z + \delta z)$$

The tension loads at the ends of the element $BDEGHJ$ differ by an amount

$$\frac{F\delta z}{I_x} \int_A y dA$$

If $\bar{y}$ is the distance of the centroid of the area A from Cx, then

$$\int_A y\,dA = A\bar{y}$$

The out-of-balance tension load is equilibrated by a shearing force over the cylindrical surface *BDEGHJ*.

This shearing force is then

$$\frac{F\delta z}{I_x}\int_A y\,dA = \frac{F\delta z}{I_x}A\bar{y}$$

and acts along the surface *BDEGHJ* and parallel to Cz. The total shearing force per unit length of the beam is

$$q = \frac{F\delta z}{I_x}A\bar{y}\;/\;\delta z = \frac{FA\bar{y}}{I_x} \qquad (10.4)$$

If b is the length of the curve *BDE*, or *GHJ*, then the average shearing stress over the surface *BDEGHJ* is

$$\bar{\tau} = \frac{FA\bar{y}}{bI_x} \qquad (10.5)$$

When b is small compared with the other linear dimensions of the cross-section we find that the shearing stress is nearly uniformly distributed over the surfaces of the type *BDEGHJ*. This is the case in thin-walled beams, such as I-sections and channel sections.

10.4 Shearing stresses in an I-beam

As an application of the general method developed in the preceding paragraph, consider the shearing stresses induced in a thin-walled I-beam carrying a concentrated load F at the free end, acting parallel to Cy, Figure 10.6. The cross-section has two axes of symmetry Cx and Cy; the flanges are of breadth b, and the distance between the centres of the flanges is h; the flanges and web are assumed to be of uniform thickness t.

Equation (10.4) gives the shearing force q per unit length of beam at any region of the cross-section. Consider firstly a point *1* of the flange at a distance s_1 from a free edge, Figure 10.6(iii); the area of flange cut off by a section through the point *1* is

$$A = s_1 t$$

The distance of the centroid of this area from the neutral axis Cx is

$$\overline{y_1} = \frac{1}{2}h$$

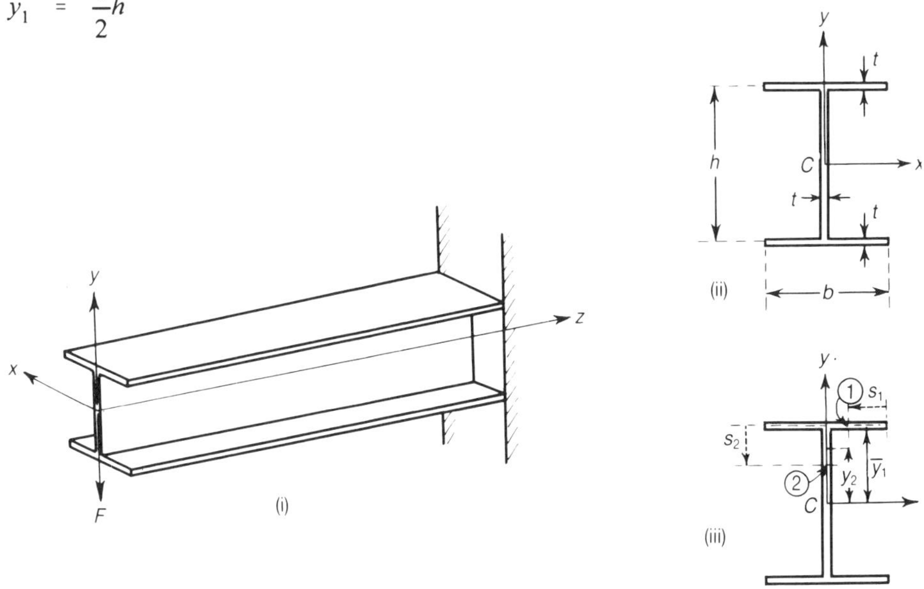

Figure 10.6 Flexural shearing stresses in an I-beam.

Then from equation (10.4), the shearing force at point *1* of the cross-section is

$$q = \frac{Fs_1th}{2I_x} \tag{10.6}$$

If the wall thickness t is small compared with the other linear dimensions of the cross-section, we may assume that q is distributed uniformly over the wall thickness t; the shearing stress is then

$$\tau = \frac{q}{t} = \frac{Fs_1h}{2I_x} \tag{10.7}$$

at point *1*. At the free edge, given by $s_1 = 0$, we have $\tau = 0$, since there can be no longitudinal shearing stress on a free edge of the cross-section. The shearing stress τ increases linearly in intensity as s_1 increases from zero to ½b; at the junction of web and flanges s_1 = ½b, and

$$\tau = \frac{Fbh}{4I_x} \tag{10.8}$$

As the cross-section is symmetrical about Cy, the shearing stress in the adjacent flange also increases linearly from zero at the free edge.

Consider secondly a section through the web at the point *2* at a distance s_2 from the junctions of the flanges and web. In evaluating $A\bar{y}$ for this section we must consider the total area cut off by the section through the point *2*. However, we can evaluate $A\bar{y}$ for the component areas cut off by the section through the point *2*; we have

$$A\bar{y} = (bt)\,\frac{1}{2}h + (s_2 t)\left(\frac{1}{2}h - \frac{1}{2}s_2\right)$$

$$= \frac{1}{2}t\left[bh + s_2\left(h - s_2\right)\right]$$

The from equation (10.4),

$$q = \frac{Ft}{2I_x}\left[bh + s_2\left(h - s_2\right)\right]$$

If this shearing force is assumed to be uniformly distributed as a shearing stress, then

$$\tau = \frac{q}{t} = \frac{F}{2I_x}\left[bh + s_2\left(h - s_2\right)\right] \qquad (10.9)$$

At the junction of web and flanges $s_2 = 0$, and

$$\tau = \frac{Fbh}{2I_x} \qquad (10.10)$$

At the neutral axis, $s_2 = \frac{1}{2}h$, and

$$\tau = \frac{Fbh}{2I_x}\left[1 + \frac{h}{4b}\right] \qquad (10.11)$$

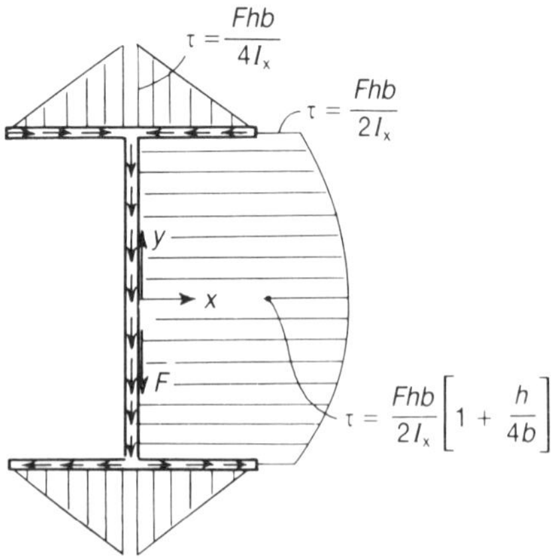

Figure 10.7 Variation of shearing stresses in an I-beam.

We notice that τ varies parabolically throughout the depth of the web, attaining a maximum value at $s_2 = \frac{1}{2}h$, the neutral axis, Figure 10.7. In any cross-section of the beam the shearing stresses vary in the form shown; in the flanges the stresses are parallel to Cx, and contribute nothing to the total force on the section parallel to Cy.

At the junctions of the web and flanges the shearing stress in the web is twice the shearing stresses in the flanges. The reason for this is easily seen by considering the equilibrium conditions at this junction. Consider a unit length of the beam along the line of the junction, Figure 10.8; the shearing stresses in the flanges are

$$\tau_f = \frac{Fbh}{4I_x} \tag{10.12}$$

while the shearing stress in the web we have estimated to be

$$\tau_w = \frac{Fbh}{2I_x} \tag{10.13}$$

For longitudinal equilibrium of a unit length of the junction of web and flanges, we have

$$2\left[\tau_f \times (t \times 1)\right] = \tau_w \times (t \times 1)$$

which gives

$$\tau_w = 2\tau_f \tag{10.14}$$

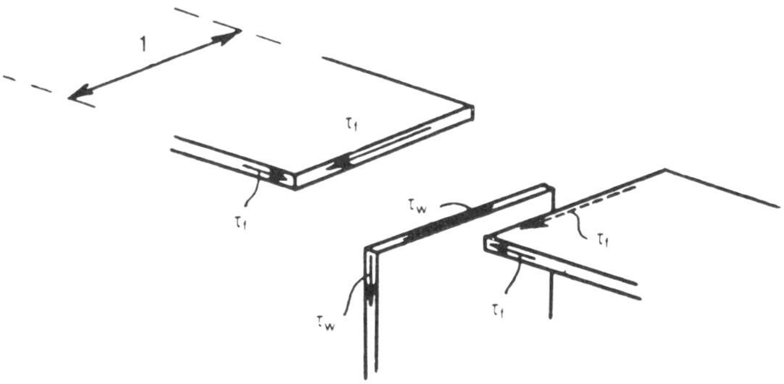

Figure 10.8 Equilibrium of shearing forces at the junction of the web and flanges of an I-beam.

This is true, in fact, for the relations we have derived above; longitudinal equilibrium is ensured at any section of the cross-section in our treatment of the problem. If the flanges and web were of different thicknesses, t_f and t_w, respectively, the equilibrium condition at the junction would be

$$2\tau_f t_f = \tau_w t_w$$

Then

$$\frac{\tau_w}{\tau_f} = \frac{2t_f}{t_w} \tag{10.15}$$

The implication of this equilibrium condition is that at a junction, such as that of the flanges and web of an I-section, the sum of the shearing forces per unit length for the components meeting at that junction is zero when account is taken of the relevant directions of these shearing forces. For a junction

$$\sum \tau t = 0 \tag{10.16}$$

where τ is the shearing stress in an element at the junction, and t is the thickness of the element; the summation is carried out for all elements meeting at the junction.

For an I-section carrying a shearing force acting parallel to the web we see that the maximum shearing stress occurs at the middle of the web, and is given by equation (10.11). Now, I_x for the section is given approximately by

$$I_x = \frac{1}{12} h^3 t + \frac{1}{2} h^2 bt = \frac{h^3 t}{12}\left(1 + \frac{6b}{h}\right) \tag{10.17}$$

Then

$$\tau_{max} = \frac{6Fb}{h^2 t}\left[\frac{1 + h/4b}{1 + 6b/h}\right] \tag{10.18}$$

The total shearing force in the web of the beam parallel to Cy is F; if this were distributed uniformly over the depth of the web the average shearing stress would be

$$\tau_{av} = \frac{F}{ht} \tag{10.19}$$

Then for the particular case when $h = 3b$, we have

$$\tau_{max} = \frac{7}{6}\left(\frac{F}{ht}\right) \tag{10.20}$$

Then τ_{max} is only one-sixth or about 17% greater than the mean shearing stress over the web.

Problem 10.1 The web of a girder of I-section is 45 cm deep and 1 cm thick; the flanges are each 22.5 cm wide by 1.25 cm thick. The girder at some particular section has to withstand a total shearing force of 200 kN. Calculate the shearing stresses at the top and middle of the web. *(Cambridge)*

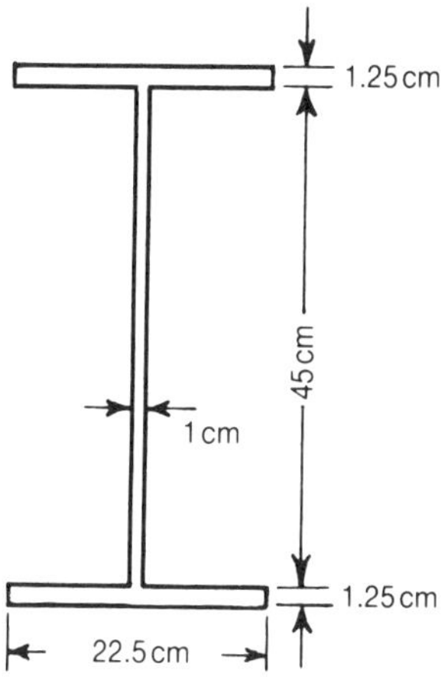

Solution

The second moment of area of the web about the centroidal axis is

$$\frac{1}{12}(0.010)(0.45)^3 = 0.0760 \times 10^{-3} \text{ m}^4$$

The second moment of are of each flange about the centroidal axis is

$$(0.225)(0.0125)(0.231)^2 = 0.150 \times 10^{-3} \text{ m}^4$$

The total second moment of area is then

$$I_x = [0.076 + 2(0.150)]\, 10^{-3} = 0.376 \times 10^{-3} \text{ m}^4$$

At a distance y above the neutral axis, the shearing stress from equation (10.9) is

$$\tau = \frac{F}{2I_x}\left[\left(bh + \frac{1}{4}h^2\right) - y^2\right]$$

$$= \frac{200 \times 10^3}{2 \times 0.376 \times 10^{-3}}\left[(0.225)(0.4625) + \frac{1}{4}(0.4625)^2 - y^2\right]$$

where $s_2 = h/2 - y$

At the top of the web, we have $y = 0.231$ m, and

$$\tau = 27.7 \text{ MN/m}^2$$

While at the middle of the web, where $y = 0$, we have

$$\tau = 41.9 \text{ MN/m}^2$$

10.5 Principal stresses in beams

We have shown how to find separately the longitudinal stress at any point in a beam due to bending moment, and the mean horizontal and vertical shearing stresses, but it does not follow that these are the greatest direct or shearing stresses. Within the limits of our present theory we can employ the formulae of Sections 5.7 and 5.8 to find the principal stresses and the maximum shearing stress.

We can draw, on a side elevation of the beam, lines showing the direction of the principal stresses. Such lines are called the *lines of principal stress*; they are such that the tangent at any point gives the direction of principal stress. As an example, the lines of principal stress have been drawn in Figure 10.9 for a simply-supported beam of uniform rectangular cross-section, carrying a uniformly distributed load. The stresses are a maximum where the tangents to the curves are parallel to the axis of the beam, and diminish to zero when the curves cut the faces of the beam at right angles. On the neutral axis, where the stress is one of shear, the principal stress curves cut the axis at 45°.

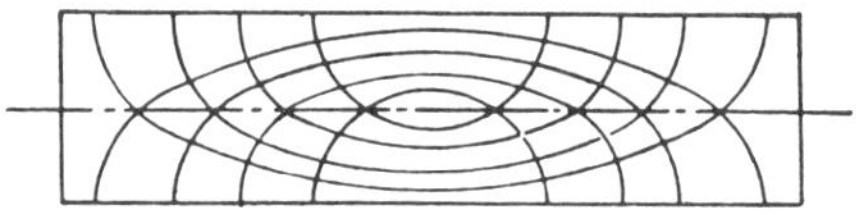

Figure 10.9 Principal stress lines in a simply-supported rectangular beam carrying a uniformly distributed load.

Problem 10.2 The flanges of an I-girder are 30 cm wide by 2.5 cm thick and the web is 60 cm deep by 1.25 cm thick. At a particular section the sagging bending moment is 500 kNm and the shearing force is 500 kN. Consider a point in the section at the top of the web and calculate for this point; (i) the longitudinal stress, (ii) the shearing stress, (iii) the principal stresses. *(Cambridge)*

<u>*Solution*</u>

First calculate the second moment of area about the neutral axis; the second moment of area of the web is

$$\frac{1}{12}(0.0125)(0.6)^3 = 0.225 \times 10^{-3} \text{ m}^4$$

The second moment of area of each flange is

$$(0.3)(0.025)(0.3125)^2 = 0.733 \times 10^{-3} \text{ m}^4$$

The total second moment of area is then

$$I_x = [0.225 + 2(0.733)]10^{-3} = 1.691 \times 10^{-3} \text{ m}^4$$

Next, for a point at the top of the web,

$$A\bar{y} = (0.3 \times 0.025)(0.3125) = 2.34 \times 10^{-3} \text{ m}^4$$

Then, for this point, with $M = 500$ kNm we have

$$\sigma = \frac{My}{I_x} = \frac{(500 \times 10^3)(0.3)}{1.691 \times 10^{-3}} = 88.6 \text{ MN/m}^2 \text{ (compressive)}$$

$$\tau = \frac{FA\bar{y}}{I_x t} = \frac{(500 \times 10^3)(2.34 \times 10^{-3})}{(1.691 \times 10^{-3})(0.0125)} = 55.3 \text{ MN.m}^2$$

The principal stresses are then

$$\frac{1}{2}\sigma \pm \left[\frac{1}{4}\sigma^2 + \tau^2\right]^{\frac{1}{2}} = (-44.3 \pm 70.9) \text{ MN/m}^2$$

$$= 26.6 \text{ and } -115.2 \text{ MN/m}^2$$

It should be noticed that the greater principal stress is about 30% greater than the longitudinal stress. At the top of the flange the longitudinal stress is -96 MN/m^2, so the greatest principal stress at the top of the web is 20% greater than the maximum longitudinal stress.

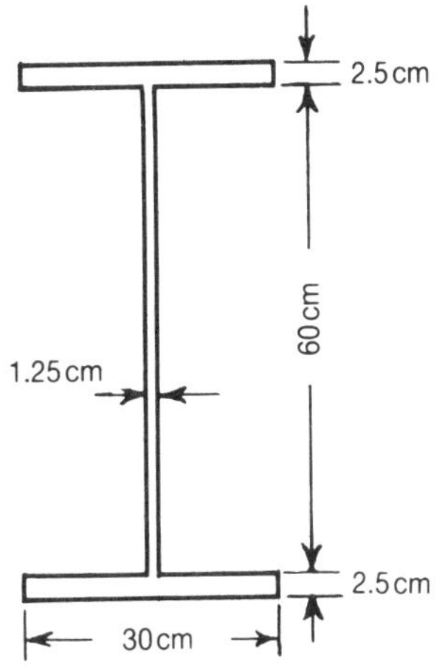

10.6 Superimposed beams

If we make a beam by placing one member on the top of another, Figure 10.10, there will be a tendency, under the action of lateral loads for the two members to slide over each other along the plane of contact AB, Figure 10.10. Unless this sliding action is prevented in some way, the one beam will act independently of the other; when there is no shearing connection between the beams along AB, the strength of the compound beam is the sum of the strengths of the separate beams.

However, if the sliding action is resisted, the compound beam behaves more nearly as a solid member; for elastic bending the permissible moment is proportional to the elastic section modulus.

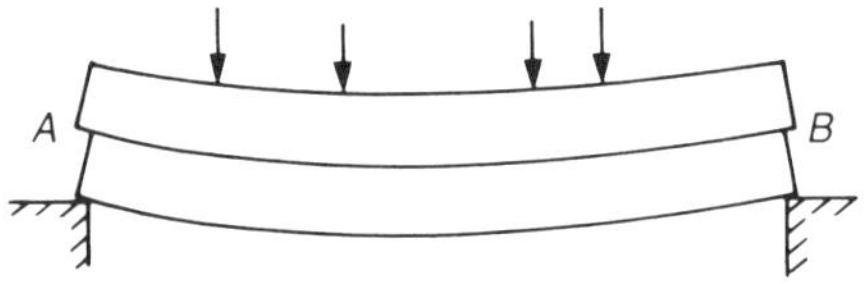

Figure 10.10 Sliding action between two beams superimposed without shearing connections.

In the case of two equal beams of rectangular cross-section, the elastic section modulus of each beam is

$$\frac{bh^2}{6}$$

where b is the breadth and h is the depth of each beam. For two such beams, placed one on the other, without shearing connection, the elastic section modulus is

$$2 \times \frac{1}{6} bh^2 = \frac{1}{3} bh^2$$

If the two beams have a rigid shearing connection, the effective depth is $2h$, and the elastic section modulus is

$$\frac{1}{6} b(2h)^2 = \frac{2}{3} bh^2$$

The elastic section modulus, and therefore the permissible bending moment, is doubled by providing a shearing connection between the two beams. In the case of steel beams, the flanges along the plane of contact AB, may be riveted, bolted, or welded together.

10.7 Shearing stresses in a channel section; shear centre

We have discussed the general case of shearing stresses in the bending of a beam having an axis of symmetry in the cross-section; we assumed that the shearing forces were applied parallel to this axis of symmetry. This is a relatively simple problem to treat because there can be no twisting of the beam when a shearing force is applied parallel to the axis of symmetry. We consider now the case when the shearing force is applied at right angles to an axis of symmetry of the cross-section. Consider for example a channel section having an axis Cx of symmetry in the cross-section, Figure 10.11; the section is of uniform wall-thickness t, b is the total breadth of each flange, and h is the distance between the flanges; C is the centroid of the cross-section. Suppose the beam is supported at one end, and that a shearing force F is applied at the free end in a direction parallel to Cy. We apply this shearing force at a point O on Cx such that no torsion of the channel occurs, Figure 10.12; if F is applied considerably to the left of C, twisting obviously will occur in a counter-clockwise direction; if F is applied considerably to the right then twisting occurs in a clockwise direction. There is some intermediate position of O for which no twisting occurs; as we shall see this position is not coincident with the centroid C.

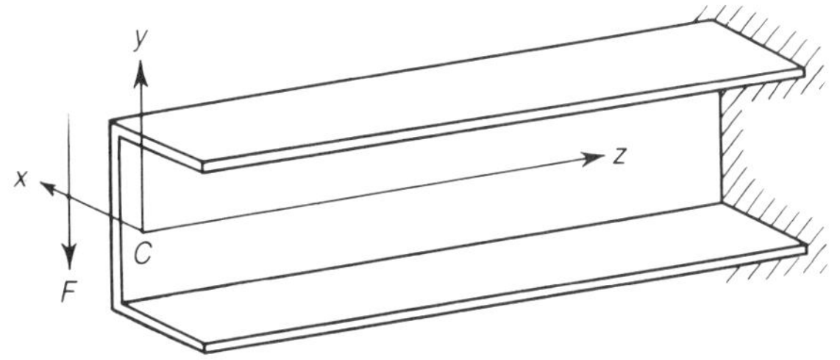

Figure 10.11 Shearing of a channel cantilever.

The problem is greatly simplified if we assume that F is applied at a point O on Cx to give no torsion of the channel; suppose O is a distance e from the centre of the web, Figure 10.12.

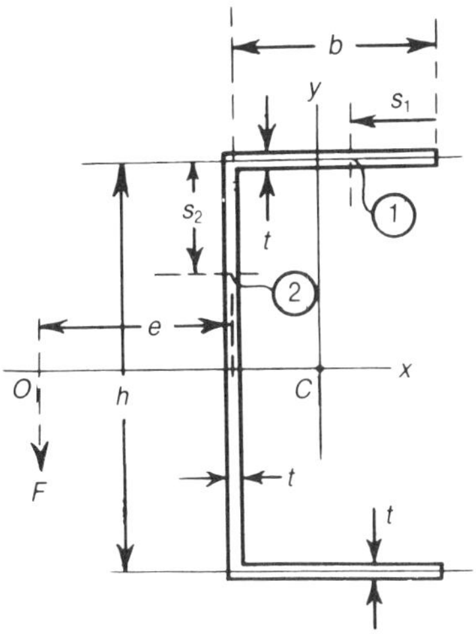

Figure 10.12 Shearing stress at any point of a channel beam.

At any section of the beam there are only bending actions present; therefore, we can again use the relation

$$q = \frac{FA\bar{y}}{I_x} \tag{10.21}$$

At a distance s_1 from the free edge of a flange

$$q_1 = \frac{Fht}{2I_x} s_1$$

At a distance s_2 along the web from the junction of web and flange

$$q_2 = \frac{Ft}{2I_x} \left[bh + s_2 (h - s_2)\right]$$

The shearing stress in flange is

$$\tau_1 = \frac{q_1}{t} = \frac{Fh}{2I_x} s_1$$

and in the web is

$$\tau_2 = \frac{q_2}{t} = \frac{F}{2I_x} \left[bh + s_2 (h - s_2)\right]$$

The shearing stress τ_1 in the flanges increases linearly from zero at the free edges to a maximum at the corners; the variation of shearing stress τ_2 in the web is parabolic in form, attaining a maximum value

$$\tau_{max} = \frac{Fbh}{2I_x} \left(1 + \frac{h}{4b}\right) \tag{10.22}$$

at the mid-depth of the web, Figure 10.13. The shearing stresses τ_1 in the flanges imply total shearing forces of amounts

$$\frac{Fht}{2I_x} \int_0^b s_1 \, ds_1 = \frac{Fb^2ht}{4I_x} \tag{10.23}$$

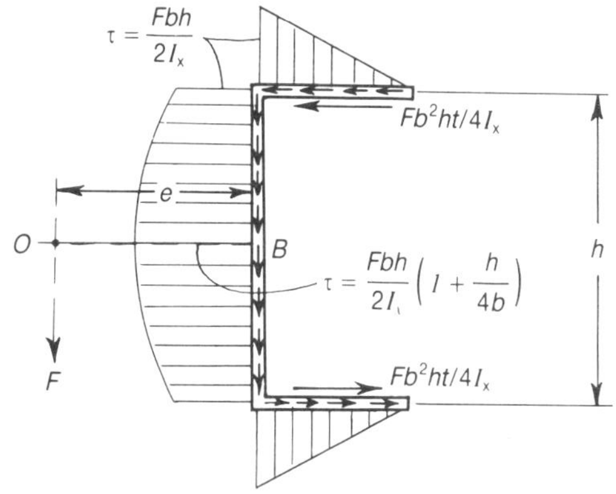

Figure 10.13 Variation of shearing stresses over the cross-section of a channel beam; e is the distance to the stress centre O.

acting parallel to the centre lines of the flanges; the total shearing forces in the two flanges are in opposite directions. If the distribution of shearing stresses τ_1 and τ_2 is statically equivalent to the applied shearing force F, we have, on taking moments about B— the centre of the web— that

$$Fe = \frac{Fb^2ht}{4I_x}.h = \frac{Fb^2h^2t}{4I_x}$$

Then $$e = \frac{b^2h^2t}{4I_x} \tag{10.24}$$

which, as we should expect, is independent of F. We note that O is remote from the centroid C of the cross-section; the point O is usually called the *shear centre;* it is the point of the cross-section through which the resultant shearing force must pass if bending is to occur without torsion of the beam.

Problem 10.3 Determine the maximum value of the shearing stress and the shear centre position 'e' for the thin-walled split tube in the figure below.

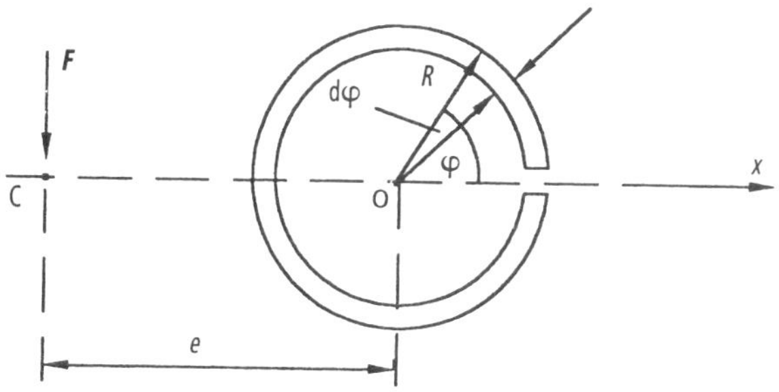

<u>*Solution*</u>

Consider an infinitesimally small element of the tube wall at an angle φ

$$\int y\,dA = \int_0^{\varphi} R\sin\varphi\,.\,(t\,.\,R\,.\,d\varphi)$$

$$= R^2\,t\left[-\cos\varphi\right]_0^{\varphi}$$

$$= R^2\,t\left[-\cos\varphi - (-1)\right]$$

$$= R^2\,t\,(1 - \cos\varphi)$$

At φ, the shearing stress τ_φ is given by

$$\tau_\varphi = \frac{F}{bI_x}\int y\,dA$$

or $$\tau_\varphi = \frac{F}{tI_x}R^2\,t\,(1 - \cos\varphi) \qquad (10.25)$$

or $$\tau_\varphi = \frac{FR^2}{I_x}(1 - \cos\varphi)$$

Now $$I_x = \int y^2\,dA$$

$$= \int_0^{2\pi}(R\sin\varphi)^2(Rt\,d\varphi)$$

$$= R^3\,t\int_0^{2\pi}\sin^2\varphi\,d\varphi$$

but $$\sin^2\varphi = \frac{1 - \cos 2\varphi}{2}$$

$$\therefore I_x = R^3\,t\int_0^{2\pi}\frac{(1 - \cos 2\varphi)}{2}\,d\varphi$$

$$= \frac{R^3\,t}{2}\left[\varphi - \frac{\sin 2\varphi}{2}\right]_o^{2\pi}$$

$$= \frac{R^3\,t}{2}\left[(2\pi - 0) - (0 - 0)\right]$$

or $$I_x = \pi R^3 t \tag{10.26}$$

Substituting equation (10.26) into (10.25), we get

$$\tau_\varphi = \frac{FR^2}{\pi R^3 t}(1 - \cos\varphi)$$
$$= \frac{F}{\pi R t}(1 - \cos\varphi) \tag{10.27}$$

$\tau_{\varphi(\max)}$ occurs when $\varphi = \pi$

$$\tau_{\varphi(\max)} = \frac{F}{\pi R t}(1 + 1) = \frac{2F}{\pi R t} \tag{10.28}$$

To determine the shear centre position, take moments about the point 'O'.

i.e. $$Fe = \int_0^{2\pi} \tau_\varphi \, (t \; R \; d\varphi) \; R$$

$$= \frac{R^2 t \, . \, F}{\pi R t} \int_0^{2\pi} (1 - \cos\varphi) \, . \, d\varphi$$

$$= \frac{FR}{\pi} \left[\varphi - \sin\varphi\right]_0^{2\pi}$$

$$= \frac{FR}{\pi} \left[(2\pi - 0) - (0 - 0)\right]$$

$$= \frac{FR}{\pi} 2\pi$$

$$\therefore e = 2R$$

Further problems (*answers on page 692*)

10.4 A plate web girder consists of four plates, in each flange, of 30 cm width. The web is 60 cm deep, 2 cm thick and is connected to the flanges by 10 cm by 10 cm by 1.25 cm angles, riveted with 2 cm diameter rivets. Assuming the maximum bending moment to be 1000 kNm, and the shearing force to be 380 kN, obtain suitable dimensions for (i) the thickness of the flange plates, (ii) the pitch of the rivets. Take the tensile stress as 100 MN/m^2, and the shearing stress in the rivets as 75 MN/m^2. *(RNEC)*

10.5 In a small gantry for unloading goods from a railway waggon, it is proposed to carry the lifting tackle on a steel joist, 24 cm by 10 cm, of weight 320 N/m, supported at the ends, and of effective length 5 m. The equivalent dead load on the joist due to the load to be raised is 30 kN, and this may act at any point of the middle 4 m. By considering the fibre stress and the shear, examine whether the joist is suitable. The flanges are 10 cm by 1.2 cm, and the web is 0.75 cm thick. The allowable fibre stress is 115 MN/m^2, and the allowable shearing stress 75 MN/m^2. *(Cambridge)*

10.6 A girder of I-section has a web 60 cm by 1.25 cm and flanges 30 cm by 2.5 cm. The girder is subjected at a bending moment of 300 kNm and a shearing force of 1000 kN at a particular section. Calculate how much of the shearing force is carried by the web, and how much of the bending moment by the flanges. *(Cambridge)*

10.7 The shearing force at a given section of a built-up I-girder is 1000 kN and the depth of the web is 2 m. The web is joined to the flanges by fillet welds. Determine the thickness of the web plate and the thickness of the welds, allowing a shearing stress of 75 MN/m^2 in both the web and welds.

10.8 A thin metal pipe of mean radius R, thickness t and length L, has its ends closed and is full of water. If the ends are simply-supported, estimate the form of the distribution of shearing stresses over a section near one support, ignoring the intrinsic weight of the pipe.

10.9 A compound girder consists of a 45 cm by 18 cm steel joist, of weight 1000 N/m, with a steel plate 25 cm by 3 cm welded to each flange. If the ends are simply-supported and the effective span is 10 m, what is the maximum uniformly distributed load which can be supported by the girder? What weld thicknesses are required to support this load?

Allowable longitudinal stress in plates = 110 MN/m^2
Allowable shearing stress in welds = 60 MN/m^2
Allowable shearing stress in web of girder = 75 MN/m^2

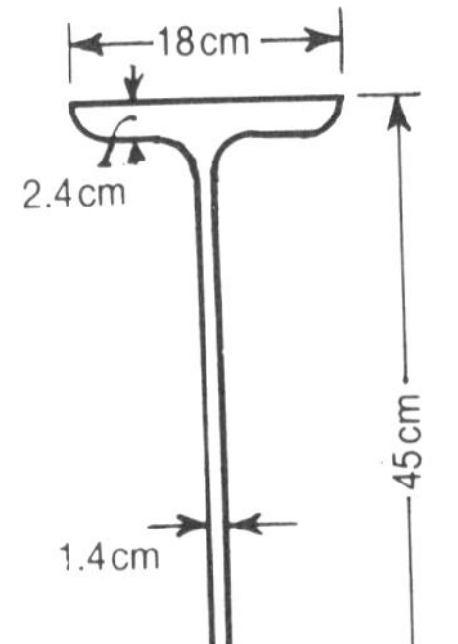

10.10 Determine the maximum value of the shearing stresses and the positions of the shear centres for the thin-walled tubes shown in the figures below.

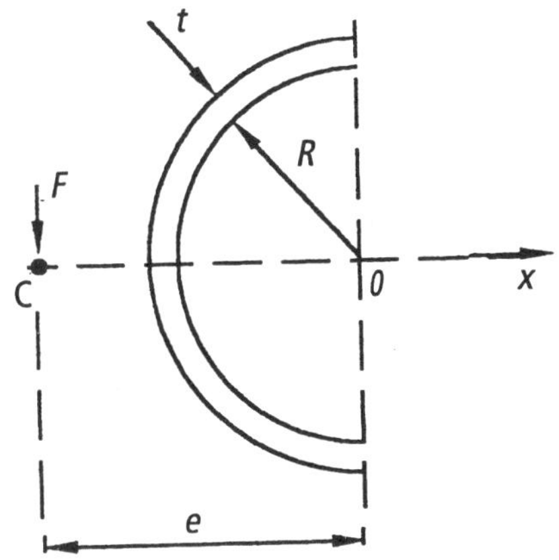

(a)

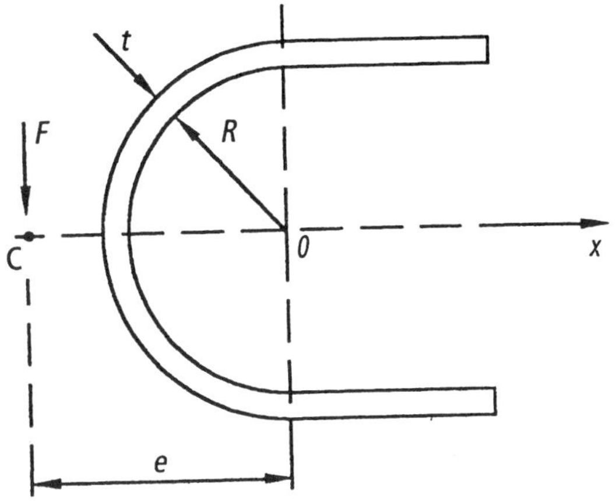

(b)

11 Beams of two materials

11.1 Introduction

Some beams used in engineering structures are composed of two materials. A timber joist, for example, may be reinforced by bolting steel plates to the flanges. Plain concrete has little or no tensile strength, and beams of this material are reinforced therefore with steel rods or wires in the tension fibres. In beams of these types there is a composite action between the two materials.

11.2 Transformed sections

The composite beam shown in Figure 11.1 consists of a rectangular timber joist of breadth b and depth h, reinforced with two steel plates of depth h and thickness t.

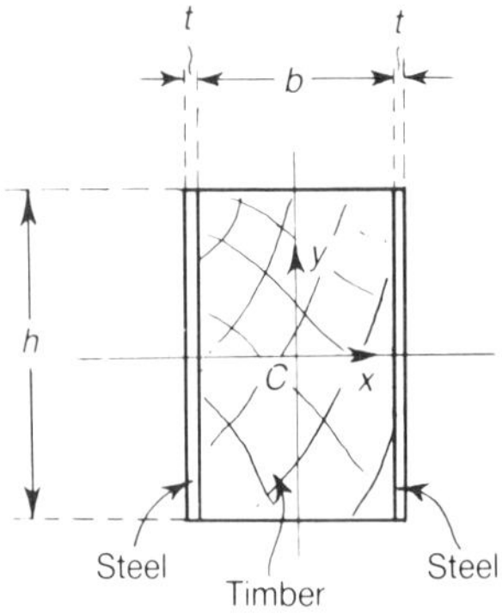

Figure 11.1 Timber beam reinforced with steel side plates.

Consider the behaviour of the composite beam under the action of a bending moment M applied about Cx; if the timber beam is bent into a curve of radius R, then, from equation (9.5), the bending moment carried by the timber beam is

$$M_t = \frac{(EI)_t}{R} \tag{11.1}$$

where $(EI)_t$ is the bending stiffness of the timber beam. If the steel plates are attached to the timber beam by bolting, or glueing, or some other means, the steel plates are bent to the same radius of curvature R as the timber beam. The bending moment carried by the two steel plates is then

$$M_s = \frac{(EI)_s}{R}$$

where $(EI)_s$ is the bending stiffness of the two steel plates. The total bending moment is then

$$M = M_t + M_s = \frac{1}{R}\left[(EI)_t + (EI)_s\right]$$

This gives

$$\frac{1}{R} = \frac{M}{(EI)_t + (EI)_s} \tag{11.2}$$

Clearly, the beam behaves as though the total bending stiffness EI were

$$EI = (EI)_t + (EI)_s \tag{11.3}$$

If E_t and E_s are the values of Young's modulus for timber and steel, respectively, and if I_t and I_s are the second moments of area about Cx of the timber and steel beams, respectively, we have

$$EI = (EI)_t + (EI)_s = E_t I_t + E_s I_s \tag{11.4}$$

Then

$$EI = E_t\left[I_t + \left(\frac{E_s}{E_t}\right) I_s\right] \tag{11.5}$$

If I_s is multiplied by (E_s/E_t), which is the ratio of Young's moduli for the two materials, then from equation (11.5) we see that the composite beam may be treated as wholly timber, having an equivalent second moment of area

$$I_t + \left(\frac{E_s}{E_t}\right) I_s \tag{11.6}$$

This is equivalent to treating the beam of Figure 11.2(i) with reinforcing plates made of timber, but having thicknesses

$$\left(\frac{E_s}{E_t}\right) \times t$$

as shown in Figure 11.2(ii); the equivalent timber beam of Figure 11.2(ii) is the *transformed section* of the beam. In this case the beam has been transformed wholly to timber. Equally the beam may be transformed wholly to steel, as shown in Figure 11.2(iii). For bending about *Cx* the *breadths* of the component beams are factored to find the transformed section; the depth *h* of the beam is unaffected.

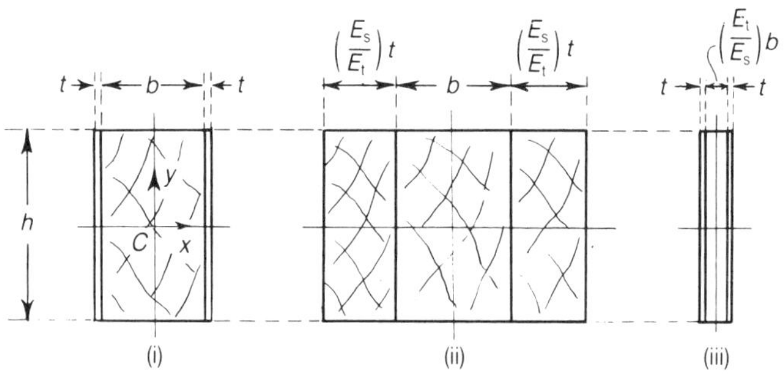

Figure 11.2 (i) Composite beam of timber and steel bent about *Cx*. (ii) Equivalent timber beam. (iii) Equivalent steel beam.

The bending stress σ_t in the fibre of the timber core of the beam a distance y from the neutral axis is

$$\sigma_t = M_t \frac{y}{I_t}$$

Now, from equations (11.1) and (11.2)

$$M_t = \frac{(EI)_t}{R}, \quad M = \frac{1}{R}\left[(EI)_t + (EI)_s\right]$$

and on eliminating R,

$$M_t = \frac{M}{1 + \dfrac{E_s I_s}{E_t I_t}} \tag{11.7}$$

Then

$$\sigma_t = \frac{My}{I_t\left(1 + \dfrac{E_s I_s}{E_t I_t}\right)} = \frac{My}{I_t + \left(\dfrac{E_s}{E_t}\right) I_s} \tag{11.8}$$

the bending stresses in the timber core are found therefore by considering the *total* bending moment M to be carried by the transformed timber beam of Figure 11.2(ii). The longitudinal strain at the distance y from the neutral axis Cx is

$$\varepsilon = \frac{\sigma_t}{E_t} = \frac{My}{E_t I_t + E_s I_s}$$

Then at the distance y from the neutral axis the stress in the steel reinforcing plates is

$$\sigma_s = E_s \varepsilon = \frac{My}{I_s + \left(\frac{E_t}{E_s}\right) I_t} \tag{11.9}$$

because the strains in the steel and timber are the same at the same distance y from the neutral axis. This condition of equal strain is implied in the assumption made earlier that the steel and the timber components of the beam are bent to the same radius of curvature R.

Problem 11.1 A composite beam consists of a timber joist, 15 cm by 10 cm, to which reinforcing steel plates, ½ cm thick, are attached. Estimate the maximum bending moment which may be applied about Cx, if the bending stress in the timber is not to exceed 5 MN/m^2, and that in the steel 120 MN/m^2. Take $E_s/E_t = 20$.

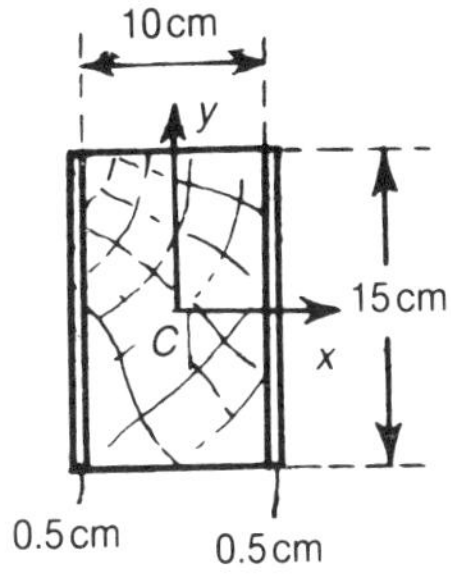

Solution

The maximum bending stresses occur in the extreme fibres. If the stress in the timber is 5 MN/m^2, the stress in the steel at the same distance from Cx is

$$5 \times 10^6 \times \frac{E_s}{E_t} = 100 \times 10^6 \text{ N/m}^2 = 100 \text{ MN/m}^2$$

Thus when the maximum timber stress is attained, the maximum steel stress is only 100 MN/m^2. If the maximum permissible stress of 120 MN/m^2 were attained in the steel, the stress in the timber would exceed 5 MN/m^2, which is not permissible. The maximum bending moment gives therefore a stress in the timber of 5 MN/m^2. The second moment of area about Cx of the equivalent timber beam is

$$I_x = \frac{1}{12}(0.10)(0.15)^3 + \frac{1}{12}(0.010)(0.15)^3 \times 20$$

$$= 0.0842 \times 10^{-3}\ \text{m}^4$$

For a maximum stress in the timber of 5 MN/m^2, the moment is

$$M = \frac{(5 \times 10^6)(0.0842 \times 10^{-3})}{0.075} = 5610\ \text{Nm}$$

11.3 Timber beam with reinforcing steel flange plates

In Section 11.2 we discussed the composite bending action of a timber be‿m reinforced with steel plates over the depth of the beam. A similar bending problem arises when the timber joist is reinforced on its upper and lower faces with steel plates, as shown if Figure 11.3(i); the timber web of the composite beams may be transformed into steel to give the equivalent steel section of Figure 11.3(ii); alternatively, the steel flanges may be replaced by equivalent timber flanges to give the equivalent timber beam of Figure 11.3(iii). The problem is then treated in the same way as the beam in Section 11.2; the stresses in the timber and steel are calculated from the second moment of area of the transformed timber and steel sections.

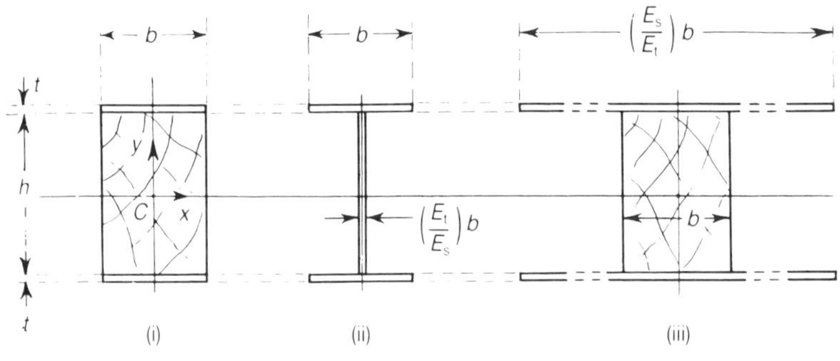

Figure 11.3 (i) Timber beam with reinforced steel flange plates. (ii) Equivalent steel I-beam. (iii) Equivalent timber I-beam.

An important difference, however, between the composite actions of the beams of Figures 11.2 and 11.3 lies in their behaviour under shearing forces. The two beams, used as cantilevers carrying end loads F, are shown in Figure 11.4; for the timber joist reinforced over the depth, Figure 11.4(i), there are no shearing actions between the timber and the steel plates, except near the loaded ends of the cantilever.

However, for the joist of Figure 11.4(ii), a shearing force is transmitted between the timber and the steel flanges at all sections of the beam. In the particular case of thin reinforcing flanges, it is sufficiently accurate to assume that the shearing actions in the cantilever of Figure 11.4(ii) are resisted largely by the timber joist; on considering the equilibrium of a unit length of the composite beam, equilibrium is ensured if a shearing force (F/h) per unit length of beam is transmitted between the timber joist and the reinforcing flanges, Figure 11.5. This shearing force must be carried by bolts, glue or some other suitable means. The end deflections of the cantilevers shown in Figure 11.4 may be difficult to estimate; this is due to the fact that account may have to be taken of the shearing distortions of the timber beams.

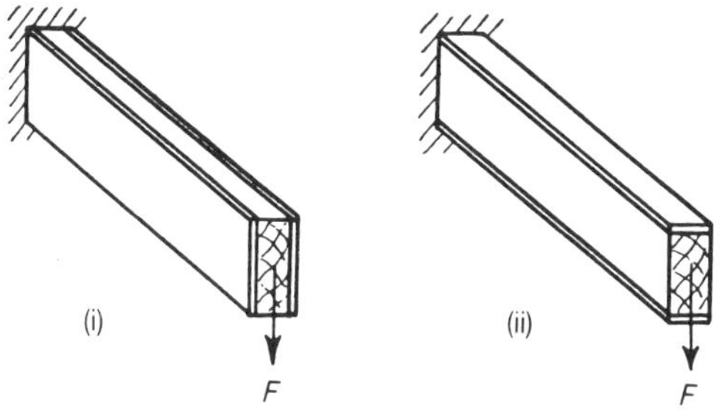

Figure 11.4 composite beams under shearing action, showing (i) steel and timber both resisting shear and (ii) timber alone resisting shear.

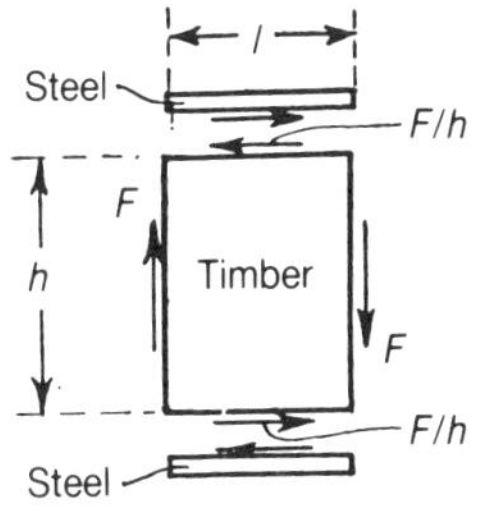

Figure 11.5 Shearing actions in a timber joist with reinforcing steel flanges.

Problem 11.2 A timber joist 15 cm by 7.5 cm has reinforcing steel flange plates 1.25 cm thick. The composite beam is 3 m long, simply-supported at each end, and carries a uniformly distributed lateral load of 10 kN. Estimate the maximum bending stresses in the steel and timber, and the intensity of shearing force transmitted between the steel plates and the timber. Take $E_s/E_t = 20$.

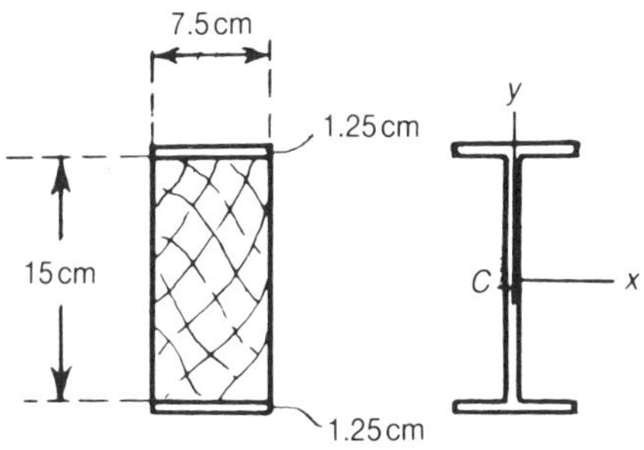

Solution

The second moment of area of the equivalent steel section is

$$I_x = \frac{1}{20}\left[\frac{1}{12}(0.075)(0.15)^3\right] + 2\left[(0.0125)(0.075)^3\right] = 11.6 \times 10^{-6}\ \text{m}^4$$

The maximum bending moment is

$$\frac{\left(10 \times 10^3\right)(3)}{8} = 3750\ \text{Nm}$$

The maximum bending stress in the steel is then

$$\sigma_s = \frac{(3750)(0.0875)}{\left(11.6 \times 10^{-6}\right)} = 28.3\ \text{MN/m}^2$$

The bending stress in the steel at the junction of web and flange is

$$\sigma_s = \frac{(3750)(0.0750)}{\left(11.6 \times 10^{-6}\right)} = 24.2\ \text{MN/m}^2$$

The stress in the timber at this junction is then

$$\sigma_t = \frac{E_t}{E_s} \times \sigma_s = \frac{1}{20}(24.2) = 1.2\ \text{MN/m}^2$$

On the assumption that the shearing forces at any section of the beam area taken largely by the timber, the shearing force between the timber and steel plates is

$$\left(5 \times 10^3\right) / (0.15) = 33.3\ \text{kN/m}$$

because the maximum shearing force in the beam is 5 kN.

11.4 Ordinary reinforced concrete

It was noted in Chapter 1 that concrete is a brittle material which is weak in tension. Consequently a beam composed only of concrete has little or no bending strength since cracking occurs in the extreme tension fibres in the early stages of loading. To overcome this weakness steel rods are embedded in the tension fibres of a concrete beam; if concrete is cast around a steel rod, on setting the concrete shrinks and grips the steel rod. It happens that the coefficients of linear expansion of concrete and steel are very nearly equal; consequently, negligible stresses are set up by temperature changes.

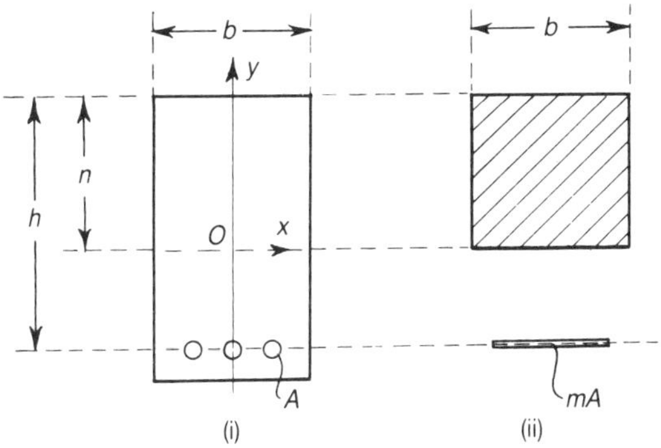

Figure 11.6 Simple rectangular concrete beam with reinforcing steel in the tension flange.

The bending of an ordinary reinforced concrete beam may be treated on the basis of transformed sections. Consider the beam of rectangular cross-section shown in Figure 11.6. The breadth of the concrete is b, and h is the depth of the steel reinforcement below the upper extreme fibres. The beam is bent so that tensile stresses occur in the lower fibres. The total area of cross-section of the steel reinforcing rods is A; the rods are placed longitudinally in the beam. The beam is now bent so that Ox becomes a neutral axis, compressive stresses being induced in the concrete above Ox. We assume that concrete below the neutral axis cracks in tension, and is therefore ineffectual; we neglect the contribution of the concrete below Ox to the bending strength of the beam. Suppose m is the ratio of Young's modulus of steel, E_s, to Young's modulus of concrete, E_c; then

$$m = \frac{E_s}{E_c} \tag{11.10}$$

If the area A of steel is transformed to concrete, its equivalent area is mA; the equivalent concrete beam then has the form shown in Figure 11.6(ii). The depth of the neutral axis Ox below the extreme upper fibres is n. The equivalent concrete area mA on the tension side of the beam is concentrated approximately at a depth h.

We have that the neutral axis of the beam occurs at the centroid of the equivalent concrete beam; then

$$bn \times \frac{1}{2}n = mA\,(h - n)$$

Thus n is the root of the quadratic equation

$$\frac{1}{2}\,bn^2 + mAn - mAh = 0 \tag{11.11}$$

The relevant root is

$$n = \frac{mA}{b}\left(\sqrt{1 + \frac{2bh}{mA}} - 1\right) \qquad (11.12)$$

The second moment of area of the equivalent concrete beam about its centroidal axis is

$$I_c = \frac{1}{3}bn^3 + mA\,(h - n)^2 \qquad (11.13)$$

The maximum compressive stress induced in the upper extreme fibres of the concrete is

$$\sigma_c = \frac{Mn}{I_c} \qquad (11.14)$$

$$\sigma_s = \frac{M(h - n)}{I_c} \times \frac{E_s}{E_c} = \frac{mM(h - n)}{I_c} \qquad (11.15)$$

Problem 11.3 A rectangular concrete beam is 30 cm wide and 45 cm deep to the steel reinforcement. The direct stresses are limited to 115 MN/m^2 in the steel and 6.5 MN/m^2 in the concrete, and the modular ratio is 15. What is the area of steel reinforcement if both steel and concrete are fully stressed? Estimate the permissible bending moment for this condition.

Solution

From equations (11.14) and (11.15)

$$\sigma_s = \frac{M(h - n)}{\dfrac{bn^3}{3m} + A\,(h - n)^2} = 115\ \text{MN/m}^2$$

and

$$\sigma_c = \frac{Mn}{\dfrac{1}{3}bn^3 + mA\,(h - n)^2} = 6.5\ \text{MN/m}^2$$

Then

$$\frac{M(h-n)}{115} = \frac{Mn}{6.5} m$$

Hence

$$h - n = 1.18n \quad \text{and} \quad \frac{n}{h} = \frac{1}{2.18} = 0.458$$

Then

$$n = 0.458 \times 0.45 = 0.206 \text{ m}$$

From equation (11.11)

$$\frac{2mA}{bh} = \frac{(n/h)^2}{1-(n/h)} = \frac{(0.458)^2}{0.542} = 0.387$$

Then

$$A = 0.387 \frac{bh}{2m} = \frac{0.387 \times 0.30 \times 0.45}{30} = 1.75 \times 10^{-3} \text{ m}^2$$

As the maximum allowable stresses of both the steel and concrete are attained, the allowable bending moment may be elevated on the basis of either the steel or the concrete stress. The second moment of area of the equivalent concrete beam is

$$I_c = \frac{1}{3} bn^3 + mA(h-n)^2$$

$$= \frac{1}{3}(0.30)(0.206)^3 + 15(0.00174)(0.244)^2 = 2.42 \times 10^{-3} \text{ m}^2$$

The permissible bending moment is

$$M = \frac{\sigma_c I_c}{y_c} = \frac{(6.5 \times 10^6)(2.42 \times 10^{-3})}{(0.206)} = 76.4 \text{ kNm}$$

Problem 11.4 A rectangular concrete beam has a breadth of 30 cm and is 45 cm deep to the steel reinforcement, which consists of two 2.5 cm diameter bars. Estimate the permissible bending moment if the stresses are limited to 115 MN/m^2 and 6.5 MN/m^2 in the steel and concrete, respectively, and if the modular ratio is 15.

Solution

The area of steel reinforcement is $A = 2(\pi/4)(0.025)^2 = 0.982 \times 10^{-3}$ m^2. From equation (11.12)

$$\frac{n}{h} = \frac{mA}{bh}\left[\sqrt{1 + \frac{2bh}{mA}} - 1\right]$$

Now

$$\frac{mA}{bh} = \frac{(15)\,(0.982) \times 10^{-3}}{(30)\,(45) \times 10^{-4}} = 0.1091$$

Then

$$\frac{n}{h} = 0.1091\left[\left(1 + \frac{2}{0.1091}\right)^{\frac{1}{2}} - 1\right] = 0.370$$

Thus

$$n = 0.370h = 0.167 \text{ m}$$

The second moment of area of the equivalent concrete beam is

$$I_c = \frac{1}{3}bn^3 + mA\,(h - n)^2$$

$$= \frac{1}{3}(0.30)\,(0.167)^3 + 15\,(0.982 \times 10^{-3})\,(0.283)^2$$

$$= (0.466 + 1.180)\,10^{-3} \text{ m}^4$$

$$= 1.646 \times 10^{-3} \text{ m}^4$$

If the maximum allowable concrete stress is attained, the permissible moment is

$$M = \frac{\sigma_c I_c}{n} = \frac{(6.5 \times 10^6)\,(1.646 \times 10^{-3})}{0.167} = 64 \text{ kNm}$$

If the maximum allowable steel stress is attained, the permissible moment is

$$M = \frac{\sigma_s I_c}{m(h - n)} = \frac{(115 \times 10^6)\,(1.646 \times 10^{-3})}{15(0.283)} = 44.6 \text{ kNm}$$

Steel is therefore the limiting material, and the permissible bending moment is

$$M = 44.6 \text{ kNm}$$

Problem 11.5 A rectangular concrete beam, 30 cm wide, is reinforced on the tension side with four 2.5 cm diameter steel rods at a depth of 45 cm, and on the compression side with two 2.5 diameter rods at a depth of 5 cm. Estimate the permissible bending moment if the stresses in the concrete are not to exceed 6.5 MN/m^2 and in the steel 115 MN/m^2. The modular ratio is 15.

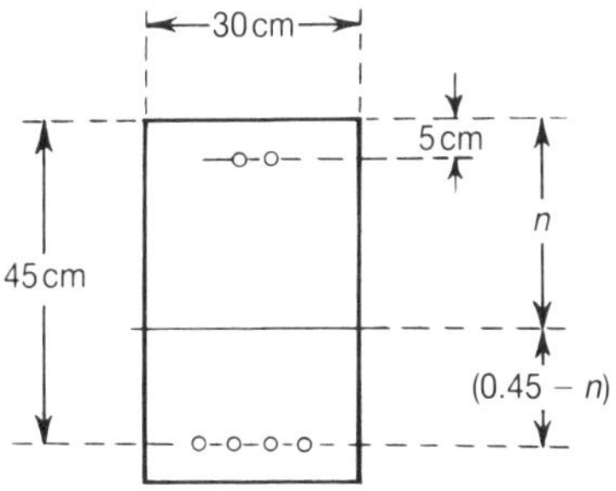

Solution

The area of steel reinforcement is 1.964×10^{-3} m^2 on the tension side, and 0.982×10^{-3} m^2 on the compression side. The cross-sectional area of the equivalent concrete beam is

$$(0.30)n + (m - 1)(0.000982) + m(0.001964) = (0.30n + 0.0433)\text{m}^2$$

The position of the neutral axis is obtained by taking moments, as follows:

$$(0.30)n\left(\frac{1}{2}n\right) + (m - 1)(0.000982)(0.05) + m(0.001964)(0.45)$$

$$= (0.30n + 0.0433)n$$

This reduces to

$$n^2 - 0.288n - 0.093 = 0$$

giving

$$n = -0.144 \pm 0.337$$

The relevant root is $n = 0.193$ m

The second moment of area of the equivalent concrete beam is

$$I_c = \frac{1}{3}(0.30)n^3 + (m-1)(0.000982)(n-0.05)^2 + m(0.001964)(0.45-n)^2$$

$$= (0.720 + 0.281 + 1.950)10^{-3}$$

$$= 2.95 \times 10^{-3} \text{ m}^4$$

If the maximum allowable concrete stress is attained, the permissible moment is

$$M = \frac{\sigma_c I_c}{n} = \frac{(6.5 \times 10^6)(2.95 \times 10^{-3})}{0.193} = 99.3 \text{ kNm}$$

If the maximum allowable steel stress is attained, the permissible moment is

$$M = \frac{\sigma_s I_c}{m(0.45 - n)} = \frac{(115 \times 10^6)(2.95 \times 10^{-3})}{15(0.257)} = 88.0 \text{ kNm}$$

Thus, steel is the limiting material, and the allowable moment is 88.0 kNm.

Problem 11.6 A steel I-section, 12.5 cm by 7.5 cm, is encased in a rectangular concrete beam of breadth 20 cm and depth 30 cm to the lower flange of the I-section. Estimate the position of the neutral axis of the composite beam, and find the permissible bending moment if the steel stress is not to exceed 115 MN/m^2 and the concrete stress 6.5 MN/m^2. The modular ratio is 15. The area of the steel beam is 0.00211 m^2 and its second moment of area about its minor axis is 5.70×10^{-6} m^4.

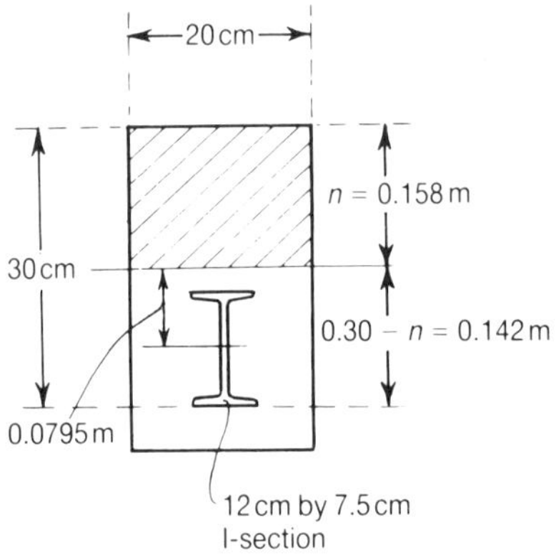

Solution

The area of the equivalent steel beam is

$$\frac{(0.20)n}{15} + 0.00211 \text{ m}^2$$

The position of the neutral axis is obtained by taking moments, as follows:

$$\left(\frac{0.20n}{15} + 0.00211\right)n = \left(\frac{0.20n}{15}\right)\left(\frac{1}{2}n\right) + (0.00211)(0.2375)$$

This reduces to

$$n^2 + 0.316n - 0.075 = 0$$

The relevant root of which is

$$n = 0.158 \text{ m}$$

The second moment of area of the equivalent steel beam is

$$I_s = \frac{1}{3}\left(\frac{0.20}{15}\right)(0.158)^3 + (0.00211)(0.0795)^2 = 0.0366 \times 10^{-3} \text{ m}^4$$

The allowable bending moment on the basis of the steel stress is

$$M = \frac{\sigma_s I_s}{(0.30 - n)} = \frac{(115 \times 10^6)(0.0366 \times 10^{-3})}{0.142} = 29.7 \text{ kNm}$$

If the maximum allowable concrete stress is 6.5 MN/m^2, the maximum allowable compressive stress in the equivalent steel beam is

$$m(6.5 \times 10^6) = 97.5 \text{ MN/m}^2$$

On this basis, the maximum allowable moment is

$$M = \frac{(97.5 \times 10^6)(0.0366 \times 10^{-3})}{0.158} = 22.6 \text{ kNm}$$

Concrete is therefore the limiting material, and the maximum allowable moment is

$$M = 22.6 \text{ kNm}$$

Problem 11.7 A reinforced concrete T-beam contains 1.25×10^{-3} m^2 of steel reinforcement on the tension side. If the steel stress is limited to 115 MN/m^2 and the concrete stress to 6.5 MN/m^2, estimate the permissible bending moment. The modular ratio is 15.

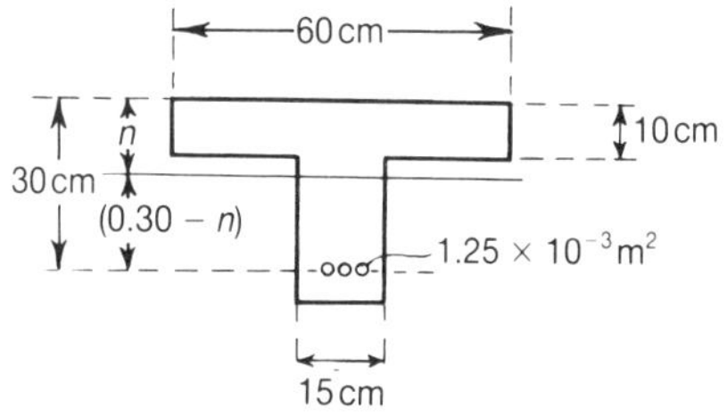

Solution

Suppose the neutral axis falls below the underside of the flange. The area of the equivalent concrete beam is

$$(0.60)n - 0.45(n - 0.10) + (0.00125)15 = 0.15n + 0.0638 \text{ m}^2$$

The position of the neutral axis is obtained by taking moments, as follows:

$$(0.60n)\left(\frac{1}{2}n\right) + (0.00125)(15)(0.30) - 0.45(n - 0.10)\left(\frac{1}{2}\right)(n + 0.10)$$

$$= (0.15n + 0.0638)n$$

This reduces to

$$n^2 + 0.850n - 0.1044 = 0$$

the relevant root of which is $n = 0.109$ m which agrees with our assumption earlier that the neutral axis lies below the flange.

The second moment of area of the equivalent concrete beam is

$$I_c = \frac{1}{3}(0.60)\left(n^3\right) - \frac{1}{3}(0.45)(n - 0.10)^3 + 0.00125\,(15)\,(0.30 - n)^2$$

$$= (0.259 + 0.000 + 0.685)10^{-3} \text{ m}^4$$

$$= 0.944 \times 10^{-3} \text{ m}^4$$

If the maximum allowable concrete stress is attained, the permissible moment is

$$M = \frac{\sigma_c I_c}{n} = \frac{\left(6.5 \times 10^6\right)\left(0.944 \times 10^{-3}\right)}{0.109} = 56.3 \text{ kNm}$$

If the maximum allowable steel stress is attained, the permissible moment is

$$M = \frac{\sigma_s I_c}{m(0.30 - n)} = \frac{(115 \times 10^6)(0.944 \times 10^{-3})}{15(0.191)} = 37.9 \text{ kNm}$$

Steel is therefore the limiting material, and the permissible bending moment is 37.9 kNm.

Further Problems *(answers on page 693)*

11.8 A concrete beam of rectangular section is 10 cm wide and is reinforced with steel bars whose axes are 30 cm below the top of the beam. Estimate the required total area of the cross-section of the steel if the maximum compressive stress in the concrete is to be 7.5 MN/m^2 and the tensile stress in the steel is 135 MN/m^2 beam is subjected to pure bending. What bending moment would the beam withstand when in this condition? Assume that Young's modulus for steel is 15 times that for concrete and that concrete can sustain no tensile stresses. *(Cambridge)*

11.9 A reinforced concrete T-beam carries a uniformly distributed super-load on a simply-supported span of 8 m. The stresses in the steel and concrete are not to exceed 125 MN/m^2 and 7 MN/m^2, respectively. The modular ratio is 15, and the density of concrete is 2400 kg/m^3. Determine the permissible super-load. *(Nottingham)*

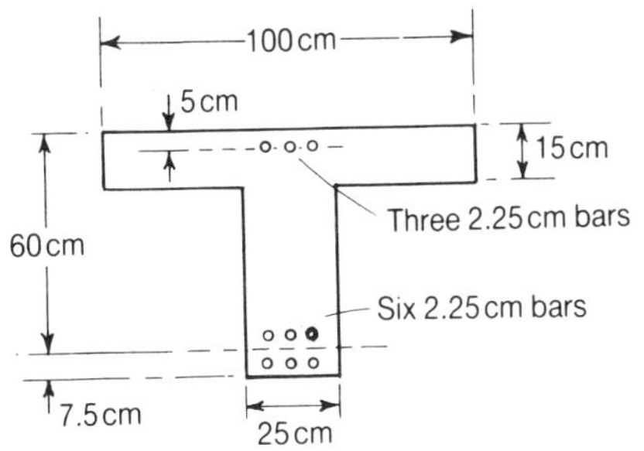

11.10 A wooden joist 15 cm deep by 7.5 cm wide is reinforced by glueing to its lower face a steel strip 7.5 cm wide by 0.3 cm thick. The joist is simply-supported over a span of 3 m, and carries a uniformly distributed load of 5000 N. Find the maximum direct stresses in the wood and steel and the maximum shearing stress in the glue. Take $E_s/E_t = 20$. *(Cambridge)*

11.11 A timber beam is 15 cm deep by 10 cm wide, and carries a central load of 30 kN at the centre of a 3 m span; the beam is simply-supported at each end. The timber is reinforced with flat steel plates 10 cm wide by 1.25 cm thick bolted to the upper and lower surfaces of the beam. Taking E for steel as 200 GN/m^2 and E for timber as 1 GN/m^2, estimate

(i) the maximum direct stress in the steel strips;
(ii) the average shearing stress in the timber;
(iii) the shearing load transmitted by the bolts;
(iv) the bending and shearing deflections at the centre of the beam.

12 Bending stresses and direct stresses combined

12.1 Introduction

Many instances arise in practice where a member undergoes bending combined with a thrust or pull. If a member carries a thrust, direct longitudinal stresses are set up; if a bending moment is now superimposed on the member at some section, additional longitudinal stresses are induced.

In this chapter we shall be concerned with the combined bending and thrust of short stocky members; in such cases the presence of a thrust does not lead to overall instability of the member. Buckling of beams under end thrust is discussed later in Chapter 18.

12.2 Combined bending and thrust of a stocky strut

Consider a short column of rectangular cross-section, Figure 12 1(i). The column carries an axial compressive load P, together with bending moment M, at some section, applied about the centroidal axis Cx.

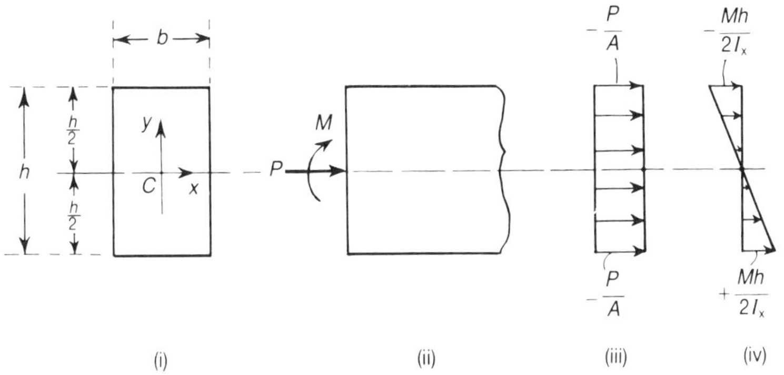

Figure 12.1 Combined bending and thrust of a rectangular cross-section beam.

The area of the column is A, and I_x is the second moment of the area about Cx. If P acts alone, the average longitudinal stress over the section is

$$-\frac{P}{A}$$

the stress being compressive. If the couple M acts alone, and if the material remains elastic, the longitudinal stress in any fibre a distance y from Cx is

$$-\frac{My}{I_x}$$

for positive values of y. We assume now that the combined effect of the thrust and the bending moment is the sum of the separate effects of P and M. The stresses due to P and M acting separately are shown in Figure 12.1(iii) and (iv). On combining the two stress systems, the resultant stress in any fibre is

$$\sigma = -\frac{P}{A} - \frac{My}{I_x} \tag{12.1}$$

Clearly the greatest compressive stress occurs in the upper extreme fibres, and has the value

$$\sigma_{max} = -\frac{P}{A} - \frac{Mh}{2I_x} \tag{12.2}$$

In the lower fibres of the beam y is negative; in the extreme lower fibres

$$\sigma = -\frac{P}{A} + \frac{Mh}{2I_x} \tag{12.3}$$

which is compressive or tensile depending upon whether $(Mh/2I_x)$ is less than or greater than (P/A). The two possible types of stress distribution are shown in Figure 12.2(i) and (ii). When $(Mh/2I_x) < (P/A)$, the stresses are compressive for all parts of the cross-section, Figure 12.2(i). When $(Mh/2I_x) > (P/A)$, the stress is zero at a distance (PI_x/AM) below the centre line of the beam, Figure 12.2(ii); this defines the position of the neutral axis of the column, or the axis of zero strain. In Figure 12.2(i) the imaginary neutral axis is also a distance (PI_x/AM) from the centre line, but it lies outside the cross-section.

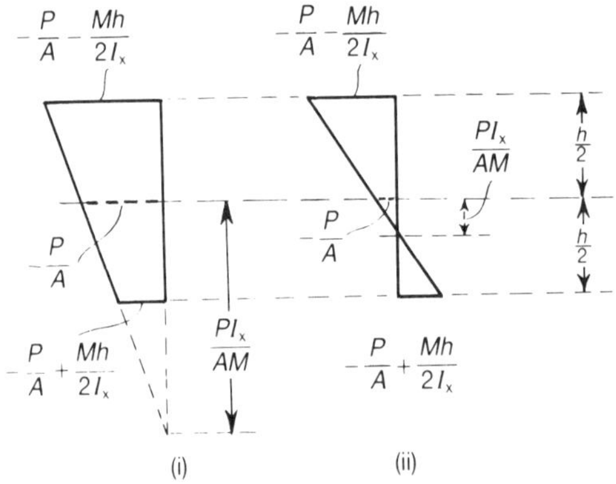

Figure 12.2 Position of the neutral axis for combined bending and thrust.

12.3 Eccentric thrust

We can use the analysis of Section 12.2 to find the stresses due to the eccentric thrust. The column of rectangular cross-section shown in Figure 12.3(i) carries a thrust P, which can be regarded as concentrated at the point D, which lies on the centroidal axis Cy, at a distance e_y from C, Figure 12.3(ii). The eccentric load P is statically equivalent to an axial thrust P and a bending moment Pe_y applied about Cx, Figure 12.3(iii). Then, from equation (12.1), the longitudinal stress any fibre is

$$\sigma = -\frac{P}{A} - \frac{Pe_y y}{I_x} = -\frac{P}{A}\left(1 + \frac{Ae_y y}{I_x}\right) \qquad (12.4)$$

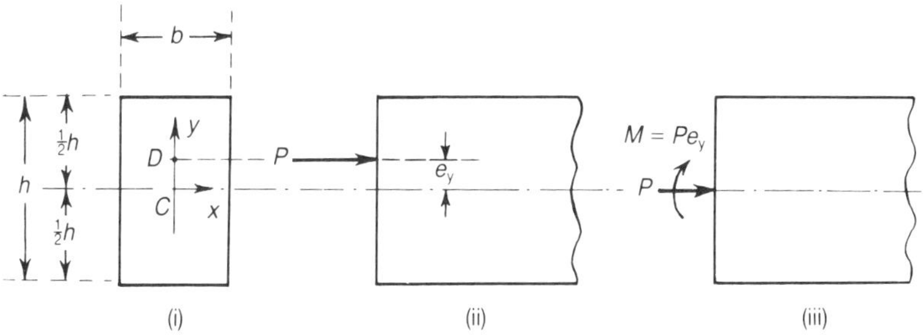

Figure 12.3 Column of rectangular cross-section carrying an eccentric thrust.

We are interested frequently in the condition that no tensile stresses occur in the column; clearly, tensile stresses are most likely to occur in the lowest extreme fibres, where

$$\sigma = -\frac{P}{A}\left(1 - \frac{Ae_y h}{2I_x}\right) \qquad (12.5)$$

This stress is tensile if

$$\frac{Ae_y h}{2I_x} > 1 \qquad (12.6)$$

that is, if

$$\frac{6e_y}{h} > 1$$

or

$$e_y = \frac{1}{6}h \qquad (12.7)$$

Now suppose the thrust P is applied eccentrically about both centroidal axes, at a distance e_x from the axis Cy and a distance e_y from the axis Cx, Figure 12.4. We replace the eccentric thrust P by an axial thrust P at C, together with couples Pe_y and Pe_x about Cx and Cy, respectively.

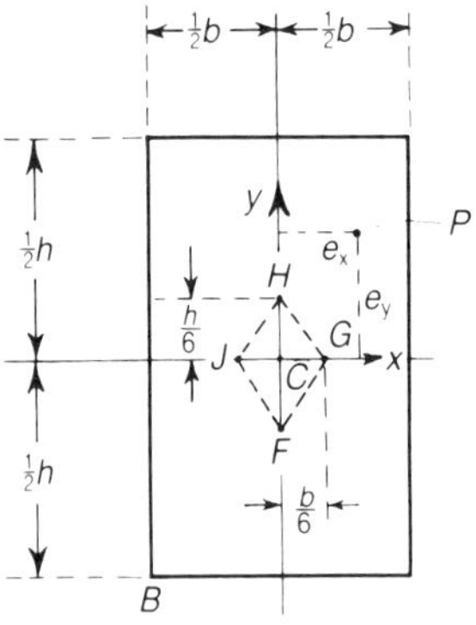

Figure 12.4 Core of a rectangular cross-section.

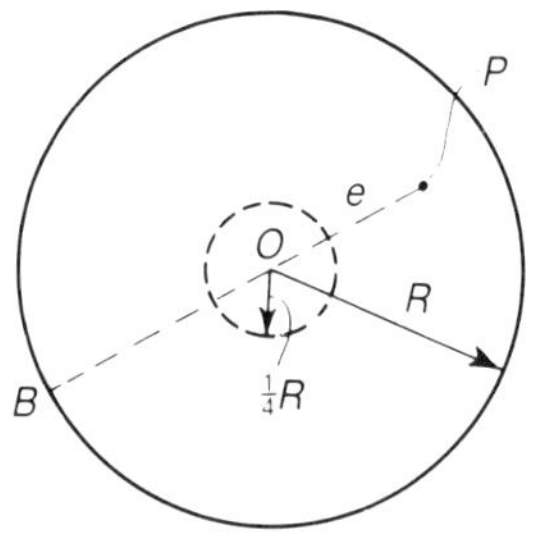

Figure 12.5 Core of a circular cross-section.

The resultant compressive stress at any fibre defined by co-ordinate (x, y) is

$$\sigma = -\frac{P}{A} - \frac{Pe_x x}{I_y} - \frac{Pe_y y}{I_x}$$

$$= -\frac{P}{A}\left[1 + \frac{Ae_x x}{I_y} + \frac{Ae_y y}{I_x}\right] \qquad (12.8)$$

Suppose e_x and e_y are both positive; then a tensile stress is more likely to occur at the corner B of the rectangle. The stress at B is tensile when

$$1 - \frac{Ae_x b}{2I_y} - \frac{Ae_y h}{2I_x} < 0 \qquad (12.9)$$

On substituting for A, I_x and I_y, this becomes

$$1 - \frac{6e_x}{h} - \frac{6e_y}{h} < 0 \qquad (12.10)$$

If P is applied at a point on the side of the line HG remote from C, this inequality is satisfied, and the stress at B becomes tensile, regardless of the value of P. Similarly, the lines HJ, JF and FG define limits on the point of application of P for the development of tensile stresses at the other three corners of the column. Clearly, if no tensile stresses are to be induced at all, the load P must not be applied outside the parallelogram $FGHJ$ in Figure 12.4; the region $FGHJ$ is known as the *core of the section*. For the rectangular section of Figure 12.4 the core is a parallelogram with diagonals of lengths ⅓h and ⅓b.

For a column with a circular cross-section of radius R, Figure 12.5, the tensile stress is most likely to develop at a point B on the perimeter diametrically opposed to the point of application of P. The stress at B is

$$\sigma = -\frac{P}{A} + \frac{PeR}{I} = -\frac{P}{A}\left(1 - \frac{AeR}{I}\right) \qquad (12.11)$$

where I is the second moment of area about a diameter. Tensile stresses are developed if

$$\frac{AeR}{I} > 1 \qquad (12.12)$$

On substituting for A and I, this becomes

$$\frac{4e}{R} > 1$$

or

$$e > \frac{R}{4} \qquad (12.13)$$

The core of the section is then a circle of radius ¼R.

Problem 12.1 Find the maximum stress on the section AB of the clamp when a pressure of 2500 N is exerted by the screw. The section is rectangular 2.5 cm by 1 cm. *(Cambridge)*

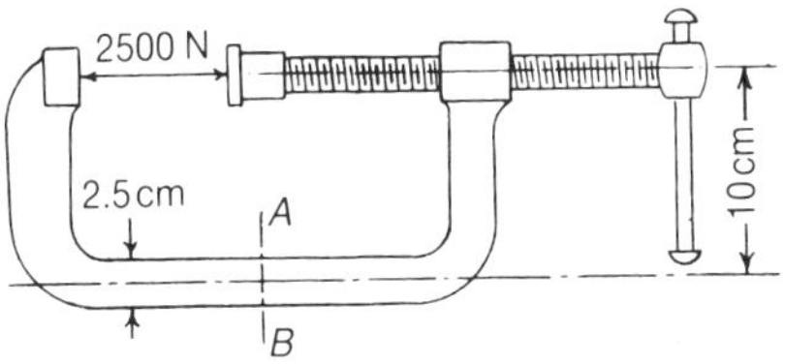

<u>*Solution*</u>

The section AB is subjected to a tension of 2500 N, and a bending moment (2500)(0.10) = 250 Nm. The area of the section = 0.25×10^{-3} m^2. The direct tensile stress = $(2500)/(0.25 \times 10^{-3})$ = 10 MN/m^2. The second moment of area = $1/12\,(0.01)(0.025)^3$ = 13.02×10^{-9} m^4.

Therefore, the maximum bending stresses due to the couple of 250 Nm are equal to

$$\frac{(250)(0.0125)}{(13.02 \times 10^{-9})} = 240 \text{ MN/m}^2$$

Hence the maximum tensile stress on the section is

$$(240 + 10) = 250 \text{ MN/m}^2$$

The maximum compressive stress is

$$(240 - 10) = 230 \text{ MN/m}^2$$

Problem 12.2 A masonry pier has a cross-section 3 m by 2 m, and is subjected to a load of 1000 kN, the line of the resultant being 1.80 m from one of the shorter sides, and 0.85 m from one of the longer sides. Find the maximum tensile and compressive stresses produced. *(Cambridge)*

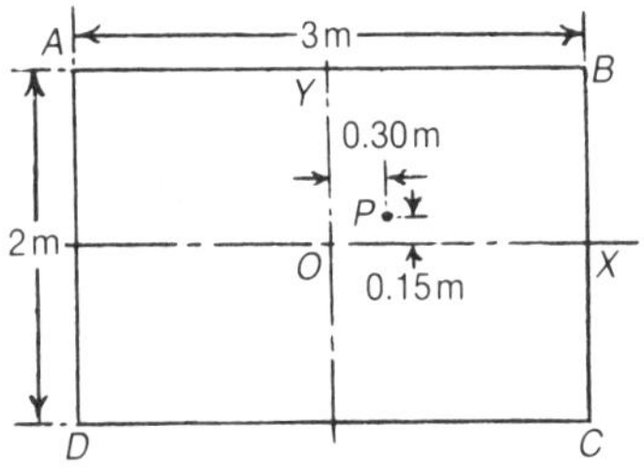

<u>*Solution*</u>

P represents the line of action of the thrust. The bending moments are

$$(0.15)(1000 \times 10^3) = 150 \text{ kNm about } OX$$

$$(0.30)(1000 \times 10^3) = 300 \text{ kNm about } OY$$

Now,

$$I_x = \frac{1}{12}(3)(2)^3 = 2 \text{ m}^4$$

$$I_y = \frac{1}{12}(2)(3)^3 = 4.5 \text{ m}^4$$

The cross-sectional area is

$$A = (3)(2) = 6 \text{ m}^2$$

For a point whose co-ordinates are (x, y) the compressive stress is

$$\sigma = -\frac{P}{A}\left(1 + \frac{Ae_x x}{I_y} + \frac{Ae_y y}{I_x}\right)$$

which gives

$$\sigma = -\frac{1000 \times 10^3}{6}\left(1 + \frac{x}{2.5} + \frac{9y}{20}\right)$$

The compressive stress is a maximum at B, where $x = 1.5$ m and $y = 1$ m. Then

$$\sigma_B = -\frac{10^6}{6}\left(1 + \frac{3}{5} + \frac{9}{20}\right) = -0.342 \text{ MN/m}^2$$

The stress at D, where $x = -1.5$ m and $y = -1$ m, is

$$\sigma_D = -\frac{10^6}{6}\left(1 - \frac{3}{5} - \frac{9}{20}\right) = +0.008 \text{ MN/m}^2$$

which is the maximum tensile stress.

12.4 Pre-stressed concrete beams

The simple analysis of Section 12.2 is useful for problems of pre-stressed concrete beams. A concrete beam, unreinforced with steel, can withstand negligible bending loads because concrete is so weak in tension. But if the beam be pre-compressed in some way, the tensile stresses induced by bending actions are countered by the compressive stresses already present. In Figure 12.6, for example, a line of blocks carries an axial thrust; if this is sufficiently large, the line of blocks can be used in the same way as a solid beam.

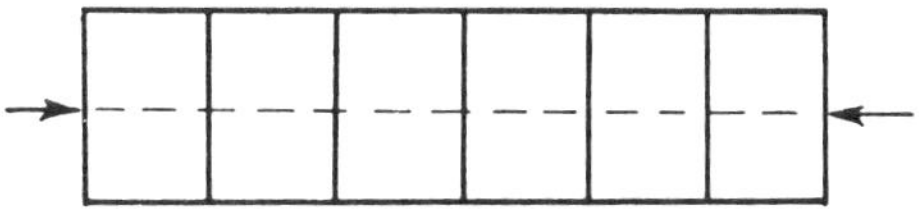

Figure 12.6 Bending strength of a pre-compressed line of blocks.

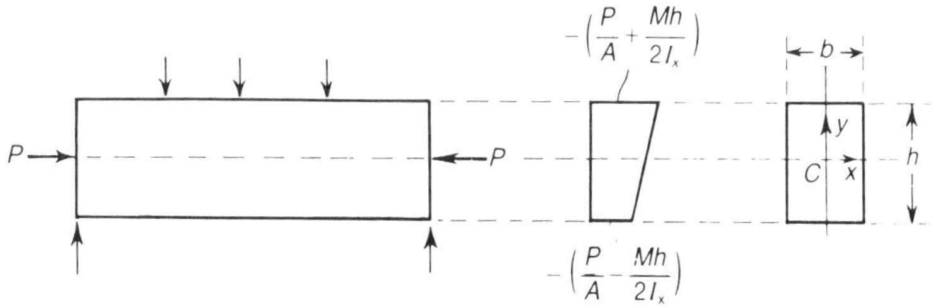

Figure 12.7 Concrete beam with axial pre-compression.

Suppose a concrete beam of rectangular cross-section, Figure 12.7, carries some system of lateral loads and is supported at its ends. An axial pre-compression P is applied at the ends. If M is the sagging moment at any cross-section, the greatest compressive stress occurs in the extreme top fibres, and has the value

$$\sigma = -\left(\frac{P}{A} + \frac{Mh}{2I_x}\right) \tag{12.14}$$

The stress in the extreme bottom fibres is

$$\sigma = -\left(\frac{P}{A} - \frac{Mh}{2I_x}\right) \tag{12.15}$$

Now suppose the maximum compressive stress in the concrete is limited to σ_1, and the maximum tensile stress to σ_2. Then we must have

$$\frac{P}{A} + \frac{Mh}{2I_x} \leq \sigma_1 \tag{12.16}$$

and

$$-\frac{P}{A} + \frac{Mh}{2I_x} \leq \sigma_2 \tag{12.17}$$

Then the design conditions are

$$\frac{Mh}{2I_x} \leq \sigma_1 - \frac{P}{A} \tag{12.18}$$

$$\frac{Mh}{2I_x} \leq \sigma_2 + \frac{P}{A} \tag{12.19}$$

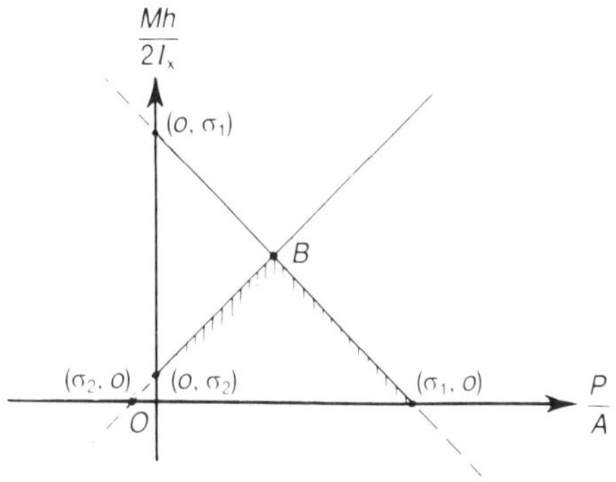

Figure 12.8 Optimum conditions for a beam with axial pre-compression.

These two inequalities are shown graphically in Figure 12.8, in which (P/A) is plotted against $(Mh/2I_x)$. Usually σ_2 is of the order of one-tenth of σ_1. The optimum conditions satisfying both inequalities occur at the point B; the maximum bending moment which can be given by

$$\frac{Mh}{I_x} = (\sigma_1 + \sigma_2) \tag{12.20}$$

that is,

$$M_{max} = \frac{I_x}{h}(\sigma_1 + \sigma_2) \tag{12.21}$$

The required axial thrust for this load is

$$P = \frac{1}{2}A(\sigma_1 - \sigma_2) \tag{12.22}$$

Some advantage is gained by pre-compressing the beam eccentrically; in Figure 12.9(i) a beam of rectangular cross-section carries a thrust P at a depth $(1/6)h$ below the centre line. As we saw in Section 12.3, this lies on the edge of the core of cross-section, and no tensile stresses are induced. In the upper extreme fibres the longitudinal stress is zero, and in the lower extreme fibres the compressive stress is $2P/A$, Figure 12.9(i).

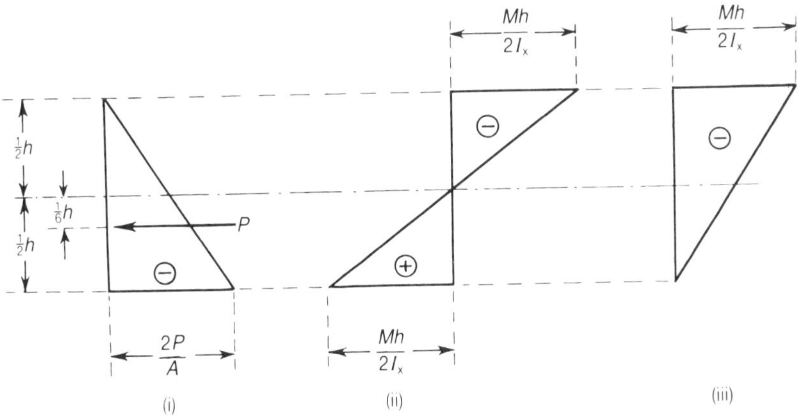

Figure 12.9 Concrete beam with eccentric pre-compression.

Now suppose a sagging bending moment M is superimposed on the beam; the extreme fibre stresses due to M are $(Mh/2I_x)$ tensile on the lower and compressive on the upper fibres, Figure 12.9(ii). If

$$\frac{2P}{A} = \frac{Mh}{2I_x} \tag{12.23}$$

then the resultant stresses, Figure 12.9(iii), are zero in the extreme lower fibres and a compressive stress of $(Mh/2I_x)$ in the extreme upper fibres. If this latter compressive stress does not exceed σ_1, the allowable stress in concrete, the design is safe. The maximum allowable value of M is

$$M = \frac{2I_x}{h}\sigma_1 \qquad (12.24)$$

As σ_2 in equation (12.21) is considerably less then σ_1, the bending moment given by equation (12.21) is approximately half that given by equation (12.24). Thus pre-compression by an eccentric load gives a considerably higher bending strength.

In practice the thrust is applied to the beam either externally through rigid supports, or by means of a stretched high-tensile steel wire passing through the beam and anchored at each end.

Further problems *(answers on page 693)*

12.3 The single rope of a cantilever crane supports a load of 200 kN and passes over two pulleys and then vertically down the axis of the crane to the hoisting apparatus. The section AB of the crane is a hollow rectangle. The outside dimensions are 37.5 cm and 75 cm and the material is 2.5 cm thick all round, and the longer dimension is in the direction AB. Calculate the maximum tensile and compressive stresses set up in the section, and locate the position of the neutral axis. *(Cambridge)*

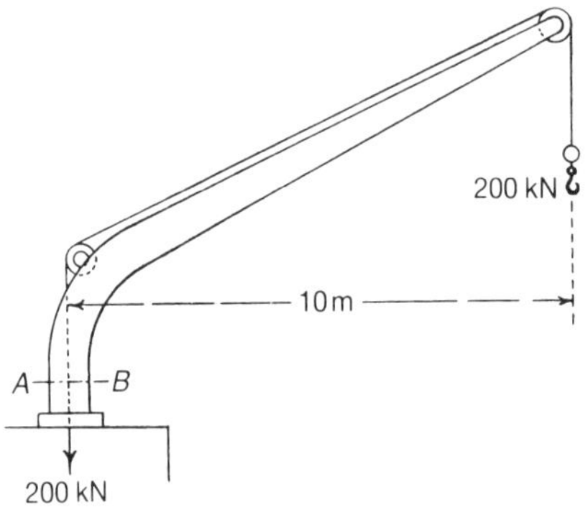

12.4 The horizontal cross-section of the cast-iron standard of a vertical drilling machine has the form shown. The line of thrust of the drill passes through P. Find the greatest value the thrust may have without the tensile stress exceeding 15 MN/m^2. What will be the stress along the face AB? *(Cambridge)*

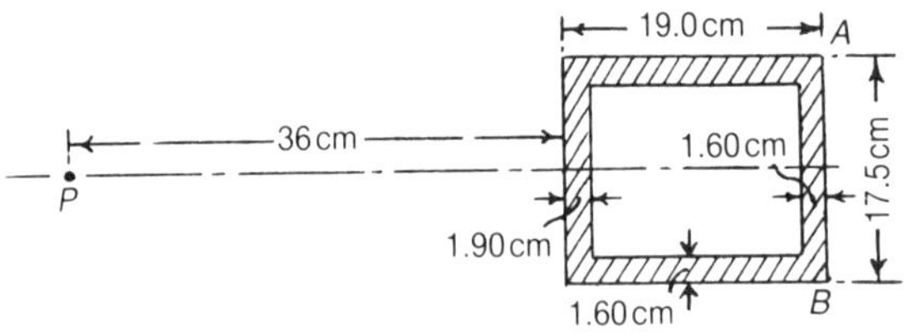

12.5 A vertical masonry chimney has a internal diameter d_1 and an external diameter d_0. The base of the chimney is given a horizontal acceleration a m/s^2, and the whole chimney moves horizontally with this acceleration. Show that at a section at depth h below the top of the chimney, the resultant normal force acts at a distance $ah/2g$ from the centre of the section. If the chimney behaves as an elastic solid, show that at a depth $g(d_0^2 + d_1^2)/4ad_0$ below the top, tensile stress will be developed in the material. *(Cambridge)*

12.6 A link of a valve gear has to be curved in one plane, for the sake of clearance. Estimate the maximum tensile and compressive stress in the link if the thrust is 2500 N. *(Cambridge)*

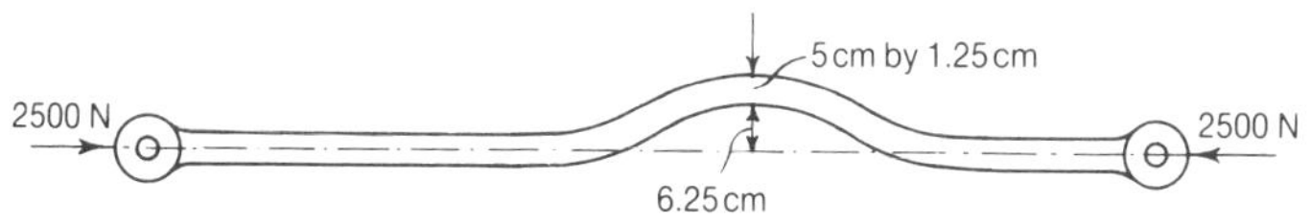

12.7 A cast-iron crank has a section on the line AB of the form shown. Show how to determine the greatest compressive and tensile stresses at AB, normal to the section, due to the thrust P of the connecting rod at the angle φ shown.

If the stresses at the section must not exceed 75 MN/m^2, either in tension or compression, find the maximum value of the thrust P. *(Cambridge)*

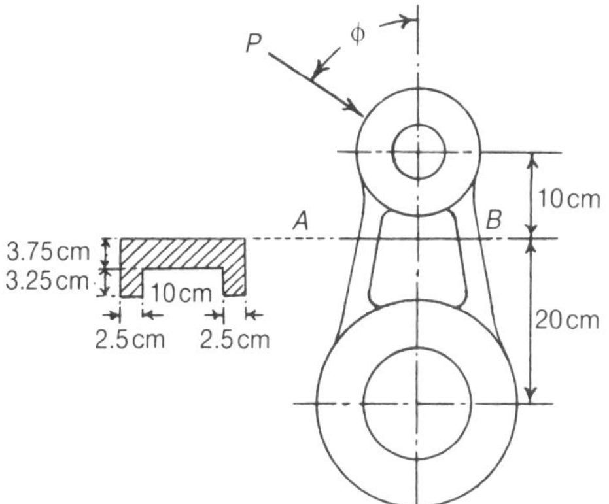

12.8 The load on the bearing of a cast-iron bracket is 5 kN. The form of the section *AB* is given. Calculate the greatest tensile stress across the section *AB* and the distance of the neutral axis of the section from the centre of gravity of the section. *(Cambridge)*

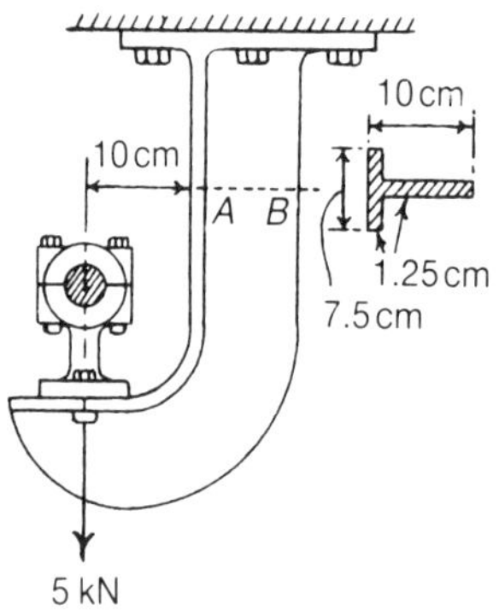

13 Deflections of beams

13.1 Introduction

In Chapter 7 we showed that the loading actions at any section of a simply-supported beam or cantilever can be resolved into a bending moment and a shearing force. Subsequently, in Chapters 9 and 10, we discussed ways of estimating the stresses due to these bending moments and shearing forces. There is, however, another aspect of the problem of bending which remains to be treated, namely, the calculation of the *stiffness* of a beam. In most practical cases, it is necessary that a beam should be not only strong enough for its purpose, but also that it should have the requisite stiffness, that is, it should not deflect from its original position by more than a certain amount. Again, there are certain types of beams, such as those carried by more than two supports and beams with their ends held in such a way that they must keep their original directions, for which we cannot calculate bending moments and shearing forces without studying the deformations of the axis of the beam; these problems are statically indeterminate, in fact.

In this chapter we consider methods of finding the deflected form of a beam under a given system of external loads and having known conditions of support.

13.2 Elastic bending of straight beams

It was shown in Section 9.2 that a straight beam of uniform cross-section, when subjected to end couples M applied about a principal axis, bends into a circular arc of radius R, given by

$$\frac{1}{R} = \frac{M}{EI} \qquad (13.1)$$

where EI, which is the product of Young's modulus E and the second moment of area I about the relevant principal axis, is the flexural stiffness of the beam; equation (13.1) holds only for *elastic* bending.

Where a beam is subjected to shearing forces, as well as bending moments, the axis of the beam is no longer bent to a circular arc. To deal with this type of problem, we assume that equation (13.1) still defines the radius of curvature at any point of the beam where the bending moment is M. This implies that where the bending moment varies from one section of the beam to another, the radius of curvature also varies from section to section, in accordance with equation (13.1).

In the unstrained condition of the beam, Cz is the longitudinal centroidal axis, Figure 13.1, and Cx, Cy are the principal axes in the cross-section. The co-ordinate axes Cx, Cy are so arranged that the y-axis is vertically downwards. This is convenient as most practical loading conditions give rise to vertically downwards deflections. Suppose bending moments are applied about axes parallel to Cx, so that bending is restricted to the yz-plane, because Cx and Cy are principal axes.

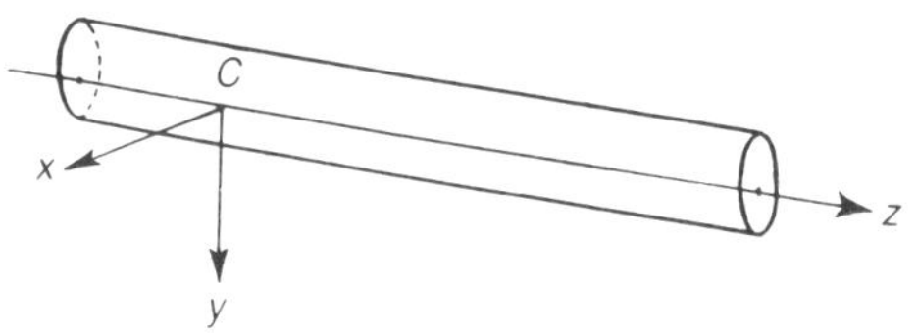

Figure 13.1 Longitudinal and principal centroidal axes for a straight beam.

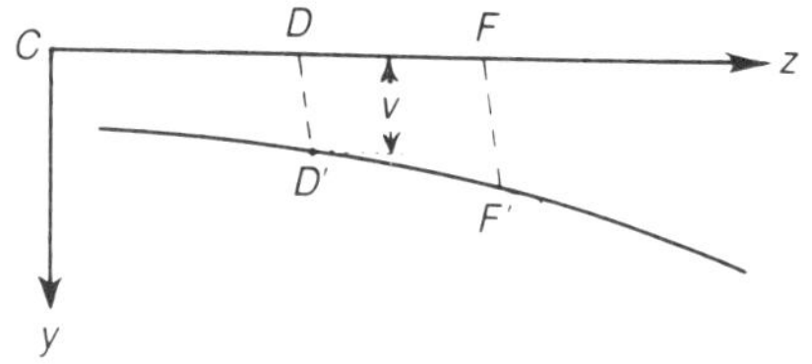

Figure 13.2 Displacements of the longitudinal axis of the beam.

Consider a short length of the unstrained beam, corresponding with *DF* on the axis *Cz*, Figure 13.2. In the strained condition *D* and *F* are displaced to *D'* and *F'*, respectively, which lies in the *yz*-plane. Any point such as *D* on the axis *Cz* is displaced by an amount *v* parallel to *Cy*; it is also displaced a small, but negligible, amount parallel to *Cz*.

The radius of curvature *R* at any section of the beam is then given by

$$\frac{1}{R} = \frac{\dfrac{d^2v}{dz^2}}{\pm\left[1 + \left(\dfrac{dv}{dz}\right)^2\right]^{3/2}} \tag{13.2}$$

We are concerned generally with only small deflections, in which *v* is small; this implies that (dv/dz) is small, and that $(dv/dz)^2$ is negligible compared with unity. Then, with sufficient accuracy, we may write

$$\frac{1}{R} = \pm\frac{d^2v}{dz^2} \tag{13.3}$$

The equations (13.1) and (13.3) give

$$\pm\, EI\frac{d^2v}{dz^2} = M \tag{13.4}$$

We must now consider whether the positive or negative sign is relevant in this equation; we have already adopted the convention in Section 7.4 that sagging bending moments are positive. When a length of the beam is subjected to sagging bending moments, as in Figure 13.3, the value of (dv/dz) along the length diminishes as *z* increases; hence a sagging moment implies that the curvature is negative. Then

$$EI\frac{d^2v}{dz^2} = -M \tag{13.5}$$

where *M* is the *sagging* bending moment.

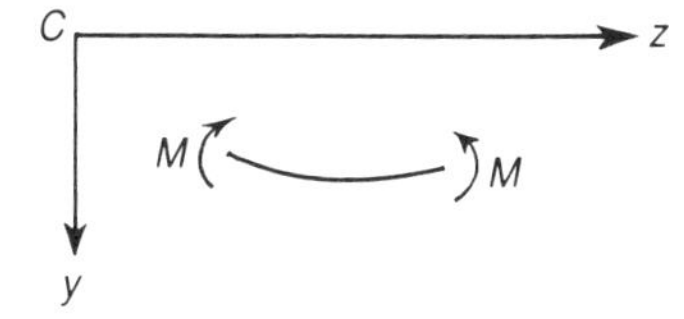

Figure 13.3 Curvature induced by sagging bending moment.

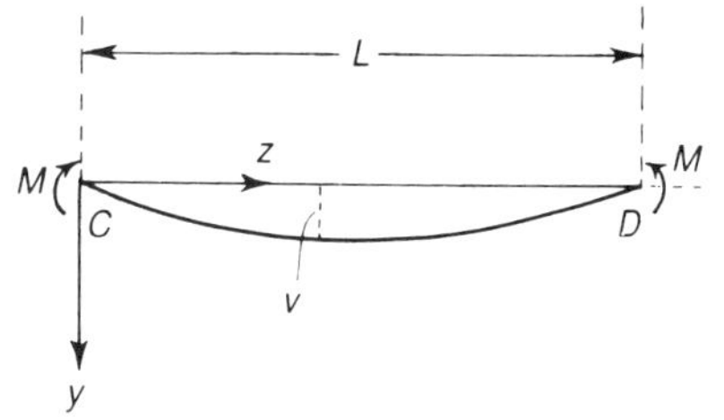

Figure 13.4 Deflected form of a beam in pure bending.

Where the beam is loaded on its axis of shear centres, so that no twisting occurs, M may be written in terms of shearing force F and intensity w of vertical loading at any section. From equation (7.9) we have

$$\frac{d^2M}{dz^2} = \frac{dF}{dz} = -w$$

On substituting for M from equation (13.5), we have

$$\frac{d^2}{dz^2}\left[-EI\frac{d^2v}{dz^2}\right] = \frac{dF}{dz} = -w \qquad (13.6)$$

This relation is true if EI varies from one section of a beam to another. Where EI is constant along the length of a beam,

$$-EI\frac{d^4v}{dz^4} = \frac{dF}{dz} = -w \qquad (13.7)$$

As an example of the use of equation (13.4), consider the case of a uniform beam carrying couples M at its ends, Figure 13.4. The bending moment at any section is M, so the beam is under a constant bending moment. Equation (13.5) gives

$$EI\frac{d^2v}{dz^2} = -M$$

On integrating once, we have

$$EI\frac{dv}{dz} = -Mz + A \qquad (13.8)$$

where A is a constant. On integrating once more

$$EIv = -\frac{1}{2} Mz^2 + Az + B \tag{13.9}$$

where B is another constant. If we measure v relative to a line CD joining the ends of the beam, v is zero at each end. Then $v = 0$, for $z = 0$ and $z = L$.

On substituting these two conditions into equation (13.9), we have

$$B = 0 \quad \text{and} \quad A = \frac{1}{2} ML$$

The equation (13.9) may be written

$$EIv = \frac{1}{2} Mz(L - z) \tag{13.10}$$

At the mid-length, $z = ½L$, and

$$v = \frac{ML^2}{8EI} \tag{13.11}$$

which is the greatest deflection. At the ends $z = 0$ and $z = L/2$,

$$\frac{dv}{dz} = \frac{ML}{2EI} \text{ at } C; \quad \frac{dv}{dz} = -\frac{ML}{2EI} \text{ at } D \tag{13.12}$$

It is important to appreciate that equation (13.3), expressing the radius of curvature R in terms of v, is only true if the displacement v is small.

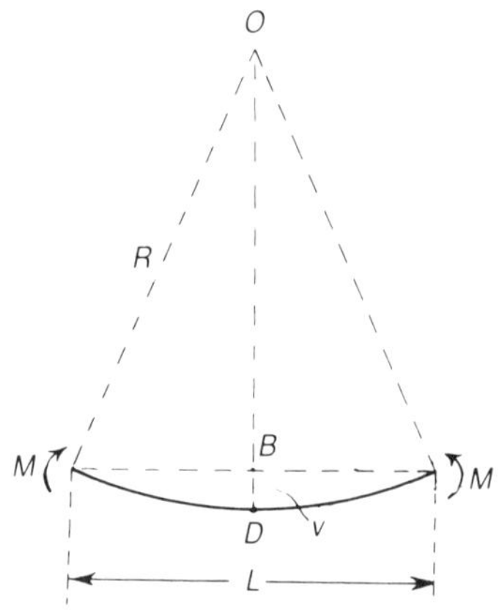

Figure 13.5 Distortion of a beam in pure bending.

We can study more accurately the pure bending of a beam by considering it to be deformed into the arc of a circle, Figure 13.5; as the bending moment M is constant at all sections of the beam, the radius of curvature R is the same for all sections. If L is the length between the ends, Figure 13.5, and D is the mid-point,

$$OB = \sqrt{R^2 - (L^2/4)}$$

Thus the central deflection v, is

$$v = BD = R - \sqrt{R^2 - (L^2/4)}$$

Then

$$v = R\left[1 - \sqrt{1 - \frac{L^2}{4R^2}}\right]$$

Suppose L/R is considerably less than unity; then

$$v = R\left[\frac{1}{2}\left(\frac{L^2}{4R^2}\right) + \frac{1}{8}\left(\frac{L^2}{4R^2}\right)^2 + \ldots\right]$$

which can be written

$$v = \frac{L^2}{8R}\left[1 + \frac{L^2}{4R^2} + \ldots\right]$$

But

$$\frac{1}{R} = \frac{M}{EI}$$

and so

$$v = \frac{ML^2}{8EI}\left[1 + \frac{M^2L^2}{4(EI)^2} + \ldots\right] \qquad (13.13)$$

Clearly, if $(L^2/4R^2)$ is negligible compared with unity we have, approximately,

$$v = \frac{ML^2}{8EI}$$

which agrees with equation (13.11). The more accurate equation (13.13) shows that, when $(L^2/4R^2)$

is not negligible, the relationship between v and M is non-linear; for all practical purposes this refinement is unimportant, and we find simple linear relationships of the type of equation (13.11) are sufficiently accurate for engineering purposes.

13.3 Simply-supported beam carrying a uniformly distributed load

A beam of uniform flexural stiffness EI and span L is simply-supported at its ends, Figure 13.6; it carries a uniformly distributed lateral load of w per unit length, which induces bending in the yz plane only. Then the reactions at the ends are each equal to $\frac{1}{2}wL$; if z is measured from the end C, the bending moment at a distance z from C is

$$M = \frac{1}{2}wLz - \frac{1}{2}wz^2$$

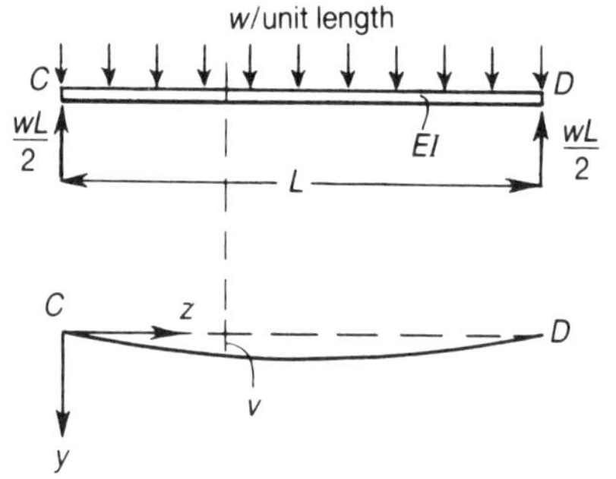

Figure 13.6 Simply-supported beam carrying a uniformly supported load.

Then from equation (13.5),

$$EI\frac{d^2v}{dz^2} = -M = -\frac{1}{2}wLz + \frac{1}{2}wz^2$$

On integrating,

$$EI\frac{dv}{dz} = -\frac{wLz^2}{4} + \frac{wz^3}{6} + A$$

and

$$EIv = -\frac{wLz^3}{12} + \frac{wz^4}{24} + Az + B \tag{13.14}$$

Suppose $v = 0$ at the ends $z = 0$ and $z = L$; then

$$B = 0, \quad \text{and} \quad A = wL^3/24$$

Then equation (13.14) becomes

$$EIv = \frac{wz}{24}\left[L^3 - 2Lz^2 + z^3\right] \qquad (13.15)$$

The deflection at the mid-length, $z = \frac{1}{2}L$, is

$$v = \frac{5wL^4}{384EI} \qquad (13.16)$$

13.4 Cantilever with a concentrated load

A uniform cantilever of flexural stiffness EI and length L carries a vertical concentrated load W at the free end, Figure 13.7. The bending moment a distance z from the built-in end is

$$M = -W(L - z)$$

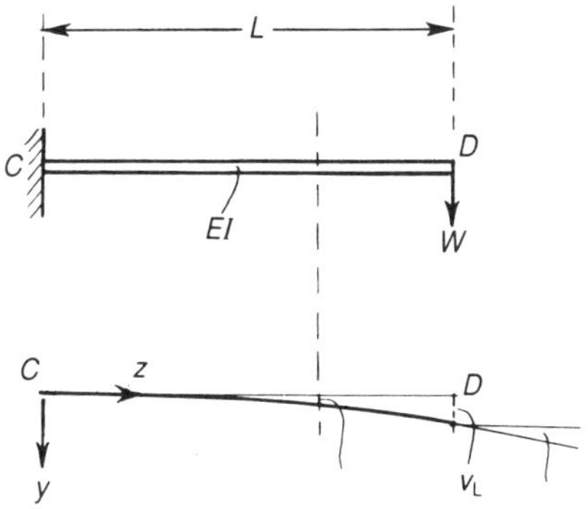

Figure 13.7 Cantilever carrying a vertical load at the remote end.

Hence equation (13.5) gives

$$EI\frac{d^2v}{dz^2} = W(L - z)$$

Then

$$EI\frac{dv}{dz} = W\left(Lz - \frac{1}{2}z^2\right) + A \qquad (13.17)$$

and

$$EIv = W\left(\frac{1}{2}Lz^2 - \frac{1}{6}z^3\right) + Az + B$$

At the end $z = 0$, there is zero slope in the deflected form, so that $dv/dz = 0$; then equation (13.17) gives $A = 0$. Furthermore, at $z = 0$ there is also no deflection, so that $B = 0$. Then

$$EIv = \frac{Wz^2}{6}(3L - z)$$

At the free end, $z = L$,

$$v_L = \frac{WL^3}{3EI} \qquad (13.18)$$

The slope of the beam at the free end is

$$\theta_L = \left(\frac{dv}{dz}\right)_{z = L} = \frac{WL^2}{2EI} \qquad (13.19)$$

When the cantilever is loaded at some point between the ends, at a distance a, say, from the built-in support, Figure 13.8, the beam between G and D carries no bending moments and therefore remains straight. The deflection at G can be deduced from equation (13.18); for $z = a$,

$$v_a = \frac{Wa^3}{3EI} \qquad (13.20)$$

and the slope at $z = a$ is

$$\theta_a = \frac{Wa^2}{2EI} \qquad (13.21)$$

Then the deflection at the free end D of the cantilever is

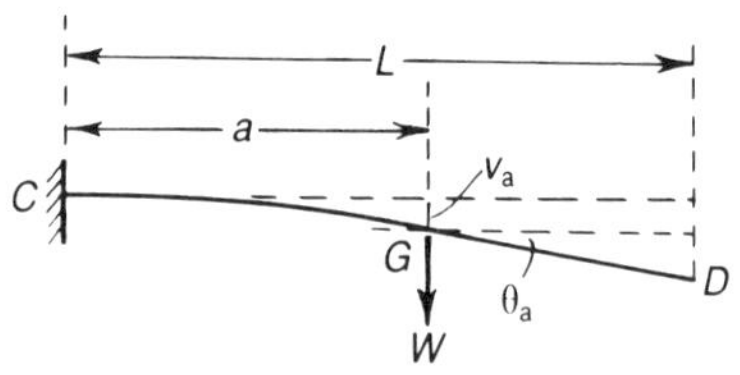

Figure 13.8 Cantilever with a load applied between the ends.

$$v_L = \frac{Wa^3}{3EI} + (L - a)\frac{Wa^2}{2EI}$$
$$= \frac{Wa^2}{6EI}(3L - a) \qquad (13.22)$$

13.5 Cantilever with a uniformly distributed load

A uniform cantilever, Figure 13.9, carries a uniformly distributed load of w per unit length over the whole of its length. The bending moment at a distance z from C is

$$M = -\frac{1}{2}w(L - z)^2$$

Then, from equation (13.5),

$$EI\frac{d^2v}{dz^2} = \frac{1}{2}w(L - z)^2 = \frac{1}{2}w(L^2 - 2Lz + z^2)$$

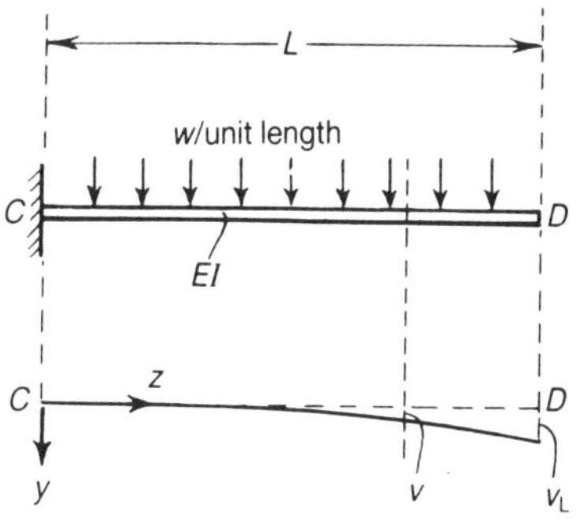

Figure 13.9 Cantilever carrying a uniformly distributed load.

Thus

$$EI\frac{dv}{dz} = \frac{1}{2}w\left(L^2z - Lz^2 + \frac{1}{3}z^3\right) + A$$

and

$$EIv = \frac{1}{2}w\left(\frac{1}{2}L^2z^2 - \frac{1}{3}Lz^3 + \frac{1}{12}z^4\right) + Az + B$$

At the built end, $z = 0$, and we have

$$\frac{dv}{dz} = 0 \quad \text{and} \quad v = 0$$

Thus $A = B = 0$. Then

$$EIv = \frac{1}{24} w \left(6L^2 z^2 - 4Lz^3 + z^4\right)$$

At the free end, D, the vertical deflection is

$$v_L = \frac{wL^4}{8EI} \tag{13.23}$$

13.6 Propped cantilever with distributed load

The uniform cantilever of Figure 13.10(i) carries a uniformly distributed load w and is supported on a rigid knife edge at the end D. Suppose P is the force on the support at D. Then we regard Figure 13.10(i) as the superposition of the effects of P and w acting separately.

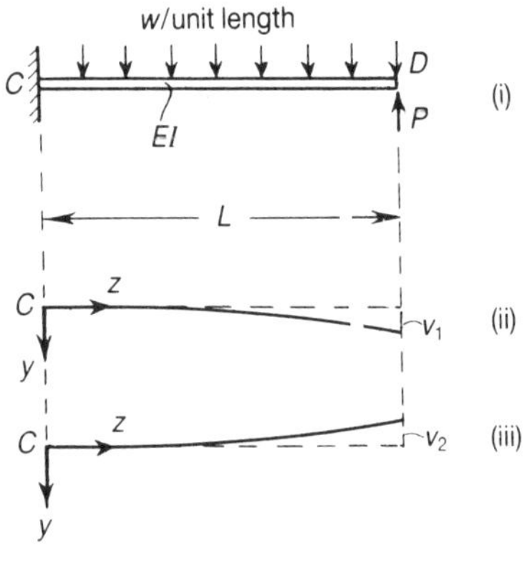

Figure 13.10 (i) Uniformly loaded cantilever propped at one end. (ii) Deflections due to w alone. (iii) Deflections due to P alone.

If w acts alone, the deflection at D is given by equation (13.23), and has the value

$$v_1 = \frac{wL^4}{8EI}$$

If the reaction P acted alone, there would be an upward deflection

$$v_2 = \frac{PL^3}{3EI}$$

at D. If the support maintains zero deflection at D,

$$v_1 - v_2 = 0$$

This gives

$$\frac{PL^3}{3EI} = \frac{wL^4}{8EI}$$

or

$$P = \frac{3wL}{8} \qquad (13.24)$$

Problem 13.1 A steel rod 5 cm diameter protrudes 2 m horizontally from a wall. (i) Calculate the deflection due to a load of 1 kN hung on the end of the rod. The weight of the rod may be neglected. (ii) If a vertical steel wire 3 m long, 0.25 cm diameter, supports the end of the cantilever, being taut but unstressed before the load is applied, calculate the end deflection on application of the load. Take $E = 200\ \mathrm{GN/m^2}$. *(RNEC)*

<u>*Solution*</u>

(i) The second moment of are of the cross-section is

$$I_x = \frac{\tau}{64}(0.050)^4 = 0.307 \times 10^{-6}\ \mathrm{m}^4$$

The deflection at the end is then

$$v = \frac{PL^3}{3EI} = \frac{(1000)(2)^3}{3(200 \times 10^9)(0.307 \times 10^{-6})} = 0.0434\ \mathrm{m}$$

(ii) Let T = tension in the wire; the area of cross-section of the wire is $4.90 \times 10^{-6}\ \mathrm{m}^2$. The elongation of the wire is then

$$e = \frac{Tl}{EA} = \frac{T(3)}{(200 \times 10^9)(4.90 \times 10^{-6})}$$

The load on the end of the cantilever is then $(1000 - T)$, and this produces a deflection of

$$v = \frac{(1000 - T)(2)^3}{3(200 \times 10^9)(0.307 \times 10^{-6})}$$

If this equals the stretching of the wire, then

$$\frac{(1000 - T)(2)^3}{3(200 \times 10^9)(0.307 \times 10^{-6})} = \frac{T(3)}{(200 \times 10^9)(4.90 \times 10^{-6})}$$

This gives $T = 934$ N, and the deflection of the cantilever becomes

$$v = \frac{(66)(2)^3}{3(200 \times 10^9)(0.307 \times 10^{-6})} = 0.00276 \text{ m}$$

Problem 13.2 A platform carrying a uniformly distributed load rests on two cantilevers projecting a distance l m from a wall. The distance between the two cantilevers is $\frac{1}{2}l$. In what ratio might the load on the platform be increased if the ends were supported by a cross girder of the same section as the cantilevers, resting on a rigid column in the centre, as shown? It may be assumed that when there is no load on the platform the cantilevers just touch the cross girder without pressure. *(Cambridge)*

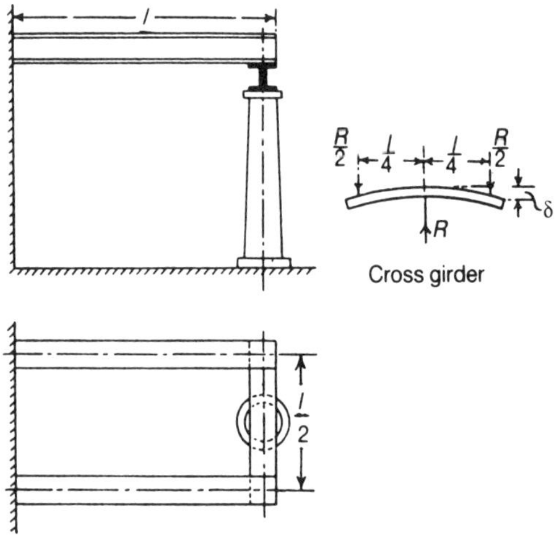

Solution

Let w_l = the safe load per unit length on each cantilever when unsupported.
Then the maximum bending moment = $\frac{1}{2}w_1 l^2$.
Let w_2 = the safe load when supported,
δ = the deflection of the end of each cantilever,
$\frac{1}{2}R$ = the pressure between each cantilever and the cross girder.
Then the pressure is

$$\frac{R}{2} = \frac{3}{8}w_2 l - \frac{3EI\delta}{l^3}$$

We see from the figure above that

$$\delta = \frac{(R/2)(l/4)^3}{3EI} = \frac{Rl^3}{384EI}$$

I having the same value for the cantilevers and cross girder. Substituting this value of δ

$$\frac{R}{2} = \frac{3w_2 l}{8} - \frac{R}{128}$$

or

$$R = \frac{48}{65} w_2 l$$

The upward pressure on the end of each cantilever is $\frac{1}{2}R = 24w_2l/65$, giving a bending moment at the wall equal to $24w_2l^2/65$. The bending moment of opposite sign due to the distributed load is $\frac{1}{2}w_2l^2$. Hence it is clear that the maximum bending moment due to both acting together must occur at the wall and is equal to $(\frac{1}{2} - 24/65)\, w_2l^2 = (17/130)\, w_2l^2$. If this is to be equal to $\frac{1}{2}\, w_1l^2$, we must have $w_2 = (65/17)\, w_1$; in other words, the load on the platform can be increased in the ratio 65/17, or nearly 4/1. The bending moment at the centre of the cross girder is $6w_2l^2/65$, which is less than that at the wall.

13.7 Simply-supported beam carrying a concentrated lateral load

Consider a beam of uniform flexural stiffness EI and length L, which is simply-supported at its ends C and G, Figure 13.11. The beam carries a concentrated lateral load W at a distance a from C. Then the reactions at C and G are

$$V_c = \frac{W}{L}(L - a) \qquad V_G = \frac{Wa}{L}$$

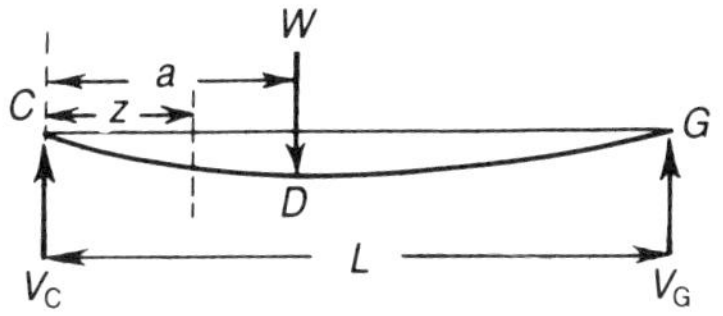

Figure 13.11 Deflections of a simply-supported beam carrying a concentrated lateral load.

Now consider a section of the beam a distance z from C; if $z < a$, the bending moment at the section is

$$M = V_c z$$

and if $z > a$,

$$M = V_c z - W(z - a)$$

Then

$$EI\frac{d^2v}{dz^2} = -V_c z \quad \text{for} \quad z < a$$

and

$$EI\frac{d^2v}{dz^2} = -V_c z + W(z - a) \quad \text{for} \quad z > a$$

On integrating these equations, we have

$$EI\frac{dv}{dz} = -\frac{1}{2}V_c z^2 + A \quad \text{for} \quad z < a \tag{13.25}$$

$$EI\frac{dv}{dz} = -\frac{1}{2}V_c z^2 + W\left(\frac{1}{2}z^2 - az\right) + A' \quad \text{for } z > a \tag{13.26}$$

and

$$EIv = -\frac{1}{6}V_c z^3 + Az + B \quad \text{for} \quad z < a \tag{13.27}$$

$$EIv = -\frac{1}{6}V_c z^3 + W\left(\frac{1}{6}z^3 - \frac{1}{2}az^2\right) + A'z + B' \text{ for } z > a \tag{13.28}$$

In these equations A, B, A' and B' are arbitrary constants. Now for $z = a$ the values of v given by equations (13.27) and (13.28) are equal, and the slopes given by equations (13.25) and (13.26) are also equal, as there is continuity of the deflected form of the beam through the point D. Then

$$-\frac{1}{6}V_c a^3 + Aa + B = -\frac{1}{6}V_c a^3 + W\left(\frac{1}{6}a^3 - \frac{1}{2}a^3\right) + A'a + B'$$

and

$$-\frac{1}{2}V_c a^2 + A = -\frac{1}{2}V_c a^2 + W\left(\frac{1}{2}a^2 - a^2\right) + A'$$

These two equations give

$$A' = A + \frac{1}{2}Wa^2$$
$$B' = B - \frac{1}{6}Wa^3 \qquad (13.29)$$

At the extreme ends of the beam $v = 0$, so that when $z = 0$ equation (13.27) gives $B = 0$, and when $z = L$, equation (13.28) gives

$$-\frac{1}{6}V_c L^3 + W\left(\frac{1}{6}L^3 - \frac{1}{2}aL^2\right) + A'L + B' = 0$$

We have finally,

$$A = \frac{1}{6}V_c L^2 - \frac{W}{6L}(L - a)^3$$
$$B = 0$$
$$A' = \frac{1}{6}V_c L^2 - \frac{W}{6L}(L - a)^3 + \frac{1}{2}Wa^2 \qquad (13.30)$$
$$B' = -\frac{1}{6}Wa^3$$

But $V_c = W(L - a)/L$, so that equations (13.30) become

$$A = \frac{Wa}{6L}(L - a)(2L - a)$$
$$B = 0$$
$$A' = \frac{Wa}{6L}\left(2L^2 + a^2\right) \qquad (13.31)$$
$$B' = -\frac{1}{6}Wa^3$$

Then equations (13.27) and (13.28) may be written

$$EIv = -\frac{W}{6L}(L - a)z^3 + \frac{Wa}{6L}\left(2L^2 - 3aL + a^2\right)z \quad \text{for } z < a \tag{13.32}$$

$$EIv = -\frac{W}{6L}(L - a)z^3 + \frac{W}{6}\left(z^3 - 3az^2\right)$$

$$+ \frac{Wa}{6L}\left(2L^2 + a^2\right)z - \frac{Wa^3}{6} \quad \text{for } z > a \tag{13.33}$$

The second relation, for $z > a$, may be written

$$EIv = -\frac{W}{6L}(L - a)z^3 + \frac{Wa}{6L}\left(2L^2 - 3aL + a^2\right)z + \frac{W}{6}(z - a)^3 \tag{13.34}$$

Then equations (13.32) and (13.33) differ only by the last term of equation (13.34); if the last term of equation (13.34) is discarded when $z < a$, then equation (13.34) may be used to define the deflected form in all parts of the beam.

On putting $z = a$, the deflection at the loaded point D is

$$v_D = \frac{Wa^2 (L - a)^2}{3EIL} \tag{13.35}$$

When W is at the centre of the beam, $a = ½L$, and

$$v_D = \frac{WL^3}{48EI} \tag{13.36}$$

This is the maximum deflection of the beam only when $a = ½L$.

13.8 Macaulay's method

The observation that equations (13.32) and (13.33) differ only by the last term of equation (13.34) leads to Macaulay's method, which ignores terms which are negative within the Macaulay brackets. That is, if the term $[z - a]$ in equation (13.34) is negative, it is ignored, so that equation (13.34) can be used for the whole beam. The method will be demonstrated by applying it to a few examples.

Consider the beam shown in Figure 13.12, which is simply-supported at its ends and loaded with a concentrated load W.

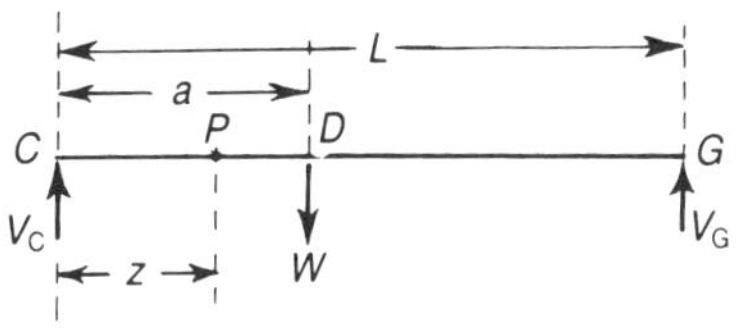

Figure 13.12 Form of step-function used in deflection analysis of a beam.

By taking moments, it can be seen that

$$V_c = W(L - a)/L \tag{13.37}$$

and the bending moment when $z < a$ is

$$M = V_c z \tag{13.38}$$

Then bending moment when $z > a$ is

$$M = V_c z - W(z - a) \tag{13.39}$$

Now

$$EI \frac{d^2v}{dz^2} = -M$$

hence, the Macaulay method allows us to express this relationship as follows

←------- $z <= a$ --------→ ←----$a < z < L$ -----→

$$EI \frac{d^2v}{dz^2} = -V_c z \qquad +W[z - a] \tag{13.40}$$

On integrating equation (13.40), we get

$$EI \frac{dv}{dz} = -\frac{V_c z^2}{2} + A \qquad +\frac{W}{2}[z - a]^2 \tag{13.41}$$

and

$$EIv = \frac{-V_c z^3}{6} + Az + B \qquad +\frac{W}{6}[z-a]^3 \tag{13.42}$$

The term on the right of equations (13.40) and (13.41) must be integrated by the manner so that the arbitrary constants A and B apply when $z < a$ and also when $z > a$. The square brackets [ ] are called Macaulay brackets and *do not apply* when the term inside them is *negative*.

The two boundary conditions are:

$$\text{at } z = 0, \quad v = 0 \quad \text{and} \quad \text{at } z = L, \quad v = 0$$

Applying the first boundary condition to equation (13.42), we get

$$B = 0$$

Applying the second boundary condition to equation (13.42), we get

$$0 = -V_c L^3/6 + AL + W(L - a)^3/6$$

or

$$AL = W(L - a)L^3/(6L) - W(L - a)^3/6$$

or

$$A = W(L - a)L/6 - W(L - a)^3/(6L)$$

$$= \frac{W(L - a)}{6}\{L - (L - a)^2/L\}$$

$$\therefore EIv = -W(L - a)z^3/(6L)$$

$$+ W(L - a)\{L - (L - a)^2/L\}x/6$$

$$+ W[z - a]^3/6$$

On putting $z = a$, we get the deflection at D, namely v_D

i.e.

$$v_D = \frac{W(L - a)}{6EI}\{-a^3/L + (L - (L - a)^2/L)a + 0\}$$

$$= \frac{W(L - a)}{6EI}\{-a^3/L + (L - (L^2 - 2aL + a^2)/L)a\}$$

$$= \frac{W(L - a)}{6EI}(-a^3/L + La - La + 2a^2 - a^3/L)$$

$$= \frac{W(L - a)}{6EI}.(2a^2 - 2a^3/L)$$

or

$$v_D = \frac{W(L - a)^2 a^2}{3EIL}$$

If W is *placed centrally*, so that $a = L/2$,

$$v_D = \frac{W(L - L/2)^2 (L/2)^2}{3EIL}$$

or $$v_D = \frac{WL^3}{48EI} \qquad (13.43)$$

13.9 Simply-supported beam with distributed load over a portion of the span

Suppose that the load is w per unit length over the portion DG, Figure 13.13; the reactions at the ends of the beam are

$$V_c = \frac{w}{2L}(L - a)^2$$

$$V_G = \frac{w}{2L}(L^2 - a^2)$$

The bending moment at a distance z from C is

$$M = V_c z - \frac{w}{2}[z - a]^2,$$

where the square brackets are Macaulay brackets, which only apply when the term inside them is positive.

i.e. $$M = \frac{w}{2L}(L-a)^2 z - \frac{w}{2}[z-a]^2$$

←---------- $z <= a$ ----------→ ←------ $a < z < L$ -----→

Hence $$EI\frac{d^2v}{dz^2} = \frac{-w}{2L}(L-a)^2 z \qquad + \frac{w}{2}[z-a]^2 \qquad (13.44)$$

so that $$EI\frac{dv}{dz} = \frac{w}{4L}(L-a)^2 z^2 + A \qquad + \frac{w}{6}[z-a]^3 \qquad (13.45)$$

and $$EIv = \frac{-w}{12L}(L-a)^2 z^3 + Az + B \qquad + \frac{w}{24}[z-a]^4 \qquad (13.46)$$

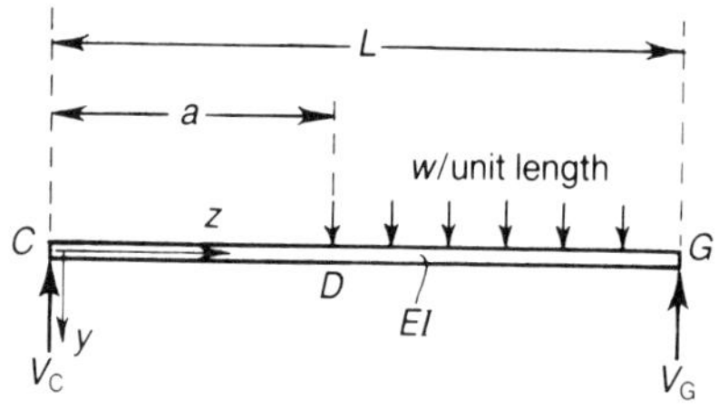

Figure 13.13 Load extending to one support.

The boundary conditions are that when

$$z = 0, \quad v = 0 \quad \text{and when} \quad z = L, \quad v = 0$$

Applying the first boundary condition to equation (13.46), we get

$$B = 0$$

Applying the second boundary condition to equation (13.46), we get

$$0 = -\frac{w}{12}(L-a)^2 L^2 + AL + \frac{w}{24}(L-a)^4$$

$$\therefore A = \frac{w}{12}(L-a)^2 L - \frac{w}{24L}(L-a)^4$$

$$= \frac{w}{24L}(L-a)^2\left\{2L^2 - (L-a)^2\right\}$$

$$= \frac{w}{24L}(L-a)^2\left\{2L^2 - L^2 - a^2 + 2aL\right\}$$

or $$A = \frac{w}{24L}(L-a)^2\left(L^2 + 2La - a^2\right)$$

The equation for the deflection curve is then:

$$EIv = \frac{-w}{2L}(L-a)^2 z^3 + \frac{w}{24L}(L-a)^2\left(L^2 + 2La - a^2\right)z$$

$$+ \frac{w}{24}[z-a]^4 \qquad (13.47)$$

where the square brackets in equation (13.47) are Macaulay brackets.

When the load does not extend to either support, Figure 13.14(i), the result of equation (13.47) may be used by superposing an upwards distributed load of w per unit length over the length GH

on a downwards distributed load of w per unit length over DH, Figure 13.14(ii). Due to the downwards distributed load alone

$$EIv = \frac{-w}{2L}(L-a)^2 z^3 + \frac{w}{24L}(L-a)^2\left(L^2 + 2La - a^2\right)z$$

$$+\frac{w}{24}[z-a]^4 \qquad (13.48)$$

where the square brackets in equation (13.48) are Macaulay brackets.

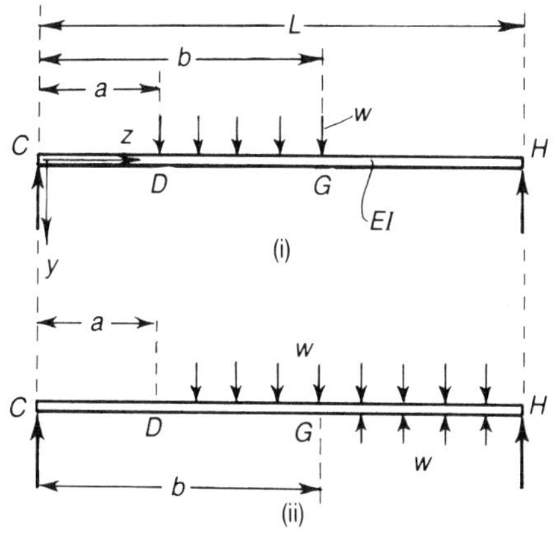

Figure 13.14 Load not extending to either support.

Due to the upwards distributed load

$$EIv = \frac{w}{2L}(L-b)^2 z^3 - \frac{w}{24L}(L-b)^2\left(L^2 + 2Lb - b^2\right)z$$

$$-\frac{w}{24}[z-b]^4 \qquad (13.49)$$

where the square brackets in equation (13.49) are Macaulay brackets.

On superposing the two deflected forms, the resultant deflection is given by

$$EIv = -\frac{wz^3}{2L}(b-a)\,(2L-a-b) + \frac{w}{24L}$$

$$\left\{(L-a)^2\left(L^2 + 2La - a^2\right) - (L-b)^2\left(L^2 + 2Lb - b^2\right)\right\} \qquad (13.50)$$

$$+\frac{w}{24}[z-a]^4 - \frac{w}{24}[z-b]^4$$

where the square brackets of equation (13.50) are Macaulay brackets and must be ignored if the term inside them becomes negative.

13.10 Simply-supported beam with a couple applied at an intermediate point

The simply-supported beam of Figure 13.15 carries a couple M_a applied to the beam at a point a distance a from C. The vertical reactions at each end are (M_a/L). The bending moment a distance z from C is

$$M = \frac{M_a z}{L} + Ma\,[z - a]^o \tag{13.51}$$

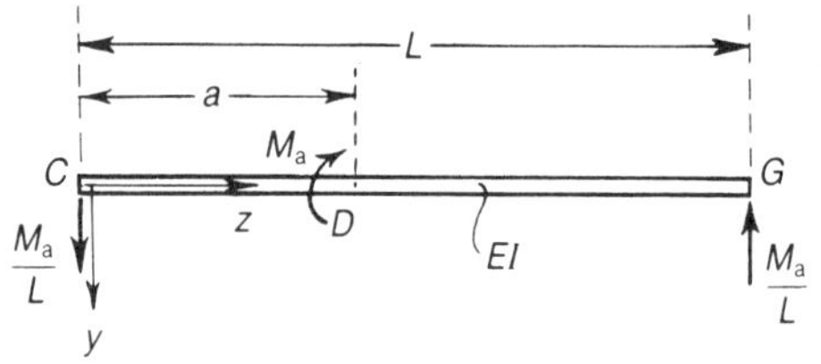

Figure 13.15 Beam with a couple applied at a point in the span.

The term on the right of equation (13.51) is so written, so that equation (13.51) applied over the whole length of the beam.

Hence,

←- - - - - - - - - $z <= a$ - - - - - - - - - - → ←- - - - - - $a < z < L$ - - - - - →

$$EI\frac{d^2v}{dz^2} = \frac{M_a z}{L} \qquad - M_a[z-a]^o$$

$$\therefore\ EI\frac{dv}{dz} = \frac{M_a z^2}{2L} + A \qquad - M_a[z-a] \tag{13.52}$$

and

$$EIv = \frac{M_a z^3}{6L} + Az + B \qquad -\frac{M_a}{2}[z-a]^2 \tag{13.53}$$

The boundary conditions are that

$$v = 0 \quad \text{at} \quad z = 0 \quad \text{and at} \quad z = L$$

From the first boundary condition, we get

$$B = 0$$

From the second boundary condition, we get

$$o = \frac{M_a L^2}{6} + AL \frac{-M_a}{2} (L - a)^2$$

$$\therefore A = \frac{-M_a L}{6} + \frac{M_a}{2L} (L - a)^2$$

$$= \frac{M_a}{6L} \left(-L^2 + 3L^2 + 3a^2 - 6aL\right)$$

$$= \frac{M_a}{6L} \left(2L^2 - 6La + 3a^2\right)$$

$$\therefore EIv = \frac{M_a z^3}{6L} + \frac{M_a z}{6L}\left(2L^2 - 6La + 3a^2\right) + \frac{-M_a}{2}\left[z - a\right]^2 \tag{13.54}$$

where the square brackets in equation (13.54) are Macaulay brackets.

The deflection at D, when $z = a$, is

$$v_D = \frac{M_a a}{3EIL} (L - a)(L - 2a) \tag{13.55}$$

Problem 13.3 A steel beam rests on two supports 6 m apart, and carries a uniformly distributed load of 10 kN per metre run. The second moment of area of the cross-section is 1×10^{-3} m^4 and $E = 200$ GN/m^2. Estimate the maximum deflection.

<u>*Solution*</u>

The greatest deflection occurs at mid-length and has the value given by equation (13.16):

$$v = \frac{5wL^4}{384EI} = \frac{5\left(100 \times 10^3\right)(6)^4}{384\left(200 \times 10^9\right)\left(1 \times 10^{-3}\right)} = 0.00844 \text{ m}$$

Problem 13.4 A uniform, simply-supported beam of span L carries a uniformly distributed lateral load of w per unit length. It is propped on a knife-edge support at a distance a from one end. Estimate the vertical force on the prop.

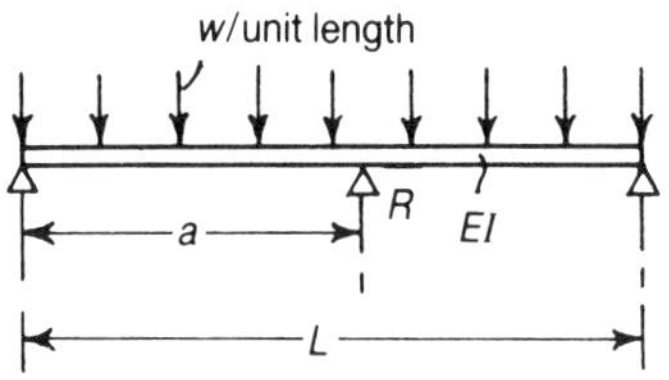

Solution

If the beam is unpropped, then, from equation (13.15), the downwards vertical deflection at the position of the prop is

$$(v)_{z=a} = \frac{wa}{24EI}\left(L^3 - 2La^2 + a^3\right)$$

If R is the reaction on the prop, then under the action of R alone the upwards vertical deflection at the prop is, from equation (13.35),

$$(v)_{z=a} = \frac{Ra^2(L-a)^2}{3EIL}$$

If there is no resultant deflection at the prop, we have

$$\frac{Ra^2(L-a)^2}{3EIL} = \frac{wa}{24EI}\left(L^3 - 2La^2 - a^3\right)$$

Thus, the reaction on the prop is

$$R = \frac{wL}{8}\left[\frac{1 - 2\left(\frac{a}{L}\right)^2 + \left(\frac{a}{L}\right)^3}{\frac{a}{L}\left(1 - \frac{a}{L}\right)^2}\right]$$

The propping force is least when the prop is at mid-span; in this case, $a/L = 0.5$ and $R = 5\,wL/8$.

Problem 13.5 A simply-supported, uniform beam, of span L and flexural stiffness EI, carries a vertical lateral load W at a distance a from one end. Calculate the greatest lateral deflection in the beam.

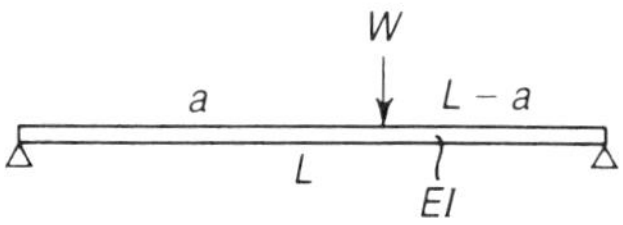

Solution

From section 13.7, the lateral deflection at any point is given by

$$EIv = -\frac{W}{6L}(L - a)z^3 + \frac{Wa}{6L}(2L^2 - 3aL + a^2)z \text{ for } z > a$$

$$EIv = -\frac{W}{6L}(L - a)z^3 + \frac{Wz^2}{6}(z - 3a) + \frac{Wa}{6L}(2L^2 + a^2)z - \frac{Wa^3}{6} \text{ for } z > a$$

Let us suppose first that $a > \frac{1}{2}L$, when we would expect the greatest deflection to occur in the range $z < a$; over this range

$$EI\frac{dv}{dz} = -\frac{W}{2L}(L - a)z^2 + \frac{Wa}{6L}(2L^2 - 3aL + a^2)$$

This is zero when

$$-\frac{W}{2L}(L - a)z^2 + \frac{Wa}{6L}(2L^3 - 3aL + a^2) = 0$$

i.e. when

$$(L - a)z^2 = \frac{1}{3}a(2L^2 - 3aL + a^2)$$

or when

$$z = \sqrt{\frac{a}{3}(2L - a)}$$

If this gives a root in the range $z < a$, then

$$\sqrt{\frac{a}{3}(2L - a)} < a$$

and $2L - a < 3a$, or $a > \frac{1}{2}L$. This is compatible with our earlier suppositions. Then, with $a > \frac{1}{2}L$, the greatest deflection occurs at the point

$$z = \left[(a/3)\,(2L - a)\right]^{\frac{1}{2}} \text{ and has the value}$$

$$v_{max} = \frac{Wa}{9LEI}\,(2L - a)\,(L - a)\,\sqrt{\frac{a}{3}\,(2L - a)}$$

If $a < \frac{1}{2}L$, the greatest deflection occurs in the range $z > a$; in this case we replace a by $(L - a)$, whence the greatest deflection occurs at the point

$$z = \sqrt{\frac{1}{3}\left(L^2 - a^2\right)}\text{ , and has the value}$$

$$v_{max} = \frac{Wa}{9LEI}\,(L^2 - a^2)\,\sqrt{\frac{a}{3}\,(2L - a)}$$

13.11 Beam with end couples and distributed load

Suppose the ends of the beam CD, Figure 13.16, rest on knife-edges, and carry couples M_C and M_D. If, in addition, the beam carries a uniformly distributed lateral load w per unit length, the bending moment a distance z from C is

$$M = \frac{M_C}{L}\,(L - z) + M_D\,\frac{z}{L} + \frac{1}{2}\,wz\,(L - z)$$

The equation of the deflection curve is then given by

$$EI\,\frac{d^2v}{dz^2} = -\frac{M_C}{L}\,(L - z) - M_D\,\frac{z}{L} - \frac{1}{2}\,wz\,(L - z)$$

Then

$$EI\,\frac{dv}{dz} = -\frac{M_C}{L}\left(Lz - \frac{1}{2}z^2\right) - \frac{M_D}{L}\left(\frac{z}{2}\right) - \frac{1}{2}w\left(\frac{Lz^2}{2} - \frac{z^3}{3}\right) + A$$

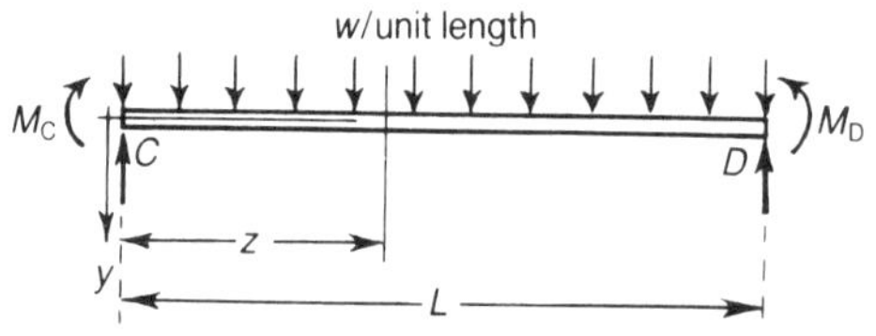

Figure 13.16 Simply-supported beam carrying a uniformly supported load.

and

$$EIv = \frac{M_C}{L}\left(\frac{Lz^2}{2} - \frac{z^3}{6}\right) - \frac{M_D}{L}\left(\frac{z^3}{6}\right) - \frac{1}{2}w\left(\frac{Lz^2}{6} - \frac{z^4}{12}\right) + Az + B \tag{13.56}$$

If the ends of the beam remain at the same level, $v = 0$ for $z = 0$ and $z = L$. Then $B = 0$ and

$$AL = \frac{1}{3}M_C L^2 + \frac{1}{6}M_D L^2 + \frac{1}{24}wL^4$$

Then

$$EIv = -\frac{M_C}{L}\left(\frac{Lz^2}{2} - \frac{z^3}{6}\right) - \frac{M_D}{L}\left(\frac{z^3}{6}\right) - \frac{1}{2}w\left(\frac{Lz^3}{6} - \frac{z^4}{12}\right)$$

$$+ z\left(\frac{M_C L}{3} + \frac{M_D L}{6} + \frac{wL^3}{24}\right)$$

The slopes at the ends are

$$\left(\frac{dv}{dz}\right)_{z=0} = \frac{L}{24EI}\left(8M_C + 4M_D + wL^2\right)$$

$$\left(\frac{dv}{dz}\right)_{z=L} = -\frac{L}{24EI}\left(4M_C + 8M_D + wL^2\right)$$

Suppose that the end D of the beam now sinks an amount δ downwards relative to C. Then at $v = L$ we have $v = \delta$, instead of $v = 0$. In equation (13.56), A is then given by

$$AL = EI\delta + \frac{1}{3}M_C L^2 + \frac{1}{6}M_D L^2 + \frac{1}{24}wL^4$$

For the slopes at the ends we have

$$\left(\frac{dv}{dz}\right)_{z=0} = \frac{L}{24EI}\left(8M_C + 4M_D + wL^2\right) + \frac{\delta}{L}$$

$$\left(\frac{dv}{dz}\right)_{z=L} = -\frac{L}{24EI}\left(4M_C + 8M_D + wL^2\right) + \frac{\delta}{L} \tag{13.57}$$

13.12 Beams with non-uniformly distributed load

When a beam carries a load which is not uniformly distributed the methods of the previous articles can still be employed if M and $\int M dz$ are both integrable functions of z, for we have in all cases

$$-EI \frac{d^2v}{dz^2} = M$$

which can be written in the form

$$\frac{d}{dz}\left(\frac{dv}{dz}\right) = -\frac{M}{EI}$$

If I is uniform along the beam the first integral of this is

$$\frac{dv}{dz} = A - \frac{1}{EI}\int M dz \qquad (13.58)$$

where A is a constant. The second integral is

$$v = Az + B - \frac{1}{EI}\int\int M dz dz \qquad (13.59)$$

If M and $\int M\,dz$ are integrable function of z the process of finding v can be continued analytically, the constants A and B being found from the terminal conditions. Failing this the integrations must be performed graphically or numerically. This is most readily done by plotting the bending-moment curve, and from that deducing a curve of areas representing $\int M\,dz$. From this curve a third is deduced representing $\int\int M\,dz\,dz$.

Problem 13.6 A uniform, simply-supported beam carries a distributed lateral load varying in intensity from w_0 at one end to $2w_0$ at the other. Calculate the greatest lateral deflection in the beam.

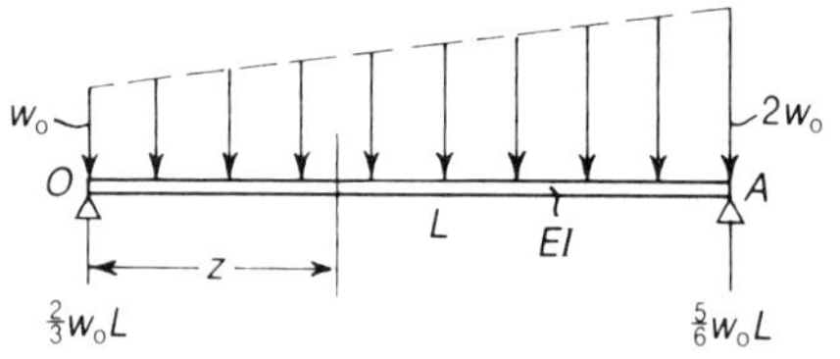

<u>*Solution*</u>

The vertical reactions at O and A are $(2/3)\,w_0L$ and $(5/6)\,w_0L$. The bending moment at any section a distance z from O is then

$$M = \frac{2}{3}w_0 Lz - \frac{1}{2}w_0 z^2 - \frac{w_0 z^3}{6L}$$

Then

$$EI\frac{d^2v}{dz^2} = -\left[\frac{2}{3}w_0 Lz - \frac{1}{2}w_0 z^2 - \frac{w_0 z^3}{6L}\right]$$

On integrating once,

$$EI\frac{dv}{dz} = -\left[\frac{w_0 Lz^2}{3} - \frac{w_0 z^3}{6} - \frac{w_0 z^4}{24L} + C_1\right]$$

where C_1 is a constant. On integrating further,

$$EIv = -\left[\frac{w_0 Lz^3}{9} - \frac{w_0 z^4}{24} - \frac{w_0 z^5}{120L} + C_1 z + C_2\right]$$

where C_2 is a further constant. If $v = 0$ at $z = L$, we have

$$C_1 = -\frac{11}{180}w_0 L^3 \quad \text{and} \quad C_2 = 0$$

Then

$$EIv = \frac{11}{180}w_0 L^3 z - \frac{w_0 Lz^3}{9} - \frac{w_0 z^4}{24} + \frac{w_0 z^5}{120L}$$

The greatest deflection occurs at $dv/dz = 0$, i.e. when

$$\frac{11}{180}w_0 L^3 - \frac{w_0 Lz^2}{3} + \frac{w_0 z^3}{6} + \frac{w_0 z^4}{24L} = 0$$

or when

$$15\left(\frac{z}{L}\right)^4 + 60\left(\frac{z}{L}\right)^3 - 120\left(\frac{z}{L}\right)^2 + 22 = 0$$

The relevant root of this equation is $z/L = 0.506$ which gives the point of maximum deflection near to the mid-length. The maximum deflection is

$$v_{max} \doteq \frac{7.03}{360}\frac{w_0 L^4}{EI} = 0.0195\frac{w_0 L^4}{EI}$$

This is negligibly different from the deflection at mid-span, which is

$$(v)_z = L/2 = \frac{5w_0 L^4}{256EI}$$

13.13 Cantilever with irregular loading

In Figure 13.17(i) a cantilever is free at *D* and built-in to a rigid wall at *C*. The bending moment curve is *DM* of Figure 13.17(ii); the bending moments are assumed to be hogging, and are therefore negative. The curve *CH* represents $\int_0^z M\,dz$, and its ordinates are drawn downwards because *M* is negative. The curve *CG* is then constructed from *CH* by finding

$$\int_0^z \int_0^z M dz dz$$

In equation (13.51), the constants *A* and *B* are both zero as $v = 0$ and $dv/dz = 0$ at $z = 0$. Then *CD* is the base line for both curves.

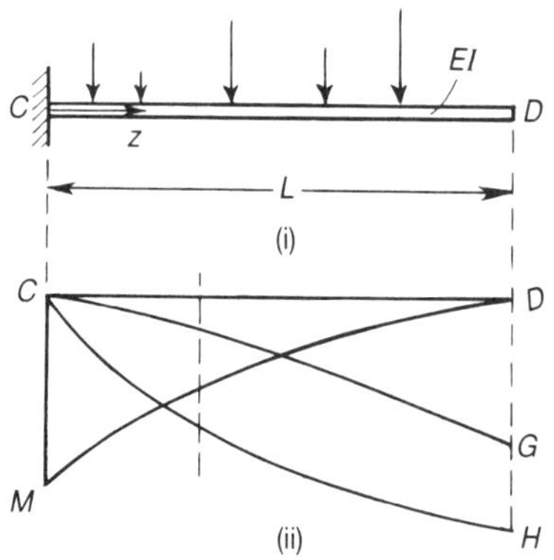

Figure 13.17 Cantilever carrying any system of lateral loads.

13.14 Beams of varying section

When the second moment of area of a beam varies from one section to another, equations (13.58) and (13.59) take the forms

$$\frac{dv}{dz} = A - \frac{1}{E}\int\frac{Mdz}{I}$$

and

$$v = Az + B - \frac{1}{E}\iint \frac{M}{I} dz\, dz$$

The general method of procedure follows the same lines as before. If (M/I) and $\int (M/I)dz$ are integrable functions of z, then (dv/dz) and v may be evaluated analytically; otherwise graphical or numerical methods must be employed, when a curve of (M/I) must be taken as the starting point instead of a curve of M.

Problem 13.7 A cantilever strip has a length L, a constant breadth b and thickness t varying in such a way that when the cantilever carries a lateral end load W, the centre line of the strip is bent into a circular arc. Find the form of variation of the thickness t.

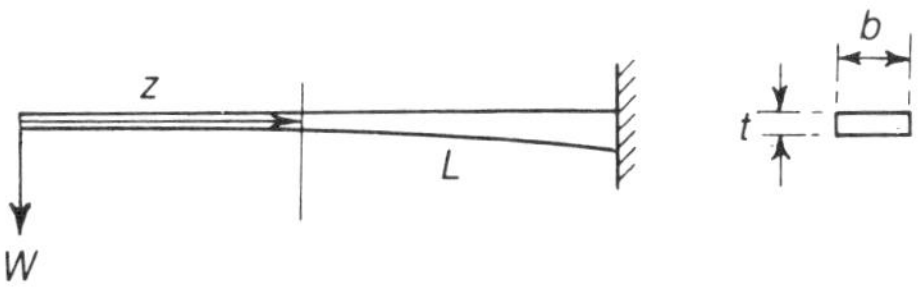

Solution

The second moment of area, I, at any section is

$$I = \frac{1}{12} bt^3$$

The bending moment at any section is $(-Wz)$, so that

$$EI\frac{d^2v}{dz^2} = Wz$$

Then

$$\frac{d^2v}{dz^2} = \frac{Wz}{EI}$$

If the cantilever is bent into a circular arc, then d^2v/dz^2 is constant, and we must have

$$\frac{Wz}{EI} = \text{constant}$$

This requires that

$$\frac{z}{I} = \text{constant}$$

or $\quad I \propto z$

Thus,

$$\frac{1}{12} bt^3 \propto z$$

or $\quad t \propto z^{\frac{1}{3}}$

Any variation of the form

$$t = t_0 \left(\frac{z}{L}\right)^{\frac{1}{3}}$$

where t_0 is the thickness at the built-in end will lead to bending in the form of a circular arc.

Problem 13.8 The curve M, below, represents the bending moment at any section of a timber cantilever of variable bending stiffness. The second moments of area are given in the table below. Taking $E = 11 \text{ GN/m}^2$, deduce the deflection curve.

z (from supported end)(m)	0	0.1	0.2	0.3	0.5	0.7	0.9	1.1	1.3	1.5	1.6	1.7
I (m^4)	50.8	27.4	17.4	12.25	5.65	3.23	1.69	0.783	0.278	0.074	0.0298	0 $\times 10^{-4}$

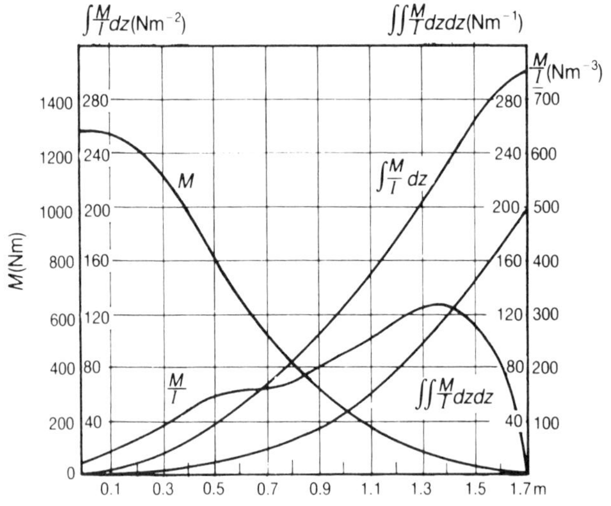

<u>Solution</u>

The first step is to calculate *M/I* at each section and to plot the *M/I* curve. We next plot the area under this curve at any section to give the curve

$$\int_0^z \frac{M}{I}\, dz$$

From this, the curve

$$\int_0^z \int \frac{M}{I}\, dz dz$$

is plotted to give the deflected form

$$v = \frac{1}{E} \int_0^z \int \frac{M}{I}\, dz dz$$

The maximum deflection at the free end of the cantilever is

$$v = \frac{1}{E}\left(300 \times 10^6\right) = \frac{300 \times 10^6}{11 \times 10^9} = 0.0272 \text{ m}$$

13.15 Non-uniformly distributed load and terminal couples; the method of moment-areas

Consider a simply-supported beam carrying end moments M_C and M_D, as in Figure 13.16, and a distributed load of varying intensity w. Suppose M_0 is the bending moment at any section due to the load w acting alone on the beam. Then

$$M = M_0 + \frac{M_C}{L}(L - z) + \frac{M_D}{L} z$$

The differential equation for the deflection curve is

$$EI\frac{d^2v}{dz^2} = -M_0 - \frac{M_C}{L}(L - z) - \frac{M_D}{L} z \qquad (13.60)$$

The integral between the limits $z = 0$ and $z = L$ is

$$EI\left[\left(\frac{dv}{dz}\right)_{z=L} - \left(\frac{dv}{dz}\right)_{z=0}\right] = -\frac{1}{2}M_C L - \frac{1}{2}M_D L - \int_0^L M_0 dz \qquad (13.61)$$

Again, on multiplying equation (13.60) by z, we have

$$EI\left(z\frac{d^2v}{dz^2}\right) = -M_0 z - \frac{M_C z}{L}(L-z) - \frac{M_D}{L}z^2 \tag{13.62}$$

But

$$EI\left(z\frac{d^2v}{dz^2}\right) = EI\frac{d}{dz}\left(z\frac{dv}{dz} - v\right)$$

Thus, on integrating equation (13.62),

$$EI\left[z\frac{dv}{dz} - v\right]_0^L = -\frac{M_D}{L}\left[\frac{z^3}{3}\right]_0^L - \frac{M_C}{L}\left[\frac{Lz^2}{2} - \frac{z^3}{3}\right]_0^L - \int_0^L M_0 z dz \tag{13.63}$$

But if $v = 0$ when $z = 0$ and $z = L$, then equation (13.63) becomes

$$EI\left[L\left(\frac{dv}{dz}\right)_{z=L}\right] = -\frac{1}{3}M_D L^2 - \frac{1}{6}M_C L^2 - \int_0^L M_0 z dz$$

Then

$$\left(\frac{dv}{dz}\right)_{z=L} = -\frac{M_D L}{3EI} - \frac{M_C L}{6EI} - \frac{1}{EIL}\int_0^L M_0 z dz \tag{13.64}$$

On substituting this value of $(dv/dz)_{z=L}$ into equation (13.61),

$$\left(\frac{dv}{dz}\right)_{z=0} = \frac{M_C L}{3EI} - \frac{M_D L}{6EI} - \frac{1}{EI}\int_0^L M_0 z dz - \frac{1}{EIL}\int_0^L M_0 z dz \tag{13.65}$$

the integral $\int_0^L M_0\, dz$ is the area of the bending moment curve due to the load w alone; $\int_0^L M_0\, z dz$ is the moment of this area about the end $z = 0$ of the beam. If A is the area of the bending moment diagram due to the lateral loads only, and $\bar{z}$ is the distance of its centroid from $z = 0$, then

$$A = \int_0^L M_0 dz, \qquad \bar{z} = \frac{1}{A}\int_0^L M_0 z\, dz$$

and equations (13.64) and (13.65) may be written

$$\left(\frac{dv}{dz}\right)_{z=0} = \frac{M_C L}{3EI} + \frac{M_D L}{6EI} + \frac{A(L - \bar{z})}{EIL} \tag{13.66}$$

$$\left(\frac{dv}{dz}\right)_{z=L} = -\frac{M_C L}{6EI} - \frac{M_D L}{3EI} - \frac{A\bar{z}}{EIL} \tag{13.67}$$

The method of analysis, making use of A and $\bar{z}$, is known as the *method of moment-areas*; it can be extended to deal with most problems of beam deflections.

When the section of the beam is not constant, equation (13.60) becomes

$$E\frac{d^2v}{dz^2} = -\frac{M_0}{I} - \frac{M_C}{I} + \frac{M_C - M_D}{L}\left(\frac{z}{I}\right)$$

The slopes at the ends of the beam are then given by

$$E\left[\left(\frac{dv}{dz}\right)_{z=L} - \left(\frac{dv}{dz}\right)_{z=0}\right] = -\int_0^L M_0\frac{dz}{I} - M_C\int_0^L \frac{dz}{I} + \frac{1}{L}(M_C - M_D)\int_0^L \frac{zdz}{I}$$

and

$$E\left[\left(\frac{dv}{dz}\right)_{z=L}\right] = \frac{1}{L}\int_0^L M_0\frac{zdz}{I} - \frac{M_C}{L}\int_0^L \frac{zdz}{I} + \frac{1}{L^2}(M_C - M_D)\int_0^L \frac{z^2dz}{I}$$

It is necessary to plot five curves of (M_0/I), $(1/I)$, (z/I), (z^2/I), (M_0z/I) and to find their areas.

As an example of the use of equations (13.66) and (13.67), consider the beam of Figure 13.18(i), which carries end couples, M_C and M_D, and a concentrated load W at a distance a from C.

The bending moment diagram for W acting alone is the triangle CBD, Figure 13.18(ii). The area of this triangle is

$$A = \frac{1}{2}L\left(\frac{Wa}{L}\right)(L - a) = \frac{Wa}{2}(L - a)$$

To evaluate its first moment about C, divide the triangle into two right-angled triangles, having centroids at G_1 and G_2, respectively. Then

$$A\bar{z} = \frac{1}{2}a\left[\frac{Wa}{L}(L-a)\right]\frac{2a}{3} + \frac{1}{2}[L-a]\left[\frac{Wa}{L}(L-a)\right]\left[\frac{1}{3}(L+2a)\right]$$

$$= \frac{1}{6}Wa\left(L^2 - a^2\right).$$

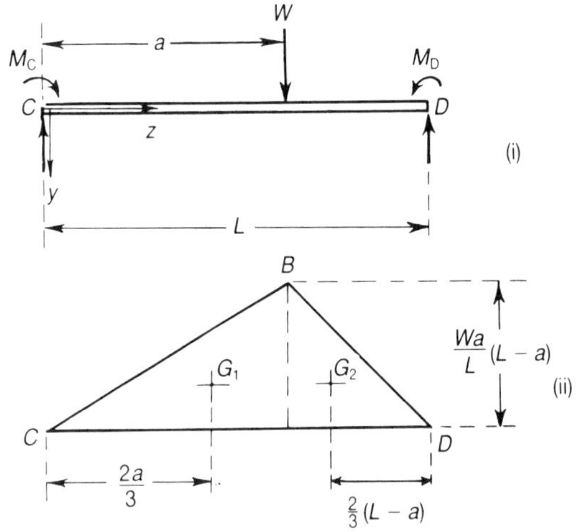

Figure 13.18 Moment-area solution of a beam carrying end couples and a concentrated load.

Then equations (13.66) and (13.67) give

$$\left(\frac{dv}{dz}\right)_{z=0} = \frac{M_C L}{3EI} + \frac{M_D L}{6EI} + \frac{Wa}{6EIL}\left(a^2 - 3aL + 2L^2\right)$$

$$\left(\frac{dv}{dz}\right)_{z=L} = -\frac{M_C L}{6EI} - \frac{M_D L}{3EI} - \frac{Wa}{6EIL}\left(L^2 - a^2\right)$$

Problem 13.9 Determine the deflection of the free end of the stepped cantilever shown in Figure 13.19(a).

Solution

The bending moment diagram is shown in Figure 13.19(b) and the *M/I* diagram is shown in Figure 13.19(c).

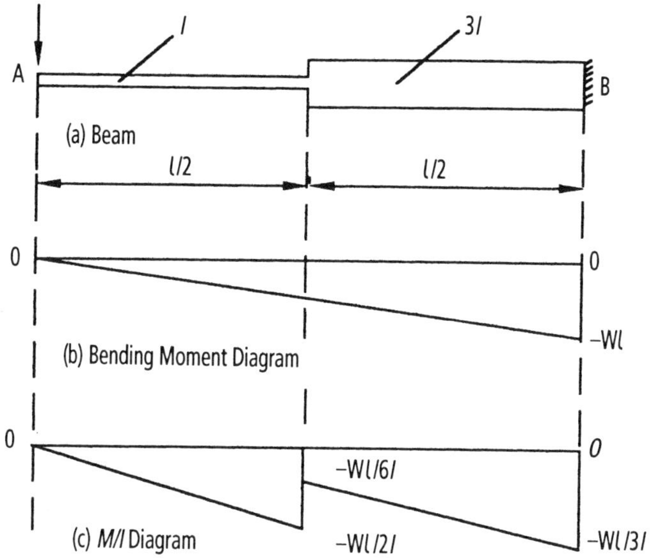

Figure 13.19 Stepped cantilever.

From equation (13.61)

$$EI\left(z\frac{dv}{dz} - v\right)_0^L = - \text{ moment of area of the bending moment diagram}$$

$$\text{or } \left(z\frac{dv}{dz} - v\right)_0^L = -\frac{1}{E} \times \text{ moment of area of the } M/I \text{ diagram}$$

Consider the moment of area of M/I about the point A, because we know that

$$\frac{dv}{dz} \text{ and } v = 0 \text{ at the point B}$$

$$\therefore \left[z\frac{dv}{dz} - v\right]_{z=L} - \left[0 \times \frac{dv}{dz} - v_A\right]_{z=0}$$

$$= \frac{1}{E}\left[\frac{WL}{2I} \times \frac{L}{4} \times \frac{2}{3} \times \frac{L}{2} + \frac{WL}{6I} \times \frac{L}{2} \times \frac{3L}{4} + \frac{WL}{6I} \times \frac{L}{4} \times \left(\frac{L}{2} + \frac{2}{3} \times \frac{L}{2}\right)\right]$$

or

$$0 + v_A = \frac{WL^3}{EI}\left[\frac{1}{24} + \frac{1}{16} + \frac{1}{24} \times \left(\frac{1}{2} + \frac{1}{3}\right)\right]$$

$$= \frac{WL^3}{EI}\left(\frac{1}{24} + \frac{1}{16} + \frac{5}{144}\right)$$

$$v_A = \frac{5WL^3}{36EI}$$

Problem 13.10 Determine the deflection of the free end of the varying depth cantilever shown in Figure 13.20(a)

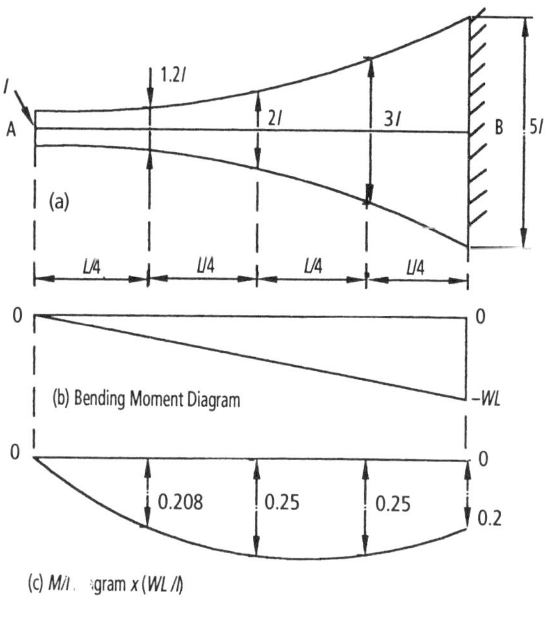

Figure 13.20 Varying depth cantilever.

Solution

Taking the moment of area of the *M/I* diagram about *A*, we eliminate v_B and dv/dz at *B*, because they are both zero. Additionally, as the *M/I* diagram is numerical, we can use numerical integration, namely Simpsons rule, as shown in Table 13.1.

Table 13.1 Numerical integration of the moment of *M/I* about *A*

Ordinate	*M/I*	*z*	*zM/I*	SM	*f(zM/I)*
1	0	0	0	1	0
2	0.208 WL/I	$L/4$	0.052 WL^2/I	4	0.208 WL^2/I
3	0.25 WL/I	$L/2$	0.125 WL^2/I	2	0.25 WL^2/I
4	0.25 WL/I	$3L/4$	0.188 WL^2/I	4	0.752 WL^2/I
5	0.2 WL/I	L	0.2 WL^2/I	1	0.2 WL^2/I
				Σ	**1.41 WL^2/I**

From Table 13.1,

$$v_A = \frac{1}{3} \times \frac{L}{4} \times 1.41\ WL^2/I$$

$$v_A = 0.1175\ WL^2/I$$

13.16 Deflections of beams due to shear

In our simple theory of bending of beams, we assumed that plane sections remain plane during bending. The effect of shearing forces in a beam is to distort plane cross-sections into curved planes. In the cantilever of Figure 13.21, the cross-section *DH* warps as the force *F* is applied, due to the shearing strains in the fibres of the beam. We assume that the shearing stresses set up by *F* are distributed in the manner already discussed in Chapter 10. This is not true strictly, because shearing distortions no longer allow sections to remain plane; however, we assume these shearing effects are secondary, and we are justified therefore in estimating them on our original theory.

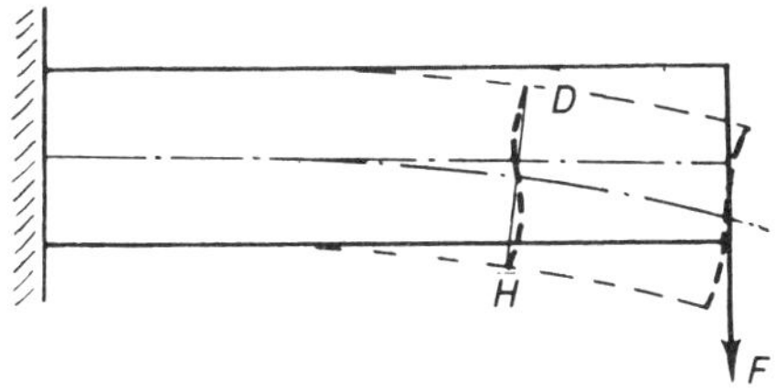

Figure 13.21 Shearing distortions in a cantilever.

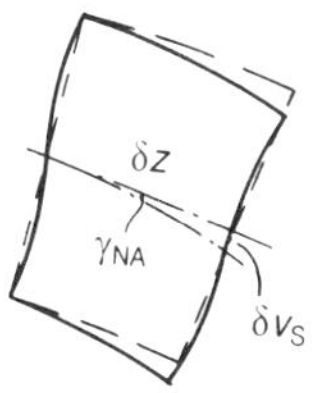

Figure 13.22 Shearing deflection at the neutral axis of a beam.

Suppose the shearing stress at the neutral axis of the beam is τ_{NA}, then the shearing strain at the neutral axis is

$$\gamma_{NA} = \frac{\tau_{NA}}{G} \tag{13.68}$$

where G is the shearing modulus. The additional deflection arising from shearing of the cross-section is then

$$\delta v_s = \gamma_{NA} \ \delta z = \frac{\tau_{NA}}{G} \ \delta z$$

Then

$$\frac{dv_s}{dz} = \frac{\tau_{NA}}{G} \tag{13.69}$$

For a cantilever of thin rectangular cross-section, Section 10.2,

$$\tau_{NA} = \frac{3F}{2ht} \tag{13.70}$$

where h is the depth of the cross-section, and t is the thickness. Then

$$\frac{dv_s}{dz} = \frac{3F}{2Ght}$$

Then

$$v_s = \frac{3Fz}{2Ght} + A \tag{13.71}$$

At $z = 0$, there is no shearing deflection, so $A = 0$. At the end $z = L$,

$$(v_s)_L = \frac{3FL}{2Ght} \tag{13.72}$$

The bending deflection at the free end, $z = L$, is

$$(v)_L = \frac{FL^3}{3EI} = \frac{4FL^3}{Eh^3t} \tag{13.73}$$

Then the total end deflection is

$$v_L = \frac{4FL^3}{Eh^3t} + \frac{3FL}{2Ght} = \frac{4FL^3}{Eh^3t}\left[1 + \frac{3E}{8G}\left(\frac{h}{L}\right)^2\right] \qquad (13.74)$$

For most materials $(3E/8G)$ is of order unity, so the contribution of the shear to the total deflection is equal approximately to $(h/L)^2$. Clearly, the shearing deflection is important only for deep beams.

Table 13.2 provides a summary of the maximum bending moments and lateral deflections for some statically determinate beams.

Problem 13.11 A 1.5 m length of the beam of Problem 11.2 is simply-supported at each end, and carries concentrated lateral load of 10 kN at the mid-span. Compare the central deflections due to bending and shearing.

Solution

From Problem 11.2, the second moment of area of the equivalent steel I-beam is 12.1×10^{-6} m^4. The central deflection due to bending is, therefore,

$$v_B = \frac{WI^3}{48E_s I_s} = \frac{(10 \times 10^3)(1.5)^3}{48(200 \times 10^9)(12.1 \times 10^{-6})} = 0.290 \times 10^{-3}\ \text{m}$$

The average shearing stress in the timber is

$$\frac{5 \times 10^3}{(0.15)(0.075)} = 0.445\ \text{MN/m}^2$$

If the shearing modulus for timber is

$$4 \times 10^9\ \text{N/m}^2$$

the shearing strain in the timber is

$$\gamma = \frac{0.445 \times 10^6}{4 \times 10^9} = 0.111 \times 10^{-3}$$

The resulting central deflection due to shearing is

$$v_s = \gamma \times 0.75 = (0.111 \times 10^{-3})(0.75) = 0.0833 \times 10^{-3} \text{ m}$$

Table 13.2 Bending moment and deflections for some simple beams

Beam type and loading – length = L	M_{max}	Maximum deflection
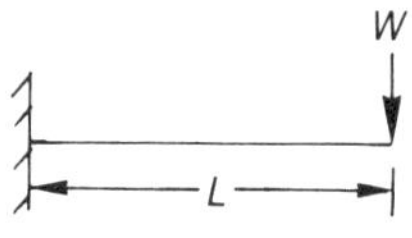	$-WL$	$WL^3/3EI$ @ $z = L$
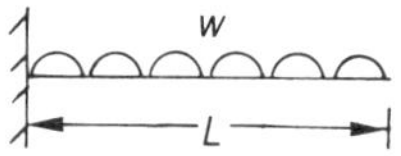	$-wL^2/2$	$wL^4/8EI$ @ $z = L$
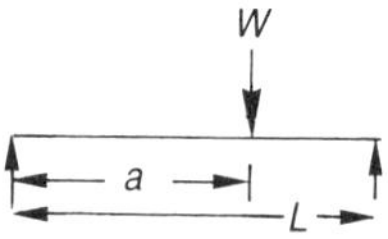	$W(L-a)\,a/L$ @ $z = a$	$Wa^2(L-a)^2/(3EIL)$ @ $z = L - \left[\frac{L^2 - a^2}{3}\right]^{\frac{1}{2}}$ when $a < L/2$
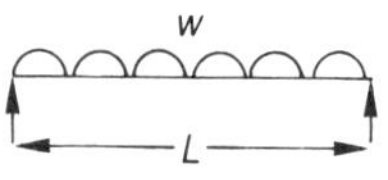	$wL^2/8$ @ $z = L/2$	$5wL^4/384EI$ @ $z = L/2$
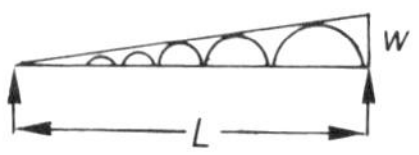	$0.0641\ wL^2$ @ $z = 0.5773\ L$	$0.00653\ wL^4/EI$ @ $z = 0.5195\ L$

Thus, the shearing deflection is nearly 30% of the bending deflection. The estimated total central deflection is

$$v = v_B + v_s = 0.373 \times 10^{-3} \text{ m}$$

Further problems *(answers on page 693)*

13.12 A straight girder of uniform section and length L rests on supports at the ends, and is propped up by a third support in the middle. The weight of the girder and its load is w per unit length. If the central support does not yield, prove that it takes a load equal to $(5/8)wL$.

13.13 A horizontal steel girder of uniform section, 15 m long, is supported at its extremities and carries loads of 120 kN and 80 kN concentrated at points 3 m and 5 m from the two ends, respectively. I for the section of the girder is 1.67×10^{-3} m^4 and $E = 200$ GN/m^2. Calculate the deflections of the girder at points under the two loads. *(Cambridge)*

13.14 A wooden mast, with a uniform diameter of 30 cm, is built into a concrete block, and is subjected to a horizontal pull at point 10 m from the ground. The wire guy A is to be adjusted so that it becomes taut and begins to take part of the load when the mast is loaded to a maximum stress of 7 MN/m^2.

Estimate the slack in the guy when the mast is unloaded. Take E for timber = 10 GN/m^2. *(Cambridge)*

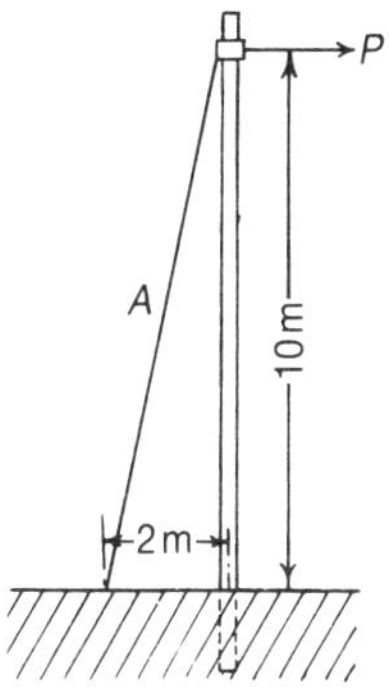

13.15 A bridge across a river has a span $2l$, and is constructed with beams resting on the banks and supported at the middle on a pontoon. When the bridge is unloaded the three supports are all at the same level, and the pontoon is such that the vertical displacement is equal to the load on it multiplied by a constant λ. Show that the load on the pontoon, due to a concentrated load W, placed one-quarter of the way along the bridge, is given by

$$\frac{11W}{16\left(1 + \dfrac{6EI\lambda}{l^3}\right)}$$

where I is the second moment of area of the section of the beams. *(Cambridge)*

13.16 Two equal steel beams are built-in at one end and connected by a steel rod as shown. Show that the pull in the tie rod is

$$P = \frac{5Wl^3}{32\left(\dfrac{6aI}{\pi d^2} + l^3\right)}$$

where d is the diameter of the rod, and I is the second moment of area of the section of each beam about its neutral axis. *(Cambridge)*

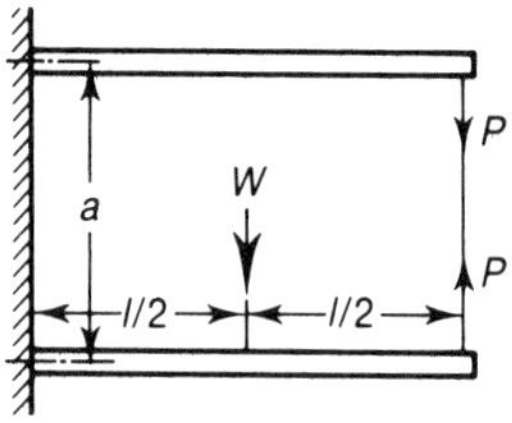

14 Built-in and continuous beams

14.1 Introduction

In all our investigations of the stresses and deflections of beams having two supports, we have supposed that the supports exercise no constraint on bending of the beam, i.e. the axis of the beam has been assumed free to take up any inclination to the line of supports. This has been necessary because, without knowing how to deal with the deformation of the axis of the beam, we were not in a position to find the bending moments on a beam when the supports constrain the direction of the axis. We shall now investigate this problem. When the ends of a beam are fixed in direction so that the axis of the beam has to retain its original direction at the points of support, the beam is said to be built-in or direction fixed.

Consider a straight beam resting on two supports A and B (Figure 14.1) and carrying vertical loads. If there is no constraint on the axis of the beam, it will become curved in the manner shown by broken lines, the extremities of the beam rising off the supports.

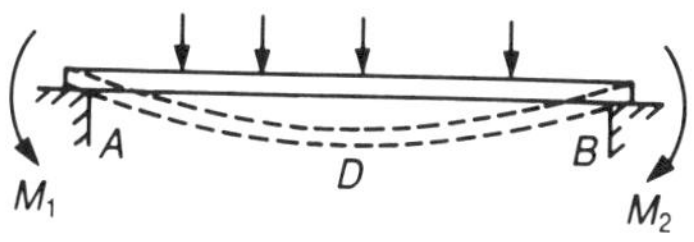

Figure 14.1 Beam with end couples.

In order to make the ends of the beam lie flat on the horizontal supports, we shall have to apply couples as shown by M_1 and M_2. If the beam is firmly built into two walls, or bolted down to two piers, or in any way held so that the axis cannot tip up at the ends in the manner indicated, the couples such as M_1 and M_2 are supplied by the resistance of the supports to deformation. These couples are termed *fixed-end moments*, and the main problem of the built-in beam is the determination of these couples; when we have found these we can draw the bending moment diagram and calculate the stresses in the usual way. The couples M_1 and M_2 in Figure 14.1 must be such as to produce curvature in the opposite direction to that caused by the loads.

14.2 Built-in beam with a single concentrated load

We may deduce the bending moments in a built-in beam under any conditions of lateral loading from the case of a beam under a single concentrated lateral load.

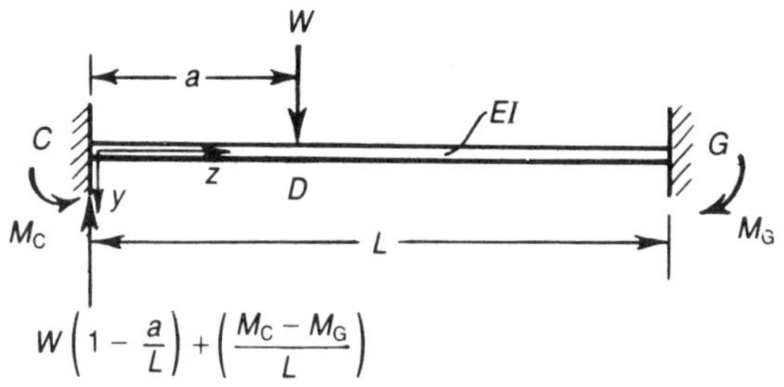

Figure 14.2 Built-in beam carrying a single lateral load.

Consider a uniform beam, of flexural stiffness EI, and length L, which is built-in to end supports C and G, Figure 14.2. Suppose a concentrated vertical load W is applied to the beam at a distance a from C. If M_C and M_G are the restraining moments at the supports, then the vertical reaction is at C is

$$W\left(1-\frac{a}{L}\right)+\frac{1}{L}\left(M_C-M_G\right)$$

The bending moment in the beam at a distance z from C is therefore

← – – – – – – – – – – – – $z <= a$ – – – – – – – – – – – – → ← – – – $a < z <= L$ – – – →

$$M=\left\{W\left(1-\frac{a}{L}\right)+\frac{1}{L}\left(M_C-M_G\right)\right\}z-M_C \qquad -W[z-a]$$

Then, for the deflected form of the beam, the displacement is given by

← – – – – – – – – – – – – $z <= a$ – – – – – – – – – – – – → ← – – – $a < z <= L$ – – – →

$$EI\frac{d^2v}{dz^2}=-\left\{W\left(1-\frac{a}{L}\right)+\frac{1}{L}\left(M_C-M_G\right)\right\}z+M_C \qquad +W\left[z-a\right] \tag{14.1}$$

or

$$EI\frac{dv}{dz}=-\left\{W\left(1-\frac{a}{L}\right)+\frac{1}{L}\left(M_C-M_G\right)\right\}\frac{z^2}{2}+M_Cz+A \qquad +\frac{W}{2}\left[z-a\right]^2 \tag{14.2}$$

and

$$EIv=-\left\{W\left(1-\frac{a}{L}\right)+\frac{1}{L}\left(M_C-M_G\right)\right\}\frac{z^3}{6}+\frac{M_Cz^2}{2}+Az+B \qquad +\frac{W}{6}\left[z-a\right]^3 \tag{14.3}$$

Two suitable boundary conditions are:

when $z = 0, \; v = dv/dz = 0$

As the Macaulay brackets will be negative when these boundary conditions are substituted, the terms on the right of equations (14.2) and (14.3) can be ignored, hence

$$A = B = 0$$

Two other boundary conditions are:

at $z = L, \; v = dv/dz = 0,$

which on substituting into equations (14.2) and (14.3) give the following two simultaneous equations:

$$-\left[W\left(1 - \frac{a}{L}\right) + \frac{1}{L}(M_C - M_G)\right]\frac{L^2}{2} + M_C L + \frac{W}{2}(L - a)^2 = 0$$

$$-\left[W\left(1 - \frac{a}{L}\right) + \frac{1}{L}(M_C - M_G)\right]\frac{L^3}{6} + \frac{M_C L^2}{6} + \frac{W}{6}(L - a)^3 = 0$$

These simultaneous equations give

$$M_C = Wa\left(\frac{L - a}{L}\right)^2 \tag{14.4}$$

$$M_G = W(L - a)\left(\frac{a}{L}\right)^2 \tag{14.5}$$

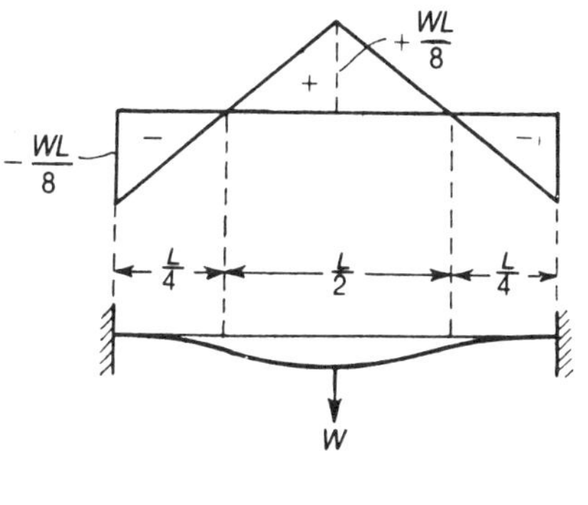

Figure 14.3 Variation in bending moment in a built-in beam carrying a concentrated load at mid-length.

M_C and M_G are referred to as the *fixed-end moments* of the beam; M_C is measured anticlockwise, and M_G clockwise.

In the particular case when the load W is applied at the mid-length, $a = \frac{1}{2}L$, and

$$M_C = M_G = \frac{WL}{8}$$

The bending moment in the beam vary linearly from hogging moments of $WL/8$ at each end to a sagging moment of $WL/8$ at the mid-length, Figure 14.3. There are points of contraflexure, or zero bending moment, at distances $L/4$ from each end.

14.3 Fixed-end moments for other loading conditions

The built-in beam of Figure 14.4 carries a uniformly distributed load of w per unit length over the section of the beam from $z = a$ to $z = b$.

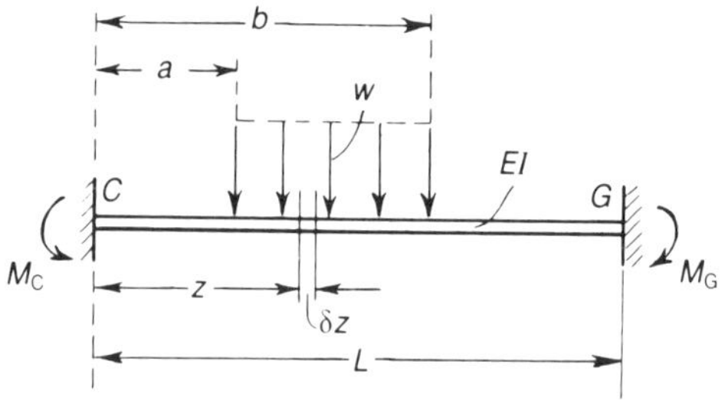

Figure 14.4 Distributed load over part of the span of a built-in beam.

Consider the loading on an elemental length δz of the beam; the vertical load on the element is $w\delta z$, and this induces a retraining moment at C of amount

$$\delta M_C = w\delta z \frac{z(L - z)^2}{L^2}$$

from equation (14.4).

The total moment at C due to all loads is

$$M_C = \int_a^b \frac{w}{L^2} z(L - z)^2 dz$$

which gives

$$M_C = \frac{w}{L^2}\left[\frac{L^2}{2}\left(b^2 - a^2\right) - \frac{2L}{3}\left(b^3 - a^3\right) + \frac{1}{4}\left(b^4 - a^4\right)\right] \qquad (14.6)$$

M_G may be found similarly. When the load covers the whole of the span, $a = 0$ and $b = L$, and equation (14.6) reduces to

$$M_C = \frac{wL^2}{12} \tag{14.7}$$

In this particular case, $M_G = M_C$; the variation of bending moment is parabolic, and of the form shown in Figure 14.5; the bending moment at the mid-length is $wL^2/24$, so the fixed-end moments are also the greatest bending moments in the beam.

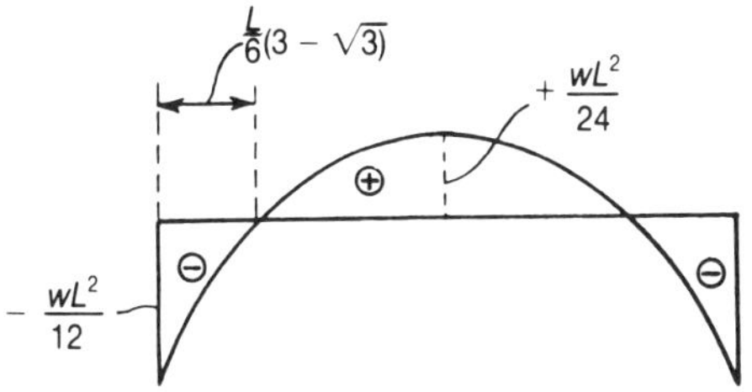

Figure 14.5 Variation of bending moment in a built-in beam carrying a uniformly distributed load over the whole span.

The points of contraflexure, or points of zero bending moment, occur at a distance

$$\frac{L}{6}\left(3 - \sqrt{3}\right) \tag{14.8}$$

from each end of the beam.

When a built-in beam carries a number of concentrated lateral loads, W_1, W_2, and W_3, Figure 14.6, the fixed-end moments are found by adding together the fixed-end moments due to the loads acting separately. For example,

$$M_C = \sum_{r = 1,2,3} W_r a_r \left(\frac{L - a_r}{L}\right)^2 \tag{14.9}$$

for the case shown in Figure 14.6.

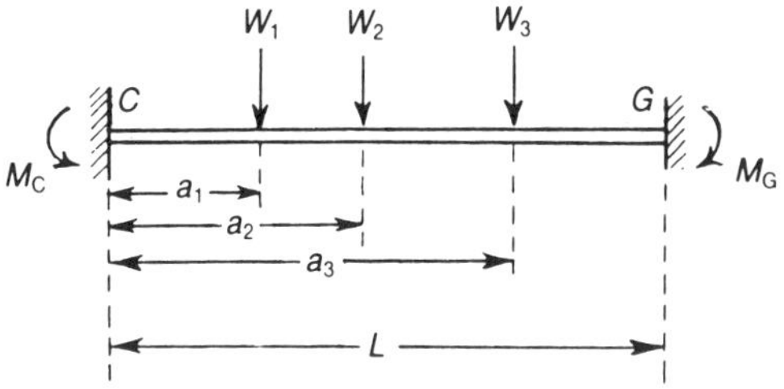

Figure 14.6 Built-in beam carrying a number of concentrated loads.

We may treat the case of a concentrated couple M_0, applied a distance a from the end C, Figure 14.7, as a limiting case of two equal and opposite loads W a small distance δa apart. The fixed-end moment at C is

$$M_C = -\frac{Wa}{L^2}(L - a)^2 + \frac{W(a + \delta a)}{L^2}(L - a - \delta a)^2$$

If δa is small,

$$M_C \doteqdot -\frac{Wa}{L^2}(L - a)^2 + \frac{W}{L^2}\left[a(L - a)^2 + \delta a\,(L - a)(L - 3a)\right]$$

which gives

$$M_C = \frac{W\delta a}{L^2}(L - a)(L - 3a)$$

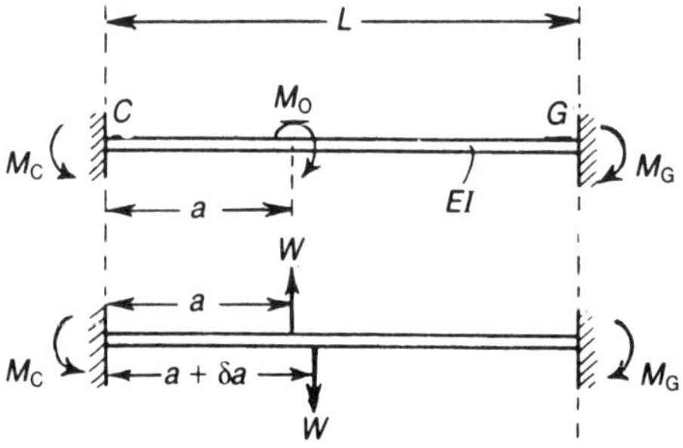

Figure 14.7 Built-in beam carrying a concentrated couple.

But if δa is small, M_0 is statically equivalent to the couple $W\delta a$, and

$$M_C = \frac{M_0}{L^2}(L - a)(L - 3a) \tag{14.10}$$

Similarly,

$$M_G = \frac{M_0}{L^2}a(2L - 3a) \tag{14.11}$$

14.4 Disadvantages of built-in beams

The results we have obtained above show that a beam which has its ends firmly fixed in direction is both stronger and stiffer than the same beam with its ends simply-supported. On this account it might be supposed that beams would always have their ends built-in whenever possible; in practice it is not often done. There are several objections to built-in beams: in the first place a small subsidence of one of the supports will tend to set up large stresses, and, in erection, the supports must be aligned with the utmost accuracy; changes of temperature also tend to set up large stresses. Again, in the case of live loads passing over bridges, the frequent fluctuations of bending moment, and vibrations, would quickly tend to make the degree of fixing at the ends extremely uncertain.

Most of these objections can be obviated by employing the double cantilever construction. As the bending moments at the ends of a built-in beam are of opposite sign to those in the central part of the beam, there must be points of inflexion, i.e. points where the bending moment is zero. At these points a hinged joint might be made in the beam, the axis of the hinge being parallel to the bending axis, because there is no bending moment to resist. If this is done at each point of inflexion, the beam will appear as a central girder freely supported by two end cantilevers; the bending moment curve and deflection curve will be exactly the same as if the beam were solid and built in. With this construction the beam is able to adjust itself to changes of temperature or subsistence of the supports.

14.5 Effect of sinking of supports

When the ends of a beam are prevented from rotating but allowed to deflect with respect to each other, bending moments are set up in the beam. The uniform beam of Figure 14.8 is displaced so that no rotations occur at the ends but the remote end is displaced downwards an amount δ relative to C.

The end reactions consist of equal couples M_C and equal and opposite shearing forces $2M_C/L$, because the system is antisymmetric about the mid-point of the beam. The half-length of the beam behaves as a cantilever carrying an end load $2M_CL$; then, from equation (13.18),

$$\frac{1}{2}\delta = \frac{(2M_C/L)(L/2)^2}{3EI} = \frac{M_C L^2}{12EI}$$

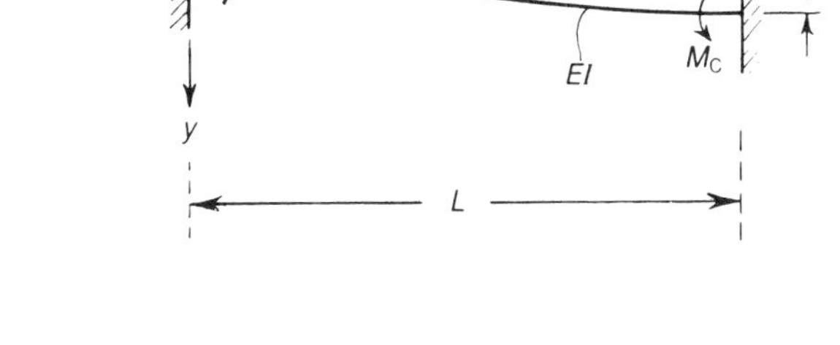

Figure 14.8 End moments induced by the sinking of the supports of a built-in beam.

Therefore

$$M_C = \frac{6EI\delta}{L^2} \qquad (14.12)$$

For a downwards deflection δ, the induced end moments are both anticlockwise; these moments must be superimposed on the fixed-end moments due to any external lateral loads on the beam.

Problem 14.1 A horizontal beam 6 m long is built-in at each end. The elastic section modulus is 0.933×10^{-3} m^3. Estimate the uniformly-distributed load over the whole span causing an elastic bending stress of 150 MN/m^2.

Solution

The maximum bending moments occur at the built-in ends, and have value

$$M_{max} = \frac{wL^2}{12}$$

If the bending stress is 150 MN/m^2,

$$M_{maz} = \frac{\sigma I}{y} = \sigma Z_e = \left(150 \times 10^6\right)\left(0.933 \times 10^{-3}\right) = 140\,\text{kNm}$$

Then

$$w = \frac{12}{L^2}\left(M_{max}\right) = 46.7 \text{ kN/m}$$

14.6 Continuous beam

When the same beam runs across three or more supports it is spoken of as a *continuous beam*. Suppose we have three spans, as in Figure 14.9, each bridged by a separate beam; the beams will bend independently in the manner shown. In order to make the axes of the three beams form a single continuous curve across the supports B and C, we shall have to apply to each beam couples acting as shown by the arrows. When the beam is one continuous girder these couples, on any bay such as BC, are supplied by the action of the adjacent bays. Thus AB and CD, bending downwards under their own loads, try to bend BC upwards, as shown by the broken curve, thus applying the couples M_B and M_C to the bay BC. This upward bending is of course opposed by the down load on BC, and the general result is that the beam takes up a sinuous form, being, in general, concave upwards over the middle portion of each bay and convex upwards over the supports.

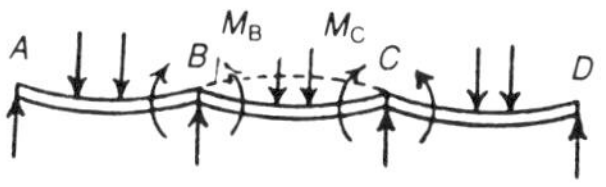

Figure 14.9 Bending moments at the supports of a continuous beam.

In order to draw the bending moment diagram for a continuous beam we must first find the couples such as M_B and M_C. In some cases there may also be external couples applied to the beam, at the supports, by the action of other members of the structure.

When the bending moments at the supports have been found, the bending moment and shearing force diagrams can be drawn for each bay according to the methods discussed in Chapter 7.

14.7 Slope-deflection equations for a single beam

In dealing with continuous beams we can make frequent use of the end slope and deflection properties of a single beam under any conditions of lateral loading. The uniform beam of Figure 14.10(i) carries any system of lateral loads; the ends are supported in an arbitrary fashion, the displacements and moments being as shown in the figure. In addition there are lateral forces at the supports. The rotations at the supports are θ_A and θ_B, respectively, reckoned positive if clockwise; M_A and M_B are also taken positive clockwise for our present purposes. The displacements δ_A and δ_B are taken positive downwards.

The loaded beam of Figure 14.10(i) may be regarded as the superposition of the loading conditions of Figures 14.10(ii) and (iii). In Figure 14.10(ii) the beam is built-in at each end; the moments at each end are easily calculable from the methods discussed in Sections 14.2 and 14.3. The fixed-end moments for this condition will be denoted by M_{FA} and M_{FB}. In Figure 14.10(iii) the beam carries no external loads between its ends, but end displacements and rotations are the same as those in Figure 14.10(i); the end couples for this condition are M_A' and M_B'. The superposition of Figures 14.10(ii) and (iii) gives the external loading and end conditions of Figure 14.10(i). We must find then the end couples in Figure 14.10(iii); from equations (13.49), putting $w = 0$, we have

$$\theta_A = \frac{M_A' L}{3EI} - \frac{M_B' L}{6EI} + \frac{1}{L}(\delta_B - \delta_A)$$

$$\theta_B = -\frac{M_A' L}{6EI} + \frac{M_B' L}{3EI} + \frac{1}{L}(\delta_B - \delta_A)$$

Then

$$\theta_A + \frac{1}{L}\left(\delta_A - \delta_B\right) = \frac{1}{6EI}\left(2M_A' - M_B'\right)$$

$$\theta_B + \frac{1}{L}\left(\delta_A - \delta_B\right) = \frac{L}{6EI}\left(2M_B' - M_A'\right)$$

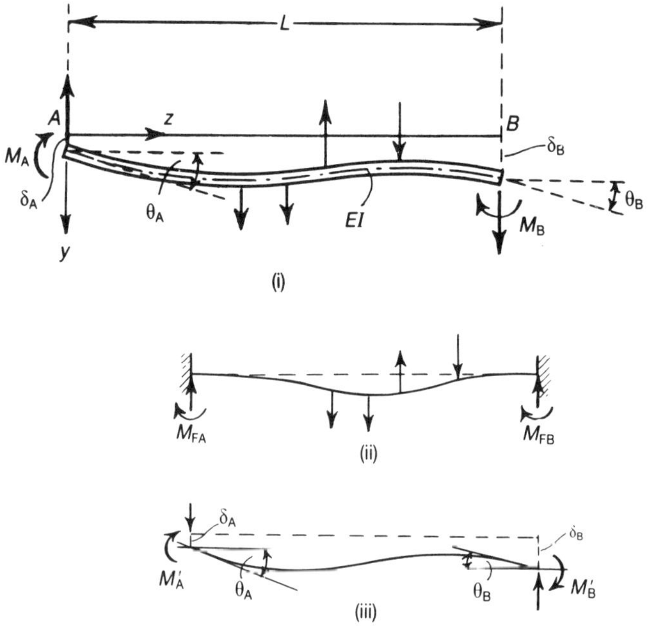

Figure 14.10 The single beam under any conditions of lateral load and end support shown in (i) can be regarded as the superposition of the built-in end beam of (ii) and the beam with end couples and end deformations of (iii).

But for the superposition we have

$$M_A' = M_A - M_{FA} \qquad M_B' = M_B - M_{FB}$$

Thus

$$\theta_A + \frac{1}{L}\left(\delta_A - \delta_B\right) = \frac{L}{6EI}\left[2\left(M_A - M_{FA}\right) - \left(M_B - M_{FB}\right)\right] \tag{14.13}$$

$$\theta_B + \frac{1}{L}\left(\delta_A - \delta_B\right) = \frac{L}{6EI}\left[2\left(M_B - M_{FB}\right) - \left(M_A - M_{FA}\right)\right] \tag{14.14}$$

These are known as the *slope-deflection equations*; they give the values of the unknown moments,

M_A and M_B. These equations will be used in the matrix displacement method of Chapter 23.

Table 14.1 provides a summary of the end fixing moments and maximum deflections for some encastré beams.

Table 14.1 End fixing moments and maximum deflections for some encastré beams

<table>
<tr><th>Beam type and loading – length = L</th><th>M_A</th><th>M_B</th><th>δ
Maximum deflection</th></tr>
<tr><td>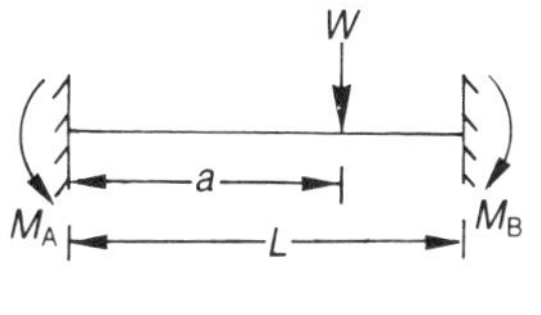
</td><td>$-Wa(L-a)^2/L^2$</td><td>$-Wa^2(L-a)/L^2$</td><td>$\frac{-2W(L-a)^2a^3}{3EI(L+2a)^2}$
@ $z = 2aL/(L+2a)$
when $a > L/2$</td></tr>
<tr><td>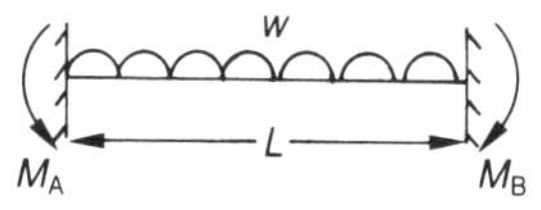
</td><td>$-wL^2/12$</td><td>$-wL^2/12$</td><td>$\frac{wL^4}{384EI}$
@ $z = L/2$</td></tr>
<tr><td>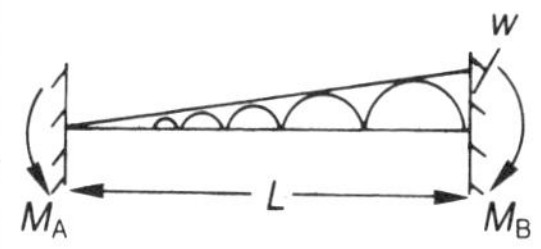
</td><td>$-wL^2/30$</td><td>$-wL^2/20$</td><td>$\frac{0{\cdot}001309wL^4}{EI}$
@ $z = 0{\cdot}525L$</td></tr>
</table>

Further problems *(answers on page 693)*

14.2 A beam 8 m span is built-in at the ends, and carries a load of 60 kN at the centre, and loads of 30 kN, 2 m from each end. Calculate the maximum bending moment and the positions of the points of inflexion.

14.3 A girder of span 7 m is built-in at each end and carries two loads of 80 kN and 120 kN respectively placed at 2 m and 4 m from the left end. Find the bending moments at the ends and centre, and the points of contraflexure. *(Birmingham)*

15 Plastic bending of mild-steel beams

15.1 Introduction

We have seen that in the bending of a beam the greatest direct stresses occur in the extreme longitudinal fibres; when these stresses attain the yield-point values, or exceed the limit of proportionality, the distribution of stresses over the depth of the beams no longer remains linear, as in the case of elastic bending.

The general problem of the plastic bending of beams is complicated; plastic bending of a beam is governed by the forms of the stress–strain curves of the material in tension and compression. Mild steel, which is used extensively as a structural material, has tensile and compressive properties which lend themselves to a relatively simple treatment of the plastic bending of beams of this material. The tensile and compressive stress–strain curves for an annealed mild steel have the forms shown in Figure 15.1; in the elastic range Young's modulus is the same for tension and compression, and of the order of 300 MN/m^2. The yield point corresponds to a strain of the order 0.0015. When the strain corresponding with the upper yield point is exceeded straining takes place continuously at a constant lower yield stress until a strain of about 0.015 is attained; at this stage further straining is accompanied by an increase in stress, and the material is said to *strain-harden*. This region of strain-hardening begins at strains about ten times larger than the strains at the yield point of the material.

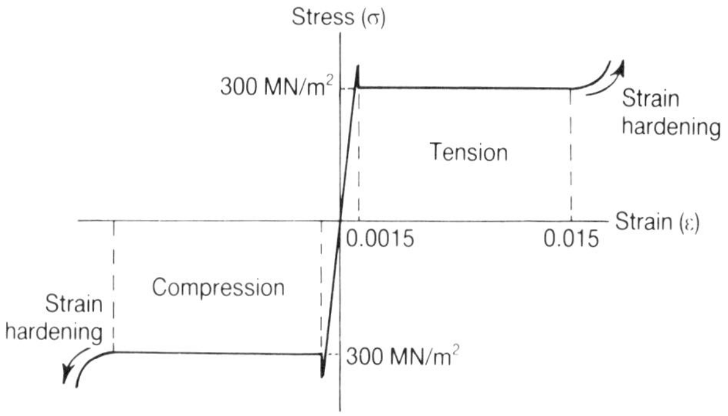

Figure 15.1 Tensile and compressive stress–strain curves of an annealed mild steel.

In applying these stress–strain curves to the plastic bending of mild-steel beams we simplify the problem by ignoring the upper yield point of the material; we assume the material is elastic, with a Young's modulus E, up to a yield stress σ_Y; Figure 15.2. We assume that the yield stress, σ_Y, and Young's modulus, E, are the same for tension and compression. These idealised stress–strain curves for tension and compression are then similar in form.

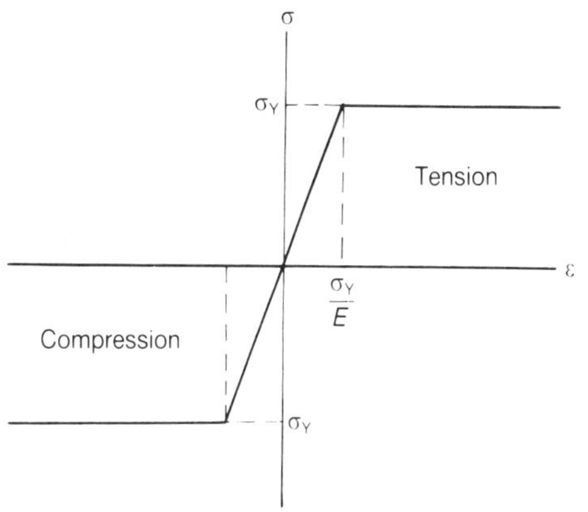

Figure 15.2 Idealized tensile and compressive stress–strain curves of annealed mild steel.

15.2 Beam of rectangular cross-section

As an example of the application of these idealised stress–strain curves for mild steel, consider the uniform bending of a beam of rectangular cross-section; b is the breadth of the cross-section and h its depth, Figure 15.3(i). Equal and opposite moments M are applied to the ends of a length of the beam. We found that in the elastic bending of a rectangular beam there is a linear distribution of direct stresses over a cross-section of the beam; an axis at the mid-depth of the cross-section is unstrained and therefore a neutral axis. The stresses are greatest in the extreme fibres of the beam; the yield stress, σ_Y, is attained in the extreme fibres, Figure 15.3(ii), when

$$M = \frac{2\sigma_Y I}{h} = M_Y \text{ (say)}$$

where I is the second moment of area of the cross-section about the axis of bending. But $I = bh^3/12$, and so

$$M_Y = \frac{1}{6} bh^2 \sigma_Y \tag{15.1}$$

As the beam is bent beyond this initial yielding condition, experiment shows that plane cross-sections of the beam remain nearly plane as in the case of elastic bending. The centroidal axis remains a neutral axis during inelastic bending, and the greatest strains occur in the extreme tension and compression fibres. But the stresses in these extreme fibres cannot exceed σ_Y, the yield stress; at an intermediate stage in the bending of the beam the central core is still elastic, but the extreme fibres have yielded and become plastic, Figure 15.3(iii).

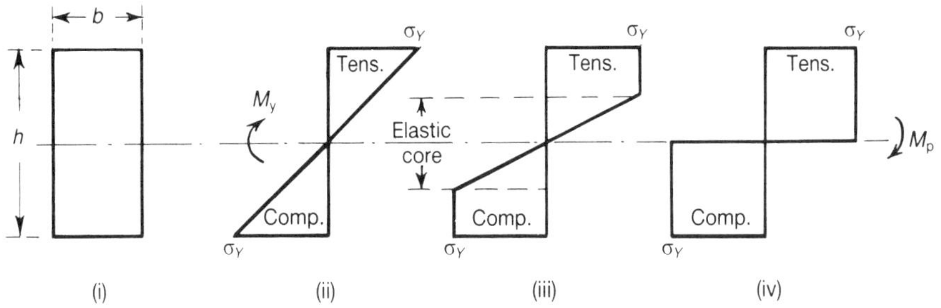

Figure 15.3 Stages in the elastic and plastic bending of a rectangular mild-steel beam.

If the curvature of the beam is increased the elastic core is diminished in depth; finally a condition is reached where the elastic core is reduced to negligible proportions, and the beam is more or less wholly plastic, Figure 15.3(iv); in this final condition there is still a central unstrained, or neutral, axis; fibres above the neutral axis are stressed to the yield point in tension, whereas fibres below the neutral axis to the yield point are in compression. In the ultimate fully plastic condition the resultant longitudinal tension in the upper half-depth of the beam is

$$\frac{1}{2} bh\sigma_Y$$

There is an equal resultant compression in the lower half-depth. There is, therefore, no resultant longitudinal thrust in the beam; the bending moment for this fully plastic condition is

$$M_P = \left(\frac{1}{2} bh\sigma_Y\right)\left(\frac{1}{2} h\right) = \frac{1}{4} bh^2\sigma_Y \tag{15.2}$$

This ultimate moment is usually called the *fully plastic moment* of the beam; comparing equations (15.1) and (15.2) we get

$$M_P = \frac{3}{2} M_Y \tag{15.3}$$

Thus plastic collapse of a rectangular beam occurs at a moment 50% greater than the bending moment at initial yielding of the beam.

15.3 Elastic–plastic bending of a rectangular mild-steel beam

In section 15.2 we introduced the concept of a fully plastic moment, M_P, of a mild-steel beam; this moment is attained when all longitudinal fibres of the beam are stressed into the plastic range of the material. Between the stage at which the yield stress is first exceeded and the ultimate stage at which the fully plastic moment is attained, some fibres at the centre of the beam are elastic and those remote from the centre are plastic. At an intermediate stage the bending is elastic–plastic.

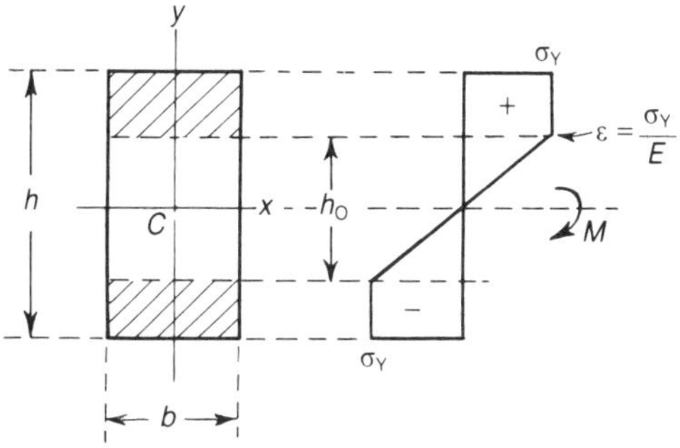

Figure 15.4 Elastic–plastic bending of a rectangular section beam.

Consider again a mild-steel beam of rectangular cross-section, Figure 15.4, which is bent about the centroidal axis Cx. In the elastic–plastic range, a central region of depth h_0 remains elastic; the yield stress σ_Y is attained in fibres beyond this central elastic core. If the central region of depth h_0 behaves as an elastic beam, the radius of curvature, R, is given by

$$\frac{2\sigma_Y}{h_0} = \frac{E}{R} \tag{15.4}$$

where E is Young's modulus in the elastic range of the material. Then

$$h_0 = \frac{2R\sigma_Y}{E} \tag{15.5}$$

Now, the bending moment carried by the elastic core of the beam is

$$M_1 = \sigma_Y \frac{bh_0^2}{6} \tag{15.6}$$

and the moment due to the stresses in the extreme plastic regions is

$$M_2 = \sigma_Y \left[\frac{bh^2}{4} - \frac{bh_0^2}{4} \right] \tag{15.7}$$

The total moment is, therefore,

$$M = M_1 + M_2 = \sigma_Y \frac{bh^2}{4} + \sigma_Y \left[\frac{bh_0^2}{6} - \frac{bh_0^2}{4} \right]$$

which gives

$$M = \sigma_Y \frac{bh^2}{4}\left[1 - \frac{h_0^2}{3h^2}\right] \tag{15.8}$$

But the fully plastic moment, M_P, of the beam is

$$M_P = \sigma_Y \frac{bh^2}{4}$$

Thus equation (15.8) may be written

$$M = M_P\left[1 - \frac{h_0^2}{3h^2}\right] \tag{15.9}$$

On substituting for h_0 from equation (15.5),

$$\frac{M}{M_P} = 1 - \frac{4}{3}\left(\frac{\sigma_Y}{E}\right)^2\left(\frac{R}{h}\right)^2 \tag{15.10}$$

At the onset of plasticity in the beam,

$$\frac{h}{R} = \frac{2\sigma_Y}{E} = \left(\frac{h}{R}\right)_Y \text{ (say)} \tag{15.11}$$

Then equation (15.10) may be written

$$\frac{M}{M_P} = 1 - \frac{1}{3}\frac{(h/R)_y^2}{(h/R)^2} \tag{15.12}$$

Values of (M/M_P) for different values of $(h/R)/(h/R)_Y$ are given in Figure 15.5; the elastic limit of the beam is reached when

$$M = \frac{2}{3}M_P = M_Y \text{ (say)}$$

As M is increased beyond M_Y, the fully plastic moment M_P is approached rapidly with increase of curvature $(1/R)$ of the beam; M is greater than 99% of the fully plastic moment when the curvature is only five times as large as the curvature at the onset of plasticity.

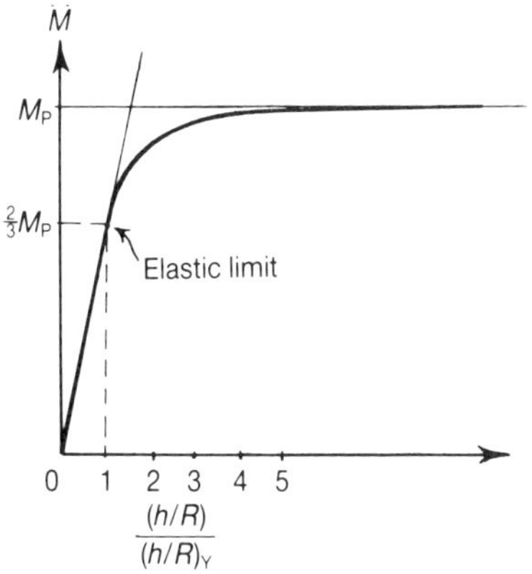

$\frac{h/R}{(h/R)_Y}$	$\frac{M}{M_P}$
1	0.667
2	0.917
3	0.963
4	0.978
5	0.992

Figure 15.5 Moment–curvature relation for the elastic–plastic bending of a rectangular mild-steel beam.

15.4 Fully plastic moment of an I-section; shape factor

The cross-sectional dimensions of an I-section are shown in Figure 15.6; in the fully plastic condition, the centroidal axis Cx is a neutral axis of bending. The tensile fibres of the beam all carry the same stress σ_Y; the total longitudinal force in the upper flange is

$$\sigma_Y b t_f$$

and its moment about Cx is

$$\sigma_Y b t_f \left(\frac{1}{2}h - \frac{1}{2}t_f \right) = \frac{1}{2}\sigma_Y b t_f (h - t_f)$$

Similarly, the total force in the tensile side of the web is

$$\sigma_Y \left(\frac{h}{2} - t_f \right) t_w$$

and its moment about Cx is

$$\frac{1}{2}\sigma_Y \left(\frac{1}{2}h - t_f \right)^2 t_w = \frac{1}{8}\sigma_Y t_w (h - 2t_f)^2$$

The compressed longitudinal fibres contribute moments of the same magnitudes. The total moment carried by the beam is therefore

$$M_P = \sigma_Y \left[bt_f(h - t_f) + \frac{1}{4} t_w (h - 2t_f)^2 \right] \qquad (15.13)$$

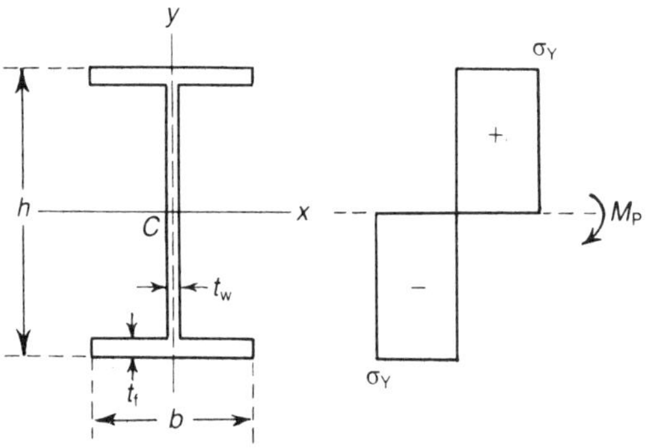

Figure 15.6 Fully plastic moment of an I-section beam.

In the case of elastic bending we defined the elastic section modulus, Z_e, as a geometrical property, which, when multiplied by the allowable bending stress, gives the allowable bending moment on the beam. In equation (15.13) suppose

$$Z_P = bt_f (h - t_f) + \frac{1}{4} t_w (h - 2t_f)^2 \qquad (15.14)$$

Then Z_P is the *plastic section modulus* of the I-beam, and

$$M_P = \sigma_Y Z_P \qquad (15.15)$$

As a particular case consider an I-section having dimensions:

$$h = 20 \text{ cm}, \qquad t_w = 0.70 \text{ cm}$$
$$b = 10 \text{ cm}, \qquad t_f = 1.00 \text{ cm}$$

Then

$$Z_P = (0.1)(0.010)(0.2 - 0.010) + \frac{1}{4}(0.007)(0.2 - 0.020)^2 = 0.247 \times 10^{-3} \text{ m}^3$$

The elastic section modulus is approximately

$$Z_e = 0.225 \times 10^{-3} \text{ m}^3$$

If M_Y is the bending moment at which the yield stress σ_Y is first reached in the extreme fibres of the beam, then

$$\frac{M_P}{M_Y} = \frac{Z_P}{Z_e} = \frac{0.247}{0.225} = 1.10 \tag{15.16}$$

Thus, in this case, the fully plastic moment is only 10% greater than the moment at initial yielding. The ratio (Z_P/Z_e) is sometimes called the *shape factor*.

15.5 More general case of plastic bending

In the case of the rectangular and I-section beams treated so far, the neutral axis of bending coincided with an axis of symmetry of the cross-section. For a section that is unsymmetrical about the axis of bending, the position of the neutral axis must be found first. The beam in Figure 15.7 has one axis of symmetry, *Oy*; the beam is bent into the fully plastic condition about *Ox*, which is perpendicular to *Oy*. The axis *Ox* is the neutral axis of bending; the total longitudinal force on the fibres above *Ox* is $A_1\sigma_Y$, where A_1 is the area of the cross-section of the beam above *Ox*. If A_2 is the area of the cross-section below *Ox*, the total longitudinal force on the fibres below *Ox* is $A_2\sigma_Y$. If there is no resultant longitudinal thrust in the beam, then

$$A_1\sigma_Y = A_2\sigma_Y$$

that is,

$$A_1 = A_2 \tag{15.17}$$

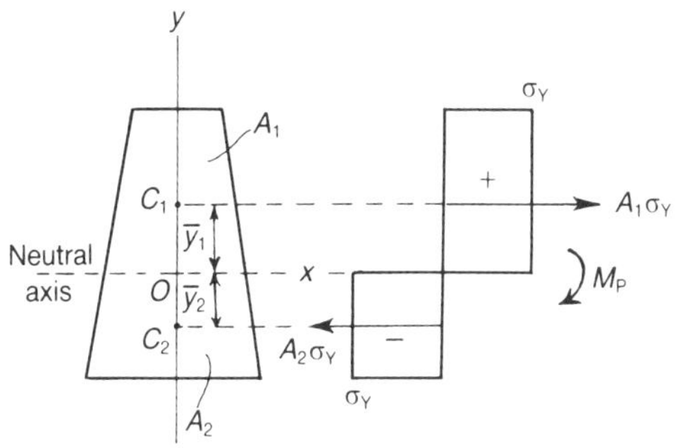

Figure 15.7 Plastic bending of a beam having one axis of symmetry in the cross-section, but unsymmetrical about the axis of bending.

The neutral axis *Ox* divides the beam cross-section into equal areas, therefore. If the total area of cross-section is A, then

$$A_1 = A_2 = \frac{1}{2}A$$

Then

$$A_1\sigma_Y = A_2\sigma_Y = \frac{1}{2}A\sigma_Y$$

Suppose C_1 is the centroid of the area A_1 and C_2 the centroid of A_2; if the centroids C_1 and C_2 are distances $\overline{y}_1$ and $\overline{y}_2$, respectively, from the neutral axis Ox, then

$$M_P = \frac{1}{2}A\sigma_Y\left(\overline{y_1} + \overline{y_2}\right) \tag{15.18}$$

The plastic section modulus is

$$Z_P = \frac{M_P}{\sigma_Y} = \frac{1}{2}A\left(\overline{y_1} + \overline{y_2}\right) \tag{15.19}$$

Problem 15.1 A 10 cm by 10 cm T-section is of uniform thickness 1.25 cm. Estimate the plastic section modulus for bending about an axis perpendicular to the web.

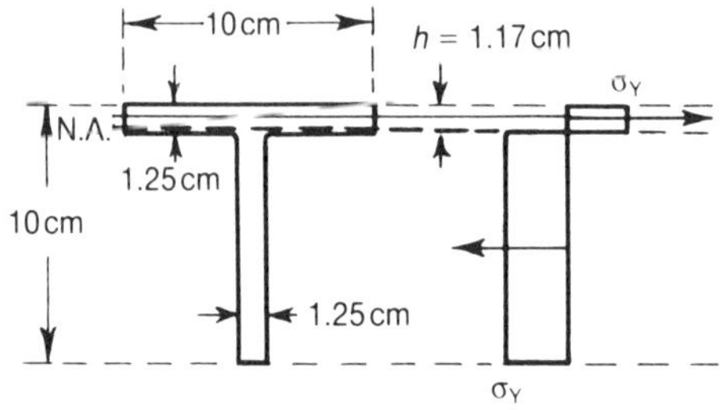

Solution

The neutral axis of plastic bending divides the section into equal areas. If the neutral axis is a distance h below the extreme edge of the flange,

$$(0.1)h = (0.0875)(0.0125) + (0.1)(0.0125 - h)$$

Then

$$h = 0.0117 \text{ m}$$

Then

$$M_P = \frac{1}{2}(0.1)(0.0117)^2\sigma_Y + \frac{1}{2}(0.0875)(0.0008)^2\sigma_Y$$

$$+ \frac{1}{2}(0.0883)^2(0.0125)\sigma_Y$$

$$= \left(0.0557 \times 10^{-3}\right)\sigma_Y$$

The plastic section modulus is then

$$Z_P = \frac{M_P}{\sigma_Y} = 0.0557 \times 10^{-3}\ \text{m}^3$$

The elastic section modulus is

$$Z_e = 0.0311 \times 10^{-3}\ \text{m}^3$$

Then

$$\frac{M_P}{M_Y} = \frac{Z_P}{Z_e} = \frac{0.0557}{0.0311} = 1.79$$

15.6 Comparison of elastic and plastic section moduli

For bending of a beam about a centroidal axis Cx, the elastic section modulus is

$$Z_e = \frac{I}{y_{\max}} \tag{15.20}$$

where I is the second moment of area of the cross-section about the axis of bending, and y_{max} is the distance of the extreme fibre from the axis of bending.

From equation (15.19) the plastic section modulus of a beam is

$$Z_P = \frac{1}{A}\left(\overline{y_1} + \overline{y_2}\right) \tag{15.21}$$

Values of Z_e and Z_p for some simple cross-sectional forms are shown in Table 15.1. In the solid rectangular and circular sections Z_p is considerably greater than Z_e; the difference between Z_p and Z_e is less marked in the case of thin-walled sections.

Table 15.1 Comparison of elastic and plastic section moduli for some simple cross-sectional forms

Cross-sectional form	Elastic section modulus, Z_e	Plastic section modulus, Z_p	Shape factor, $\frac{Z_p}{Z_e}$
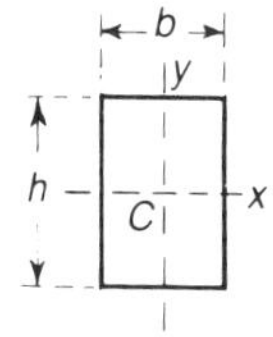 Solid rectangular section	Axis $Cy: \frac{1}{6}b^2h$ Axis $Cx: \frac{1}{6}bh^2$	Axis $Cy: \frac{1}{4}b^2h$ Axis $Cx: \frac{1}{4}bh^2$	1·5 1·5
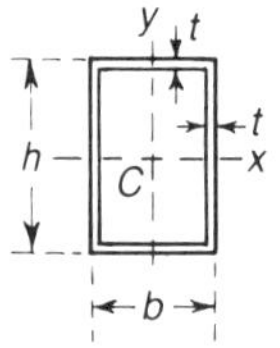 Thin-walled rectangular box of uniform wall-thickness, t	$t \ll h$; $t \ll b$ Axis $Cy: bt(h + \frac{1}{3}b)$ Axis $Cx: ht(b + \frac{1}{3}h)$	Axis $Cy: bt(h + \frac{1}{2}b)$ Axis $Cx: ht(b + \frac{1}{2}h)$	$\frac{h + \frac{1}{2}b}{h + \frac{1}{3}b}$ $\frac{b + \frac{1}{2}h}{b + \frac{1}{3}h}$
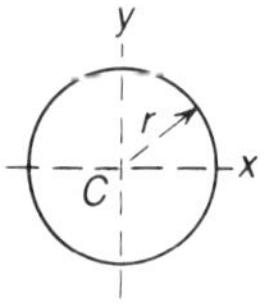 Soild circular section	Axis Cy or Cx: $\frac{\pi r^3}{4}$	Axis Cy or Cx: $\frac{4r^3}{3}$	$\frac{16}{3\pi}$
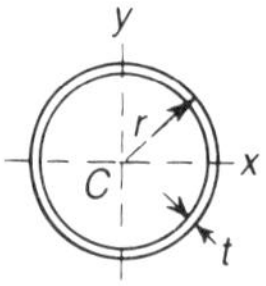 Thin-walled circular tube	$t \ll r$ Axis Cy or $Cx: \pi r^2 t$	$4r^2t$	$\frac{4}{\pi}$
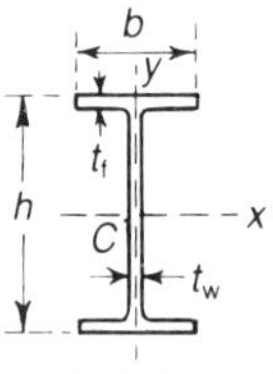 Thin-walled I-section	$t_f \ll b$; $t_w \ll h$ Axis $Cy: \frac{1}{3}b^2t$ Axis $Cx: h[bt_f + \frac{1}{6}ht_w]$	Axis $Cy: \frac{1}{2}b^2t$ Axis $Cx: h[bt_f + \frac{1}{4}ht_w]$	1·5 $\frac{bt_f + \frac{1}{4}ht_w}{bt_f + \frac{1}{6}ht_w}$

15.7 Regions of plasticity in a simply-supported beam

The mild-steel beam shown in Figure 15.8 has a rectangular cross-section; it is simply-supported at each end, and carries a central lateral load W. The variation of bending moment has the form shown in Figure 15.8(ii); the greatest bending moment occurs under the central load and has the value $WL/4$. From the preceding analysis we see that a section may take an increasing bending moment until the fully plastic moment M_P of the section is reached. The ultimate strength of the beam is reached therefore when

$$M_P = \frac{WL}{4} \tag{15.22}$$

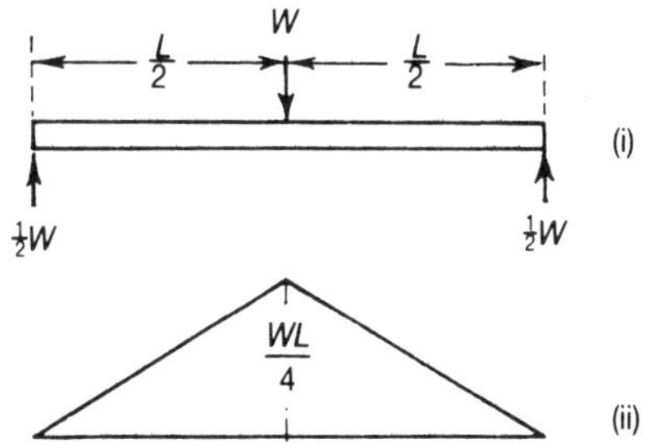

Figure 15.8 Plastic bending of a simply-supported beam.

If b is the breadth and h the depth of the rectangular cross-section, the bending moment, M_Y, at which the yield stress, σ_Y, is first attained in the extreme fibres is

$$M_Y = \sigma_Y \frac{bh^2}{6} = \frac{2}{3} M_P$$

At the ultimate strength of the beam

$$W = \frac{4M_P}{L} = \frac{4}{L}\left[\sigma_Y \frac{bh^2}{4}\right] \tag{15.23}$$

The beam is wholly elastic for a distance of

$$\frac{2}{3}\left(\frac{L}{2}\right) = \frac{1}{3}L \tag{15.24}$$

from each end support, Figure 15.9, as the bending moments in these regions are not greater than M_Y. The middle-third length of the beam is in an elastic–plastic state; in this central region consider a transverse section a–a of the beam, a distance z from the mid-length. The bending moment at this section is

$$M = \frac{1}{2}W\left(\frac{1}{2}L - z\right) \tag{15.25}$$

If W has attained its ultimate value given by equation (15.22),

$$M = \frac{2M_P}{L}\left(\frac{1}{2}L - z\right) \tag{15.26}$$

Suppose the depth of the elastic core of the beam at this section is h_0, Figure 15.9; then from equation (15.9),

$$M = M_P\left(1 - \frac{h_0^2}{3h^2}\right)$$

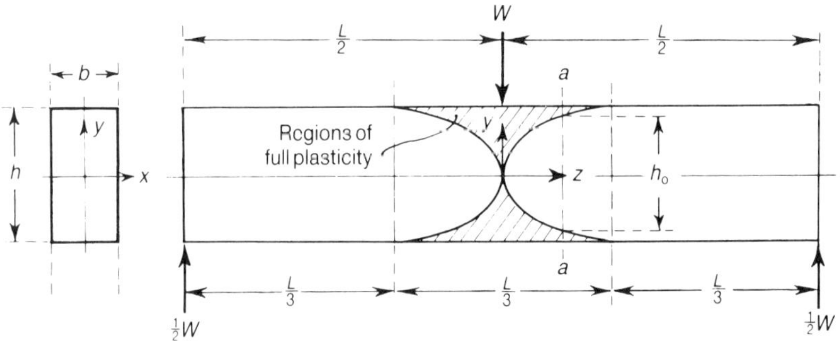

Figure 15.9 Regions of plasticity in a simply-supported beam carrying a distributed load; in the figure the depth of the beam is exaggerated.

On substituting this value of M into equation (15.26), we have

$$1 - \frac{h_0^2}{3h^2} = 1 - \frac{2z}{L} \tag{15.27}$$

and thus

$$h_0^2 = \frac{6h^2}{L} z \tag{15.28}$$

The total depth h_0, of the elastic core varies parabolically with z, therefore; from equation (15.28), $h_0 = h$ when $z = 1/6L$. The regions of full plasticity are wedge-shaped; the shapes of the regions developed in an actual mild-steel beam may be affected by, first, the stress-concentrations under the central load W, and, second, the presence of shearing stresses on sections such as a–a, Figure 15.9; equation (15.28) is true strictly for conditions of pure bending only.

For a simply-supported rectangular beam carrying a total uniformly distributed load W, Figure 15.10, the bending moment at the mid-length is

$$M_p = \frac{WL}{8}$$

at the ultimate load-carrying capacity of the beam. At a transverse section a–a, a distance z from the mid-length, the moment is

$$M = \frac{W}{8L}\left(L^2 - 4z^2\right) = \frac{M_p}{L^2}\left(L^2 - 4z^2\right) = M_p\left[1 - 4\left(\frac{z}{L}\right)^2\right] \tag{15.29}$$

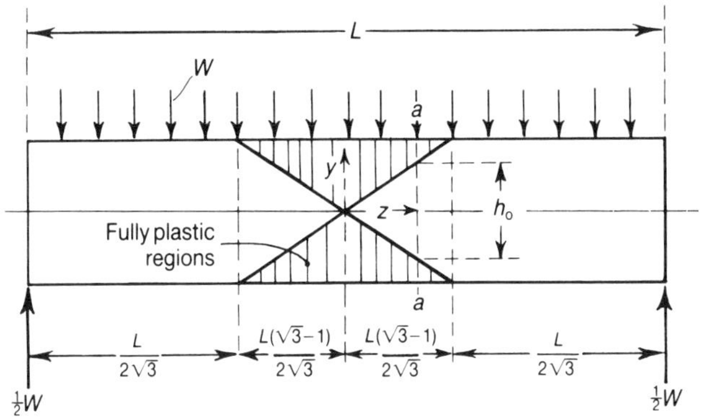

Figure 15.10 Regions of plasticity in a simply-supported beam carrying a distributed load; in the figure the depth of the beam is exaggerated.

From equation (15.9), the depth h_0 of the elastic core at the section a–a is given by

$$M = M_p\left[1 - \frac{h_0^2}{3h^2}\right]$$

Then

$$h_0^2 = 12h^2\left(\frac{z}{L}\right)^2$$

or

$$h_0 = 2\sqrt{3}\, h\left(\frac{z}{L}\right) \tag{15.30}$$

The limit of the wholly elastic length of the beam is given by $h = h_0$, or $z = L/(2\sqrt{3})$. The regions of plasticity near the mid-section are triangular-shaped, Figure 15.10.

15.8 Plastic collapse of a built-in beam

A uniform beam of length L is built-in at each end to rigid walls, and carries a uniformly distributed load w per unit length, Figure 15.11. If the material remains elastic, the bending moment at each end is $wL^2/12$, and at the mid-length $wL^2/24$. The bending moment is therefore greatest at the end supports; if yielding occurs first at a bending moment M_Y, then the lateral load at this stage is given by

$$M_Y = \frac{wL^2}{12} \tag{15.31}$$

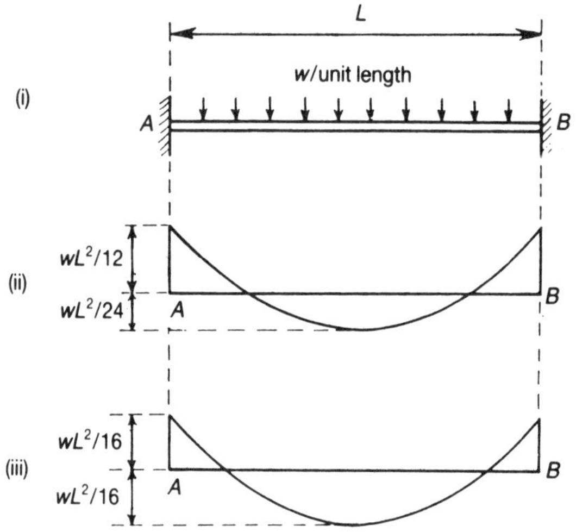

Figure 15.11 Plastic regions of a uniformly loaded built-in beam.

or

$$wL = \frac{12M_y}{L} \tag{15.32}$$

If the load w is increased beyond the limit of elasticity, plastic hinges first develop at the remote ends. The beam only becomes a mechanism when a third plastic hinge develops at the mid-length. On considering the statical equilibrium of a half-span of the beam we find that the moments at the ends and the mid-length, for plastic hinges at these sections, are

$$M_P = \frac{wL^2}{16} \tag{15.33}$$

or

$$wL = \frac{16M_P}{L} \tag{15.34}$$

Clearly, the load causing complete collapse is at least one-third greater than that at which initial yielding begins because M_P is greater than M_Y.

Another method of plastically analysing the beam of Figure 15.11 is by the *principle of virtual work* described in Chapter 17. In this case the beam is assumed to collapse in the form of a mechanism, when three plastic hinges form, as shown in Figure 15.12.

As the beam is encastré at both ends, it is statically indeterminate to the *second degree*, therefore *three hinges* are required to change it from a beam structure to a mechanism.

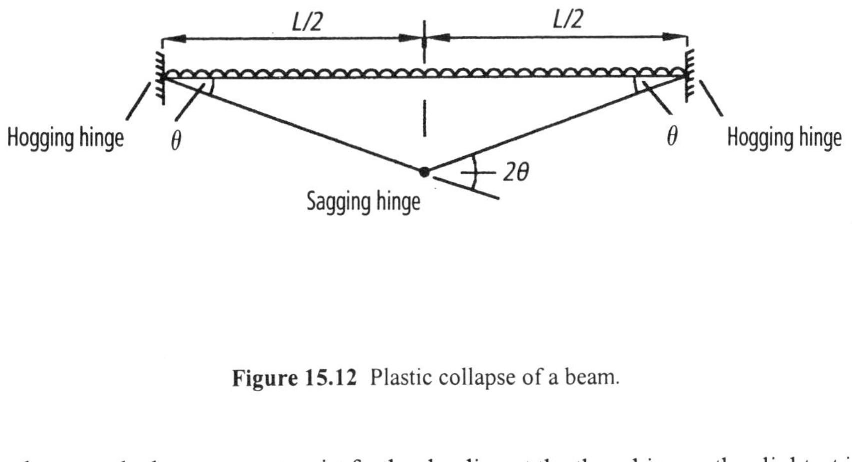

Figure 15.12 Plastic collapse of a beam.

Thus, because the beam cannot resist further loading at the three hinges, the slightest increase in load causes the hinges to rotate like 'rusty' hinges. Additionally, as the bending moment distribution is constant during this collapse, the curvature of the beam remains constant during collapse. Hence, for the purpose of analysis, the beam's two sections can be assumed to remain straight during collapse.

Work done by the three hinges during collapse

$$= M_p\theta + M_p\ 2\theta + M_p\theta \tag{15.35}$$

Work done by the distributed load

$$wL \times \frac{L}{4}\theta \tag{15.36}$$

Equating (15.35) and (15.36)

$$4M_p\theta = \frac{wL^2}{4}\theta \quad \text{or} \quad M_p = \frac{wL^2}{16} \tag{15.37}$$

which is identical to equation (15.33). This method of solution is discussed in greater detail in Chapter 17.

Further problems *(answers on page 693)*

15.2 A uniform mild-steel beam *AB* is 4 m long; it is built-in at *A* and simply-supported at *B*. It carries a single concentrated load at a point 1.5 m from *A*. if the plastic section modulus of the beam is 0.433×10^{-3} m^3, and the yield stress of the material is 235 MN/m^2, estimate the value of the concentrated load causing plastic collapse.

15.3 A uniform mild-steel beam is supported on four knife edges equally spaced a distance 8 m apart. Estimate the intensity of uniformly distributed lateral load over the whole length causing collapse, if the plastic section modulus of the beam is 1.690×10^{-3} m^3, and the yield stress of the material is 235 MN/m^2.

15.4 A uniform beam rests on three supports *A, B* and *C* with two spans each 5 m long. The collapse load is to be 100 kN per metre, and σ_Y = 235 MN/m^2. What will be a suitable mild-steel section using a shape factor 1.15?

15.5 If, in Problem 15.4, *AB* is 8 m and *BC* is 7 m, and the collapse loads are to be 100 kN/m on *AB*, 50 kN/m on *BC*, find a suitable mild-steel section I-beam, with σ_Y = 235 MN/m^2.

15.6 A continuous beam *ABCD* has spans each 8 m long, it is 45 cm by 15 cm, with flanges 2.5 cm thick and web 1 cm thick. Find the collapse load if the whole beam carries a uniformly-distributed load. Which spans collapse? σ_Y = 235 MN/m^2.

15.7 A mild-steel beam 5 cm square section is subjected to a thrust of 200 kN acting in the plane of one of the principal axes, but may be eccentric. What eccentricity will cause the whole section to become plastic if σ_Y = 235 MN/m^2?

16 Torsion of circular shafts and thin-walled tubes

16.1 Introduction

In Chapter 3 we introduced the concepts of shearing stress and shearing strain; these have an important application in torsion problems. Such problems arise in shafts transmitting heavy torques, in eccentrically loaded beams, in aircraft wings and fuselages, and many other instances. These problems are very complex in general, and at this elementary stage we can go no further than studying uniform torsion of circular shafts, thin-walled tubes, and thin-walled open sections.

16.2 Torsion of a thin circular tube

The simplest torsion problem is that of the twisting of a uniform thin circular tube; the tube shown in Figure 16.1 is of thickness t, and the mean radius of the wall is r, L is the length of the tube. Shearing stresses τ are applied around the circumference of the tube at each end, and in opposite directions.

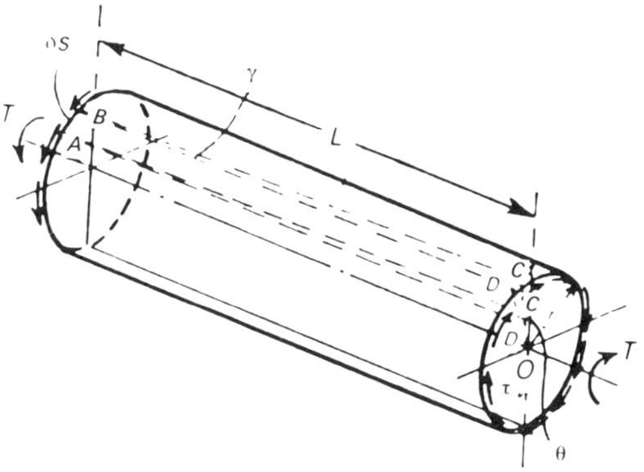

Figure 16.1 Torsion of a thin-walled circular tube.

If the stresses τ are uniform around the boundary, the total torque T at each end of the tube is

$$T = (2\pi r t)\,\tau r = 2\pi r^2 t\,\tau \tag{16.1}$$

Thus the shearing stress around the circumference due to an applied torque T is

$$\tau = \frac{T}{2\pi r^2 t} \tag{16.2}$$

We consider next the strains caused by these shearing stresses. We note firstly that complementary shearing stresses are set up in the wall parallel to the longitudinal axis of the tube. If δs is a small length of the circumference then an element of the wall *ABCD*, Figure 16.1, is in a state of pure shearing stress. If the remote end of the tube is assumed not to twist, then the longitudinal element *ABCD* is distorted into the parallelogram *ABC'D'*, Figure 16.1, the angle of shearing strain being

$$\gamma = \frac{\tau}{G} \tag{16.3}$$

if the material is elastic, and has a shearing (or rigidity) modulus G. But if θ is the angle of twist of the near end of the tube we have

$$\gamma L = r\theta \tag{16.4}$$

Hence

$$\theta = \frac{\gamma L}{r} = \frac{\tau L}{Gr} \tag{16.5}$$

It is sometimes more convenient to define the twist of the tube as the rate of change of twist per unit length; this is given by (θ/L), and from equation (16.5) this is equal to

$$\frac{\theta}{L} = \frac{\tau}{Gr} \tag{16.6}$$

16.3 Torsion of solid circular shafts

The torsion of a thin circular tube is a relatively simple problem as the shearing stress may be assumed constant throughout the wall thickness. The case of a solid circular shaft is more complex because the shearing stresses are variable over the cross-section of the shaft. The solid circular shaft of Figure 16.2 has a length L and radius a in the cross-section.

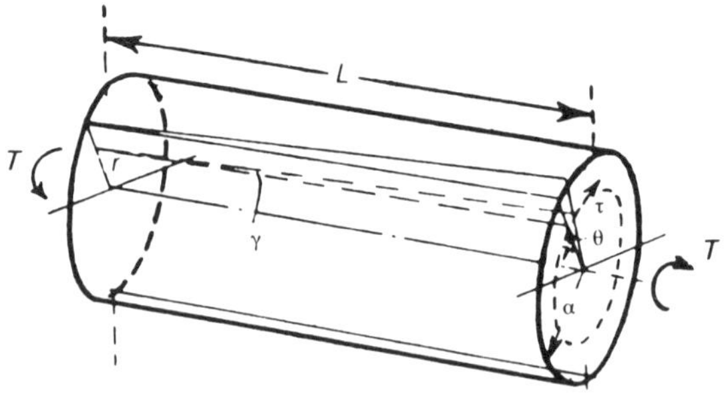

Figure 16.2 Torsion of a solid circular shaft.

When equal and opposite torques T are applied at each end about a longitudinal axis we assume that

(i) the twisting is uniform along the shaft, that is, all normal cross-sections the same distance apart suffer equal relative rotation;

(ii) cross-sections remain plane during twisting; and

(iii) radii remain straight during twisting.

If θ is the relative angle of twist of the two ends of the shaft, then the shearing strain γ of an elemental tube of thickness δr and at radius r is

$$\gamma = \frac{r\theta}{L} \tag{16.7}$$

If the material is elastic, and has a shearing (or rigidity) modulus G, Section 3.4, then the circumferential shearing stress on this elemental tube is

$$\tau = G\gamma = \frac{Gr\theta}{L} \tag{16.8}$$

The thickness of the elemental tube is δr, so the total torque on this tube is

$$(2\pi r \delta r)\tau r = 2\pi r^2 \tau \delta r$$

The total torque on the shaft is then

$$T = \int_0^a 2\pi r^2 \tau \, dr$$

On substituting for τ from equation (16.8), we have

$$\mathrm{T} = 2\pi\left(\frac{G\theta}{L}\right)\int_0^a r^3 \, dr \tag{16.9}$$

Now

$$2\pi \int_0^a r^3 \, dr = \frac{\pi a^4}{2} \tag{16.10}$$

This is the polar second moment of area of the cross-section about an axis through the centre, and is usually denoted by J. Then equation (16.9) may be written

$$T = \frac{GJ\theta}{L} \tag{16.11}$$

We may combine equations (16.8) and (16.11) in the form

$$\frac{T}{J} = \frac{\tau}{r} = \frac{G\theta}{L} \tag{16.12}$$

We see from equation (16.8) that τ increases linearly with r, from zero at the centre of the shaft to $Ga\theta/L$ at the circumference. Along any radius of the cross-section, the shearing stresses are normal to the radius and in the plane of the cross-section, Figure 16.3.

16.4 Torsion of a hollow circular shaft

It frequently arises that a torque is transmitted by a hollow circular shaft. Suppose a_1 and a_2 are the internal and external radii, respectively, of such a shaft, Figure 16.4. We make the same general assumptions as in the torsion of a solid circular shaft. If τ is the shearing stress at radius r, the total torque on the shaft is

$$T = \int_{a_1}^{a_2} 2\pi r^2 \tau dr \tag{16.13}$$

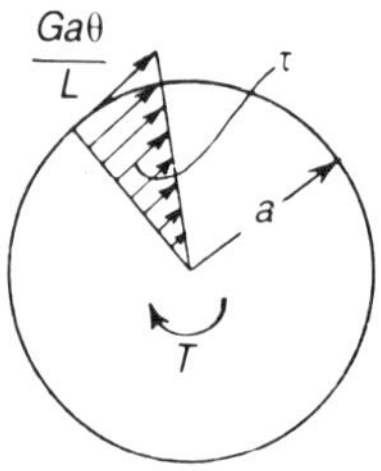

Figure 16.3 Variation of shearing stresses over the cross-section for elastic torsion of a solid circular bar.

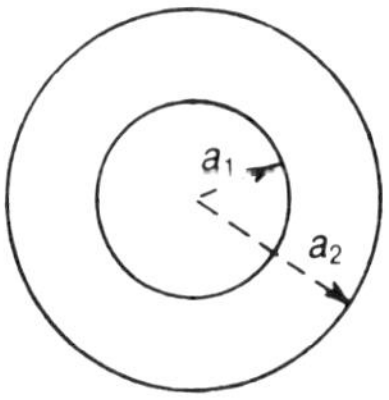

Figure 16.4 Cross-section of a hollow circular shaft.

If we assume, as before, that radii remain straight during twisting, and that the material is elastic, we have

$$\tau = \frac{Gr\theta}{L}$$

Then equation (16.13) becomes

$$T = \int_{a_1}^{a_2} \left(\frac{G\theta}{L}\right) 2\pi r^3 dr = \frac{GJ\theta}{L} \tag{16.14}$$

where

$$J = \int_{a_1}^{a_2} 2\pi r^3 \, dr \qquad (16.15)$$

Here, J is the polar second moment of area or, more generally, the *torsion constant* of the cross-section about an axis through the centre; J has the value

$$J = \int_{a_1}^{a_2} 2\pi r^3 \, dr = \frac{\pi}{2}\left(a_2^4 - a_1^4\right) \qquad (16.16)$$

Thus, for both hollow and solid shafts, we have the relationship

$$\frac{T}{J} = \frac{\tau}{r} = \frac{G\theta}{L}$$

Problem 16.1 What torque, applied to a hollow circular shaft of 25 cm outside diameter and 17.5 cm inside diameter will produce a maximum shearing stress of 75 MN/m^2 in the material *(Cambridge)*

Solution

We have

$$r_1 = 12.5 \text{ cm}, \qquad r_2 = 8.75 \text{ cm}$$

Then

$$J = \frac{\pi}{2}\left[(0.125)^4 - (0.0875)^4\right] = 0.292 \times 10^{-3} \text{ m}^4$$

If the shearing stress is limited to 75 MN/m^2, the torque is

$$T = \frac{J\tau}{r_1} = \frac{\left(0.292 \times 10^{-3}\right)\left(75 \times 10^6\right)}{(0.125)} = 175.5 \text{ kNm}$$

Problem 16.2 A ship's propeller shaft has external and internal diameters of 25 cm and 15 cm. What power can be transmitted at 110 rev/minute with a maximum shearing stress of 75 MN/m^2, and what will then be the twist in degrees of a 10 m length of the shaft? G = 80 GN/m^2. *(Cambridge)*

Solution

In this case

$$r_1 = 0.125 \text{ m}, \quad r_2 = 0.075 \text{ m}, \quad l = 10 \text{ m}$$

$$J = \frac{\pi}{2}\left[(0.125)^4 - (0.075)^4\right] = 0.335 \times 10^{-3} \text{ m}^4$$

and

$$\tau = 75 \text{ MN/m}^2$$

Then

$$T = \frac{J\tau}{r_1} = \frac{(0.335 \times 10^{-3})(75 \times 10^6)}{0.125} = 201 \text{ kNm}$$

At 110 rev/min the power generated is

$$(201 \times 10^3)\left(2\pi \times \frac{110}{60}\right) = 2.32 \times 10^6 \text{ Nm/s}$$

The angle of twist is

$$\theta = \frac{TL}{GJ} = \frac{(201 \times 10^3)(10)}{(80 \times 10^9)(0.335 \times 10^{-3})} = 0.075 \text{ radians} = 4.3°$$

Problem 16.3 A solid circular shaft of 25 cm diameter is to be replaced by a hollow shaft, the ratio of the external to internal diameters being 2 to 1. Find the size of the hollow shaft if the maximum shearing stress is to be the same as for the solid shaft. What percentage economy in mass will this change effect? *(Cambridge)*

Solution

Let r be the inside radius of the new shaft; then = $2r$ the outside radius of the new shaft

$$J \text{ for the new shaft} = \frac{\pi}{2}\left(16r^4 - r^4\right) = 7.5\pi r^4$$

$$J \text{ for the old shaft} = \frac{\pi}{2} \times (0.125)^4 = 0.384 \times 10^{-3} \text{ m}^4$$

If T is the applied torque, the maximum shearing stress for the old shaft is

$$\frac{T(0.125)}{0.384 \times 10^{-3}}$$

and that for the new one is

$$\frac{T(2r)}{7.5\pi r^4}$$

If these are equal,

$$\frac{T(0.125)}{0.384 \times 10^{-3}} = \frac{T(2r)}{7.5\pi r^4}$$

Then

$$r^3 = 0.261 \times 10^{-3} \text{ m}^3$$

or $$r = 0.640 \text{ m}$$

Hence the internal diameter will be 0.128 m and the external diameter 0.256 m.

$$\frac{\text{area of new cross-section}}{\text{area of old cross-section}} = \frac{(0.128)^2 - (0.064)^2}{(0.125)^2} = 0.785$$

Thus, the saving in mass is about 21%.

Problem 16.4 A ship's propeller shaft transmits 7.5×10^6 W at 240 rev/min. The shaft has an internal diameter of 15 cm. Calculate the minimum permissible external diameter if the shearing stress in the shaft is to be limited to 150 MN/m^2. *(Cambridge)*

Solution

If T is the torque on the shaft, then

$$T\left(\frac{2\pi \times 240}{60}\right) = 7.5 \times 10^6$$

Thus

$$T = 298 \text{ kNm}$$

If d_1 is the outside diameter of the shaft, then

$$J = \frac{\pi}{32}\left(d_1^4 - 0.150^4\right) \text{ m}^4$$

If the shearing stress is limited to 150 MN/m^2, then

$$\frac{Td_1}{2J} = 150 \times 10^6$$

Thus,

$$Td_1 = \left(300 \times 10^6\right)J$$

On substituting for J and T

$$\left(298 \times 10^3\right)d_1 = \left(300 \times 10^6\right)\left(\frac{\pi}{32}\right)\left(d_1^4 - 0.150^4\right)$$

This gives

$$\left(\frac{d_1}{0.150}\right)^4 - 3\left(\frac{d_1}{0.150}\right) - 1 = 0$$

On solving this by trial-and-error, we get

$$d_1 = 1.54(0.150) = 0.231 \text{ m}$$

or $$d_1 = 23.1 \text{ cm}$$

16.5 Principal stresses in a twisted shaft

It is important to appreciate that uniform torsion of circular shafts, of the form discussed in Section 16.3, involves no shearing between concentric elemental tubes of the shaft. Shearing stresses τ occur in a cross-section of the shaft, and complementary shearing stresses parallel to the longitudinal axis, Figure 16.5.

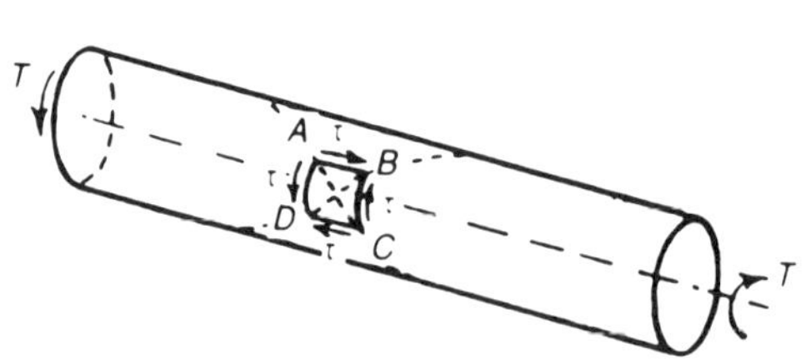

Figure 16.5 Principal stresses in the outer surface of a twisted circular shaft.

An element $ABCD$ in the surface of the shaft is in a state of pure shear. The principal plane makes angles of 45° with the axis of the shaft, therefore, and the principal stresses are $\pm\tau$. If the element $ABCD$ is square, then the principal planes are AC and BD. The direct stress on AC is compressive and of magnitude τ; the direct stress on BD is tensile and of the same magnitude. Principal planes such as AC cut the surface of the shaft in a helix; for a brittle material, weak in tension, we should expect breakdown in a torsion test to occur by tensile fracture along planes such as BD. The failure of a twisted bar of a brittle material is shown in Figure 16.6.

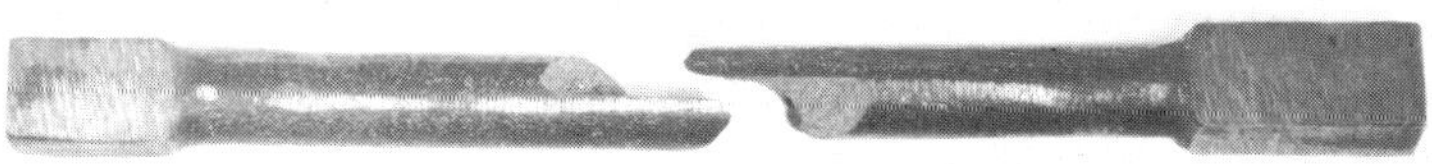

Figure 16.6 Failure in torsion of a circular bar of brittle cast iron, showing a tendency to tensile fracture across a helix on the surface of the specimen.

The torsional failure of ductile materials occurs when the shearing stresses attain the yield stress of the material. The greatest shearing stresses in a circular shaft occur in a cross-section and along the length of the shaft. A circular bar of a ductile material usually fails by breaking off over a normal cross-section, as shown in Figure 16.7.

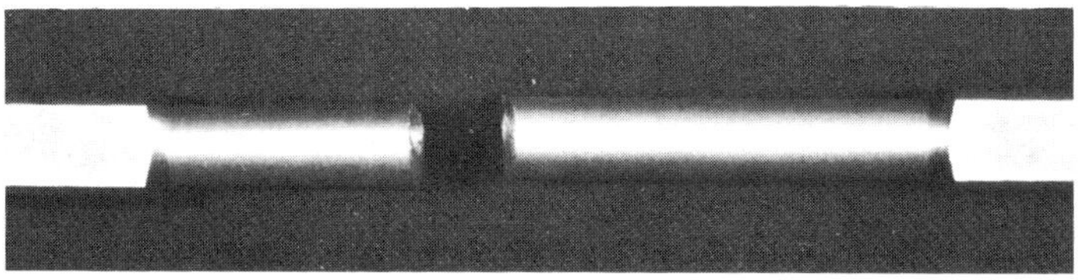

Figure 16.7 Failure of torsion of a circular bar of ductile cast iron, showing a shearing failure over a normal cross-section of the bar.

16.6 Torsion combined with thrust or tension

When a circular shaft is subjected to longitudinal thrust, or tension, as well as twisting, the direct stresses due to the longitudinal load must be combined with the shearing stresses due to torsion in order to evaluate the principal stresses in the shaft. Suppose the shaft is axially loaded in tension so that there is a longitudinal direct stress σ at all points of the shaft.

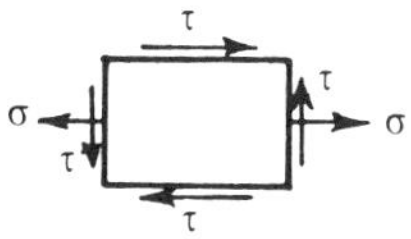

Figure 16.8 Shearing and direct stresses due to combined torsion and tension.

If τ is the shearing stress at any point, then we are interested in the principal stresses of the system shown in Figure 16.8; for this system the principal stresses, from equations (5.12), have the values

$$\frac{1}{2}\sigma \pm \frac{1}{2}\sqrt{\sigma^2 + 4\tau^2} \tag{16.17}$$

and the maximum shearing stress, from equation (5.14), is

$$\tau_{max} = \frac{1}{2}\sqrt{\sigma^2 + 4\tau^2} \tag{16.18}$$

Problem 16.5 A steel shaft, 20 cm external diameter and 7.5 cm internal, is subjected to a twisting moment of 30 kNm, and a thrust of 50 kN. Find the shearing stress due to the torque alone and the percentage increase when the thrust is taken into account. *(RNC)*

Solution

For this case, we have

$$r_1 = 0.100 \text{ m}, \qquad r_2 = 0.0375 \text{ m}$$

$$A = \pi\left(r_1^2 - r_2^2\right) = 0.0270 \text{ m}^2$$

The compressive stress is

$$\sigma = -\frac{P}{A} = -\frac{50 \times 10^3}{0.0270} = -1.85 \text{ MN/m}^2$$

Now

$$J = \frac{\pi}{2}\left(r_1^4 - r_2^4\right) = 0.00247 \text{ m}^4$$

The shearing stress due to torque alone is

$$\tau = \frac{Tr_1}{J} = \frac{\left(30 \times 10^3\right)(0.100)}{0.00247} = 1.22 \text{ MN/m}^2$$

The maximum shearing stress due to the combined loading is

$$\tau_{max} = \frac{1}{2}\left[\sigma^2 + 4\tau^2\right]^{\frac{1}{2}} = 1.53 \ \text{MN/m}^2$$

Problem 16.6 A thin steel tube of 2.5 cm diameter and 0.16 cm thickness has an axial pull of 10 kN, and an axial torque of 23.5 Nm applied to it. Find the magnitude and direction of the principal stresses at any point. *(Cambridge)*

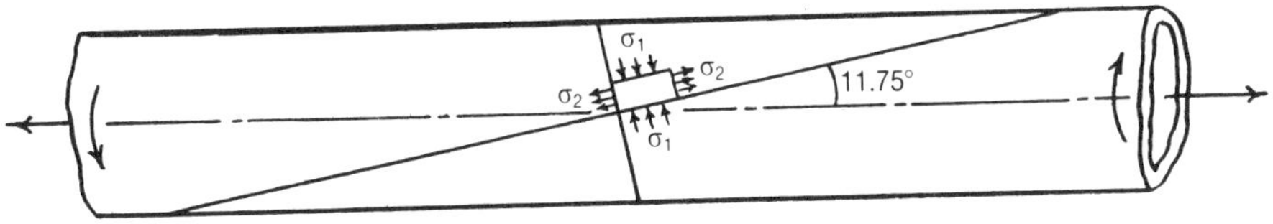

Solution

It will be easier, and sufficiently accurate, to neglect the variation in the shearing stress from the inside to the outside of the tube. Let

τ = the mean shearing stress due to torsion

r = the mean radius = 0.0109 m

t = the thickness = 0.016 m

then the moment of the total resistance to shear

$$= 2\,\pi r^2\,\tau t = (1.19 \times 10^{-6})\,\tau \ \text{Nm}$$

If this is equal to 23.5 Nm, then

$$\tau = 19.75 \ \text{MN/m}^2$$

The area of the cross-section is approximately

$$2\pi rt = 0.1098 \times 10^{-3} \ \text{m}^2$$

Hence, the tensile stress is

$$\sigma = \frac{10 \times 10^3}{0.1098 \times 10^{-3}} = 91.1 \ \text{MN/m}^2$$

The principal stresses are

$$\frac{1}{2}\left\{\sigma \pm \sqrt{\sigma^2 + 4\tau^2}\right\} = \frac{1}{2}(91.1 \pm 99.3)\ \mathrm{MN/m^2}$$

Then

$$\sigma_1 = -4.1\ \mathrm{MN/m^2}, \qquad \sigma_2 = +95.2\ \mathrm{MN/m^2}$$

the positive sign denoting tension. The planes across which they act make angles θ and $(\theta + \pi/2)$ with the axis, where

$$\tan 2\theta = \frac{2\tau}{\sigma} = \frac{39.5}{91.1} = 0.434$$

giving $\theta = 11.75°$.

16.7 Strain energy of elastic torsion

In Section 16.3 we found that the torque–twist relationship for a circular shaft has the form

$$T = \frac{GJ\theta}{L}$$

This shows that the angle of twist, θ, of one end relative to the other, increases linearly with T. If one end of the shaft is assumed to be fixed, then the work done in twisting the other end through an angle θ is the area under the T–θ relationship, Figure 16.9. This work is conserved in the shaft as strain energy, which has the value

$$U = \frac{1}{2}T\theta \tag{16.19}$$

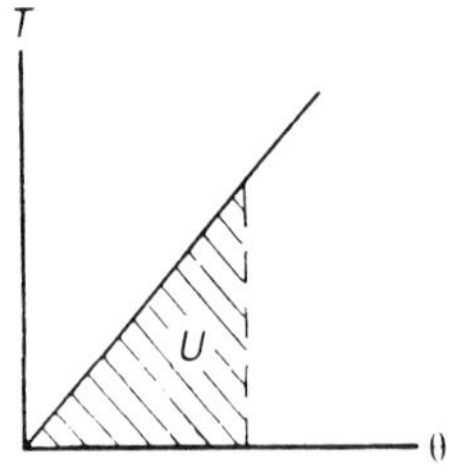

Figure 16.9 Linear torque–twist relationship and strain energy of elastic torsion.

On using equation (16.11) we may eliminate either θ or T, and we have

$$U = \left(\frac{L}{2GJ}\right) T^2 = \left(\frac{GJ}{2L}\right) \theta^2 \tag{16.20}$$

16.8 Plastic torsion of a circular shaft

When a circular shaft is twisted the shearing stresses are greatest in the surface of the shaft. If the limit of proportionality of the material in shear is at a stress τ_Y, then this stress is first attained in the surface of the shaft at a torque

$$T = \frac{J\tau_Y}{a} \tag{16.21}$$

where J is the polar second moment of area, and a is the radius of the cross-section.

Suppose the material has the idealised shearing stress–strain curve shown in Figure 16.10; behaviour is elastic up to a shearing stress τ_Y, the shearing modulus being G. Beyond the limit of proportionality shearing proceeds at a constant stress τ_Y. This behaviour is nearly true of mild steel with a well-defined yield point.

If we are dealing with a solid circular shaft, then after the onset of plasticity in the surface fibres the shearing stresses vary radially in the form shown in Figure 16.11. The material within a radius b is still elastic; the material beyond a radius b is plastic and is everywhere stressed to the yield stress τ_Y

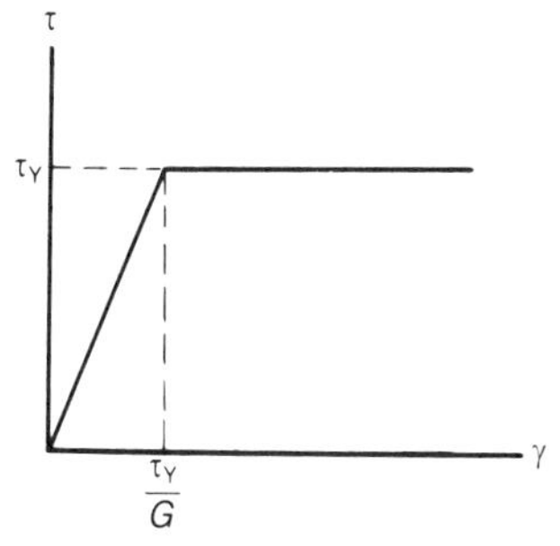

Figure 16.10 Idealized shearing stress–strain curve of mild steel.

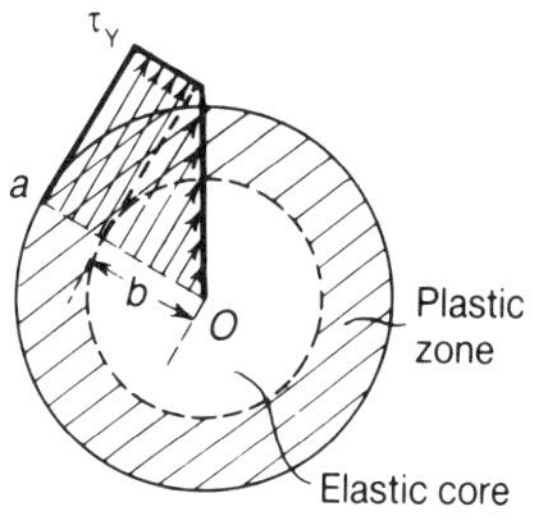

Figure 16.11 Elastic-plastic torsion of a solid circular shaft.

The torque sustained by the elastic core is

$$T_1 = \frac{J_1 \tau_Y}{b} = \frac{\pi}{2} b^3 \tau_Y \tag{16.22}$$

where subscripts 1 refer to the elastic core. The torque sustained by the outer plastic zone is

$$T_2 = \int_b^a 2\pi r^2 \tau_Y dr = \frac{2\pi}{3} \tau_Y \left[a^3 - b^3\right] \qquad (16.23)$$

The total torque on the shaft is

$$T = T_1 + T_2 = \pi\tau_Y\left(\frac{2}{3}a^3 - \frac{1}{6}b^3\right) = \frac{2\pi a^3}{3}\tau_Y\left[1 - \frac{b^3}{4a^3}\right] \qquad (16.24)$$

The angle of twist of the elastic core is

$$\theta = \frac{\tau_Y L}{Gb} \qquad (16.25)$$

where L is the length of the shaft. We assume that the outer plastic region suffers the same angle of twist; this is tantamount to assuming that radii remain straight during plastic torsion of the shaft.

Equation (16.25) gives

$$b = \frac{\tau_Y L}{G\theta} \qquad (16.26)$$

Then the torque becomes

$$T = \frac{2}{3}\pi a^3 \tau_Y \left[1 - \frac{(\tau_Y/G)^3}{4(a/L)^3}\left(\frac{1}{\theta^3}\right)\right] \qquad (16.27)$$

At the onset of plasticity

$$\theta = \frac{\tau_Y L}{Ga} = \theta_Y \text{ (say)} \qquad (16.28)$$

Then, for any other condition of torsion,

$$\theta = \frac{\tau_Y L}{Gb} = \theta_Y\left(\frac{a}{b}\right) \qquad (16.29)$$

which gives

$$\left(\frac{b}{a}\right) = \frac{\theta_Y}{\theta} \qquad (16.30)$$

and equation (16.27) becomes

$$T = \frac{2\pi a^3}{3} \tau_Y \left[1 - \frac{1}{4} \left(\frac{\theta_Y}{\theta} \right)^3 \right] \qquad (16.31)$$

When θ becomes very large, T approaches the value

$$\frac{2\pi a^3}{3} \tau_Y = T_P \text{ (say)} \qquad (16.32)$$

which is the torque on the shaft when it is fully plastic. For smaller values of θ, we have then

$$\frac{T}{T_P} = 1 - \frac{1}{4} \left(\frac{\theta_Y}{\theta} \right)^3 \qquad (16.33)$$

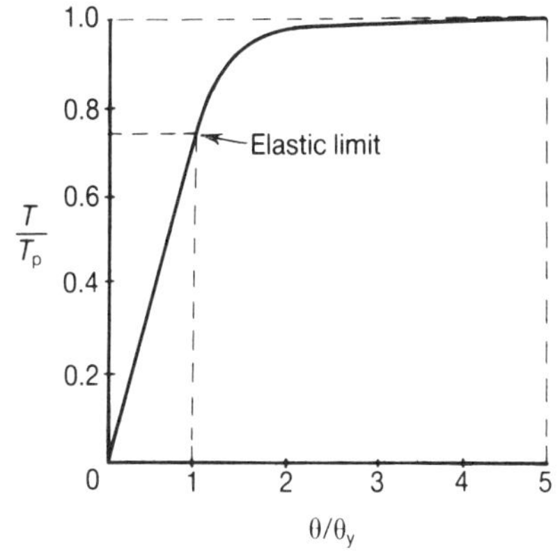

Figure 16.12 Development of full plasticity in the torsion of a solid circular shaft.

This relationship, which is plotted in Figure 16.12 for values of θ/θ_P up to 5, shows that the fully plastic torque T_Y is approached rapidly after the elastic limit is exceeded. The torque T_Y at the elastic limit is

$$T_Y = \frac{3}{4} T_P \qquad (16.34)$$

If a torsion test is carried out on a thin-walled circular tube of mean radius r and thickness t, the average shearing stress due to a torque T is

$$\tau = \frac{T}{2\pi r^2 t}$$

from equation (16.2). If θ is the angle of twist of a length L of the tube, the shearing strain is

$$\gamma = \frac{r\theta}{L}$$

from equation (16.4). Thus, from a torsion test, in which values of T and θ are measured, the shearing stress τ and strain γ can be deduced. The resulting variation of τ and γ is called the *shearing stress–strain curve* of the material; the forms of these stress–strain curves are similar to tensile and compressive stress–strain curves, as shown in Figure 16.13. In the elastic range of a material

$$\tau = G\gamma$$

where G is the *shearing modulus* of the material (Section 3.4).

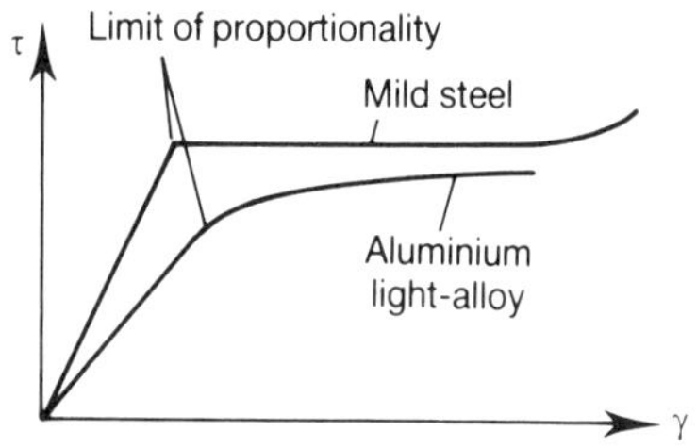

Figure 16.13 Forms of shearing stress–strain curves for mild steel and for aluminium light alloys.

It is important to appreciate that the shearing stress–strain curve cannot be directly deduced from a torsion test of a *solid* circular bar, although the limit of proportionality can be estimated reasonably accurately.

16.9 Torsion of thin tubes of non-circular cross-section

In general the problem of the torsion of a shaft of non-circular cross-section is a complex one; in the particular case when the shaft is a hollow thin tube we can develop, however, a simple theory giving results that are sufficiently accurate for engineering purposes.

Consider a thin-walled closed tube of uniform section throughout its length. The thickness of the wall at any point is t, Figure 16.14, although this may vary at points around the circumference of the tube. Suppose torques T are applied to each end so that the tube twists about a longitudinal axis Cz. We assume that the torque T is distributed over the end of the tube in the form of shearing stresses which are parallel to the tangent to the wall at any point, Figure 16.14, and that the ends of the tube are free from axial restraint. If the shearing stress at any point of the circumference is τ, then an equal complementary shearing stress is set up along the length of the tube. Consider the equilibrium of the section $ABCD$ of the wall: if the shearing stress τ at any point is uniform throughout the wall thickness then the shearing force transmitted over the edge BC is τt per unit length.

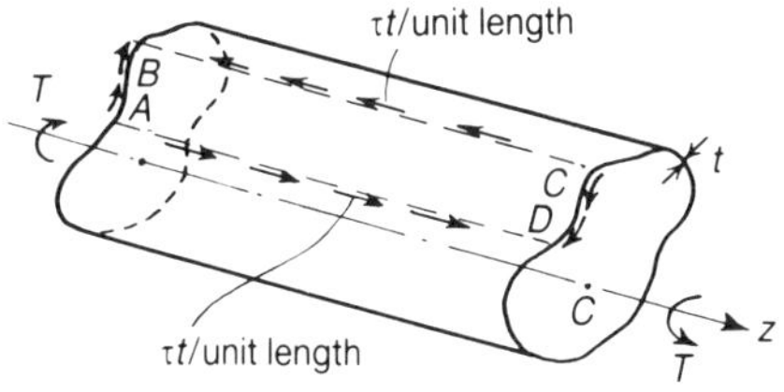

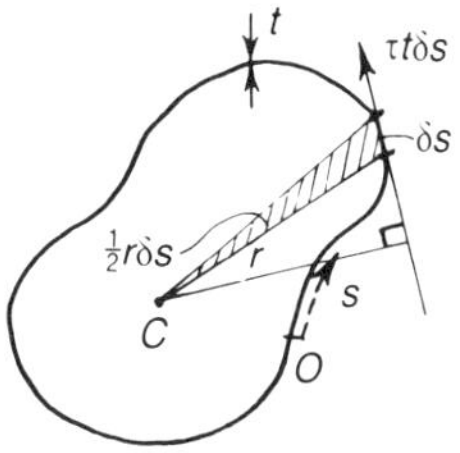

Figure 16.14 Torsion of a thin-walled tube of any cross-section.

For longitudinal equilibrium of $ABCD$ we must have that τt on BC is equal and opposite to τt on AD; but the section $ABCD$ is an arbitrary one, and we must have that τt is constant for all parts of the tube. Suppose this constant value of τt is

$$\tau t = q \tag{16.35}$$

The symbol 'q' is called the *shear flow*; it has the units of a load per unit length of the circumference of the tube.

Suppose we measure a distance s round the tube from some point O on the circumference, Figure 16.14. The force acting along the tangent to an element of length δs in the cross-section is $\tau t \delta s$. Suppose r is the length of the perpendicular from the centre of twist C onto the tangent. Then the moment of the force $\tau t \delta s$ about C is

$$\tau t r \delta s$$

The total torque on the cross-section of the tube is therefore

$$T = \oint \tau t r ds \tag{16.36}$$

where the integration is carried out over the whole of the circumference. But τt is constant and equal to q for all values of s. Then

$$T = \tau t \oint r ds = q \oint r ds \tag{16.37}$$

Now $\oint r ds$ is twice the area, A, enclosed by the centre line of the wall of the tube, and so

$$T = 2Aq \tag{16.38}$$

The shearing stress at any point is then

$$\tau = \frac{q}{t} = \frac{T}{2At} \tag{16.39}$$

To find the angle of twist of the tube we consider the strain energy stored in the tube, and equate this to the work done by the torques T in twisting the tube. When a material is subjected to shearing stresses τ the strain energy stored per unit volume of material is, from equation (3.5),

$$\frac{\tau^2}{2G}$$

where G is the shearing (or rigidity) modulus of the material. In the tube the shearing stresses are varying around the circumference but not along the length of the tube. Then the strain energy stored in a longitudinal element of length L, width δs and thickness t is

$$\left(\frac{\tau^2}{2G}\right) Lt\delta s$$

The total strain energy stored in the tube is therefore

$$U = \oint \frac{\tau^2}{2G} \, Lt\, ds \tag{16.40}$$

where the integration is carried out over the whole circumference of the tube. But τt is constant, and equal to q, and we may write

$$U = \oint \frac{\tau^2}{2G} \, Lt^2 \frac{ds}{t} = \frac{q^2 L}{2G} \oint \frac{ds}{t} \tag{16.41}$$

If the ends of the tube twist relative to each other by an angle θ, then the work done by the torques T is

$$W = \frac{1}{2} T\theta \tag{16.42}$$

On equating U and W, we have

$$\theta = \frac{q^2 L}{GT} \int \frac{ds}{t} \tag{16.43}$$

But from equation (16.38) we have

$$q = \frac{T}{2A} \tag{16.44}$$

Then equation (16.43) may be written

$$\theta = \frac{TL}{2A^2G}\int \frac{ds}{t} \tag{16.45}$$

For a tube of uniform thickness t,

$$\theta = \frac{TL}{4A^2G}\left(\frac{S}{t}\right) \tag{16.46}$$

where S is the total circumference of the tube.

Equation (16.45) can be written in the form

$$\theta = \frac{TL}{GJ}$$

where

$$J = \frac{4A^2}{\int \frac{ds}{t}}$$

J is the *torsion constant* for the section; for circular cross-sections J is equal to the polar second moment of area, but this is not true in general.

16.10 Torsion of a flat rectangular strip

A long flat strip of rectangular cross-section has a breadth b, thickness t, and length L. For uniform torsion about the centroid of the cross-section, the strip may be treated as a set of concentric thin hollow tubes, all twisted by the same amount.

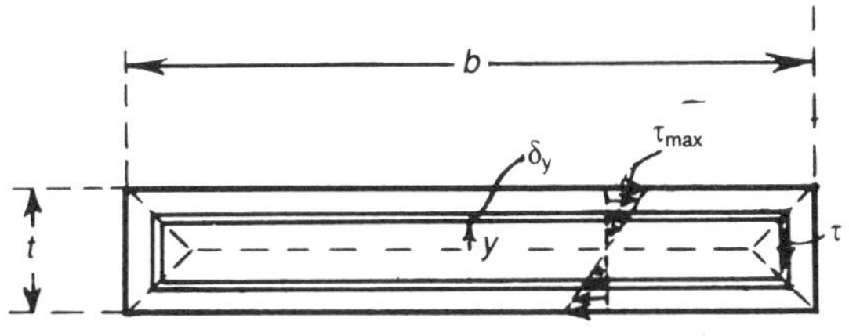

Figure 16.15 Torsion of a thin strip.

Consider such an elemental tube which is rectangular in shape the longer sides being a distance y from the central axis of the strip; the thickness of the tube is δy, Figure 16.15.

If δT is the torque carried by this elemental tube then the shearing stress in the longer sides of the tube is

$$\tau = \frac{\delta T}{4by\,\delta y} \tag{16.47}$$

where b is assumed very much greater than t. This relationship gives

$$\frac{dT}{dy} = 4by\tau \tag{16.48}$$

For the angle of twist of the elemental tube we have, from equation (16.46),

$$\theta = \frac{2bL\delta T}{16b^2y^2G\delta y} \tag{16.49}$$

where L is the length of the strip. This gives the further relationship

$$\frac{dT}{dy} = 8by^2G\frac{\theta}{L} \tag{16.50}$$

On comparing equations (16.48) and (16.50), we have

$$\tau = 2yG\left(\frac{\theta}{L}\right) \tag{16.51}$$

This shows that the shearing stress τ varies linearly throughout the thickness of the strip, having a maximum value in the surface of

$$\tau_{max} = Gt\left(\frac{\theta}{L}\right) \tag{16.52}$$

An important feature is that the shearing stresses τ act parallel to the longer side b of the strip, and that their directions reverse over the thickness of the strip. This approximate solution gives an inexact picture of the shearing stresses near the corners of the cross-section.

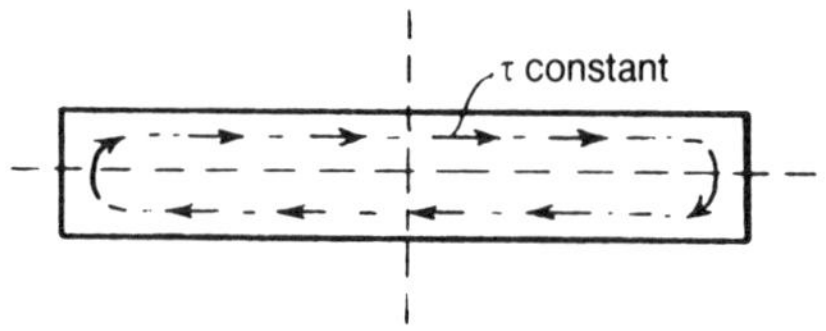

Figure 16.16 Directions of shearing stresses in the torsion of a thin strip.

We ought to consider not rectangular elemental tubes but flat tubes with curved ends. The contours of constant shearing stress are then continuous curves, Figure 16.16.

The total torque on the cross-section is

$$T = \int_0^{\frac{1}{2}t} 8by^2 G\left(\frac{\theta}{L}\right) dy = \frac{1}{3} bt^3 G \frac{\theta}{L} \tag{16.53}$$

The polar second moment of area of the cross-section about its centre is

$$J = \frac{1}{12}\left(bt^3 + b^3 t\right) \tag{16.54}$$

If b is very much greater than t, then, approximately,

$$J = \frac{1}{12} b^3 t \tag{16.55}$$

The geometrical constant occurring in equation (16.53) is $bt^3/3$; thus, in the torsion of a thin strip we cannot use the polar second moment of area for J in the relationship

$$\frac{T}{J} = \frac{G\theta}{L} \tag{16.56}$$

Instead we must use

$$J = \frac{1}{3} bt^3 \tag{16.57}$$

16.11 Torsion of thin-walled open sections

We may extend the analysis of the preceding section to the uniform torsion of thin-walled open-sections of any cross-sectional form.

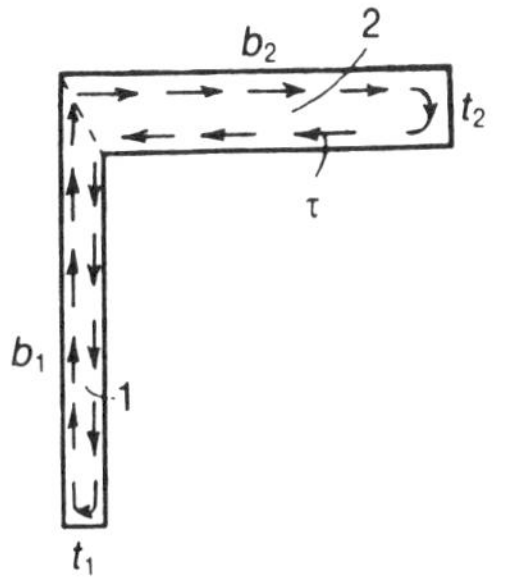

Figure 16.17 Torsion of an angle section.

In the angle section of Figure 16.17, we take elemental tubes inside the two limbs of the section. If t_1 and t_2 are small compared with b_1 and b_2, the maximum shearing stresses in limbs 1 and 2 are

$$\tau_1 = Gt_1\left(\frac{\theta}{L}\right) \qquad \tau_2 = Gt_2\left(\frac{\theta}{L}\right) \tag{16.58}$$

where the angle of twist per unit length, θ/L, is common to both limbs.

The greatest shearing stress occurs then in the surface of the thicker limb of the cross-section. The total torque is the summation of the torques carried by the two limbs, and has the value

$$T = \frac{1}{3}\left(b_1 t_1^3 + b_2 t_2^3\right) G\left(\frac{\theta}{L}\right) \tag{16.59}$$

In general, for a thin-walled open-section of any shape the shearing stress in the surface of a section of thickness t is

$$\tau = Gt\left(\frac{\theta}{L}\right) \tag{16.60}$$

The total torque on the section is

$$T = G\left(\frac{\theta}{L}\right)\sum\frac{1}{3}bt^3 \tag{16.61}$$

where the summation is carried out for all limbs of the cross-section.

Further problems *(answers on page 693)*

16.7 Find the maximum shearing stress in a propeller shaft 40 cm external, and 20 cm internal diameter, when subjected to a torque of 450 kNm. If $G = 80$ GN/m^2, what is the angle of twist in a length of 20 diameters? What diameter would be required for a solid shaft with the same maximum stress and torque? *(RNC)*

16.8 A propeller shaft, 45 m long, transmits 10 MW at 80 rev/min. The external diameter of the shaft is 57 cm, and the internal diameter 24 cm. Assuming that the maximum torque is 1.19 times the mean torque, find the maximum shearing stress produced. Find also the relative angular movement of the ends of the shaft when transmitting the average torque. Take $G = 80$ GN/m^2. *(RNC)*

16.9 A steel tube, 3 m long, 3.75 cm diameter, 0.06 cm thick, is twisted by a couple of 50 Nm. Find the maximum shearing stress, the maximum tensile stress, and the angle through which the tube twists. Take $G = 80$ GN/m^2. *(Cambridge)*

16.10 Compare the mass of a solid shaft with that of a hollow one to transmit a given power at a given speed with a given maximum shearing stress, the inside diameter of the hollow shaft being two-thirds of the outside diameter. *(Cambridge)*

16.11 A 2.5 cm circular steel shaft is provided with enlarged portions *A* and *B*. On to this enlarged portion a steel tube 0.125 cm thick is shrunk. While the shrinking process is going on, the 2.5 cm shaft is held twisted by a couple of magnitude 50 Nm. When the tube is firmly set on the shaft this twisting couple is removed. Calculate what twisting couple is left on the shaft, the shaft and tube being made of the same material. *(Cambridge)*

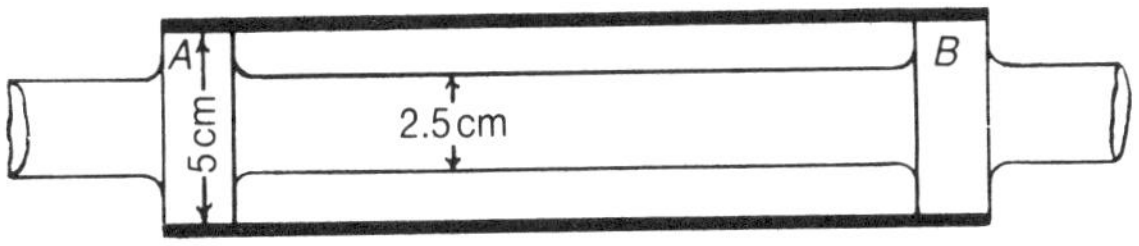

16.12 A thin tube of mean diameter 2.5 cm and thickness 0.125 cm is subjected to a pull of 7.5 kN, and an axial twisting moment of 125 Nm. Find the magnitude and direction of the principal stresses. *(Cambridge)*

17 Energy methods

17.1 Introduction

Energy methods are very useful for analysing structures, especially for those that are statically indeterminate. This chapter introduces the principle of virtual work and applies it to statically determinate and statically indeterminate frameworks. The chapter also shows how the method can be used for the plastic design of beams and rigid-jointed plane frames.

The chapter then introduces strain energy and complementary strain energy, and through the use of worked examples, shows how these methods can be used for analysing structures.

In Chapters 24 and 25, energy methods are used for developing the finite element method, which is one of the most powerful methods for analysing massive and complex structures with the aid of digital computers.

17.2 Principle of virtual work

In its simplest form the *principle of virtual work* is that

> *For a system of forces acting on a particle, the particle is in statical equilibrium if, when it is given any virtual displacement, the net work done by the forces is zero.*

A virtual displacement is any arbitrary displacement of the particle. In the virtual displacement the forces are assumed to remain constant and parallel to their original lines of actions. Consider a particle under the action of three forces, F_1, F_2 and F_3, Figure 17.1.

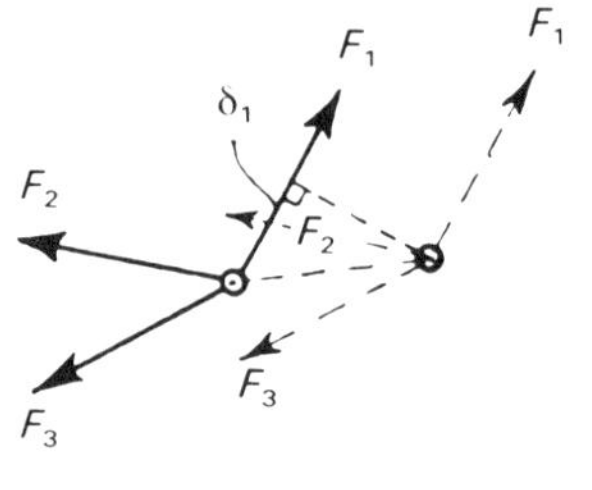

Figure 17.1 System of forces in statical equilibrium acting on a particle.

Imagine the particle to be given a virtual displacement of any magnitude in any direction. Suppose the displacements of the particle along the lines of action of the forces F_1, F_2 and F_3, are δ_1, δ_2 and δ_3, respectively; these are known as *corresponding* displacements. Then the forces form a system in statical equilibrium if

$$F_1\delta_1 + F_2\delta_2 + F_3\delta_3 = 0 \tag{17.1}$$

On the basis of the principle of virtual work we can show that the resultant of the forces acting on a particle in statical equilibrium is zero. Suppose the forces F_1, F_2 and F_3, acting on the particle of Figure 17.1, have a resultant of magnitude R in some direction; then by giving the particle a suitable virtual displacement, Δ, say, in the direction of R, the net work is

$$R\Delta$$

But by the principle of virtual work the net work is zero, so that

$$R\Delta = 0 \tag{17.2}$$

As Δ can be non-zero, R must be zero. Hence, by adopting the principle of virtual work as a basic concept, we can show that the resultant of a system of forces in statical equilibrium is zero.

17.3 Deflections of beams

In a pin-jointed frame subjected to loads applied to the joints only the tensile load in any member is constant throughout the length of that member. In the case of a beam under lateral loads the bending moments and shearing forces may vary from one section to another, so that the state of stress is not uniform along the length of the beam. In applying the principle of virtual work to problems of beams we must consider the loading actions on the virtual displacement of an elemental length of the beam.

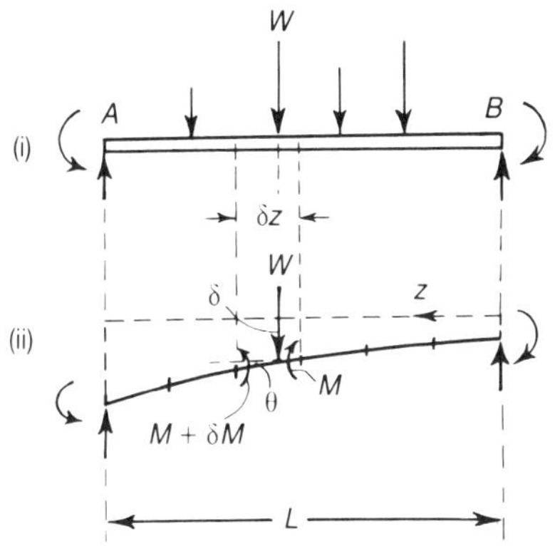

Figure 17.2 Deflections of a straight beam.

Consider a straight beam AB, Figure 17.2, which is in statical equilibrium under the action of a system of external forces and couples. The beam is divided into a number of short lengths; the loading actions on a short length such as δz consist of bending moments M and $(M + \delta M)$, an external lateral load W, and lateral shearing forces at the ends of the short length. Now suppose the short lengths of the beam are given small virtual displacements, θ. If the elements remain connected to each other, then for given values of θ the external forces, such as W, suffer certain displacements, such as δ. Then the values of θ and δ form a *compatible system* of rotations and displacements, and the virtual work of any system of forces and couples in statical in equilibrium on these rotations and displacements is zero. Then

$$\sum \delta M \times \theta + \sum W \times \delta = 0 \tag{17.3}$$

because the net work of the internal shearing forces is zero. The summation $\Sigma\delta M \times \theta$ is carried out for all short lengths of the beam, whereas the summation $\Sigma W \times \delta$ is carried out for all external loads, including couples and force reactions at points of support. If the virtual rotations θ are small, the virtual displacements δ can be found easily. If the lengths δz of the beam are infinitesimally small,

$$\sum \delta M \times \theta = \int_{z=0}^{z=L} \theta \, dM \tag{17.4}$$

where the integration is carried out over the whole length L of the beam. But

$$\int_{z=0}^{z=L} \theta \, dM = \left[M\theta - \int M d\theta \right]_{z=0}^{z=L}$$

Now

$$\left[M\theta \right]_{z=0}^{z=L} = \left[M\theta \right]_{z=L} - \left[M\theta \right]_{z=0}$$

and is the work of the end couples on their respective virtual displacements; this work has already been taken account of in the summation $\Sigma W \times \delta$, so that equation (17.3) becomes

$$\sum W \times \delta = \int_{z=0}^{z=L} M d\theta = \int_{0}^{L} M \left(\frac{d\theta}{dz} \right) dz \tag{17.5}$$

Now $(d\theta/dz)$ is the curvature of the beam when it is given the virtual rotations and displacements. If we put

$$\frac{d\theta}{dz} = \frac{1}{R} \tag{17.6}$$

where R is the radius of curvature of the beam, then

$$\sum W \times \delta = \int_{0}^{L} M \left(\frac{1}{R} \right) dz \tag{17.7}$$

As an example of the application of equation (17.7), consider the cantilever shown in Figure 17.3; having a uniform flexural stiffness EI. The cantilever carries a vertical load W at the free end; the bending moment at any section due to W is Wz, so that, if the beam remains elastic, the corresponding curvature at any section is

$$\frac{1}{R} = \frac{Wz}{EI}$$

Suppose the corresponding deflection of W is δ, Figure 17.3; then the values of $1/R$ and δ form a system of compatible curvature and displacements.

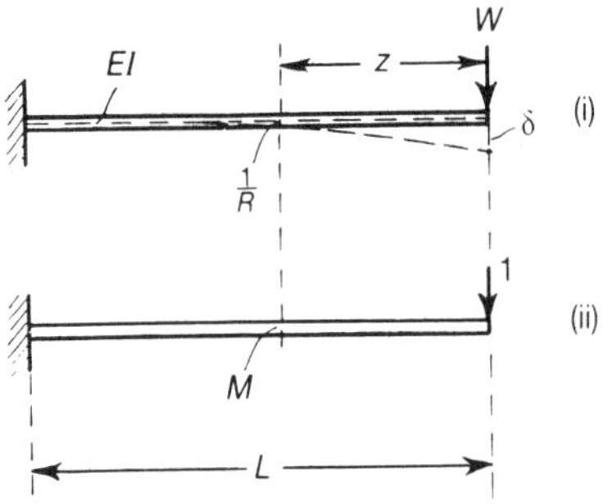

Figure 17.3 Deflections of a cantilever with an end load.

We derive a simple system of forces and couples in statical equilibrium by applying a unit vertical load at the end of the cantilever; the bending moment at any section due to this unit load is

$$M = 1 \times z = z$$

Then, from equation (17.7),

$$1 \times \delta = \int_0^L M\left(\frac{1}{R}\right) dz = \int_0^L \frac{Wz^2}{EI} dz$$

Then

$$\delta = \frac{WL^3}{3EI}$$

Problem 17.1 A simply-supported beam, of uniform flexural stiffness EI, carries a lateral load W at a distance a from the end A. Estimate the vertical deflection of W.

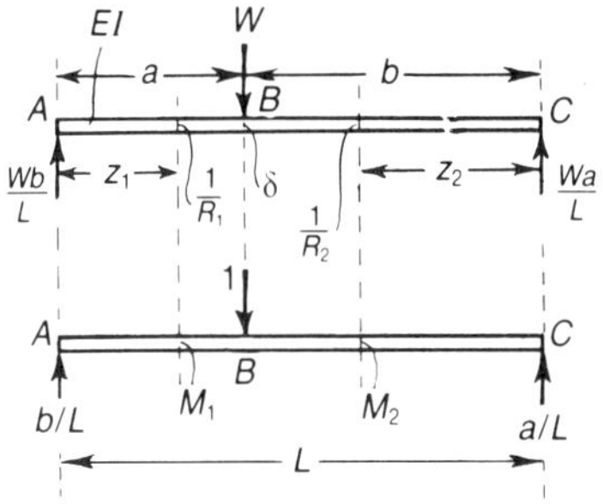

Solution

The bending moment a distance z_1 from A, for the section AB, is

$$\frac{Wbz_1}{L}$$

The curvature for AB if therefore

$$\frac{1}{R_1} = \frac{Wbz_1}{EIL}$$

Similarly, the curvature at any section in BC is

$$\frac{1}{R_2} = \frac{Waz_2}{EIL}$$

Now consider the beam with a unit vertical load at B; the bending moments at sections in AB and BC are, respectively,

$$M_1 = \frac{bz_1}{L}, \qquad M_2 = \frac{az_2}{L}$$

Then, equation (17.7) gives

$$\delta = \int_0^a M_1\left(\frac{1}{R_1}\right)dz_1 + \int_0^b M_2\left(\frac{1}{R_2}\right)dz_2$$

$$= \int_0^a \frac{Wb^2}{EIL^2}z_1^2\,dz_1 + \int_0^b \frac{Wa^2}{EIL^2}z_2^2\,dz_2$$

Therefore

$$\delta = \frac{Wa^2b^2}{3EIL^2}(a + b) = \frac{Wa^2b^2}{3EIL}$$

Problem 17.2 A cantilever of uniform flexural stiffness EI carries a uniformly-distributed load of intensity w. Estimate the vertical deflection of the free end.

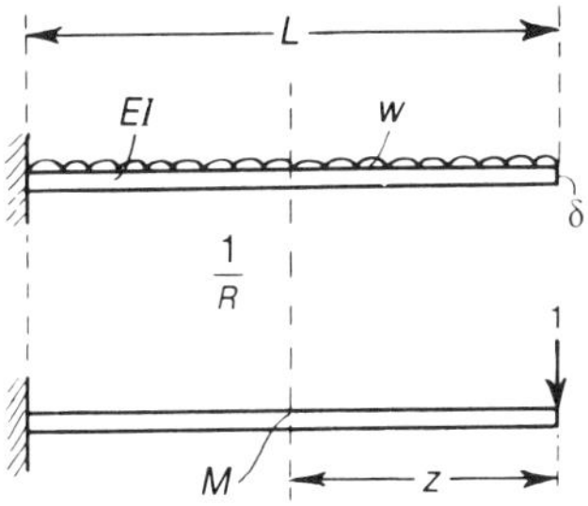

Solution

Due to the distribution load, the curvature at any section is

$$\frac{1}{R} = \frac{wz^2}{2EI}$$

For a unit vertical load at the free end, the bending moment at any section is

$$M = z$$

Then equation (17.7) gives

$$\delta = \int_0^L M\left(\frac{1}{R}\right) dz = \int_0^L \frac{wz^3}{2EI} dz$$

Then

$$\delta = \frac{wL^4}{8EI}$$

Problem 17.3 A semicircular thin ring has a radius r and uniform flexural stiffness EI. The ring carries equal and opposite loads W at the ends. Find the increase in distance between the loaded points.

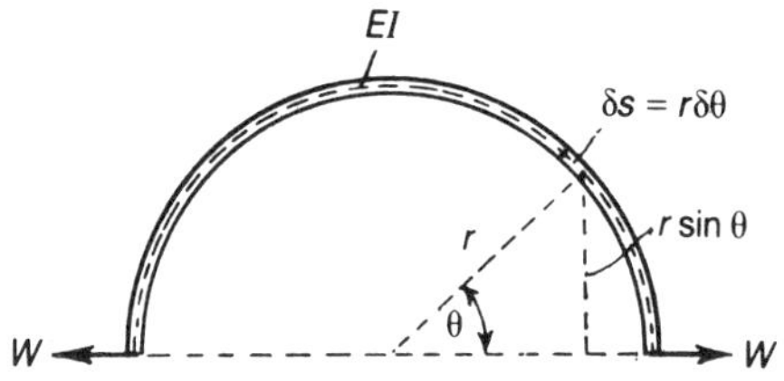

<u>*Solution*</u>

The bending moment at any angular position θ is

$$M = Wr \sin\theta$$

If the ring is thin, the change of curvature at any section is

$$\frac{1}{R} = \frac{M}{EI}$$

Now consider the virtual work of the forces and couples on their resulting displacements; if δ is the increase in distance between the loaded points

$$W \times \delta = \int_{\theta = 0}^{\theta = \pi} M\left(\frac{1}{R}\right) ds = \int_0^{\pi} \frac{M^2 r}{EI} d\theta = \frac{W^2 r^3}{EI} \int_0^{\pi} \sin^2\theta \, d\theta$$

Then

$$\delta = \frac{\pi W r^3}{2EI}$$

17.4 Statically indeterminate beam problems

The principle of virtual work may also be used in solving statically indeterminate beam problems. Consider, for example, the beam of Figure 17.4, which is built-in at *A* and supported on a roller at *B*; the beam is of uniform flexural stiffness *EI*, and carries a uniformly distributed lateral load

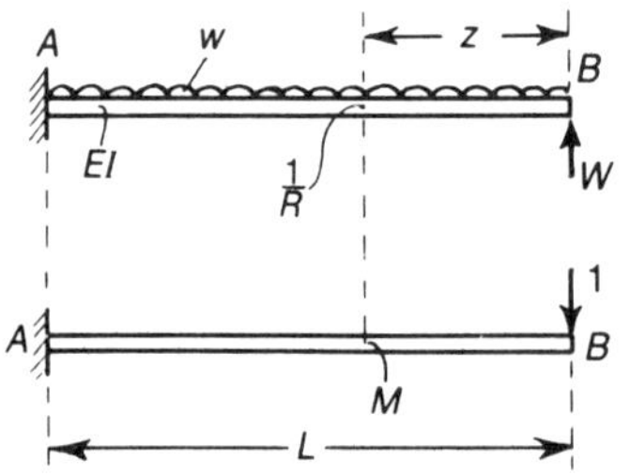

Figure 17.4 Propped cantilever under uniform lateral loading.

of intensity w. Suppose the statically indeterminate reaction at B is W; then the bending moment at any section is

$$\frac{1}{2}wz^2 - Wz$$

and if the beam remains elastic the resulting curvature at an any section is

$$\frac{1}{R} = \frac{1}{EI}\left(\frac{1}{2}wz^2 - Wz\right)$$

The bending moment at any section due to a unit lateral load at B is

$$M = z$$

Then, for no deflection at B in Figure 17.4,

$$1 \times 0 = \int_0^L M\left(\frac{1}{R}\right)dz = \int_0^L \frac{z}{EI}\left(\frac{wz^2}{2} - Wz\right)dz$$

Then

$$\int_0^L \frac{1}{2}wz^3\,dz = \int_0^L Wz^2 dz$$

Thus

$$W = \frac{3wL}{8}$$

17.5 Plastic bending of mild-steel beams

The principle of virtual work is not limited in its application to linear problems of the type discussed in the preceding problems. It is useful, for example, in solving problems of plastic bending; the uniform mild-steel beam of Figure 17.5 has a fully-plastic moment M_P. At collapse of the beam, plastic hinges develop at A and B. Suppose the point B is now given a virtual displacement δ; if δ is small, AB rotates through an angle (δ/a), and BC through an angle $[\delta/(L - a)]$. The work of the system of forces and couples of Figure 17.5(ii) on the virtual displacements and rotations of Figure 17.5(iii) is zero. Then

$$W\delta = M_P\left[\frac{2\delta}{a} + \frac{\delta}{(L - a)}\right]$$

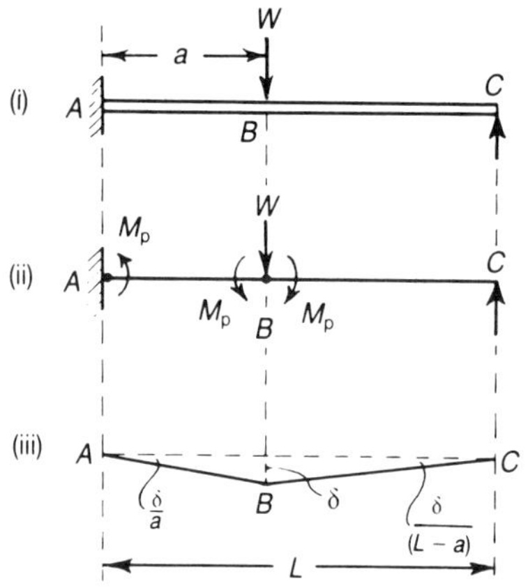

Figure 17.5 Plastic bending of a mild-steel beam.

Then

$$W = \frac{M_P(2L - a)}{a(L - a)}$$

This is the value of W at plastic collapse of the beam.

Problem 17.4 A uniform mild-steel beam has a fully-plastic moment M_P. Find the intensity of uniformly distributed loading at collapse of the beam.

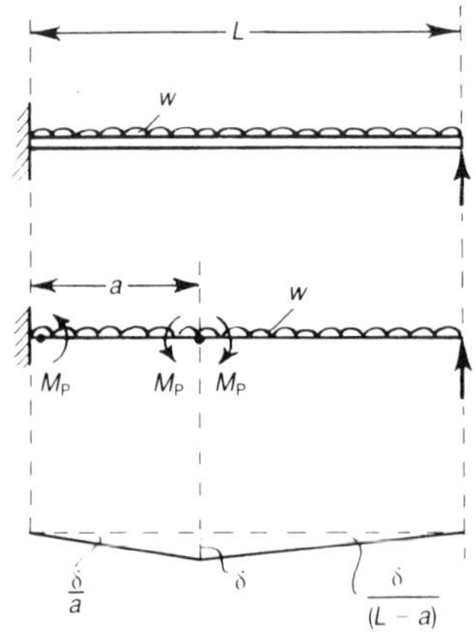

Solution

Suppose that, at plastic collapse, hinges develop at the built-in end, and at a distance a from that end. Then

$$\frac{1}{2}wa\delta + \frac{1}{2}w(L - a)\delta = M_P\left[\frac{2\delta}{a} + \frac{\delta}{(L - a)}\right]$$

Thus,

$$w = \frac{2\left(2 - \frac{a}{L}\right)}{\left(\frac{a}{L}\right)\left(1 - \frac{a}{L}\right)}\frac{M_P}{L^2}$$

This is a minimum with respect to (a/L) when

$$\frac{a}{L} = \left(2 - \sqrt{2}\right)$$

Then the relevant value of w is

$$w = \frac{2M_P}{L^2}\left(3 + 2\sqrt{2}\right)$$

An alternative method of solving the above beam problem is to consider rotations of the hinges, as shown in the figure below

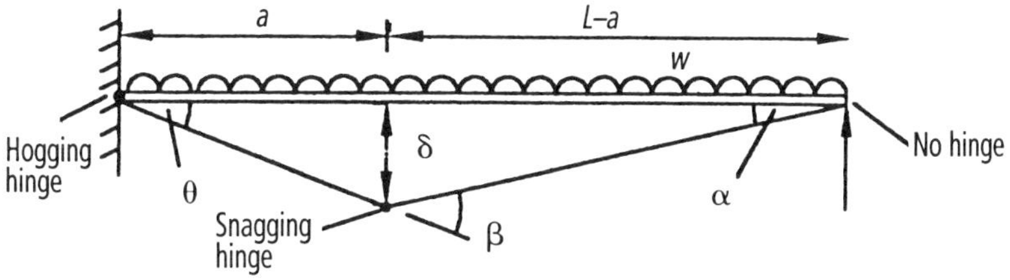

$$\delta = \theta a = \alpha(L - a)$$

$$\therefore \alpha = \theta \,.\, a/(L - a) \tag{17.8}$$

$$\beta = \alpha + \theta = \theta \; a/(L - a) + \theta$$

$$= \theta \; a/(L - a) + \theta \; (L - a) / (L - a)$$

$$= \theta \; (a + L - a) / (L - a)$$

$$\beta = \theta \; L/(L - a) \tag{17.9}$$

Now work done by the hinges

$$= M_P \; \theta + M_P \; \beta$$

$$= M_P \; \theta + M_P \; \theta \; L / (L - a)$$

$$= M_P \; \theta \; (L - a) / (L - a) + M_P \; \theta \; L / (L - a)$$

$$= M_P\theta \; (L - a + L) / (L - a)$$

$$M_P\theta \; (2L - a) / (L - a) \tag{17.10}$$

Work done by the load 'w'

$$w \times L \times \delta/2 = wL \; \theta \; a/2 \tag{17.11}$$

Equating (17.10) and (17.11)

$$M_P \; \theta \; (2L - a) / (L - a) = wL\theta \; a/2$$

$$\text{or } w = 2(2L - a) / [aL \; (L - a)] \; M_P$$

$$= \frac{2L \; (2 - a/L) \; M_P}{aL^2 \; (1 - a/L)}$$

Dividing the top and bottom by L, we get

$$w = \frac{2 \; (2 - a/L) \; M_P}{L^2 \left(\frac{a}{L}\right) \; (1 - a/L)} \tag{17.12}$$

which is the same result as before.

17.6 Plastic design of frameworks

For this case, let us make the following definitions:

λ = load or safety factor

M_P = plastic moment of resistance of the cross-section of a member of the framework

M_Y = the elastic moment of resistance of the cross-section of a member of the framework at first yield

S = shape factor = M_P/M_Y

σ_Y = yield stress

Problem 17.5 Obtain a suitable sectional modulus for the portal frame below, given that:

λ = 2.7

S = 1.15

σ_Y = 300 MPa

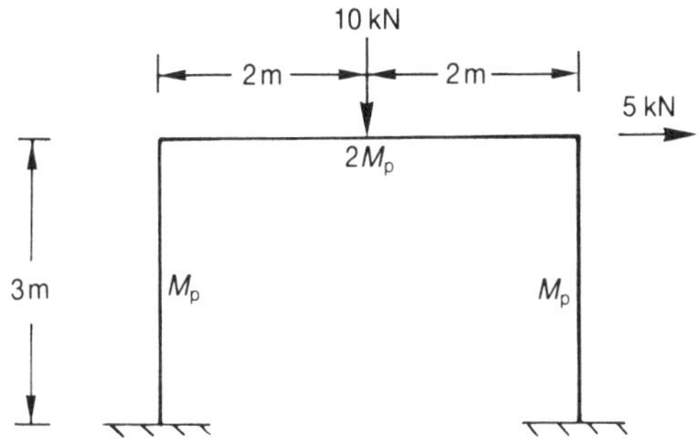

<u>*Solution*</u>

Experiments have shown[4] that this framework can fail by any of the following modes:

(a) beam mechanism
(b) sway mechanism
(c) combined beam and sway mechanism.

---

[4]Baker J F - *A Review of Recent Investigations into the Behaviour of Steel Frames in the Plastic Range*, JICE, 31, 188, 1949, and Baker J F, Horne M R and Heyman J - *The Steel Skeleton*, Cambridge University Press, 1956.

(a) Beam mechanism

This mode of failure, which was discussed in the previous section, is shown below. Applying the principle of virtual work to do this failure mechanism, we get work done by the plastic hinges when rotating = work done by the 10 kN load

or $$M_p\,\theta + 2M_p \times 2\theta + M_p\,\theta = 10 \times 2\theta$$

$$6M_p = 20\theta$$

$$M_p = 3.33 \text{ kNm}$$

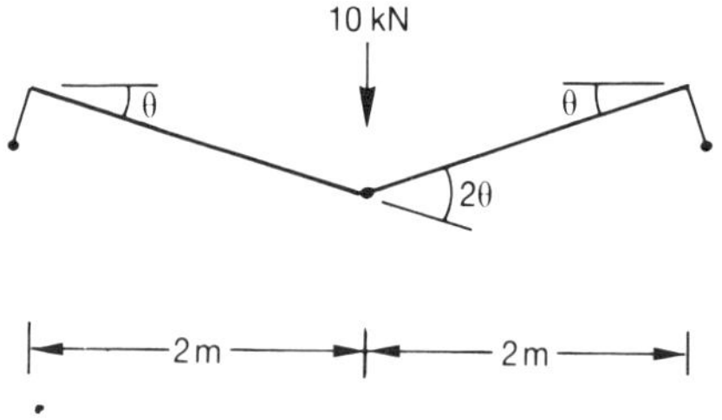

(b) Sway mechanism

This mode of failure is shown below. Applying the principle of virtual work to this failure mechanism, we get

$$M_p\ (\theta + \theta + \theta + \theta) = 5 \times 3\theta$$

or $$4M_p = 15$$

$$M_p = 3.75 \text{ kNm}$$

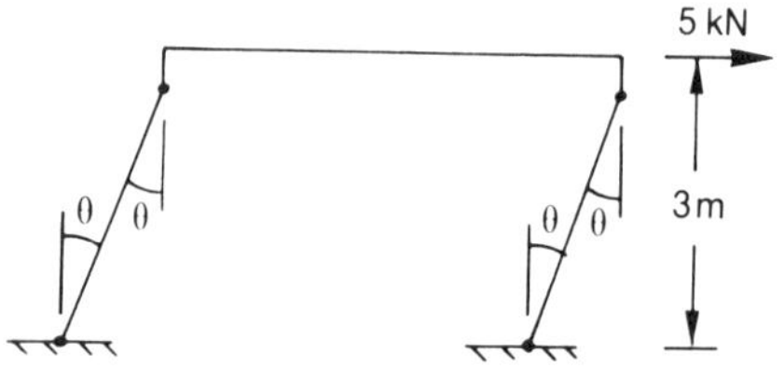

(c) Combined mechanism

This mode of failure is shown below.

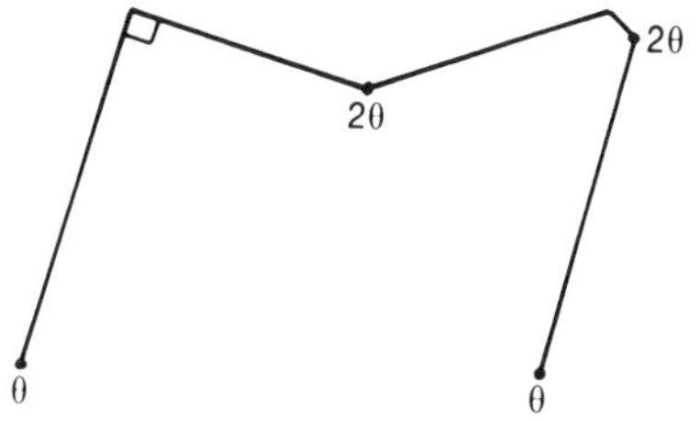

From the principle of virtual work,

$$M_P\theta + 2M_P \times 2\theta + M_P \times 2\theta + M_P\theta = 10 \times 2\theta + 5 \times 3\theta$$

or $$8M_P = 35$$

$$M_P = 4.375 \text{ kNm}$$

The design M_P is obtained from the largest of these values, as this is the value of M_P which will just prevent failure.

$$\therefore \text{design } M_P = 4.375 \times \lambda = 4.375 \times 2.7$$

$$\text{design } M_P = 11.81 \text{ KNm}$$

$$\text{Now } \frac{M_P}{M_Y} = S$$

$$\therefore M_Y = \frac{M_P}{S} = \frac{11.81}{1.15} = 10.27 \text{ kNm}$$

$$Z = \text{sectional modulus} = \frac{M_Y}{\sigma_Y}$$

$$= \frac{10.27 \times 10^3}{300 \times 10^6}$$

$$Z = 3 \times 10^{-5} \text{ m}^3 \text{ (verticals)}$$

$$Z = 6 \times 10^{-5} \text{ m}^3 \text{ (horizontal beam)}$$

Problem 17.6 Determine a suitable sectional modulus for the portal frame below, assuming that the frame has two mechanical hinges at its base, and that the following apply:

λ = 2.7

S = 1.15

σ = 300 MPa

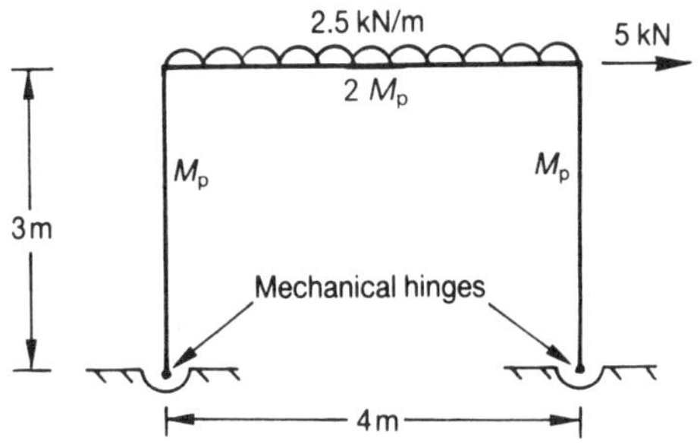

Solution

The *beam mechanism* is shown below

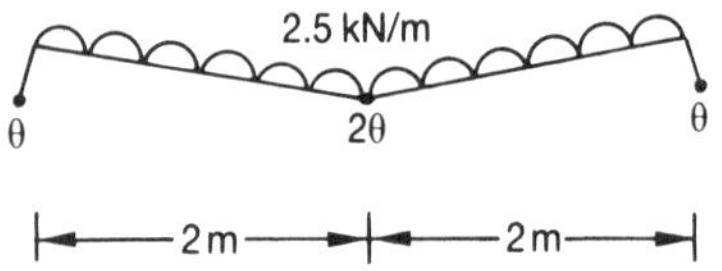

For this case

$$M_p\,\theta + 2\,M_p \times 2\theta + M_p\,\theta = 2.5 \times 4 \times 2\theta/2$$

or

$$6M_P = 10$$

$$M_P = 1.67 \text{ kNm}$$

The *sway mechanism* is shown as follows, where it must be noted that the mechanical hinge does no work during failure.

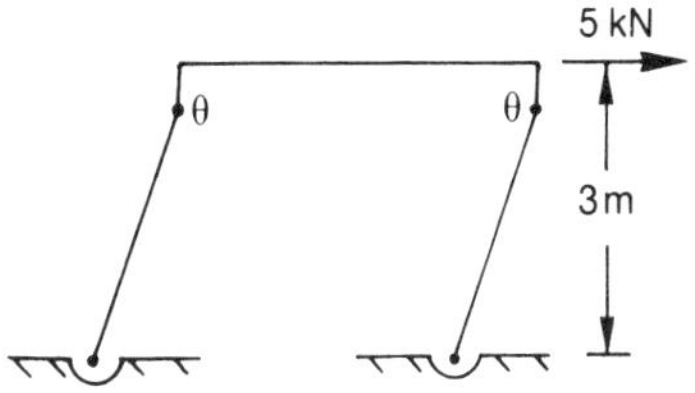

For this case

$$M_p(\theta + \theta) = 5 \times 3\theta$$

or

$$2M_p = 15$$

$$M_p = 7.5 \text{ kNm}$$

The *combined mechanism* is shown below, where it can be seen that the sagging hinge on the beam does not necessarily occur at mid-span.

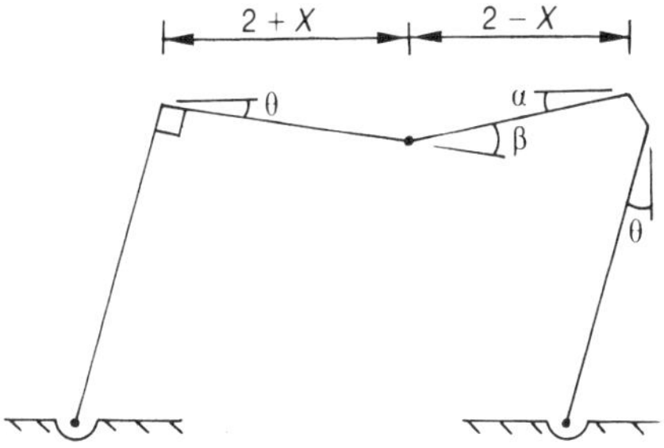

For this case,

$$2M_p\beta + M_p(\alpha + \theta) = 2.5 \times 4 \times \left(\frac{2+X}{2}\right)\theta + 5 \times 3\theta$$

$$= 5(2 + X)\theta + 15\theta \qquad (17.13)$$

but

$$(2+X)\,\theta = (2-X)\alpha$$

$$\therefore \alpha = \left(\frac{2+X}{2-X}\right)\theta \qquad (17.14)$$

$$\beta = \alpha + \theta = \left(\frac{2+X}{2-X}\right) + \theta$$

$$= \left(\frac{2 + X + 2 - X}{2 - X} \right)$$

or $$\beta = \frac{4}{(2 - X)}\theta \qquad (17.15)$$

Substituting equations (17.14) and (17.15) into equation (17.13), we get

$$2 \times M_P \times \frac{4}{(2 - X)} + M_P\left(\frac{2 + X}{2 - X} \right) + M_P = 5(2 + X) + 15$$

$$\text{or } M_P \frac{8 + 2 + X + 2 - X}{2 - X} = 5(2 + X) + 15$$

$$\text{or } M_P = [5(2 + X) + 15]\frac{(2 - X)}{12}$$

$$= \frac{1}{12}(10 + 5X + 15)(2 - X)$$

$$= \frac{1}{12}(25 + 5X)(2 - X)$$

$$= \frac{1}{12}(50 - 25X + 10X - 5X^2)$$

$$\text{or } M_P = \frac{1}{12}(50 - 15X - 5X^2) \qquad (17.16)$$

For maximum

$$M_P \frac{dM_P}{dX} = 0$$

$$\therefore \frac{dM_P}{dX} = -15 - 10X \qquad (17.17)$$

$$\text{or } X = -1.5 \text{ m}$$

Substituting equation (17.17) into equation (17.16)

$$M_P = \frac{1}{12}(50 + 22.5 - 11.25)$$

$$M_P = 5.1 \text{ kNm}$$

Design $\qquad M_P = 2.7 \times 5.1$

$$= 13.77 \text{ kNm}$$

$$M_Y = \frac{13.77}{1.15} = 11.97 \text{ kNm}$$

$$Z = \frac{11.97 \times 10^3}{300 \times 10^6}$$

$$Z = 8 \times 10^{-5} \text{ m}^3 \text{ (horizontal beam)}$$

The method will now be applied to *two-storey* and *two-bay* frameworks.

Problem 17.17 Determine a suitable sectional modulus for the two storey framework below, given that

$$\lambda = 3, \quad S = 1.16, \quad \sigma_Y = 316 \text{ MPa}$$

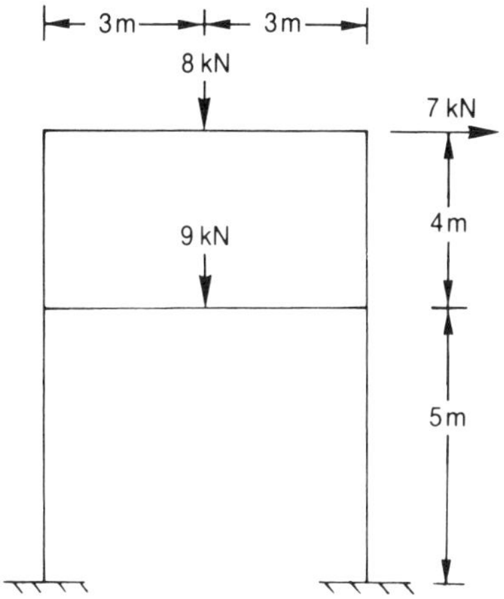

<u>*Solution*</u>

The possible mechanisms are as follows:

(a) Top beam

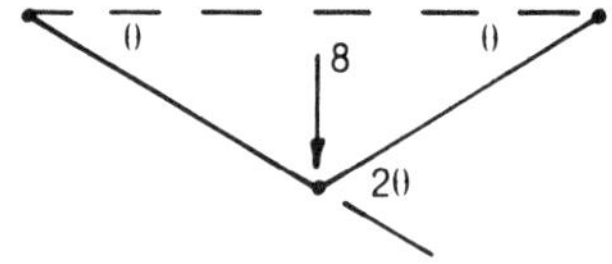

(b) Bottom beam

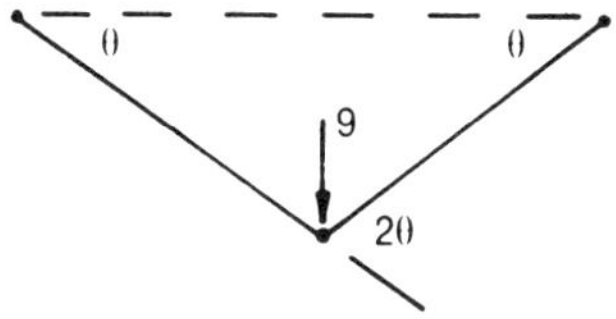

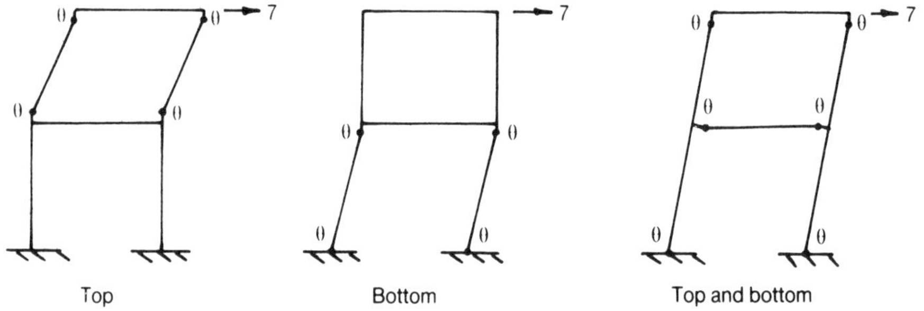

(c) Sway mechanisms (3 types)

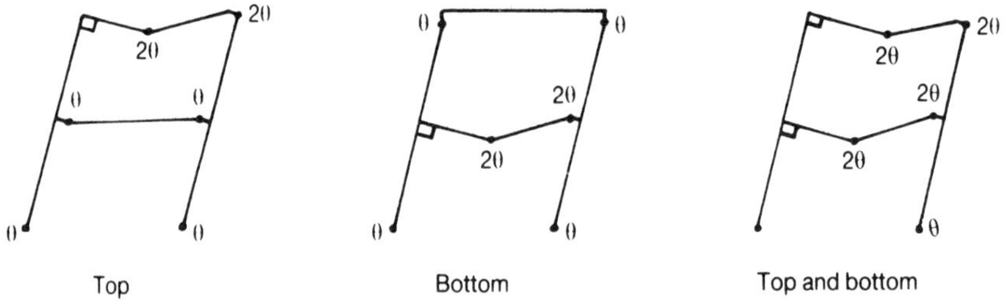

(d) Combined mechanisms (3 types)

(a) *Top beam mechanism*

$$M_P(\theta + 2\theta + \theta) = 8 \times 3\theta$$

or

$$4M_P = 24$$

$$M_P = 6 \text{ kNm}$$

(b) *Bottom beam mechanism*

$$M_P(\theta + 2\theta + \theta) = 9 \times 3\theta$$

or

$$4M_P = 27$$

$$M_P = 6.75 \text{ kNm}$$

(c) *Top sway mechanism*

$$M_P(\theta + \theta + \theta + \theta) = 7 \times 4\theta$$

$$M_P = 7 \text{ kNm}$$

(d) *Bottom sway mechanism*

$$M_P(\theta + \theta + \theta + \theta) = 7 \times 5\theta$$

$$M_P = 8.75 \text{ kNm}$$

(e) *Top and bottom sway mechanisms*

$$M_P \times 6\theta = 7 \times 9\theta$$

$$M_P = 10.5 \text{ kNm}$$

(f) *Combined top mechanism*

$$M_P(\theta + \theta + 2\theta + 2\theta + \theta + \theta) = 8 \times 3\theta + 7 \times 9\theta$$

or

$$8M_P = 87$$

$$M_P = 10.88 \text{ kNm}$$

(g) *Combined bottom mechanism*

$$M_P(\theta + \theta + 2\theta + \theta + 2\theta + \theta) = 9 \times 3\theta + 7 \times 9\theta$$

or

$$8M_P = 90$$

$$M_P = 11.25 \text{ kNm}$$

(h) *Combined top and bottom mechanisms*

$$M_P(\theta + 2\theta + 2\theta + 2\theta + 2\theta + 2\theta) = 8 \times 3\theta + 9 \times 3\theta + 7 \times 9\theta$$

or

$$10M_P = 114$$

$$M_P = 11.4 \text{ kNm}$$

Design

$$M_P = 11.4 \times 3 = 34.2 \text{ kNm}$$

$$M_Y = \frac{34.2}{1.16} = 29.48 \text{ kNm}$$

$$Z = \frac{29.48 \times 10^3}{316 \times 10^6} = 9 \times 10^{-5} \text{ m}^3$$

Problem 17.18 Determine suitable sectional moduli for the two-bay framework below, given that

$$\lambda = 3 \qquad S = 1.15 \qquad \sigma_Y = 316 \text{ MPa}$$

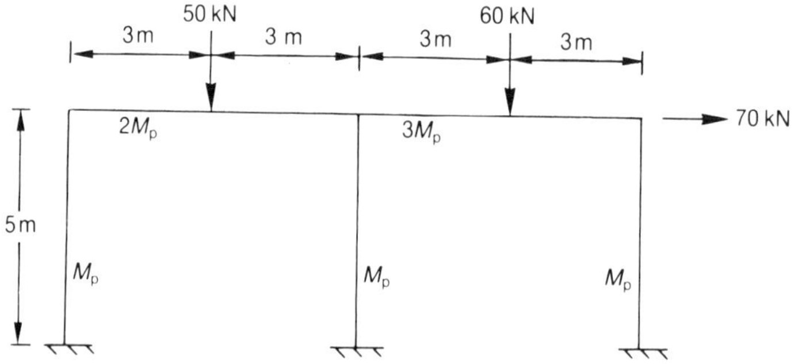

<u>*Solution*</u>

The various possible mechanisms are given below:

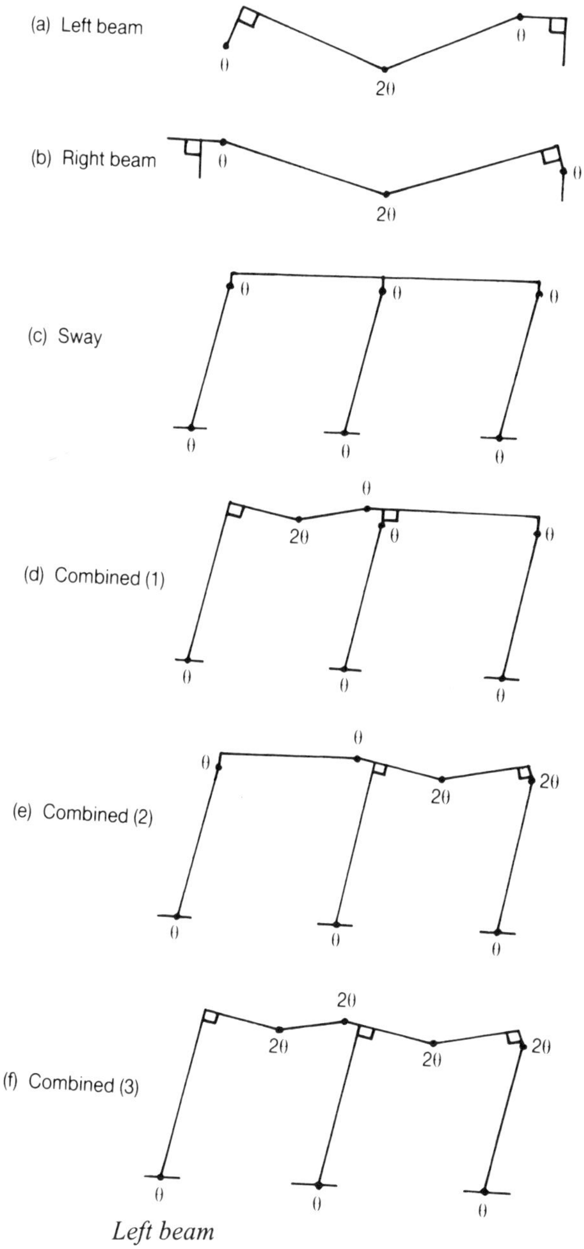

(a) *Left beam*

$$M_P(\theta + 4\theta + 2\theta) = 50 \times 3\theta$$

$$7M_P = 150$$

$$M_P = 21.4 \text{ kNm}$$

(b) *Right beam*

$$M_P(3\theta + 6\theta + \theta) = 60 \times 3\theta$$

$$10M_P = 180$$

$$M_P = 18 \text{ kNm}$$

(c) *Sway*

$$M_P \times 6\theta = 70 \times 5\theta$$

$$6M_P = \text{`}350$$

$$M_P = 58.3 \text{ kNm}$$

(d) *Combined (1)*

$$M_P(\theta + 4\theta + 2\theta + \theta + \theta + \theta + \theta) = 70 \times 5\theta + 50 \times 3\theta$$

$$11M_P = 500$$

$$M_P = 45.5 \text{ kNm}$$

(e) *Combined (2)*

$$M_P(\theta + \theta + 2\theta + \theta + 6\theta + 2\theta + \theta)$$

$$= 70 \times 5\theta + 60 \times 3\theta$$

or $$14M_P = 530$$

$$M_P = 37.86 \text{ kNm}$$

(f) *Combined (3)*

$$M_P(\theta + 4\theta + 4\theta + \theta + 6\theta + 2\theta + \theta)$$

$$= 70 \times 5\theta + 50 \times 3\theta + 60 \times 3\theta$$

or $$19M_P = 680$$

$$M_P = 35.8 \text{ kNm}$$

$$\text{Design } M_P = 58.3 \times 3 = 174.9 \ kNm$$

$$M_Y = \frac{174.9}{1.15} = 152.1 \ kNm$$

$$Z = \frac{152.1 \times 10^3}{316 \times 10^6} = 4.8 \times 10^{-4} \text{ m}^3 \text{ (verticals)}$$

$$Z = 9.6 \times 10^{-4} \text{ m}^3 \text{ (left beam)}$$

$$Z = 1.44 \quad 10^{-3} \text{ m}^3 \text{ (right beam)}$$

17.7 Complementary energy

The principle of virtual work leads also to a concept of wider application in stress–strain analysis than that of strain energy; this other property of a structure is known as *complementary energy*.

Consider the statically determinate pin-jointed frame shown in Figure 17.6; the frame is pinned to a rigid foundation at A and B, and carries external loads W_1 and W_2 at joints C and D, respectively. Suppose the corresponding displacements of the joints C and D are δ_1, and δ_2, respectively; the tensile force induced in a typical member, such as BC, is P, and its resulting extension is e. The forces W_1, W_2, P etc. are a system of forces in statical equilibrium, whereas the extensions, e, etc., are compatible with the displacements δ_1 and δ_2 of the joints. Thus by the principle of virtual work

$$W_1\delta_1 + W_2\delta_2 = \sum_m Pe \qquad (17.18)$$

where the summation is carried out for all member of the frame.

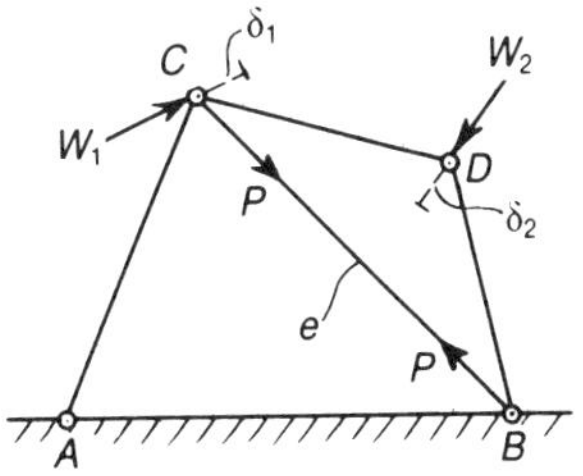

Fig. 17.6 Statically determinate plane frame under any system of external load.

Now suppose the external load W_1 is increased in magnitude by a small amount δW_1, the external load W_2 remaining unchanged; due to change in W_1 small changes occur in the forces in the members of the frame P, for example, increasing to $(P + \delta P)$. Now consider the virtual work of the modified system of forces on the original set of displacements and extensions; we have

$$(W_1 + \delta W_1)\delta_1 + W_2\delta_2 = \sum_m (P + \delta P)e$$

where the summation is carried out for all members of the frame. Now suppose the external load W_2 is increased in magnitude by a small amount δW_1, the external load W_2 remaining unchanged; due to change in W_1 small changes occur in the forces in the members of the frame, P, for example, increasing to $(P + \delta P)$.

Now consider the virtual work of the modified system of forces on the original set of displacement and extensions; we have

$$(W_1 + \delta W_1)\delta_1 + W_2\delta_2 = \sum_m (P + \delta P)e \qquad (17.19)$$

On subtracting equations (17.18) and (17.19), we have

$$\delta_1 \times \delta W_1 = \sum_m e\delta P \qquad (17.20)$$

The quantity $e\delta P$ for a member is the shaded clcmental area shown on the load-extension diagram of Figure 17.7, this is an element of the area C shown in Figure 17.8.

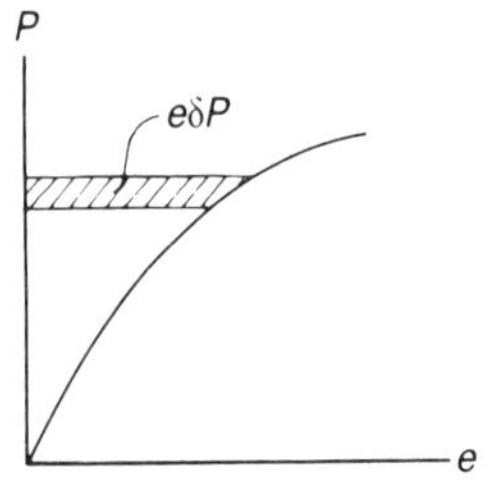

Figure 17.7 Increment of complementary energy of a single member.

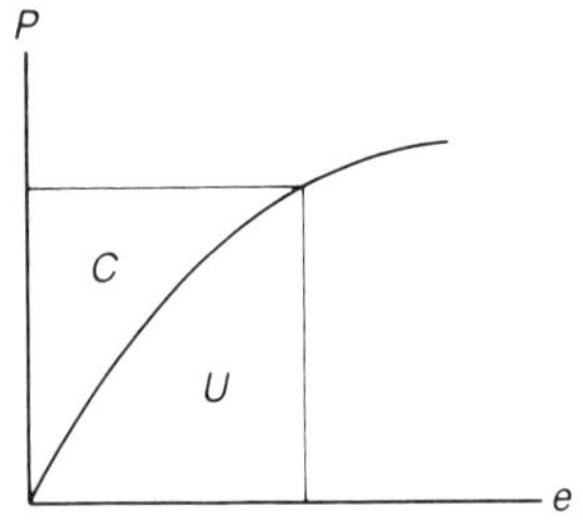

Figure 17.8 Strain energy and complementary energy of a single member.

When a bar is extended the work done on the bar is the area below the P–e curve of Figure 17.7, for a conservative structural member this work is stored as strain energy, which we have already referred to as U. We define the area above the P–e curve of Figure 17.7 as the *complementary energy, C*, of the member; we have that

$$U + C = Pe \qquad (17.21)$$

and

$$\delta C = e\delta P \tag{17.22}$$

In equation (17.18) we may write, therefore,

$$\delta_1 \times \delta W_1 = \delta C \tag{17.23}$$

where C is the complementary energy of all members of the frame. If δW_1 is infinitesimally small

$$\frac{\partial C}{\partial W_1} = \delta_1 \tag{17.24}$$

Then the partial derivative of the complementary energy function C with respect to the external load W_1 gives the corresponding displacement δ_1 of that load.

17.8 Complementary energy in problems of bending

The complementary energy method may be used to considerable advantage in the solution of problems of bending of straight and thin curved beams. In general we suppose that the moment–curvature relationship for an element of a beam is of the form shown in Figure 17.9. The complementary energy of bending of an elemental length δs due to a bending moment M is

$$\int_0^M \left(\frac{\delta s}{R}\right) dM$$

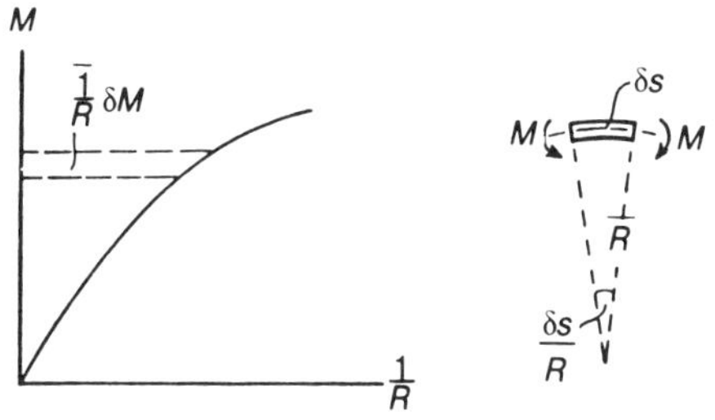

Figure 17. 9 Complementary energy of bending of the element of a beam.

For a linear-elastic beam of flexural stiffness EI

$$\frac{1}{R} = \frac{M}{EI}$$

and so the complementary energy is

$$\int_0^M \frac{M}{EI}\, dM\delta s = \frac{M^2\delta s}{2EI} \tag{17.25}$$

For a length L of the beam, the complementary energy is therefore

$$C = \int_0^L \frac{M^2\,ds}{2EI} \tag{17.26}$$

As in the case of pin-jointed frames, the partial derivative of C with respect to any external load is the corresponding displacement of that load. For statically indeterminate beams, the partial derivative of the complementary energy with respect to a redundant force or couple is zero.

Problem 17.9 Estimate the vertical displacement of the free end of the uniform cantilever shown.

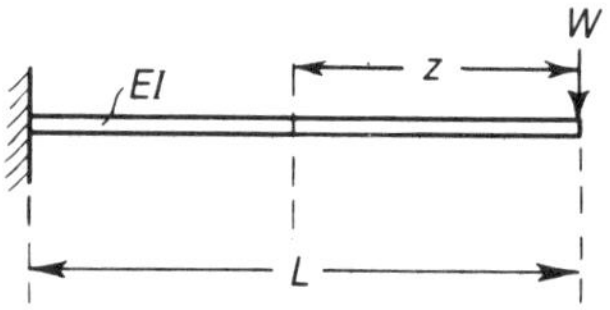

<u>*Solution*</u>

The complementary energy of bending is

$$C = \int_0^L \frac{M^2\,dz}{2EI} = \int_0^L \frac{W^2 z^2\,dz}{2EI} = \frac{W^2 L^3}{6EI}$$

The corresponding displacement of W is

$$\delta_w = \frac{\partial C}{\partial W} = \frac{WL^3}{3EI}$$

Problem 17.10 A cantilever has a uniform flexural stiffness EI. Estimate the vertical deflection at the free end if the cantilever carries a uniformly distributed lateral load of intensity w.

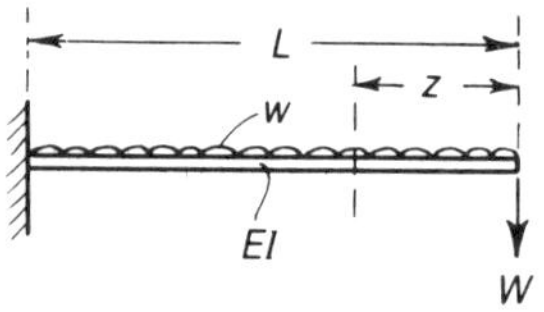

<u>*Solution*</u>

Introduce a vertical load W at the free end; the bending moment at any section is then

$$M = \frac{1}{2}wz^2 + Wz$$

The complementary energy of bending is

$$C = \frac{1}{2EI}\int_0^L \left(\frac{1}{2}wz^2 + Wz\right)^2 dz$$

The corresponding displacement of W is

$$\delta_w = \frac{\partial C}{\partial W} = \frac{1}{EI}\int_0^L \left(\frac{1}{2}wz^2 + Wz\right) zdz$$

Now put $W = 0$; then

$$\delta_W \quad \frac{1}{EI}\int_0^L \frac{1}{2}wz^3 dz = \frac{wL^4}{8EI}$$

Problem 17.11 A cantilever of uniform flexural stiffness EI carries a moment M at the remote end. Estimate the angle of rotation at that end of the beam.

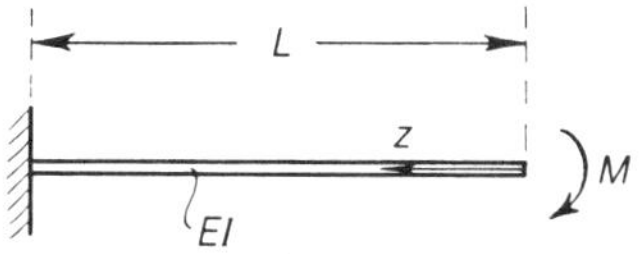

Solution

All sections of the beam carry the same bending moment M, so the complementary energy is

$$C = \int_0^L \frac{M^2 dz}{2EI} = \frac{M^2 L}{2EI}$$

The corresponding displacement of M is

$$\theta_M = \frac{ML}{EI}$$

which is the angle of rotation at the remote end.

Problem 17.12 Solve the problem discussed in Section 17.4, using complementary energy.

Solution

The bending moment at any section in terms of w and the redundant force W is $\frac{1}{2}wz^2 - Wz$. Then

$$C = \int_0^L \left(\frac{1}{2}wz^2 - Wz\right)^2 \frac{dz}{2EI}$$

The property $\partial C/\partial W = 0$ gives

$$\int_0^L \frac{1}{2}wz^3\,dz = \int_0^L Wz^2\,dz$$

Then

$$W = \frac{3wL}{8}$$

Problem 17.13 Solve Problem 17.3 using complementary energy.

Solution

The bending moment at any angular position θ is

$$M = Wr\sin\theta$$

Then

$$C = \int_0^\pi \frac{M^2}{2EI}\,r\,d\theta$$

Thus

$$\delta_w = \frac{\partial C}{\partial W} = \frac{\partial C}{\partial M}\frac{\partial M}{\partial W} = \int_0^\pi M\frac{\partial M}{\partial W}\frac{r\,d\theta}{EI}$$

$$= \int_0^\pi \frac{Wr^3\sin^2\theta}{EI}\,d\theta = \frac{\pi Wr^3}{2EI}$$

Problem 17.14 A thin circular ring of radius r and uniform flexural stiffness carries two radial loads W applied along a diameter. Estimate the maximum bending moment in the ring.

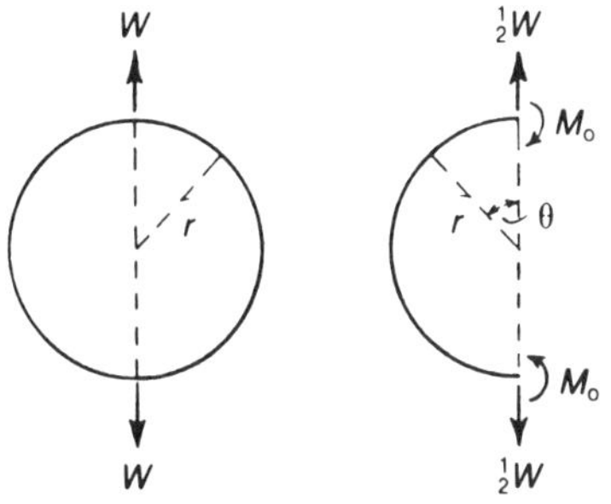

<u>*Solution*</u>

By symmetry the loading action on a half-ring are ½W and M_0. The bending moment at any angular position θ is

$$M = M_0 - \frac{1}{2}Wr\sin\theta$$

Then

$$C = \int_0^{\pi}\left(M_0 - \frac{1}{2}Wr\ \sin\theta\right)^2 \frac{rd\theta}{2EI}$$

But

$$\partial C/\partial M_0 = 0, \quad \text{so that}$$

$$\int_0^{\pi} M_0\, d\theta = \frac{1}{2}Wr \int_0^{\pi} \sin\theta\, d\theta$$

Then

$$M_0 = Wr/\pi$$

17.9 The Raleigh–Ritz method

This method is also known as the *method of minimum potential*, and in Chapters 24 and 25, it is used in the finite element method.

In mathematical terms, it can be stated, as follows:

$$\frac{\partial \pi_p}{\partial W} = 0$$

where

π_p = total potential = $U_e + WD$

U_e = strain energy

WD = the potential of the load system

W = load

The method will be applied to problem 17.12 to determine an expression for δ_w.

Now

$$U_e = \int \frac{M^2}{2EI} dz = \text{the bending strain energy of a beam}$$

As

$$M = Wz = \text{bending moment at } z,$$

$$U_e = \frac{1}{2EI} \int_0^l W^2 z^2 dz$$

or

$$U_e = \frac{W^2 l^3}{6EI}$$

By inspection

$$WD = \text{potential of the load system}$$

$$= -W \ \delta_w$$

$$\therefore \pi_p = \frac{W^2 l^3}{6EI} - W\delta_w$$

Now,

$$\frac{\partial \pi_p}{\partial W} = 0 = \frac{Wl^3}{3EI} - \delta_w$$

$$\therefore \delta_w = \frac{Wl^3}{3EI} \text{ as required}$$

Further problems *(answers on page 693)*

17.15 A thin semicircular bracket, *AB*, of radius *R* is built-in at *A*, and has at *B* a rigid horizontal arm *BC* of length *R*. the arm carries a vertical load *W* at *C*. Show that the vertical deflection at *C* is $\pi WR^3/2EI$, where *EI* is the flexural rigidity of the strip, and determine the horizontal deflection. *(Nottingham)*

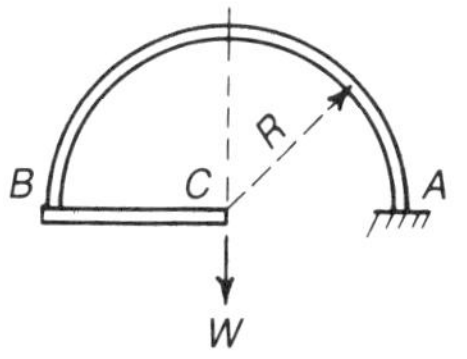

17.16 A beam has a second moment of area of $2I$ over one-half of the span and I over the other half. Find the fixed-end moments when a load of 100 kN is carried at the mid-length.

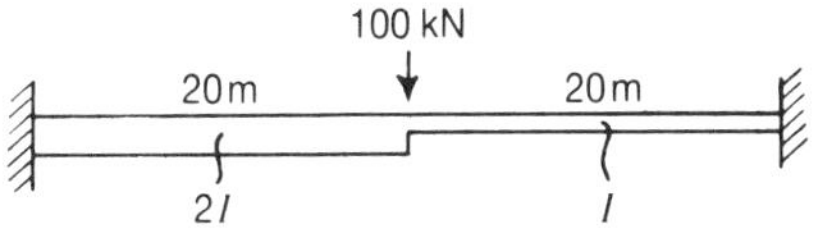

17.17 A ring radius *R* and uniform cross-section hangs from a single support. Find the position and magnitude of the maximum bending moment due to its own weight. *(London)*

17.18 An 'S' hook follows part of the outline of two equal circles of radius *R* that just touch. It embraces 5/6ths of one circle and 2/3rds of the other. If the ends are pulled apart by a force, *P*, by how much will they be moved if the hook has a constant rigidity *EI*? *(London)*

17.19 Using the plastic hinge theory determine a suitable sectional modulus for the rigid-jointed framework shown below. The following may be assumed to apply to the framework

$\lambda = 4$ $\quad\quad$ $\sigma_Y = 300$ MPa $\quad\quad$ $S = 1.15$

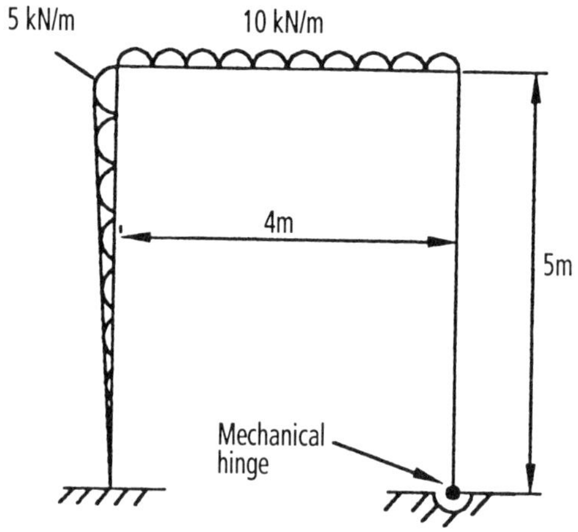

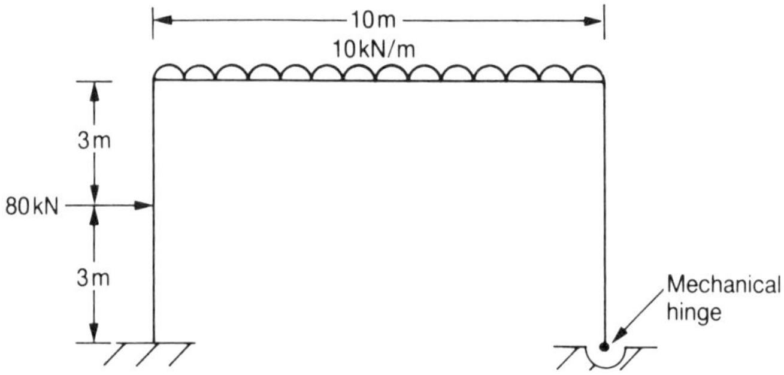

17.20 A portal frame of uniform section is subjected to the loading above. Using the plastic hinge theory, determine a suitable section modulus for the frame, based on a load factor of 4, a shape factor of 1.15 and a yield stress of 275 MPa. *(Portsmouth, Standard 1989)*

17.21 Using the plastic hinge theory, determine a suitable section modulus for the two bay rigid-jointed plane frame below.

The following assumptions should be made:-

load factor = 4

shape factor = 1.15

yield stress = 275 MPa

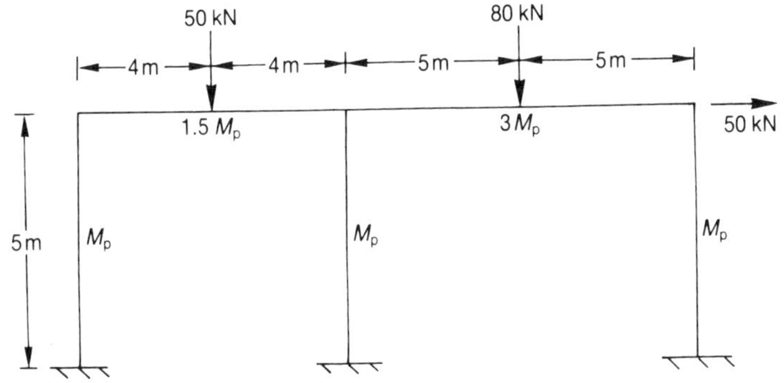

(Portsmouth, Honours 1989)

18 Buckling of columns and beams

18.1 Introduction

In all the problems treated in preceding chapters, we were concerned with the small strains and distortions of a stressed material. In certain types of problems, and especially those involving compressive stresses, we find that a structural member may develop relatively large distortions under certain critical loading conditions. Such structural members are said to *buckle*, or become *unstable*, at these critical loads.

As an example of elastic buckling, we consider firstly the buckling of a slender column under an axial compressive load.

18.2 Flexural buckling of a pin-ended strut

A perfectly straight bar of uniform cross-section has two axes of symmetry Cx and Cy in the cross-section on the right of Figure 18.1. We suppose the bar to be a flat strip of material, Cx being the weakest axis of the cross-section. End thrusts P are applied along the centroidal axis Cz of the bar, and EI its uniform flexural stiffness for bending about Cx.

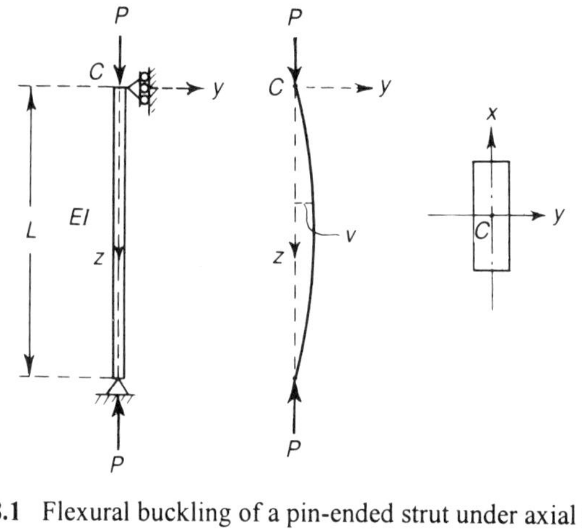

Figure 18.1 Flexural buckling of a pin-ended strut under axial thrust.

Now Cx is the weakest axis of bending of the bar, and if bowing of the compressed bar occurs we should expect bending to take place in the yz-plane. Consider the possibility that at some value of P, the end thrust, the strut can buckle laterally in the yz-plane. There can be no lateral deflections at the ends of the strut; suppose v is the displacement of the centre line of the bar parallel to Cy at any point. There can be no forces at the hinges parallel to Cy, as these would imply bending moments at the ends of the bar. The only two external forces are the end thrusts P, which are assumed to maintain their original line of action after the onset of buckling. The bending moment at any section of the bar is then

$$M = Pv \tag{18.1}$$

which is a sagging moment in relation to the axes *Cz* and *Cy*, in the sense of Section 13.2. But the moment–curvature relationship for the beam at any section is

$$M = -EI\frac{d^2v}{dz^2}$$

provided the deflection v is small. Thus

$$-EI\frac{d^2v}{dz^2} = Pv$$

Then

$$EI\frac{d^2v}{dz^2} + Pv = 0 \tag{18.2}$$

Put

$$\frac{P}{EI} = k^2 \tag{18.3}$$

Then

$$\frac{d^2v}{dz^2} + k^2v = 0 \tag{18.4}$$

The general solution of this differential equation is

$$v = A\cos kz + B\sin kz \tag{18.5}$$

where A and B are arbitrary constants. We have two boundary conditions to satisfy: at the ends $z = 0$ and $z = L$, $v = 0$. Then

$$A = 0 \quad \text{and} \quad B\sin kL = 0$$

Now consider the implications of the equation

$$B\sin kL = 0$$

which is derived from the boundary conditions. If $B = 0$, then both A and B are zero, and obviously the strut is undeflected.

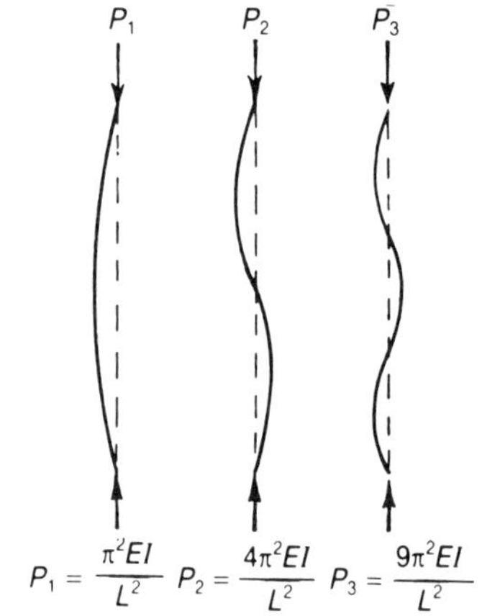

Figure 18.2 Modes of buckling of a pin-ended strut.

If, however, $\sin kL = 0$, B is indeterminate, and the strut may assume the form

$$v = B \sin kz$$

This is called a buckled condition of the strut. Obviously B is indeterminate when kL, assumes the values,

$$kL = \pi,\ 2\pi,\ \ldots \text{ etc.} \tag{18.6}$$

We need not consider the solution $kL = 0$, which implies $k = 0$, because the solution of the differential equation is not trigonometric in form when $k = 0$. Instability occurs when $kL\pi, = 2\pi$, etc.

$$\therefore P = k^2 EI = \frac{\pi^2 EI}{L^2},\ 4\pi^2 \frac{EI}{L^2} \text{ etc} \tag{18.7}$$

There are infinite number of values of P for instability, corresponding with various modes of buckling, Figure 18.2. The fundamental mode occurs at the lowest critical load

$$P_e = \pi^2 \frac{EI}{L^2} = \text{Euler load for pin-ended struts} \tag{18.8}$$

This is known as the *Euler formula* and corresponds with buckling in a single longitudinal half-wave. The critical load

$$P = 2^2\pi^2 \frac{EI}{L^2} = 4\pi^2 \frac{EI}{L^2} \tag{18.9}$$

corresponds with buckling in two longitudinal half-waves, and so on for higher modes. In practice the critical load P_e is never exceeded because high stresses develop at this load and collapse of the strut ensues. We are not therefore concerned with buckling loads higher than the lowest buckling load. For all practical purposes the buckling load of a pin-ended strut is given by equation (18.8).

At this load a perfectly straight pin-ended strut is in a state of *neutral equilibrium*; the small deflection

$$v = B \sin kz$$

is indeterminate, because B itself is indeterminate. Theoretically, the strut is in equilibrium at the load $\pi^2 EI/L^2$ for any small value of B, corresponding with a condition of *neutral equilibrium*; at this buckling load we should expect to be able to push the strut into any sinusoidal wave of small amplitude. This can be verified experimentally by compressing a long slender strip of material which remains elastic during bending.

At values of P less than $\pi^2 EI/L^2$ the strut is in a condition of *unstable equilibrium*; any small lateral disturbance produces motion and finally collapse of the strut. This, however, is a hypothetical situation as, in practice, the load $\pi^2 EI/L^2$ cannot be exceeded if the loads are static, and not applied suddenly.

The condition of neutral equilibrium at

$$P_e = \pi^2 \frac{EI}{L^2}$$

is only attained for small lateral displacements of the strut. When these displacements become large, the moment-curvature relation

$$M = -EI \frac{d^2 v}{dz^2}$$

is no longer valid; theoretically the problem becomes more difficult. The effect of large lateral displacements is to increase the flexural stiffness of the strut; in this case, provided the material remains elastic, end thrusts greater than $\pi^2 EI/L^2$ are attainable. If the thrust P is plotted against the lateral displacement v at any section, the P - v relation for a perfectly straight strut has the form shown in Figure 18.3(i), when account is taken of large deflections. Lateral deflections become possible only when

$$P \geq \frac{\pi^2 EI}{L^2}$$

This analysis is restricted to the hypothetical case of a perfectly straight strut. When the strut has small imperfections, displacements v are possible for all values of P (Figure 18.3(ii)), and the hypothetical condition of neutral equilibrium at

$$P = \frac{\pi^2 EI}{L^2}$$

is never attained. All materials have a limit of proportionality; when this is attained the flexural stiffness of the strut usually falls off rapidly. On the *P–v* diagram for the strut this corresponds with the development of a region of unstable equilibrium.

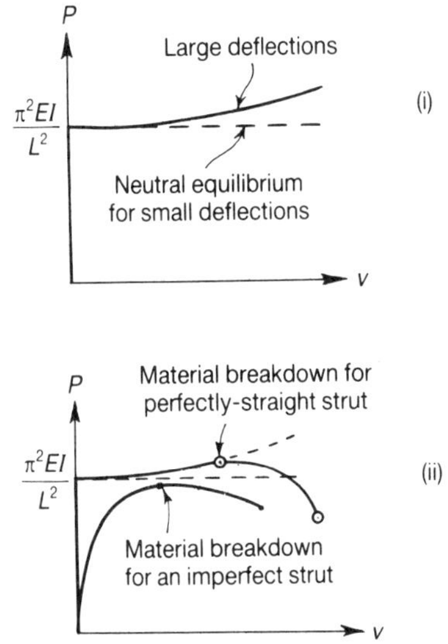

Figure 18.3 Large deflections and material breakdown of struts.

18.3 Rankine–Gordon formula

Predictions of buckling loads by the Euler formula is only reasonable for very long and slender struts that have very small geometrical imperfections. In practice, however, most struts suffer plastic knockdown and the experimentally obtained buckling loads are much less than the Euler predictions. For struts in this category, a suitable formula is the Rankine–Gordon formula which is a semi-empirical formula, and takes into account the crushing strength of the material, its Young's modulus and its slenderness ratio, namely L/k, where

L = length of the strut

k = least radius of gyration of the strut's cross-section

$$P_c = \sigma_c A \tag{18.10}$$

where

A = cross-sectional area

σ_c = crushing stress

Then

$$\frac{1}{P_R} = \frac{1}{P_e} + \frac{1}{P_c} \tag{18.11}$$

where

P_R = Rankine–Gordon buckling load
P_e = Euler buckling load

$$= \frac{\pi^2 EI}{L^2} \text{ for a pin-ended strut}$$

$$\therefore \frac{1}{P_R} = \frac{L_0^2}{\pi^2 EI} + \frac{1}{\sigma_{yc} \times A}$$

$$= \frac{L_0^2}{\pi^2 EAk^2} + \frac{1}{\sigma_{yc} A} \tag{18.12}$$

$$= \frac{L_0^2 \sigma_{yc} + \pi^2 Ek^2}{\pi^2 EAk^2 \sigma_{yc}}$$

or

$$P_R = \frac{\pi^2 EAk^2 \sigma_{yc}}{L_0^2 \sigma_{yc} + \pi^2 Ek^2}$$

$$= \frac{\sigma_{yc}}{L_0^2 \sigma_{yc} / \pi^2 EAk^2 + \pi^2 Ek^2 / \pi^2 EAk^2} \tag{18.13}$$

$$P_R = \frac{\sigma_{yc} \times A}{\left(\sigma_{yc} / \pi^2 E\right)\left(L_0 / k\right)^2 + 1}$$

Let

$$a = \frac{\sigma_{yc}}{\pi^2 E} \tag{18.14}$$

Then

$$P_R = \frac{\sigma_{yc} A}{1 + a(L_0 / K)^2} \tag{18.15}$$

where a is the denominator constant in the Rankine–Gordon formula, which is dependent on the boundary conditions and material properties.

A comparison of the Rankine–Gordon and Euler formulae, for geometrically perfect struts, is given in Figure 18.4. Some typical values for $1/a$ and σ_{yc} are given in Table 18.1. Where L_0 is the effective length of the strut; see Section 18.4.

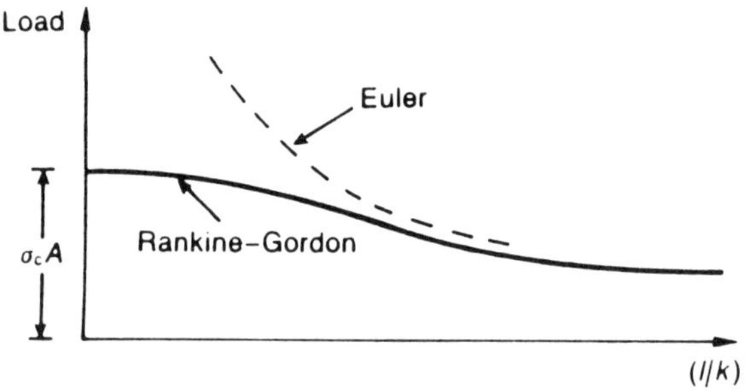

Figure 18.4 Comparison of Euler and Rankine–Gordon formulae.

Table 18.1 Rankine Constants

Material	$1/a$	σ_{yc}
Mild Steel	7500	300
Wrought Iron	8000	250
Cast Iron	18000	560
Timber	1000	35

18.4 Effects of geometrical imperfections

For intermediate struts with geometrical imperfections, the buckling load is further decreased, as shown in Figure 18.5.

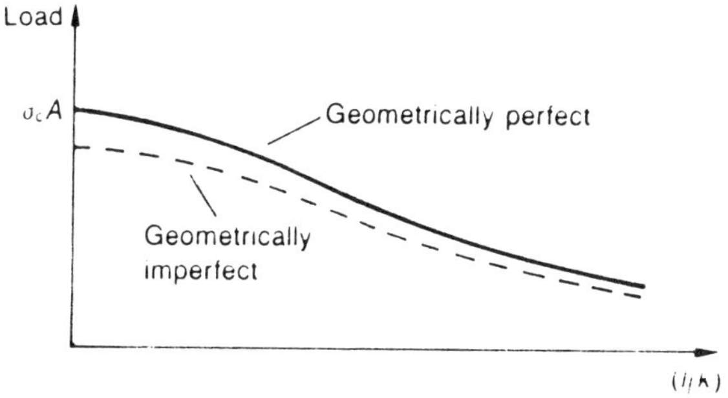

Figure 18.5 Rankine–Gordon loads for perfect and imperfect struts.

18.5 Effective lengths of struts

The theoretical buckling load for a pinned-ended strut is one-quarter of the theoretical buckling load of a fixed-ended strut and four times the theoretical buckling load for a strut fixed at one end and free at the other end; see Sections 18.10 to 18.12.

Table 18.2 Effective lengths of struts (L_0)

Type of Strut	Euler	BS449
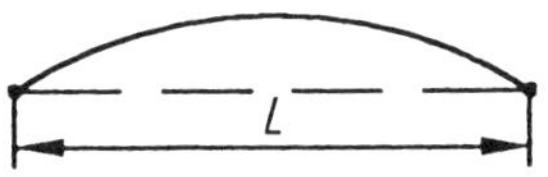	$L_0 = L$	$L_0 = L$
	$L_0 = L$	$L_0 = L$
	$L_0 = 0.5L$	$L_0 = 0.7L$
	$L_0 = 0.7L$	$L_0 = 0.85L$
	$L_0 = 2L$	$L_0 = 2L$

Table 18.2 gives the effective lengths of struts (L_0), which have actual lengths of L, for different boundary conditions, where BS449 allows for elastic relaxation at the ends of the strut.

18.6 Pin-ended strut with eccentric end thrusts

In practice it is difficult, if not impossible, to apply the end thrusts along the longitudinal centroidal axis Cz of a strut. We consider now the effect of an eccentrically applied compressive load P on a uniform strut of flexural stiffness EI and length L.

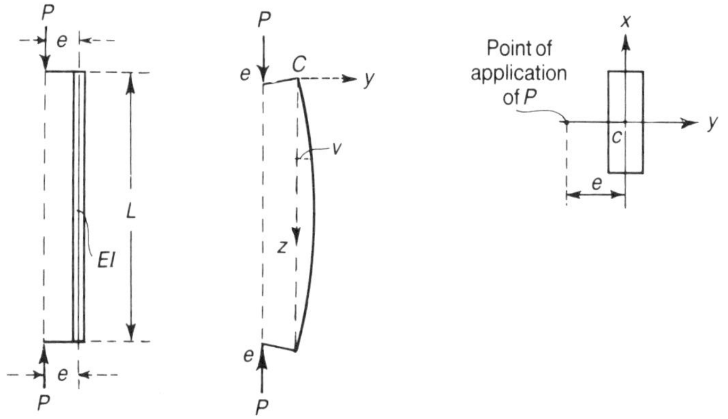

Figure 18.6 Eccentric loading of a strut.

Suppose the end thrusts are applied at a distance e from the centroid and on the axis Cy of the cross-section. We assume again that the cross-section is that of a flat rectangular strip, Cx being the weaker axis of bending. The end thrusts P are applied to rigid arms attached to the ends of the strut.

An end load P causes the straight strut to bend; suppose v is the displacement of any point on Cz from its original position. The bending moment at that section is

$$M = P(e + v)$$

which is a sagging moment in relation to the axes Cz and Cy. If v is small we have

$$M = -EI\frac{d^2v}{dz^2}$$

Thus

$$-EI\frac{d^2v}{dz^2} = P(e+v)$$

Then

$$EI\frac{d^2v}{dz^2} + Pv = -Pe$$

When $e = 0$, this differential equation reduces to that already derived for an axially loaded strut.

As before, put

$$k^2 = \frac{P}{EI}$$

Then

$$\frac{d^2v}{dz^2} + k^2 v = -k^2 e$$

The complete solution is

$$v = A \cos kz + B \sin kz - e$$

Now $v = 0$ at $z = 0$ and $z = L$, so that

$$A - e = 0, \quad \text{and} \quad A \cos kL + B \sin kL - e = 0$$

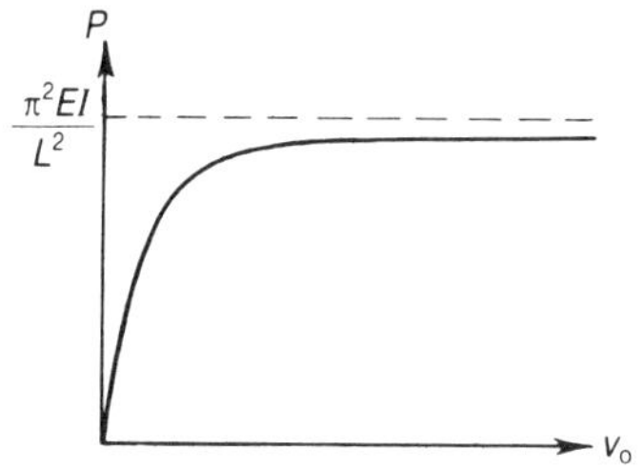

Figure 18.7 Deflections of an eccentrically loaded strut.

The first of these equations gives $A = e$, and the second gives

$$B = \frac{e(1 - \cos kL)}{\sin kL}$$

Then

$$v = e(\cos kz - 1) + \frac{e(1 - \cos kL)}{\sin kL} \sin kz \tag{18.16}$$

The displacement v at the mid-length, $z = \frac{1}{2}L$, is

$$v_0 = e\left[\left(\cos\frac{kL}{2} - 1\right) + \frac{1 - \cos kL}{\sin kL}\sin\frac{1}{2}kL\right]$$

$$= e\left[\frac{2\sin\frac{1}{2}kL\left(1 - \cos\frac{1}{2}kL\right)}{\sin kL}\right] \tag{18.17}$$

If $\sin \frac{1}{2} kL \neq 0$, we have

$$v_0 = e\left(\sec\frac{1}{2}kL - 1\right) \tag{18.18}$$

When $P = 0$,

$$\frac{1}{2}kL = \frac{L}{2}\sqrt{\frac{P}{EI}} = 0$$

and $v_0 = 0$. As P approaches $\pi^2 EI/L^2$, $\frac{1}{2}kL$ approaches $\pi/2$, and

$$\sec\frac{1}{2}kL \rightarrow \infty$$

Thus values of v_0 are possible from the onset of loading; the values of v_0 increase non-linearly with increases of P. The value of $P = \pi^2 EI/L^2$ is not attainable as this would imply an infinitely large value of v_0, and material breakdown would occur at some smaller value of P.

It is interesting to evaluate the longitudinal stresses at the mid-length of the strut; the largest lateral deflection occurs at this section, and the greatest bending moment also occurs at this section, therefore. The bending moment is

$$M = P(v_0 + e) = Pe\sec\frac{1}{2}kL \tag{18.19}$$

Suppose c is the distance from the centroidal axis Cx to the extreme fibres of the strut. Then the longitudinal bending stress set up by M is

$$\sigma_1 = \frac{Mc}{I} = \frac{Pec\sec\frac{1}{2}kL}{I} \tag{18.20}$$

The average longitudinal compressive stress set up by P is

$$\sigma_2 = \frac{P}{A} \tag{18.21}$$

where A is the cross-sectional area of the strut. Then the maximum longitudinal compressive stress is

$$\sigma_{max} = \sigma_2 + \sigma_1 = \frac{P}{A} + \frac{Pec}{I} \sec \frac{1}{2}kL \tag{18.22}$$

Suppose $I = Ar^2$, where r is the radius of gyration of the cross-section about Cx. Then

$$\sigma_{max} = \frac{P}{A}\left[1 + \frac{ec}{r^2} \sec \frac{1}{2}kL\right] \tag{18.23}$$

The minimum compressive stress is

$$\sigma_{min} = \frac{P}{A}\left[1 - \frac{ec}{r^2} \sec \frac{1}{2}kL\right] \tag{18.24}$$

The value of P giving rise to a maximum compressive stress σ is

$$P = \frac{A\sigma}{1 + \dfrac{ec}{r^2} \sec \dfrac{1}{2}kL} \tag{18.25}$$

However,

$$\frac{1}{2}kL = \frac{L}{2}\sqrt{\frac{P}{EI}}$$

and is therefore a function of P, so that the above equations must be solved by trial and error. A good approximation is derived as follows: let $\frac{1}{2}kL = \theta$, then for $0 < \theta < \frac{1}{2}\pi$

$$\sec \theta \fallingdotseq \frac{1 + 0.26\left(\dfrac{2\theta}{\pi}\right)^2}{1 - \left(\dfrac{2\theta}{\pi}\right)^2} = \frac{P_e + 0.26P}{P_e - P}$$

which leads to the following equation for P:

$$P^2\left(1 - 0.26\frac{ec}{r^2}\right) - P\left[P_e\left(1 + \frac{ec}{r^2}\right) + \sigma_A\right] + \sigma A P_e = 0$$

If $e = 0$, this has the roots $P = P_e$ or σA.

18.7 Initially curved pin-ended strut

In practice a strut cannot be made perfectly straight, and our analysis for the flexure of a compressed bar would become more realistic if account could be taken of the slight deviations from straightness of the centroidal axis of a strut.

Consider again a strut consisting of a flat strip of material. Suppose the centroidal longitudinal axis is initially curved, the lateral displacement at any point being v_0 from the axis Oz, Figure 18.8. Thrusts P are now applied at the ends of the strut and at the centroids of the end cross-sections.

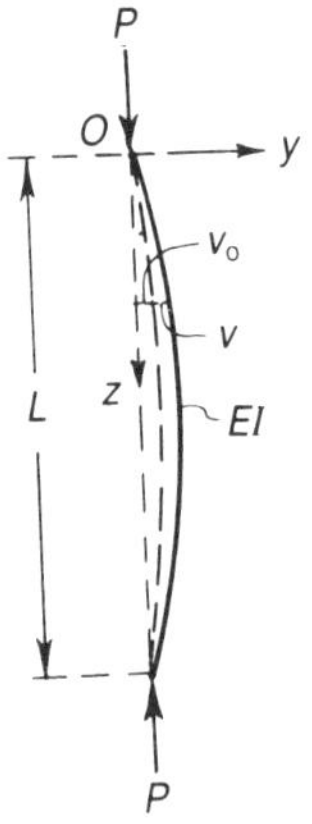

Figure 18.8 Initially curved strut.

The strut then bends further from its initial unloaded position. Suppose v is the additional lateral displacement at any section due to the application of P. If the ends of the strut are pinned there can be no lateral forces at the ends. The bending moment at any section of the strut is

$$M = P(v + v_0)$$

If the strut is initially unstressed then the bending moment at any section is proportioned to the change of curvature at that section. Then

$$M = -EI\frac{d^2v}{dz^2}$$

because the change of curvature is due only to the additional displacement v of the strut and not the total displacement $(v + v_0)$. Then

$$EI\frac{d^2v}{dz^2} + P(v + v_0) = 0$$

Put $P/EI = k^2$, as before. Then

$$\frac{d^2v}{dz^2} + k^2 v = -k^2 v_0$$

Suppose for the sake of simplicity that v_0 is sinusoidal in form; take

$$v_0 = a \sin \frac{\pi z}{L} \tag{18.26}$$

where a is a constant, and is the initial lateral displacement at the centre of the strut. Then

$$\frac{d^2v}{dz^2} + k^2 v = -k^2 a \sin \frac{\pi z}{L}$$

The general solution is

$$v = A \cos kz + B \sin kz + \frac{k^2 a}{\dfrac{\pi^2}{L^2} - k^2} \sin \frac{\pi z}{L}$$

If the ends are pinned we have

$$v = 0 \quad \text{at} \quad z = 0 \quad \text{and} \quad z = L$$

Then

$$A = 0 \quad \text{and} \quad B \sin kL = 0$$

If k is to assume any non-zero value we must have $B = 0$, so the relationship for v reduces to

$$v = \frac{k^2 a}{\dfrac{\pi^2}{L^2} - k^2} \sin \frac{\pi z}{L} \tag{18.27}$$

This may be written

$$v = \frac{a \sin \dfrac{\pi z}{L}}{\dfrac{\pi^2}{k^2 L^2} - 1} \tag{18.28}$$

But $k^2 = P/EI$, so on putting $\pi^2 EI/L^2 = P_e$, we have

$$v = \frac{a \sin \dfrac{\pi z}{L}}{\dfrac{P_e}{P} - 1} = \frac{v_0}{\dfrac{P_e}{P} - 1} \tag{18.29}$$

Now P_e is the buckling load for the perfectly straight strut. The relation for v, which is the additional lateral displacement of the strut, shows that the effect of the end thrust P is to increase v_0 by the factor $1/[(P_e/P) - 1]$. Obviously as P approaches P_e,v tends to infinity. The additional displacement at the mid-length of the strut is

$$v_c = \frac{a}{\dfrac{P_e}{P} - 1} \tag{18.30}$$

This relation between P and v_c has the form shown in Figure 18.9(i); v_c increases rapidly as P approaches P_e. Theoretically, the load P_e can only be attained at an infinitely large deflection. In practice material breakdown would occur before P_e could be attained, and at a finite displacement. We may write the relation for v_c in the form

$$P_e \frac{v_c}{P} - v_c = a \tag{18.31}$$

This gives a linear relation between (v_c/P) and v_c, Figure 18.9. The negative intercept on the axis of v_c is equal to $(-a)$. If values of (v_c/P) and v_c are plotted in a strut test, it will be found that as the critical condition is approached these variables are related by a straight-line equation of the type discussed above. The slope of this straight line defines P_e, the buckling load for a perfectly-straight strut.

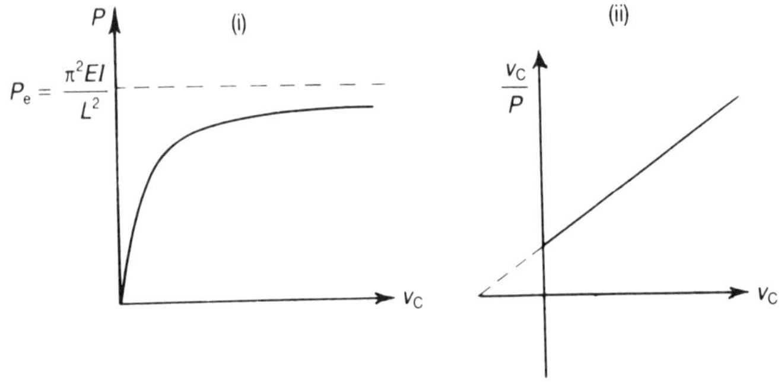

Figure 18.9 Deflections of an initially curved strut.

The P–v_c curve is asymptotic to the line $P = P_e$ if the material remains elastic. It is of considerable interest to evaluate the maximum longitudinal compressive stress in the strut. The maximum bending moment occurs at the mid-length, and has the value

$$M = P\left(a + v_c\right) = Pa\left[1 + \frac{1}{\frac{P_e}{P} - 1}\right] = Pa\left[\frac{P_e}{P_e - P}\right] \tag{18.32}$$

The maximum compressive stress occurs in an extreme fibre, and has the value

$$\sigma_{\max} = \frac{P}{A} + \frac{Pa\ P_e}{P_e - P}\left(\frac{c}{I}\right) = \frac{P}{A}\left[1 + \frac{P_e}{P_e - P}\left(\frac{ac}{r^2}\right)\right] \tag{18.33}$$

where A is the area of the cross-section, c is the distance from the centroidal axis to the extreme fibres, and r is the relevant radius of gyration of the cross-section. Now P/A is the average stress on the strut; if this is equal to σ, then

$$\sigma_{\max} = \sigma\left[1 + \frac{\sigma_e}{\sigma_e - \sigma}\left(\frac{ac}{r^2}\right)\right] \tag{18.34}$$

where

$$\sigma_e = \frac{P_e}{A} = \pi^2 E\left(\frac{r}{L}\right)^2 \tag{18.35}$$

Suppose $\frac{ac}{r^2} = \eta$. Then

$$\sigma_{\max} = \sigma\left[1 + \frac{\eta\sigma_e}{\sigma_e - \sigma}\right] \tag{18.36}$$

Then

$$\sigma_{\max} = \left(\sigma_e - \sigma\right) = \sigma\left[\left(1 + \eta\right)\sigma_e - \sigma\right]$$

Thus,

$$\sigma^2 - \sigma\left[\sigma_{\max} + \left(1 + \eta\right)\sigma_e\right] + \sigma_{\max}\,\sigma_e = 0$$

Then

$$\sigma = \frac{1}{2}\left[\sigma_{max} + (1+\eta)\sigma_e\right] - \sqrt{\frac{1}{4}\left[\sigma_{max} + (1+\eta)\sigma_e\right]^2 - \sigma_{max}\sigma_e} \qquad (18.37)$$

We need not consider the positive square root, since we are only interested in the smaller of the two roots of the equation. This relation gives the value of average stress, σ, at which a maximum compressive stress σ_{max} would be attained for any value of η. If we are interested in the value of σ at which yield stress σ_Y of a mild-steel strut is attained, we have

$$\sigma = \frac{1}{2}\left[\sigma_y + (1+\eta)\sigma_e\right] - \sqrt{\frac{1}{4}\left[\sigma_y + (1+\eta)\sigma_e\right]^2 - \sigma_y\sigma_e} \qquad (18.38)$$

18.8 Design of pin-ended struts

A commonly used structural material is mild steel. It has been found from tests on mild-steel pin-ended struts that failure of an initially-curved member takes place when the yield stress is first attained in one of the extreme fibres. From a wide range of tests Robertson concluded that the failing loads of mild-steel struts could be estimated if η is taken to be proportional to (L/r) the slenderness ratio of the strut; Robertson suggests that

$$\eta = 0.003\left(\frac{L}{r}\right) \qquad (18.39)$$

This value of η gives

$$\sigma = \frac{1}{2}\left[\sigma_y + \left(1 + 0.003\frac{L}{r}\right)\sigma_e\right] - \sqrt{\frac{1}{4}\left[\sigma_y + \left(1 + 0.003\frac{L}{r}\right)\sigma_e\right]^2 - \sigma_y\sigma_e} \qquad (18.40)$$

This represents a transition curve between yielding of the material for low slenderness ratios, Figure 18.10, and buckling at high slenderness ratios.

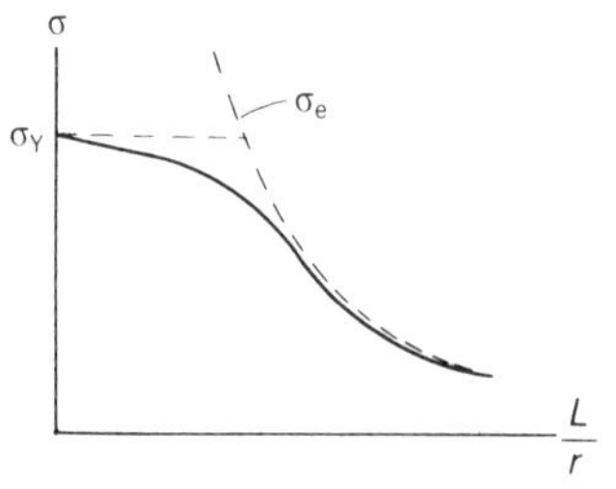

Figure 18.10 Effect of material breakdown on the buckling of a strut.

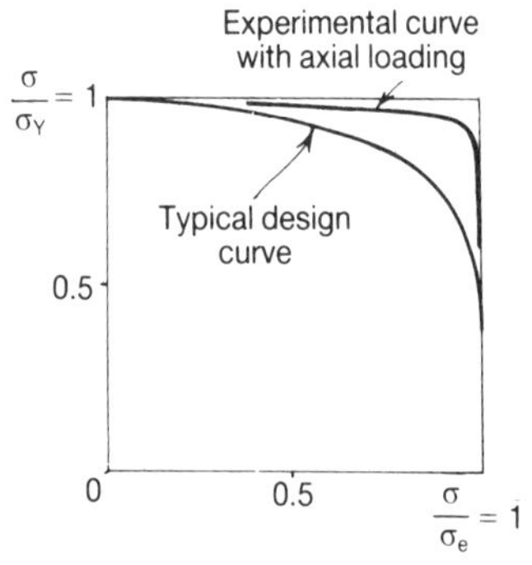

Figure 18.11 'Interaction' curves for practical struts.

In the case of mild-steel struts under true axial loading buckling occurs at σ_e the elastic buckling load or at σ_Y the yield stress. If true axial loading could be achieved in practice, all struts would fail at stresses that could be represented either by $\sigma/\sigma_Y = 1$, or $\sigma/\sigma_e = 1$. In a series of strut tests it is found that the test results are usually defined by a curve on the σ/σ_Y - σ/σ_e diagram, Figure 18.11, and not by the two straight lines $\sigma/\sigma_Y = 1$ and $\sigma/\sigma_e = 1$. if the experimental technique is improved to give better axial-loading conditions the curve approaches these two straight lines. Any convenient transition curve on this diagram may be taken as a design curve for practical conditions of axial loading.

18.9 Strut with uniformly distributed lateral loading

In the preceding sections we considered the effects of end eccentricities and initial curvatures on the lateral bending of compressed struts; these produce lateral bending of the strut from the onset of compression.

A similar problem arises when a compressed strut carries a lateral load. Consider a pin-ended strut length L and uniform flexural stiffness EI, Figure 18.12. Suppose the axial thrust on the strut is P, and that there is a lateral load of uniform intensity w per unit length. At the ends of the strut there are lateral shearing forces $\frac{1}{2}wL$.

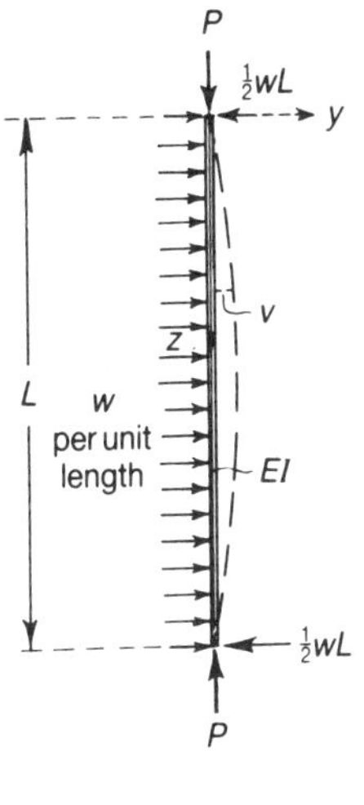

Figure 18.12 Laterally loaded struts.

If v is the lateral deflection at any point of the centroidal axis, then the bending moment at any section is

$$M = -EI\frac{d^2v}{dz^2} = Pv + \frac{1}{2}wLz - \frac{1}{2}wz^2$$

Then

$$\frac{d^2v}{dz^2} + \frac{Pv}{EI} = -\frac{w}{2EI}\left(Lz - z^2\right)$$

If $P/EI = k^2$, then

$$\frac{d^2v}{dz^2} + k^2v = -\frac{wk^2}{2P}\left(Lz - z^2\right)$$

The complete solution of this equation is

$$v = A\cos kz + B\sin kz - \frac{w}{2P}\left(Lz - z^2 + \frac{2}{k^2}\right)$$

in which A and B are arbitrary constants. Now, at $z = 0$ and $z = L$ we have $v = 0$, so

$$A - \frac{w}{Pk^2} = 0$$

and

$$A\cos kL + B\sin kL - \frac{w}{Pk^2} = 0$$

Then

$$A = \frac{w}{Pk^2}, \qquad B = \frac{w}{Pk^2}\left[\frac{1 - \cos kL}{\sin kL}\right]$$

Thus

$$v = \frac{w}{Pk^2}\left[\cos kz + \left(\frac{1-\cos kL}{\sin kL}\right)\sin kz - 1 - \frac{1}{2}k^2\left(Lz - z^2\right)\right] \tag{18.41}$$

The maximum value of v occurs at the mid-length, $z = \frac{1}{2}L$, and is given by

$$v_{max} = \frac{w}{Pk^2}\left[\cos\frac{1}{2}kL + \left(\frac{1-\cos kL}{\sin kL}\right)\sin\frac{1}{2}kL - 1 - \frac{1}{8}k^2L^2\right] \tag{18.42}$$

This may be written

$$v_{\max} = \frac{w}{Pk^2}\left[\sec\frac{1}{2}kL - 1 - \frac{1}{8}k^2L^2\right] \tag{18.43}$$

The maximum bending moment also occurs at the mid-length, and has the value

$$M_{\max} = Pv_{\max} + \frac{1}{8}wL^2 \tag{18.44}$$

On substituting for $v_{\max}$, we have

$$M_{\max} = \frac{w}{k^2}\left[\sec\frac{1}{2}kL - 1 - \frac{1}{8}k^2L^2\right] + \frac{1}{8}wL^2 = \frac{w}{k^2}\left[\sec\frac{1}{2}kL - 1\right] \tag{18.45}$$

When P is small, k is also small, and

$$\sec\frac{1}{2}kL = \frac{1}{\cos\frac{1}{2}kL} \fallingdotseq \left[1 - \frac{1}{2}\left(\frac{1}{2}kL\right)^2 + \frac{1}{24}\left(\frac{1}{2}kL\right)^4\right]^{-1}$$

Thus, approximately,

$$\sec\frac{1}{2}kL \fallingdotseq 1 + \left[\frac{1}{8}(kL)^2 - \frac{1}{384}(kL)^4\right] + \left[\frac{1}{8}(kL)^2 - \frac{1}{384}(kL)^4\right]^2$$

$$= 1 + \frac{1}{8}(kL)^2 + \frac{5}{384}(kL)^4 \tag{18.46}$$

Then

$$v_{\max} = \frac{w}{Pk^2}\left[\frac{5}{384}k^4L^4\right] = \frac{5}{384}\frac{wL^4}{EI} \tag{18.47}$$

This agrees with the value of the central deflection of a laterally loaded beam without end thrust. Similarly, when k is small,

$$M_{\max} = \frac{wL^2}{8}\left[\frac{8\left(\sec\frac{1}{2}kL - 1\right)}{k^2L^2}\right] \tag{18.48}$$

the term in square brackets is the factor by which the bending moment due to w alone must be multiplied to give the correct bending moment.

18.10 Buckling of a strut with built-in ends

In the elastic buckling of struts, we have assumed so far that the ends of the strut are always hinged to some foundation. When the ends are supported so that no rotations can occur, Figure 18.13, then the relevant mode of instability for the lowest critical load involves points of contra flexure at the quarter points. The buckling load is therefore the same as that of a pin-ended strut of half the length. Then

$$P_{cr} = \frac{\pi^2 EI}{\left(\frac{1}{2}L\right)^2} = 4\pi^2 \frac{EI}{L^2}, \text{ where } L_0 = 0.5L \tag{18.49}$$

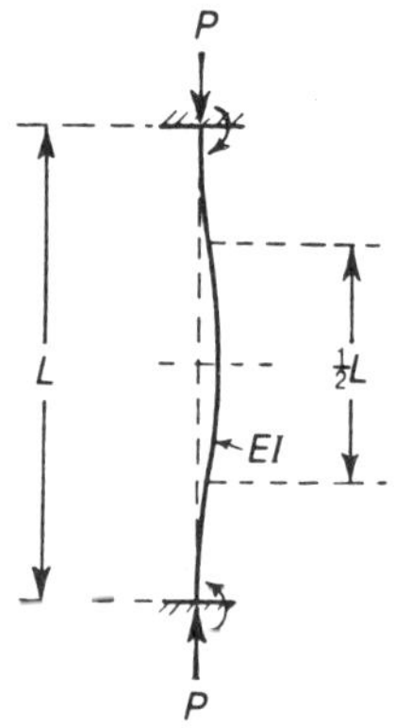

Figure 18.13 Buckling of a strut with built-in ends.

When the ends of the strut are built-in, no restraining moments are induced at the ends until the strut develops a buckled form.

18.11 Buckling of a strut with one end fixed and the other end free

When a vertical load P is applied to the free end of a vertical cantilever, AB, at the lowest critical load the laterally deflected form of the strut is a sinusoidal wave of length $2L$. If we consider the reflection of the buckled strut about A, Figure 18.14, then the strut of length $2L$ behaves as a pin-ended strut. The buckling load is

$$P_{cr} = \frac{\pi^2 EI}{(2L)^2} = \frac{\pi^2 EI}{4L^2}, \text{ where } L_0 = 2L \tag{18.50}$$

An important assumption in the preceding analysis is that the load at the free end of the cantilever is maintained in a vertical direction. If the load is always directed at A, that is its line of action is

BA, Figure 18.15 in the buckled form, then there is no restraining moment at *A*, and the cantilever behaves as a pin-ended strut. The buckling load is

$$P_{cr} = \pi^2 \frac{EI}{L^2} \tag{18.51}$$

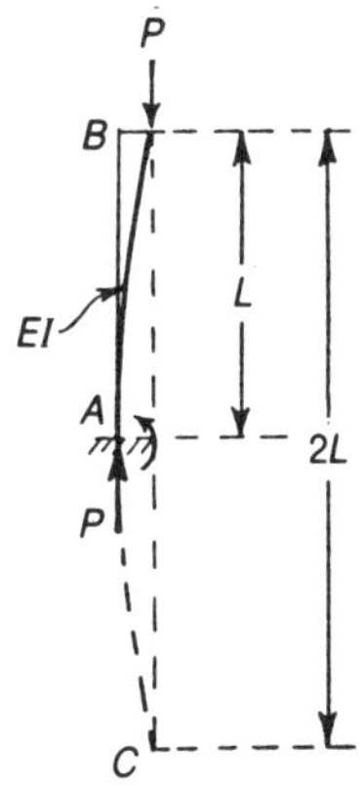

Figure 18.14 Buckling of a strut with one end free and the other built in.

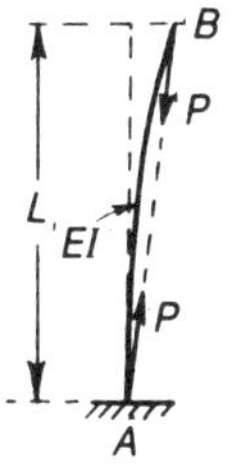

Figure 18.15 Thrust inclined to its original direction.

18.12 Buckling of a strut with one end pinned and the other end fixed

For other combinations of end conditions we are usually led to more involved calculations. A strut is pinned at its upper end and built-in to a rigid foundation at the lower end, Figure 18.16. In the buckled form of the strut a lateral shearing force *F* is induced at the upper end.

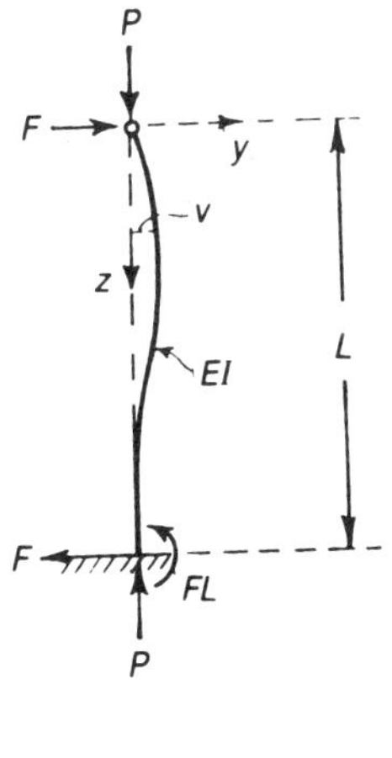

Figure 18.16 Strut with one end pinned and the other end fixed.

If v is the deflection of the central axis of the strut parallel to the y-axis, the bending moment at any section is

$$M = Pv - Fz$$

But

$$M = -EI \frac{d^2v}{dz^2}$$

Thus

$$-EI \frac{d^2v}{dz^2} = Pv - Fz$$

Put $k^2 = P/EI$. Then

$$\frac{d^2v}{dz^2} + k^2v = \frac{Fk^2z}{P}$$

The general solution is

$$v = A\cos kz + B\sin kz + \frac{F}{P}z$$

where A and B are arbitrary constants; the value of F is also unknown as yet, so there are three unknown constants in this equation. The boundary conditions are

$$v = 0, \quad \text{at} \quad z = 0$$

and $$v = 0 \text{ and } \frac{dv}{dz} = 0, \quad \text{at} \quad z = L$$

These give

$$A = 0$$

$$B\sin kL + \frac{FL}{P} = 0$$

$$Bk\cos kL + \frac{F}{P} = 0$$

The last two of these equations give

$$\frac{B}{F} = -\frac{L}{P\sin kL} = -\frac{1}{Pk\cos kL}$$

Thus

$$kL \cos kL = \sin kL \tag{18.52}$$

This equation gives the values of kL at which B and F are indeterminate, that is, at a condition of neutral equilibrium. The equation may be written

$$kL = \tan kL \tag{18.53}$$

The smallest non-zero value of kL satisfying this equation is approximately equal to 4.49 (see Figure 18.17). This gives

$$P_{cr} = k^2 EI = 4.49^2 \frac{EI}{L^2} = 20.2 \frac{EI}{L^2}$$

We may derive an approximate value of kL in the following way: suppose kL is less than $3\pi/2$ by a small amount ε, then

$$kL = \frac{3\pi}{2} - \varepsilon$$

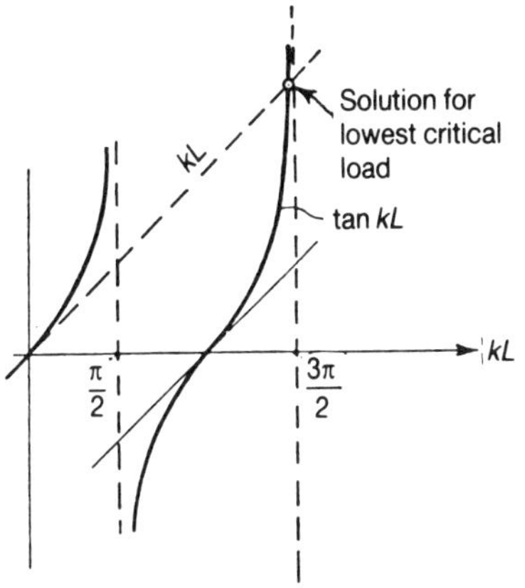

Figure 18.17 Graphical determination of buckling load.

Then we are interested in the roots of the equation

$$\frac{3\pi}{2} - \varepsilon = \tan\left(\frac{3\pi}{2} - \varepsilon\right)$$

If ε is small, then

$$\frac{3\pi}{2} - \varepsilon \doteqdot \cot \varepsilon \doteqdot \frac{1}{\varepsilon}\left(1 - \frac{1}{3}\varepsilon^2\right)$$

Approximately

$$\frac{3\pi}{2} = \frac{1}{\varepsilon}, \quad \text{or} \quad \varepsilon = \frac{2}{3\pi}$$

Then

$$kL = \frac{3\pi}{2} - \frac{2}{3\pi} = \frac{9\pi^2 - 4}{6\pi}$$

and

$$P_{cr} = k^2 EI = \left(\frac{9\pi^2 - 4}{6\pi}\right)^2 \frac{EI}{L^2} = 20.3 \frac{EI}{L^2} \qquad (18.54)$$

where

$$L_0 = \sqrt{\pi^2 / 20.3} = 0.7$$

18.13 Flexural buckling of struts with other cross-sectional forms

In Section 18.2 we considered the strut to be in the form of a flat rectangular strip. We considered buckling to involve bending about the major axis *Cx* only, Figure 18.18. In the case of a flat rectangular strip the axis *Cx* is clearly the weaker axis of bending. In practice, structural sections rarely have this simple cross-sectional form, but frequently have I-sections, or angle sections, or circular sections.

In general, if the cross-sectional form of a strut has two axes of symmetry, we can consider flexural instability about these two axes independently. If an I-section has two axes of symmetry in the cross-section, Figure 18.19, flexural instability occurs usually about the axis of smaller stiffness, usually *Cx*. In a rectangular strut, Figure 18.19, the weaker bending axis is parallel to the longer sides. Circular cross-sectional forms have the property that any two mutually perpendicular diameters are principal centroidal axes; for these sections flexural instability is equally likely about any principal centroidal axis, Figure 18.19; when buckling occurs it is usually restricted to one plane. In making these statement we assume the ends of the strut are hinged about both axes *Cy* and *Cz*; this can be achieved in practice by loading through ball-ends. When the ends are not supported in the same way about *Cy* and *Cx*, then torsional effects may become important in the buckling behaviour.

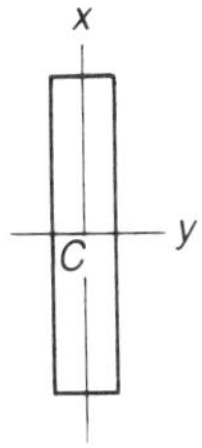

Figure 18.18 Narrow strip cross-section.

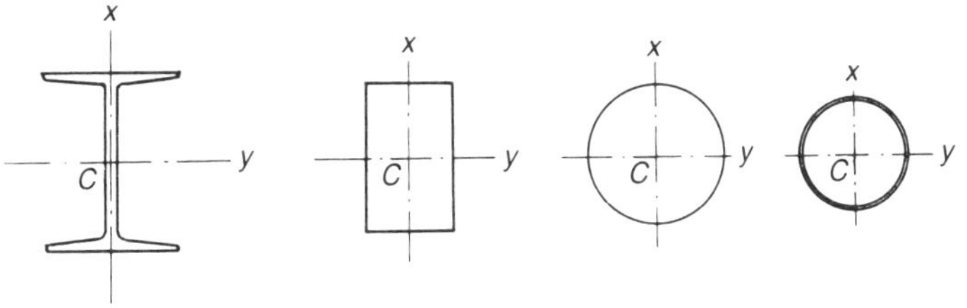

Figure 18.19 Cross-section with two axes of symmetry.

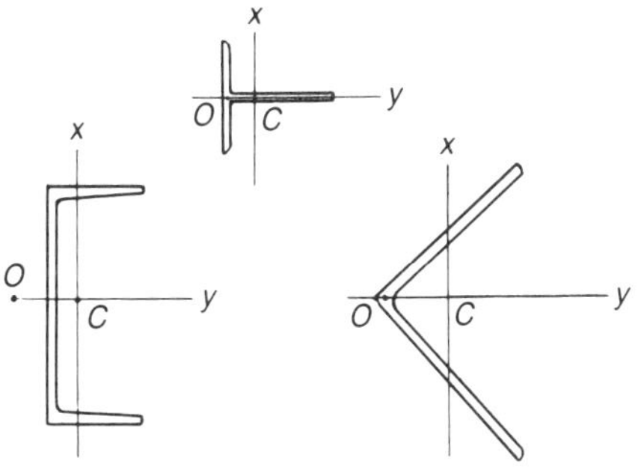

Figure 18.20 Cross-sections with only one axis of symmetry.

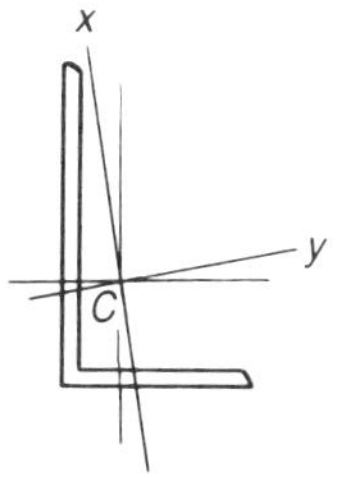

Figure 18.21 Unequal angle strut.

In the case of cross-sectional forms with only one axis of symmetry, Cy, say (Figure 18.20), torsional effects become important if the shear centre is not coincident with the centroid. This is true of channel sections, T-sections, and equal angle sections. Although for certain struts flexural instability occurs about the weaker principal axis Cz, in general twisting also occurs.

In the case of cross-sectional forms with no axes of symmetry, Figure 18.21, the buckled form always involves torsion, and the flexural buckling load has little meaning. This is true of unequal angle struts.

Problem 18.1 What thrust will a round steel rod take without buckling if it is 1.25 cm diameter, 2 m long, perfectly straight, and pin-jointed at the ends, the load being applied exactly along the axis of the rod?

Solution

We have

$$I = \frac{\pi(0.0125)^4}{64} = 1.20 \times 10^{-9}\ \mathrm{m}^4, \quad L = 2\ \mathrm{m}$$

Taking $E = 200\ \mathrm{GN/m^2}$, we have

$$P_e = \frac{\pi^2 EI}{L^2} = 591\ N$$

18.14 Torsional buckling of a cruciform strut

We mentioned above that some struts are prone to torsional buckling effects. A cross-sectional form in which torsional instability occurs independently of any other form of buckling is a symmetrical cruciform section.

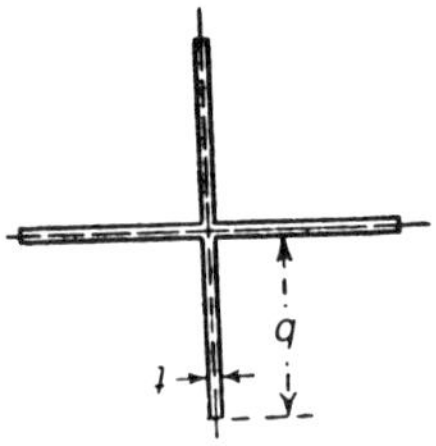

Figure 18.22 Cross-section of a cruciform strut.

The cruciform has four equally spaced limbs each of breadth b and uniform thickness t, Figure 18.22. Consider the section under a uniform compressive stress σ, Figure 18.23(i). We consider the possibility that the section may become unstable by twisting about the longitudinal axis Cz, Figure 18.23(ii); the stresses σ over the ends remain parallel to Cz during buckling.

Over any cross-section of the cruciform the stress is σ, acting parallel to Cz. Consider an elemental area δA of one limb at a distance x from the axis Cz, Figure 18.23(iii). If the relative twist between two cross-sections a distance δz apart is $\delta\theta$, then the force

$$\sigma \delta A$$

on the elemental area is statically equivalent to a force $\sigma\delta A$ acting along the twisted form of the strut and a small force

$$\sigma \delta A x \frac{d\theta}{dz}$$

acting in the plane of the cross-section. The inclined forces $\sigma\delta A$ on the two cross-sections are in equilibrium with each other, but the two forces $\sigma\delta A x\ (d\theta/dz)$ give rise to a resultant torque at any cross-section. This torque is

$$4\int_0^b \sigma x^2 \frac{d\theta}{dx}\, dA = 4\sigma \frac{d\theta}{dz}\int_0^b x^2\, dA$$

since there are four limbs.

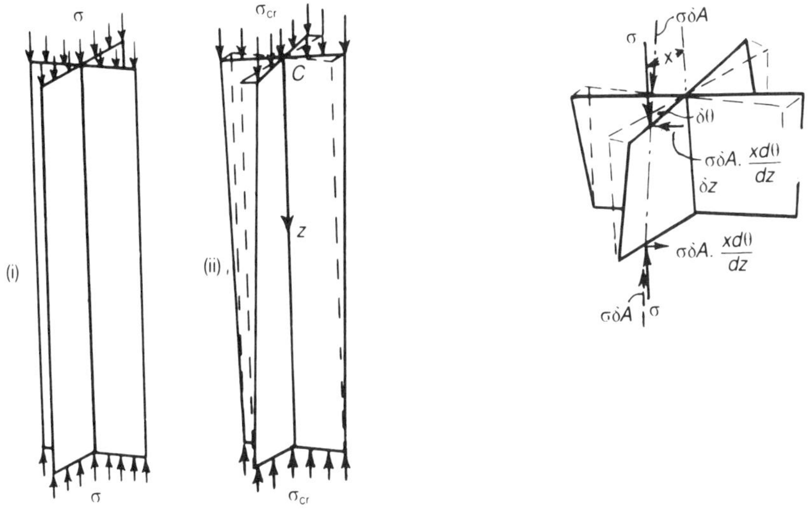

Figure 18.23 Torsional buckling of a cruciform column.

The geometrical quantity

$$4 \int_0^b x^2 dA$$

is the polar second moment of area of the cross-section about Cz. The resultant torque at any cross-section is then

$$\sigma \frac{d\theta}{dz} J_0$$

where

$$J_0 = 4 \int_0^b x^2 \, dA = 4t \int_0^b x^2 \, dz = \frac{4}{3} b^3 t$$

Now, we found in Chapter 16 that the torque-twist relation for a cruciform section is

$$\text{Torque} = GJ \frac{d\theta}{dz} = \frac{4}{3} Gbt^3 \frac{d\theta}{dz}$$

In the case of the compressed cruciform, the twisted form can be maintained if

$$\sigma \frac{d\theta}{dz} J_0 = GJ \frac{d\theta}{dz}$$

Then

$$\sigma = G\left(\frac{J}{J_0}\right) = G\left(\frac{\frac{4}{3}bt^3}{\frac{4}{3}b^3t}\right) = G\left(\frac{t}{b}\right)^2 \tag{18.55}$$

18.15 Modes of buckling of a cruciform strut

With a knowledge of the torsional and flexural buckling loads of a cruciform strut, we can estimate the range of struts, we can estimate the range of struts for which buckling is likely in the two modes.

If b is very much greater than t, and if all the limbs are similar in form, flexural buckling of a pin-ended strut is possible about any axis through the junction of the limbs, since the flexural stiffness is the same for all axes. For flexural instability the critical stress is

$$\sigma_f = \pi^2 \frac{EI}{AL^2} \tag{18.56}$$

Now $I = 1/12\ t(2b)^3 = ⅔b^3t$ and $A = 4bt$, and so

$$\sigma_f = \frac{\pi^2}{6}\frac{Eb^2}{L^2} \tag{18.57}$$

Now, as we have seen, the torsional buckling stress is independent of L, and has the value

$$\sigma_t = G\left(\frac{t}{b}\right)^2 \tag{18.58}$$

Then $\sigma_f > \sigma_t$ when

$$\frac{\pi^2}{6}\frac{Eb^2}{L^2} > G\left(\frac{t}{b}\right)^2$$

i.e. when

$$\frac{b^4}{L^2t^2} > \frac{6G}{\pi^2 E} = \frac{6}{2\pi^2(1+\nu)} = \frac{3}{\pi^2(1+\nu)} \tag{18.59}$$

If $\nu = 0.3$, then

$$\frac{b^4}{L^2t^2} > \frac{3}{1.3\pi^2} = 0.234 \tag{18.60}$$

Thus torsional buckling takes place when

$$\frac{b^2}{Lt} > \sqrt{0.234} = 0.484$$

i.e. when

$$\frac{Lt}{b^2} < 2.07$$

This condition may be written

$$\left(\frac{L}{b}\right) < 2.07\left(\frac{b}{t}\right) \tag{18.61}$$

We can show the domains of flexural and torsional instability by plotting (L/b) against (b/t), Figure 18.24. For a practical material, yielding or material breakdown occurs when L/b and b/t approach zero; the lower left-hand corner is therefore the yielding domain. Above the straight line

$$\left(\frac{L}{b}\right) = 2.07\left(\frac{b}{t}\right)$$

buckling is by flexure, whereas below this line buckling is by torsion.

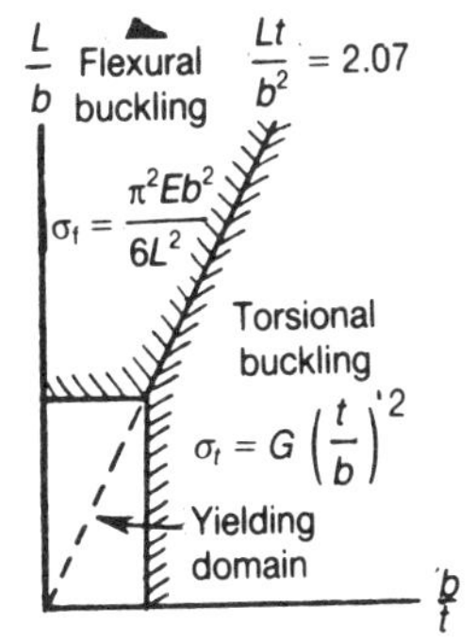

Figure 18.24 Modes of buckling of a cruciform strut.

18.16 Lateral buckling of a narrow beam

We have seen that the axial compression of a slender strut can lead to a condition of neutral equilibrium, in which at a certain compressive load a flexural form of deformation becomes possible. In the case of a cruciform strut we have shown that a form of neutral equilibrium involving torsion is possible under certain conditions.

Problems of structural instability are not restricted entirely to compression members, although there are many problems of this type. In the case of deep beams, for example, lateral buckling may occur, involving torsion and bending perpendicular to the plane of the depth of the beam. In general this problem is a complex one; however, we can determine some of the factors involved by studying the relatively simple case of the bending of a narrow deep beam.

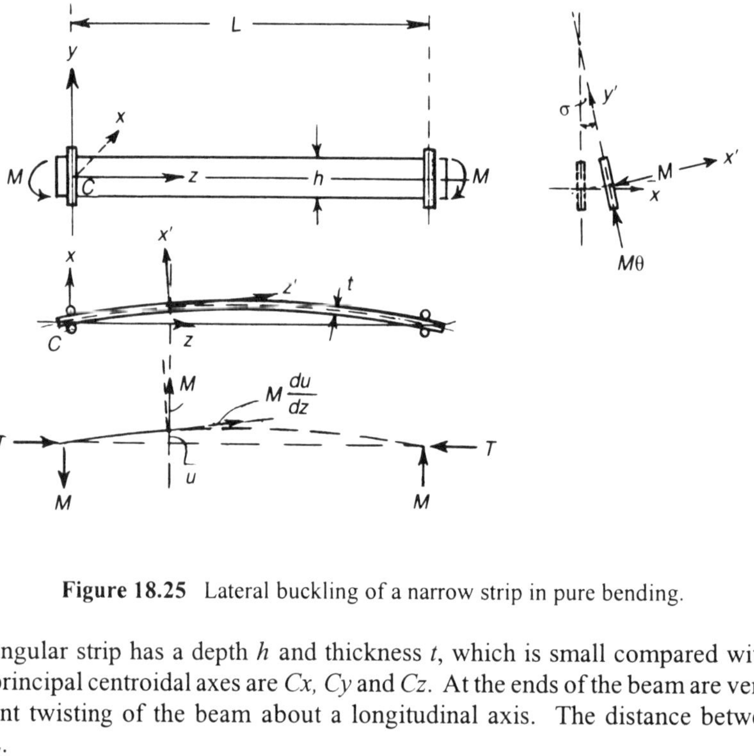

Figure 18.25 Lateral buckling of a narrow strip in pure bending.

A long rectangular strip has a depth h and thickness t, which is small compared with h, Figure 18.25. The principal centroidal axes are Cx, Cy and Cz. At the ends of the beam are vertical rollers which prevent twisting of the beam about a longitudinal axis. The distance between the end supports is L.

The beam is loaded with moments M applied at each end about axes parallel to Cx. Consider the possibility that the beam may become laterally unstable at some critical value of M. If $h >> t$ then bending displacements in the yz plane may be neglected. Suppose in the buckled form the principal centroidal axes at any cross-section are Cx', Cy' and Cz'. The lateral displacements parallel to Cx is u, and θ is the angle of twist about Cz at any cross-section. The moments M are assumed to be maintained along their original lines of action; the only other forces which may be induced at the ends are equal and opposite longitudinal torques T. The bending moment about the axis Cy' is then

$$M\theta$$

and as this gives rise to the curvature of the beam in the *xz* plane we have

$$M\theta = -EI_y \frac{d^2u}{dz^2}$$

Where EI_y is the bending stiffness of the beam about *Cy*. Again, for twisting about *Cz'*

$$T + M\frac{du}{dz} = GJ\frac{d\theta}{dz}$$

where *GJ* is the torsional stiffness about *Cx*. Differentiation of the second equations gives

$$M\frac{d^2U}{dz^2} = GJ\frac{d^2\theta}{dz^2}$$

Thus, on eliminating *u*,

$$M\theta = -EI_y \frac{GJ}{M}\frac{d^2\theta}{dz^2}$$

Then

$$\frac{d^2\theta}{dz^2} + \frac{M^2}{GJEI_y}\theta = 0$$

Put

$$k^2 = \frac{M^2}{GJEI_y} \qquad (18.62)$$

Then

$$\frac{d^2\theta}{dz^2} + k^2\theta = 0$$

The general solution is

$$\theta = A\cos kz + B\sin kz$$

where A and B are arbitrary constants. If $\theta = 0$ at $z = 0$, then $A = 0$. Further if $\theta = 0$ at $z = L$,

$$B \sin kL = 0$$

If $B = 0$, then both A and B are zero, and no buckling occurs; but B can be non-zero if

$$\sin kL = 0$$

We can disregard the root $kL = 0$, since the general solution is only valid if k is non-zero. The relevant roots are

$$kL = \pi, \quad 2\pi, \quad 3\pi\ldots \tag{18.63}$$

The smallest value of critical moment is

$$M_{cr} = k\sqrt{(GJ)(EI_y)} = \frac{\pi}{L}\sqrt{(GJ)(EI_y)}$$

Now, for a beam of rectangular cross-section,

$$GJ = \frac{1}{3}Ght^3, \quad EI_y = \frac{1}{12}Eht^3 \tag{18.64}$$

Then

$$M_{cr} = \frac{\pi}{L}\sqrt{\frac{1}{36}GEh^2t^6} = \frac{\pi}{L}\frac{ht^3}{6}\sqrt{GE} \tag{18.65}$$

If $G = E/2(1+\nu)$ then

$$\sqrt{GE} = \sqrt{E^2/2(1+\nu)} = \frac{E}{\sqrt{2(1+\nu)}} \tag{18.66}$$

The maximum bending stress at the bending moment M_{cr} is

$$\sigma_{cr} = \frac{M_{cr}}{I_x}\frac{h}{2} = \frac{6M_{cr}}{h^2t} = \frac{\pi E}{\sqrt{2(1+\nu)}}\frac{t^2}{hL} \tag{18.67}$$

For a strip of given depth h and thickness t, the buckling stress σ_{cr} is proportional to the inverse of (L/t), which is sometimes referred to as the slenderness ratio of the beam.

Further problems *(answers on page 694)*

18.2 Calculate the buckling load of a pin-joined strut made of round steel rod 2 cm diameter and 4 m long.

18.3 Find the thickness of a round steel tubular strut 3.75 cm external diameter, 2 m long, pin-joined at the end, to withstand an axial load of 10 kN.

18.4 Calculate the buckling load of a strut built-in at both ends, the cross-section being a square 1 cm by 1 cm, and the length 2 m. Take $E = 200\ \text{GN/m}^2$.

18.5 A steel scaffolding pole acts as a strut, but the load is applied eccentrically at 7.5 cm distance from the centre line with leverages in the same direction at top and bottom. The pole is tubular, 5 cm external diameter and 0.6 cm thick, 3 m in length between its ends which are not fixed in direction. If the steel has a yield stress of 300 MN/m² and $E = 200\ \text{GN/m}^2$, estimate approximately the load required to buckle the strut. *(RNEC)*

18.6 Two similar members of the same dimensions are connected together at their ends by two equal rigid links, the links being pin-jointed to the members. At the middle the members are rigidly connected by a distance piece. Equal couples are applied to the links, the axes of the couples being parallel to the pins of the joints. Show that buckling will occur in the top member if the couples M exceed a value given by the root of the equation

$$\tan \frac{1}{2}kL = \tanh \frac{1}{2}kL$$

where $k^2 = M/EId$. *(Cambridge)*

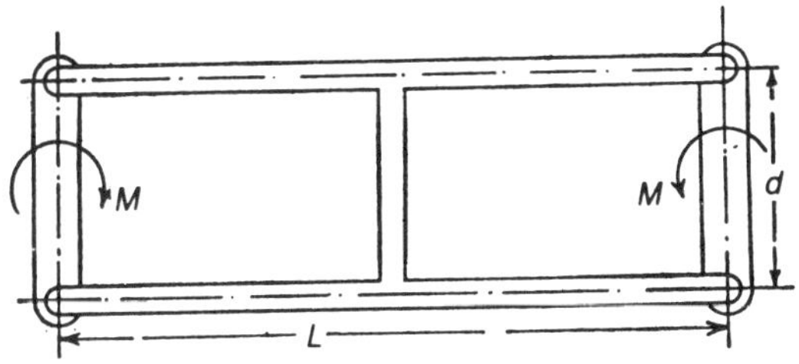

19 Lateral deflections of circular plates

19.1 Introduction

In this chapter, consideration will be made of three classes of plate problem, namely

(i) small deflections of plates, where the maximum deflection does not exceed half the plate thickness, and the deflections are mainly due to the effects of flexure;

(ii) large deflections of plates, where the maximum deflection exceeds half the plate thickness, and membrane effects become significant; and

(iii) very thick plates, where shear deflections are significant.

Plates take many and various forms from circular plates to rectangular ones, and from plates on ships' decks to ones of arbitrary shape with cut-outs etc; however, in this chapter, considerations will be made mostly of the small deflections of circular plates.

19.2 Plate differential equation, based on small deflection elastic theory

Let, w be the out-of-plane deflection at any radius r, so that,

$$\frac{dw}{dr} = \theta$$

and

$$\frac{d^2w}{dr^2} = \frac{d\theta}{dr}$$

Also let

R_t = tangential or circumferential radius of curvature at r = AC (see Figure 19.1).

R_r = radial or meridional radius of curvature at r = BC.

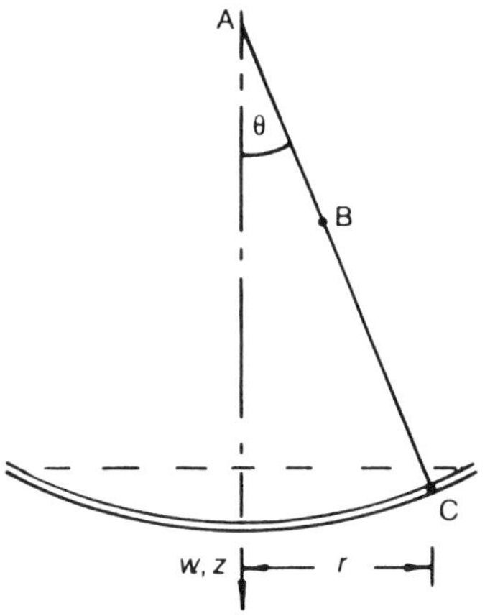

Figure 19.1 Deflected form of a circular plate.

From standard small deflection theory of beams (see Chapter 13) it is evident that

$$R_r = 1\Big/\frac{d^2w}{dr^2} = 1\Big/\frac{d\theta}{dr} \tag{19.1}$$

or

$$\frac{1}{R_r} = \frac{d\theta}{dr} \tag{19.2}$$

From Figure 19.1 it can be seen that

$$R_t = AC = r/\theta \tag{19.3}$$

or

$$\frac{1}{R_t} = \frac{1}{r}\frac{dw}{dr} = \frac{\theta}{r} \tag{19.4}$$

Let z = the distance of any fibre on the plate from its neutral axis, so that

$$\varepsilon_r = \text{radial strain} = \frac{z}{R_r} = \frac{1}{E}\left(\sigma_r - \nu\sigma_t\right) \tag{19.5}$$

and

$$\varepsilon_t = \text{circumferential strain} = \frac{z}{R_t} = \frac{1}{E}\left(\sigma_t - v\sigma_r\right) \tag{19.6}$$

From equations (19.1) to (19.6) it can be shown that

$$\sigma_r = \frac{Ez}{\left(1 - v^2\right)}\left(\frac{1}{R_r} + \frac{v}{R_t}\right) = \frac{Ez}{1 - v^2}\left(\frac{d\theta}{dr} + \frac{v\theta}{r}\right) \tag{19.7}$$

$$\sigma_t = \frac{Ez}{\left(1 - v^2\right)}\left(\frac{1}{R_t} + \frac{v}{R_r}\right) = \frac{Ez}{\left(1 - v^2\right)}\left(\frac{\theta}{r} + \frac{d\theta}{dr}\right) \tag{19.8}$$

where,

σ_r = radial stress due to bending

σ_t = circumferential stress due to bending

The tangential of circumferential bending moment per unit radial length is

$$M_t = \int_{-t/2}^{+t/2} \sigma_t z \; dz$$

$$= \int_{-t/2}^{+t/2} \frac{E}{\left(1 - v^2\right)}\left(\frac{\theta}{r} + \frac{d\theta}{dr}\right) z^2 dz$$

$$= \frac{E}{\left(1 - v^2\right)}\left(\frac{\theta}{r} + v\frac{d\theta}{dr}\right)\left[\frac{z^3}{3}\right]_{-t/2}^{+t/2}$$

$$= \frac{Et^3}{12\left(1 - v^2\right)}\left(\frac{\theta}{r} + v\frac{d\theta}{dr}\right)$$

therefore

$$M_t = D\left(\frac{\theta}{r} + v\frac{d\theta}{dr}\right) = D\left(\frac{1}{r}\frac{dw}{dr} + v\frac{d^2w}{dr^2}\right) \tag{19.9}$$

where,

t = plate thickness

and

$$D = \frac{Et^3}{12(1 - v^2)} = \text{flexural rigidity}$$

Similarly, the radial bending moment per unit circumferential length,

$$M_r = D\left(\frac{d\theta}{dr} + \frac{v\theta}{r}\right) = D\left(\frac{d^2w}{dr^2} + \frac{vdw}{r\,dr}\right) \tag{19.10}$$

Substituting equation (19.9) and (19.10) into equations (19.7) and (19.8), the bending stresses could be put in the following form:

$$\sigma_t = 12\ M_t \times z / t^3$$

and

$$\sigma = 12\ M_r \times z / t^3 \tag{19.11}$$

and the maximum stresses $\hat{\sigma}_t$ and $\hat{\sigma}_r$ will occur at the outer surfaces of the plate (ie, @$z = \pm t/2$). Therefore

$$\hat{\sigma}_t = 6M_t / t^2 \tag{19.12}$$

and

$$\hat{\sigma}_r = 6M_r / t^2 \tag{19.13}$$

The plate differential equation can now be obtained by considering the equilibrium of the plate element of Figure 19.2.

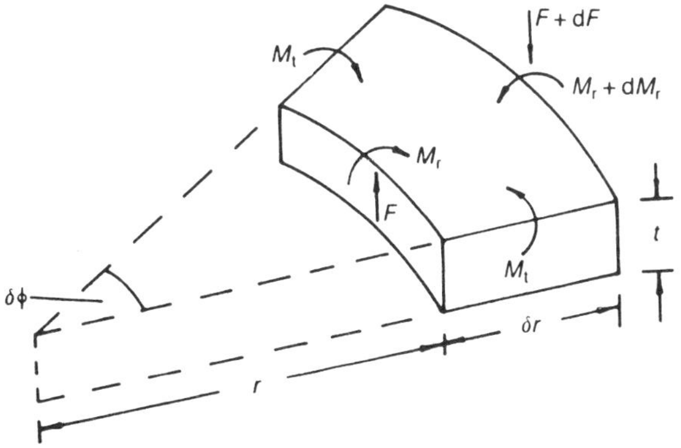

Figure 19.2 Element of a circular plate.

Taking moments about the outer circumference of the element,

$$(M_r + \delta M_t)(r + \delta r)\,\delta\varphi - M_r\, r\delta\varphi - 2M_t\ \delta r \sin\frac{\delta\varphi}{2} - F\ \ r\,\delta\varphi\delta r = 0$$

In the limit, this becomes

$$M_r + r \cdot \frac{dM_r}{dr} - M_t - Fr = 0 \tag{19.14}$$

Substituting equation (19.9) and (19.10) into equation (19.14),

$$\left(\frac{d\theta}{dr} + \frac{v\theta}{r}\right) + \left(r \cdot \frac{d^2\theta}{dr^2} + r \cdot \frac{v}{r}\frac{d\theta}{dr} - r \cdot \frac{v\theta}{r^2}\right) - \frac{\theta}{r} - v\frac{d\theta}{dr} = \frac{Fr}{D}$$

or

$$\frac{d^2\theta}{dr^2} + \left(\frac{1}{r}\right)\frac{d\theta}{dr} - \frac{\theta}{r^2} = \frac{F}{D}$$

which can be re-written in the form

$$\frac{d}{dr}\left[(1/r) \cdot \frac{d(r\theta)}{dr}\right] = \frac{F}{D} \tag{19.15}$$

where F is the shearing force / unit circumferential length.

Equation (19.15) is known as the plate differential equation for circular plates.

For a horizontal plate subjected to a lateral pressure p per unit area and a concentrated load W at the centre, F can be obtained from equilibrium considerations. Resolving 'vertically',

$$2\pi r F = \pi r^2 p + W$$

therefore

$$F = \frac{pr}{2} + \frac{W}{2\pi r} \text{ (except at } r = 0) \tag{19.16}$$

Substituting equation (19.16) into equation (19.15),

$$\frac{d}{dr}\left[(1/r)\frac{d(r\theta)}{dr}\right] = \frac{1}{D}\left[\frac{pr}{2} + \frac{W}{2\pi r}\right]$$

therefore

$$\frac{1}{r}\frac{d(r\theta)}{dr} = \frac{1}{D}\left[\frac{pr^2}{4} + \frac{W}{2\pi}\ln r\right] + C_1$$

$$\frac{d(r\theta)}{dr} = \frac{1}{D}\left[\frac{pr^3}{4} + \frac{Wr}{2\pi}\ln r\right] + C_1 r$$

$$r\theta = \frac{1}{D}\left[\frac{pr^4}{16} + \frac{Wr^2}{4\pi}\ln r - \frac{Wr^2}{8\pi}\right] + \frac{C_1 r^2}{2} + C_2$$

$$\theta = \frac{1}{D}\left[\frac{pr^3}{16} + \frac{Wr}{4\pi}\ln r - \frac{Wr}{8\pi}\right] + \frac{C_1 r}{2} + \frac{C_2}{r} \tag{19.17}$$

since,

$$\frac{dw}{dr} = \theta$$

$$w = \int \theta \, dr + C_3$$

hence,

$$w = \frac{pr^4}{64D} + \frac{Wr^2}{8\pi D}(\ln r - 1) + \frac{C_1 r^2}{4} + C_2 \ln r + C_3 \tag{19.18}$$

Note that

$$\int r \ln r \, dr = \int \frac{\ln r}{2} d(r^2)$$

$$= \frac{r^2}{2}\ln r - \int \frac{r^2}{2} d(\ln r) = \frac{r^2}{2}\ln r - \int \frac{r}{2} dr$$

$$= \frac{r^2}{2}\ln r - \frac{r^2}{4} + \text{a constant} \tag{19.19}$$

Problem 19.1 Determine the maximum deflection and stress in a circular plate, clamped around its circumference, when it is subjected to a centrally placed concentrated load W.

<u>*Solution*</u>

Putting $p = 0$ into equation (19.18),

$$w = \frac{Wr^2}{8\pi D}(\ln r - 1) + \frac{C_1 r^2}{4} + C_2 \ln r + C_3$$

$$\frac{dw}{dr} = \frac{Wr}{4\pi D}(\ln r - 1) + \frac{Wr}{8\pi D} + \frac{C_1 r}{2} + \frac{C_2}{r}$$

as dw/dr cannot equal ∞ at $r = 0$, $C_2 = 0$

$$\text{at } r = R, \quad \frac{dw}{dr} = w = 0$$

therefore

$$0 = \frac{WR^2}{8\pi D}\ln r - \frac{WR^2}{8\pi D} + \frac{C_1 R^2}{4} + C_3$$

and

$$0 = \frac{WR}{4\pi D}\ln R - \frac{WR}{4\pi D} + \frac{WR}{8\pi D} + \frac{C_1 R}{2}$$

Hence,

$$C_1 = \frac{W}{4\pi D}(1 - 2\ln R)$$

$$C_3 = -\frac{WR^2}{8\pi D}\ln R + \frac{WR^2}{8\pi D} - \frac{WR^2}{16\pi D} + \frac{WR^2}{8\pi D}\ln R = \frac{WR^2}{16\pi D}$$

$$w = \frac{WR^2}{8\pi D}\ln r - \frac{Wr^2}{8\pi D} + \frac{Wr^2}{16\pi D} - \frac{Wr^2}{8\pi D}\ln R + \frac{WR^2}{16\pi D}$$

or

$$w = \frac{WR^2}{16\pi D}\left|1 - \frac{r^2}{R^2} + \frac{2R^2}{R^2}\ln\left(\frac{r}{R}\right)\right|$$

The maximum deflection ($\hat{w}$) occurs at $r = 0$

$$\hat{w} = \frac{WR^2}{16\pi D}$$

Substituting the derivatives of w into equations (19.9) and (19.10),

$$M_r = \frac{W}{4\pi}\left[1 + \ln\left(\frac{r}{R}\right)(1 + \nu)\right]$$

$$M_t = \frac{W}{4\pi}\left[\nu + (1 + \nu)\ln\left(\frac{r}{R}\right)\right]$$

Problem 19.2 Determine the maximum deflection and stress that occur when a circular plate clamped around its external circumference is subjected to a uniform lateral pressure p.

Solution

From equation (19.18),

$$w = \frac{pr^4}{64D} + \frac{C_1 r^2}{4} + C_2 \ln r + C_3$$

$$\frac{dw}{dr} = \frac{pr^3}{16D} + \frac{C_1 r}{2} + \frac{C_2}{r}$$

and

$$\frac{d^2w}{dr^2} = \frac{3pr^2}{16D} + \frac{C_1}{2} - \frac{C_2}{r^2}$$

$$\text{at } r = 0, \quad \frac{dw}{dr} \neq \infty \text{ therefore } C_2 = 0$$

$$\text{at } r = R, \quad w = \frac{dw}{dr} = 0$$

therefore

$$0 = \frac{pr^4}{64D} + \frac{C_1 R^2}{4} + C_3$$

$$0 = \frac{pR^3}{16D} + \frac{C_1 R}{2}$$

therefore

$$C_1 = \frac{-pR^2}{8D}$$

$$C_3 = \frac{pR^4}{64D}$$

therefore

$$w = \frac{pR^4}{64D}\left(1 - \frac{r^2}{R^2}\right)^2 \tag{19.20}$$

Substituting the appropriate derivatives of w into equations (19.9) and (19.10),

$$M_r = \frac{pR^2}{16}\left[-(1 + v) + (3 + v)\frac{r^2}{R^2}\right] \tag{19.21}$$

$$M_t = \frac{pR^2}{16}\left[-(1 + v) + (1 + 3v)\frac{r^2}{R^2}\right] \tag{19.22}$$

Maximum deflection ($\hat{w}$) occurs at $r = 0$

$$\hat{w} = \frac{pR^4}{64D} \tag{19.23}$$

By inspection it can be seen that the maximum bending moment is obtained from (19.21), when $r = R$, i.e.

$$\hat{M}_r = pR^2/8$$

and $$\hat{\sigma} = 6\hat{M}_t / t^2$$

$$= 0.75\, pR^2 / t^2$$

Problem 19.3 Determine the expression for M_r and M_t in an annular disc, simply-supported around its outer circumference, when it is subjected to a concentrated load W, distributed around its inner circumference, as shown in Figure 19.3.

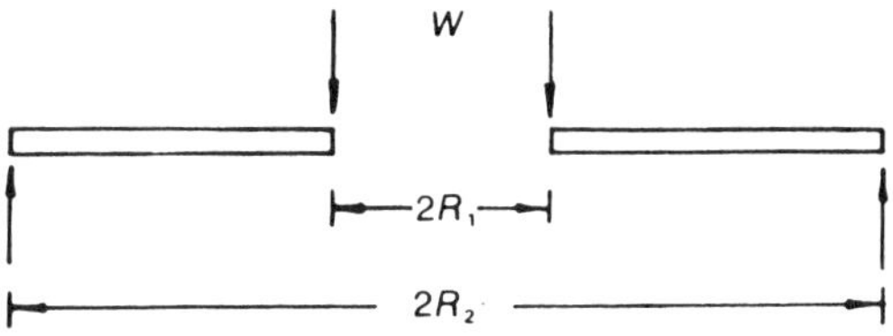

Figure 19.3 Annular disc.

W = total load around the inner circumference.

<u>Solution</u>

From equation (19.18),

$$w = \frac{Wr^2}{8\pi D}(\ln r - 1) + \frac{C_1 r^2}{4} + C_2 \ln r + C_3$$

at $r = R_2$, $w = 0$

or

$$0 = \frac{WR_2^2}{8\pi D}\left(\ln R_2 - 1\right) + \frac{C_1}{4}R_2^2 + C_2 \ln R_2 + C_3 \tag{19.24}$$

Now,

$$\frac{dw}{dr} = \frac{Wr}{4\pi D}(\ln r - 1) + \frac{Wr}{8\pi D} + \frac{C_1 r}{2} + \frac{C_2}{r} \tag{19.25}$$

and,

$$\frac{d^2w}{dr^2} = \frac{W}{4\pi D}\left(\ln r - 1\right) + \frac{W}{4\pi D} + \frac{W}{8\pi D} + \frac{C_1}{2} - \frac{C_2}{r^2} \tag{19.26}$$

A suitable boundary condition is that

$$M_r = 0 \text{ at } r = R_1 \text{ and at } r = R_2$$

but

$$M_r = D\left(\frac{d^2w}{dr^2} + \frac{v}{r}\frac{dw}{dr}\right)$$

therefore

$$\frac{W}{4\pi D}\left(\ln R_1 - 1\right) + \frac{3W}{8\pi D} + \frac{C_1}{2} - \frac{C_2}{R_1^2} + \frac{v}{R_1}\left\{\frac{WR_1}{4\pi D}\left(\ln R_1 - 1\right) + \frac{WR_1}{8\pi D} + \frac{C_1 R_1}{2} + \frac{C_2}{R_1}\right\} = 0 \tag{19.27}$$

and

$$\frac{W}{4\pi D}\left(\ln R_2 - 1\right) + \frac{3W}{8\pi D} + \frac{C_1}{2} - \frac{C_2}{R_2^2} + \frac{v}{R_2}\left\{\frac{WR_2}{4\pi D}\left(\ln R_2 - 1\right) + \frac{WR_2}{8\pi D} + \frac{C_1 R_2}{2} + \frac{C_2}{R_2}\right\} = 0 \tag{19.28}$$

Solving equations (19.27) and (19.28) for C_1 and C_2,

$$C_1 = \frac{-W}{4\pi D}\left\{\frac{2\left(R_2^2 \ln R_2 - R_1^2 \ln R_1\right)}{\left(R_2^2 - R_1^2\right)} + \frac{(1 - v)}{(1 + v)}\right\} \tag{19.29}$$

and

$$C_2 = \frac{-W}{4\pi D} \times \frac{(1 + v)}{(1 - v)} \times \frac{\left(R_2^2 R_1^2\right)}{\left(R_2^2 - R_1^2\right)} \ln\left(\frac{R_2}{R_1}\right) \tag{19.30}$$

C_3 is not required to determine expressions for M_r and M_t. Hence,

$$M_r = D(W/8\pi D)\left\{(1 + v)2 \ln r + (1 - v)\right\} + \left(C_1/2\right)(1 + v) - \left(C_2/r^2\right)(1 + v) \tag{19.31}$$

and

$$M_t = D(W/8\pi D)\{(1+\nu)2\ln r - (1-\nu)\} + (C_1/2)(1+\nu) + (C_2/r^2)(1-\nu) \qquad (19.32)$$

Problem 19.4 A flat circular plate of radius R_2 is simply-supported concentrically by a tube of radius R_1, as shown in Figure 19.4. If the 'internal' portion of the plate is subjected to a uniform pressure p, show that the central deflection δ of the plate is given by

$$\delta = \frac{pR_1^4}{64D}\left\{3 + 2\left(\frac{R_1}{R_2}\right)^2\left(\frac{1-\nu}{1+\nu}\right)\right\}$$

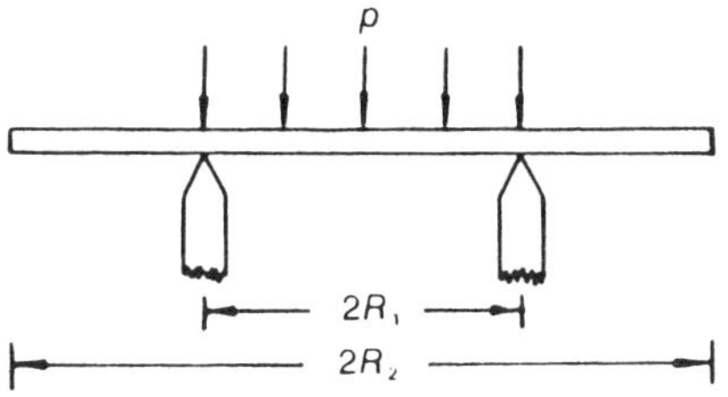

Figure 19.4 Circular plate with a partial pressure load.

<u>Solution</u>

Now the shearing force per unit length F for $r > R_1$ is zero, and for $r < R_1$,

$$F = pr/2$$

so that the plate differential equation becomes

←---------- $r < R_1$ ---------→ ←--- $r > R_1$ ---→

$$\frac{d}{dr}\left\{\frac{1}{r}\frac{d}{dr}\left(r\frac{dw}{dr}\right)\right\} = \frac{pr}{2D} \qquad\qquad = 0$$

$$\frac{1}{r}\frac{d}{dr}\left(r\frac{dw}{dr}\right) = \frac{pr^2}{4D} + A \qquad\qquad = B \qquad (19.33)$$

For continuity at $r = R_1$, the two expressions on the right of equation (19.33) must be equal, i.e.

$$\frac{pR_1^2}{4D} + A = B$$

or

$$B = \frac{pR_1^2}{4D} + A \tag{19.34}$$

or

$$\frac{1}{r}\frac{d}{dr}\left(r\frac{dw}{dr}\right) = \frac{pr^2}{4D} + A \qquad = \frac{pR_1^2}{4D} + A$$

or

$$\frac{d}{dr}\left(r\frac{dw}{dr}\right) = \frac{pr^3}{4D} + Ar \qquad = \frac{pR_1^2 r}{4D} + Ar$$

which on integrating becomes,

$$r\frac{dw}{dr} = \frac{pr^4}{16D} + \frac{Ar^2}{2} + C \qquad = \frac{pR_1^2 r^2}{8D} + \frac{Ar^2}{2} + F$$

$$\frac{dw}{dr} = \frac{pr^3}{16D} + \frac{Ar}{2} + \frac{C}{r} \qquad = \frac{pR_1^2 r}{8D} + \frac{Ar}{2} + \frac{F}{r} \tag{19.35}$$

$$\text{at } r = 0, \quad \frac{dw}{dr} \neq \infty \quad \text{therefore } C = 0$$

For continuity at $r = R_1$, the value of the slope must be the same from both expressions on the right of equation (19.35), i.e.

$$\frac{pR_1^3}{16D} + \frac{AR_1}{2} = \frac{pR_1^3}{8D} + \frac{AR_1}{2} + \frac{F}{R_1}$$

therefore

$$F = -pR_1^4 / (16D) \tag{19.36}$$

therefore

$$\frac{dw}{dr} = \frac{pr^3}{16D} + \frac{Ar}{2} \qquad = \frac{pR_1^2 r}{8D} + \frac{Ar}{2} - \frac{pR_1^4}{16Dr} \tag{19.37}$$

which on integrating becomes

$$w = \frac{pr^4}{64D} + \frac{Ar^2}{4} + G \qquad = \frac{pR_1^2 r^2}{16D} + \frac{Ar^2}{4} - \frac{R_1^4}{16D} \ln r + H \tag{19.38}$$

Now, there are three unknowns in equation (19.38), namely A, G and H, and therefore, three simultaneous equations are required to determine these unknowns. One equation can be obtained by considering the continuity of w at $r = R_1$ in equation (19.38), and the other two equations can be obtained by considering boundary conditions.

One suitable boundary condition is that at $r = R_2$, $M_r = 0$, which can be obtained by considering that portion of the plate where $R_2 > r > R_1$, as follows:

$$\frac{dw}{dr} = \frac{pR_1^2 r}{8D} + \frac{Ar}{2} - \frac{pR_1^4}{16Dr}$$

$$\frac{d^2w}{dr^2} = \frac{pR_1^2}{8D} + \frac{A}{2} + \frac{pR_1^4}{16Dr^2}$$

Now

$$M_r = D\left(\frac{d^2w}{dr^2} + \frac{\nu}{r}\frac{dw}{dr}\right)$$

$$= \left\{\left(\frac{pR_1^2}{8D} + \frac{A}{2} + \frac{pR_1^4}{16Dr^2}\right) + \frac{\nu}{r}\left(\frac{pR_1^2 r}{8D} + \frac{Ar}{2} - \frac{pR_1^4}{16Dr}\right)\right\} \tag{19.39}$$

$$= D\left\{\frac{pR_1^2}{8D}(1 + \nu) + \frac{A}{2}(1 + \nu) + \frac{pR_1^4}{16Dr^2}(1 - \nu)\right\}$$

Now, at $r = R_2$, $M_r = 0$; therefore

$$\frac{A}{2}(1 + \nu) = -\frac{pR_1^2}{8D}(1 + \nu) - \frac{pR_1^4}{16DR_2^2}(1 - \nu)$$

or

$$A = -\frac{pR_1^2}{4D} - \frac{pR_1^4}{8DR_2^2}\left(\frac{1 - \nu}{1 + \nu}\right) \tag{19.40}$$

Another suitable boundary condition is that

$$\text{at } r = R_1, \quad w = 0$$

In this case, it will be necessary to consider only that portion of the plate where $r < R_1$, as follows:

$$w = \frac{pr^4}{64D} + \frac{Ar^2}{4} + G$$

$$\text{at } r = R_1, \quad w = 0$$

Therefore

$$0 = \frac{pR_1^4}{64D} + \frac{AR_1^2}{4} + G$$

or

$$G = -\frac{pR_1^4}{64D} + \frac{pR_1^4}{16D} + \frac{pR_1^6}{32DR_2^2}\left(\frac{1 - \nu}{1 + \nu}\right)$$

$$= \frac{-pR_1^4}{64D} + \left\{\frac{pR_1^2}{4D} + \frac{pR_1^4}{8DR_2^2}\left(\frac{1 - \nu}{1 + \nu}\right)\right\}\frac{R_1^2}{4}$$

or

$$G = \frac{pR_1^4}{64D}\left\{3 + 2\left(\frac{R_1}{R_2}\right)^2\left(\frac{1 - \nu}{1 + \nu}\right)\right\} \tag{19.41}$$

The central deflection δ occurs at $r = 0$; hence, from (19.41),

$$\delta = G$$

$$\delta = \frac{pR_1^4}{64D}\left\{3 + 2\left(\frac{R_1}{R_2}\right)^2\left(\frac{1-\nu}{1+\nu}\right)\right\} \tag{19.42}$$

$$\left\{0.115\,WR^2/(ET^3);\ \frac{W}{t^2}\left[0.621\ln\left(\frac{R}{r}\right) - 0.436 + 0.0224\left(\frac{R}{r}\right)^2\right]\right\}$$

Problem 19.5 A flat circular plate of outer radius R_2 is clamped firmly around its outer circumference. If a load W is applied concentrically to the plate, through a tube of radius R_1, as shown in Figure 19.5, show that the central deflection δ is

$$\delta = \frac{W}{16\pi D}\left\{R_1^2\ln\left(\frac{R_1}{R_2}\right)^2 + \left(R_2^2 - R_1^2\right)\right\}$$

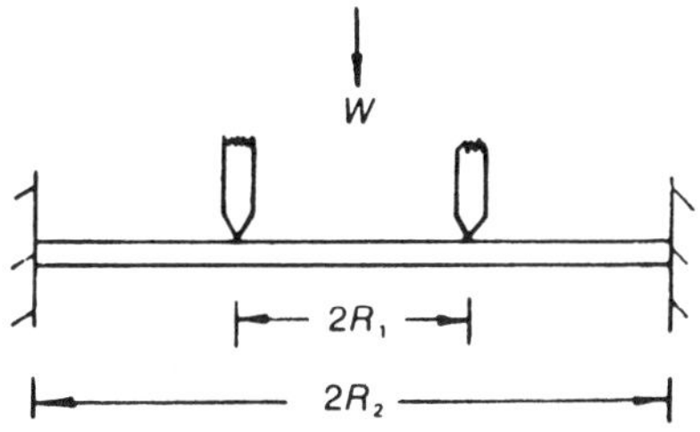

Figure 19.5 Plate under an annular load.

Solution

When $r < R_1$, $F = 0$, and when $R_2 > r > R_1$, $F = W/(2\pi r)$, so that the plate differential equation becomes

$$\longleftarrow r < R_1 \longrightarrow \qquad \longleftarrow r > R_1 \longrightarrow$$

$$\frac{d}{dr}\left\{\frac{1}{r}\,d\left(r\frac{dw}{dr}\right)\right\} = 0 \qquad\qquad = \frac{W}{2\pi D}$$

or

$$\frac{1}{r}\, d\left(r\frac{dw}{dr}\right) = A \qquad\qquad = \frac{W}{2\pi D}\ln r + B$$

or

$$d\left(r\frac{dw}{dr}\right) = Ar \qquad\qquad = \frac{Wr\ln r}{2\pi D} + Br \tag{19.43}$$

From continuity considerations at $r = R_1$, the two expressions on the right of equation (19.43) must be equal, i.e.

$$A = \frac{W}{2\pi D}\ln R_1 + B \tag{19.44}$$

On integrating equation (19.43),

$$r\frac{dw}{dr} = \frac{Ar^2}{2} + C \qquad\qquad \frac{W}{2\pi D}\left(\frac{r^2}{2}\ln r - \frac{r^2}{4}\right) + \frac{Br^2}{2} + F$$

or

$$\frac{dw}{dr} = \frac{Ar}{2} + \frac{C}{r} \qquad\qquad \frac{Wr}{4\pi D}\left(\ln r - \frac{r}{2}\right) + \frac{Br}{2} + \frac{F}{r} \tag{19.45}$$

$$\text{at } r = 0, \quad \frac{dw}{dr} \neq \infty \text{ therefore } C = 0$$

From continuity considerations for dw/dr, at $r = R_1$,

$$\frac{AR_1}{2} = \frac{WR_1}{4\pi D}\left(\ln R_1 - \frac{R_1}{2}\right) + \frac{BR_1}{2} + \frac{F}{R_1} \tag{19.46}$$

On integrating equation (19.46)

$$w = \frac{Ar^2}{2} + G \qquad\qquad = \frac{W}{2\pi D}\left(\frac{r^2}{4}\ln r - \frac{r^2}{8} - \frac{r^2}{8}\right) + \frac{Br^2}{4} + F\ln r + H$$

or

$$w = \frac{Ar^2}{2} + G \qquad\qquad \frac{Wr^2}{8\pi D}\left(\ln r - 1\right) + \frac{Br^2}{4} + F\ln r + H \tag{19.47}$$

From continuity considerations for w, at $r = R_1$,

$$\frac{Ar_1^2}{2} + G = \frac{WR_1^2}{8\pi D}\left(\ln R_1 - 1\right) + \frac{BR_1^2}{4} + F \ln R_1 + H \tag{19.48}$$

In order to obtain the necessary number of simultaneous equations to determine the arbitrary constants, it will be necessary to consider *boundary considerations*.

$$\text{at } r = R_2, \quad \frac{dw}{dr} = 0$$

therefore

$$0 = \frac{WR_2}{4\pi D}\left(\ln R_2 - \frac{R_2}{2}\right) + \frac{BR_2}{2} + \frac{F}{R_2} \tag{19.49}$$

Also, at $r = R_2$, $w = 0$; therefore

$$0 = \frac{WR_2^2}{8\pi D}\left(\ln R_2 - 1\right) + \frac{BR_2^2}{4} + F \ln\left(R_2\right) + H \tag{19.50}$$

Solving equations (19.46), (19.48), (19.49) and (19.50),

$$F = \frac{WR_1^2}{8\pi D} \tag{19.51}$$

$$H = -\frac{W}{8\pi D}\left\{-R_2^2/2 - R_1^2/2 + R_1^2 \ln\left(R_2\right)\right\}$$

and

$$G = -\frac{WR_1^2}{8\pi D} + \frac{WR_1^2}{8\pi D}\ln(R_1) + H$$

$$= -WR_1^2 + \frac{WR_1^2\ln(R_1)}{8\pi D} - \frac{W}{8\pi D}\left\{-\frac{R_2^2}{2} - \frac{R_1^2}{2} + R_1^2\ln(R_2)\right\}$$

$$= \frac{W}{16\pi D}\left\{-2R_1^2 + 2R_1^2\ln(R_1) + R_2^2 + R_1^2 - 2R_1^2\ln(R_2)\right\}$$

$$G = \frac{W}{16\pi D}\left\{R_1^2\ln\left(\frac{R_1}{R_2}\right)^2 + \left(R_2^2 - R_1^2\right)\right\}$$

δ occurs at $r = 0$, i.e.

$$\delta = G = \frac{W}{16\pi D}\left\{R_1^2\ln\left(\frac{R_1}{R_2}\right)^2 + \left(R_2^2 - R_1^2\right)\right\}$$

19.3 Large deflections of plates

If the maximum deflection of a plate exceeds half the plate thickness, the plate changes to a shallow shell, and withstands much of the lateral load as a membrane, rather than as a flexural structure.

For example, consider the membrane shown in Figure 19.6, which is subjected to uniform lateral pressure p.

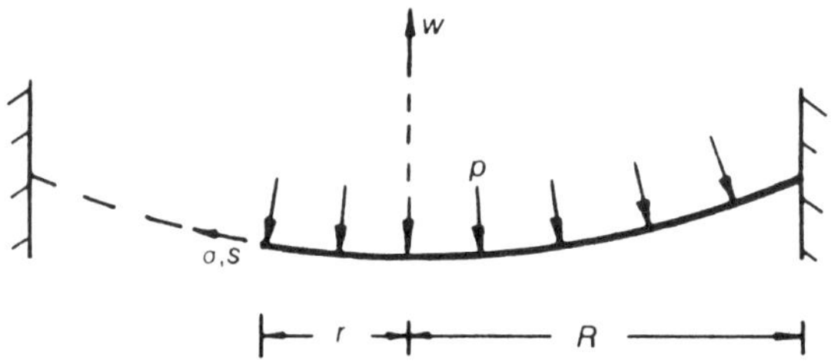

Figure 19.6 Portion of circular membrane.

Let

w = out-of-plane deflection at any radius r

σ = membrane tension at a radius r

t = thickness of membrane

Resolving vertically,

$$\sigma \times t \times 2\pi r \times \frac{dw}{dr} = p \times \pi r^2$$

or

$$\frac{dw}{dr} = \frac{pr}{2\sigma t} \tag{19.52}$$

or

$$w = \frac{pr^2}{4\sigma t} + A$$

at $r = R, \; w = 0$; therefore

$$A = \frac{-pR^2}{4\sigma t}$$

i.e.

$\hat{w}$ = maximum deflection of membrane

$$\hat{w} = -pR^2/(4\sigma t)$$

The change of meridional (or radial) length is given by

$$\delta l = \int ds - \int dr$$

where s is any length along the meridian

Using Pythagoras' theorem,

$$\delta l = \int \left(dw^2 + dr^2\right)^{1/2} - \int dr$$

$$= \int \left[1 + \left(\frac{dw}{dr}\right)^2\right]^{1/2} - \int dr$$

Expanding binomially and neglecting higher order terms,

$$\begin{aligned} \delta l &= \int\left[1 + \frac{1}{2}\left(\frac{dw}{dr}\right)^2\right]dr - \int dr \\ &= \frac{1}{2}\int\left(\frac{dw}{dr}\right)^2 dr \end{aligned} \qquad (19.53)$$

Substituting the derivative of w, namely equation (19.52) into equation (19.53),

$$\begin{aligned} \delta l &= \frac{1}{2}\int_0^R \left(\frac{pr}{2\sigma t}\right)^2 dr \\ &= p^2 R^3/(24\sigma^2 t^2) \end{aligned} \qquad (19.54)$$

but

$$\varepsilon = \text{strain} = \frac{\delta l}{R} = \frac{1}{E}(\sigma - \nu\sigma)$$

or

$$\sigma^3 = \frac{E}{(1 - \nu)}\left(\frac{p^2 R^2}{24\sigma^2 t^2}\right)$$

i.e.

$$\sigma = \sqrt[3]{\left\{\frac{E}{1 - \nu}\left(\frac{p^2 R^2}{24 t^2}\right)\right\}} \qquad (19.55)$$

but

$$\sigma = pR^2/(4t\hat{w}) \qquad (19.56)$$

From equations (19.55) and (19.56),

$$p = \frac{8E}{3(1 - \nu)}\left(\frac{t}{R}\right)\left(\frac{\hat{w}}{R}\right)^3 \qquad (19.57)$$

According to small deflection theory of plates (19.23)

$$p = \frac{64D}{R^3}\left(\frac{\hat{w}}{R}\right) \qquad (19.58)$$

Thus, for the large deflections of clamped circular plates under lateral pressure, equations (19.57) and (19.58) should be added together, as follows:

$$p = \frac{64D}{R^3}\left(\frac{\hat{w}}{R}\right) + \frac{8}{3(1 - \nu)}\left(\frac{t}{R}\right)\left(\frac{\hat{w}}{R}\right)^3 \tag{19.59}$$

If $\nu = 0.3$, then (19.59) becomes

$$\frac{pR^4}{64Dt} = \left(\frac{\hat{w}}{t}\right)\left\{1 + 0.65\left(\frac{\hat{w}}{t}\right)\right\} \tag{19.60}$$

where the second term in (19.60) represents the membrane effect, and the first term represents the flexural effect.

When $\hat{w}/t = 0.5$, the membrane effect is about 16.3% of the bending effect, but when $\hat{w}/t = 1$, the membrane effect becomes about 65% of the bending effect. The bending and membrane effects are about the same when $\hat{w}/t = 1.24$. A plot of the variation of $\hat{w}$ due to bending and due to the combined effects of bending plus membrane stresses, is shown in Figure 19.7.

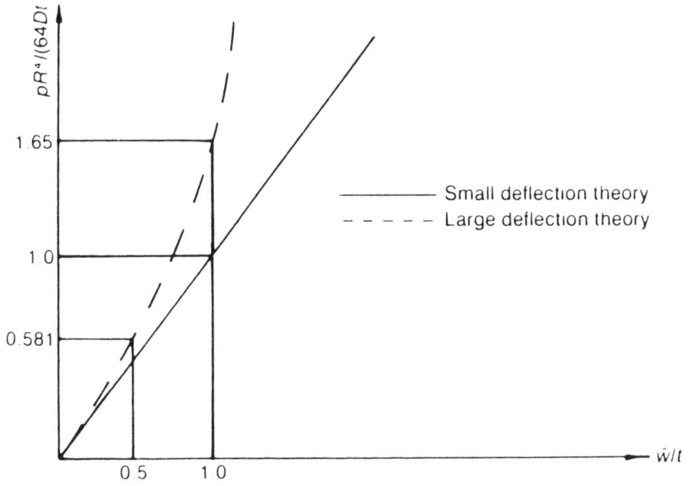

Figure 19.7 Small and large deflection theory.

19.3.1 Power series solution

This method of solution, which involves the use of data sheets, is based on a power series solution of the fundamental equations governing the large deflection theory of circular plates.

For a circular plate under a uniform lateral pressure p, the large deflection equations are given by (19.61) to (19.63).

$$D\frac{d}{dr}\left\{\frac{1}{r}\frac{d}{dr}\left(r\frac{dw}{dr}\right)\right\} = \sigma_r\frac{dw}{dr} + \frac{pr}{2} \tag{19.61}$$

$$\frac{d}{dr}\left(r\sigma_r\right) - \sigma_t = 0 \tag{19.62}$$

$$r\frac{d}{dr}\left(\sigma_r + \sigma_t\right) + \frac{E}{2}\left(\frac{dw}{dr}\right)^2 = 0 \tag{19.63}$$

Way[5] has shown that to assist in the solution of equations (19.61) to (19.63), by the power series method, it will be convenient to introduce the dimensionless ratio ζ, where

$$\zeta = r/R$$

or

$$r = \zeta R$$

$$R = \text{outer radius of disc}$$

$$r = \text{any value of radius between 0 and } R$$

Substituting for r int (19.61):

$$\frac{1}{12\left(1 - v^2\right)}\frac{d}{d(\zeta R)}\left\{\frac{1}{\zeta R} \times \frac{d(\zeta R\theta)}{d(\zeta R)}\right\} = \frac{\sigma_r\theta}{Et^2} + \frac{p\zeta R}{2Et^3}$$

or

$$\frac{1}{12\left(1 - v^2\right)}\frac{d}{d\zeta}\left\{\frac{1}{\zeta} \times \frac{d(\zeta\theta)}{d\zeta}\right\} = \frac{\sigma_r R^2\theta}{Et^2} + \frac{pR^3\zeta}{2Et^3} \tag{19.64}$$

Inspecting (19.64), it can be seen that the LHS is dependent only on the slope θ.

Now

$$\theta = \frac{dw}{dr} = \frac{dw}{d(\zeta R)}$$

---

[5] Way, S., *Bending of circular plates with large deflections*, *A.S.M.E.*, *APM-56-12*, 56, 1934.

which, on substituting into (19.64), gives:

$$\frac{1}{12\left(1-\nu^2\right)}\frac{d}{d\zeta}\left\{\frac{1}{\zeta}\frac{d}{d\zeta}\left[\frac{\zeta d(w/t)}{d\zeta}\right]\right\} = \frac{\sigma_r}{E}\left(\frac{R}{t}\right)^2\frac{d(w/t)}{d\zeta}+\frac{P}{E}\left(\frac{R}{t}\right)^4\frac{\zeta}{2} \tag{19.65}$$

but

$$\left(\frac{w}{t}\right) \quad \frac{\sigma_r}{E}\left(\frac{R}{t}\right)^2 \quad \text{and} \quad \frac{p}{E}\left(\frac{R}{t}\right)^4$$

are all dimensionless, and this feature will be used later on in the present chapter.

Substituting *r*, in terms of ζ into equation (19.62), equation (19.66) is obtained:

$$\frac{d}{d\zeta}\left\{\frac{\zeta\sigma_r}{E}\left(\frac{R}{t}\right)^2\right\} = \frac{\sigma_t}{E}\left(\frac{R}{t}\right)^2 \tag{19.66}$$

Similarly, substituting *r* in terms of ζ equation (19.63), equation (19.67) is obtained:

$$\zeta\frac{d}{d\zeta}\left\{\frac{\sigma_r}{E}\left(\frac{R}{t}\right)^2+\frac{\sigma_t}{E}\left(\frac{R}{t}\right)^2\right\}+\frac{E}{2}\theta^2 = 0 \tag{19.67}$$

Equation (19.67) can be seen to be dependent only on the deflected form of the plate.

The fundamental equations, which now appear as equations (19.65) to (19.67), can be put into dimensionless form by introducing the following dimensionless variables:

$$X = r/t = \zeta R/t$$

$$W = w/t$$

$$U = u/t$$

$$M_r' = M_r t/D$$

$$S_r = \sigma_r/E$$

$$S_t = \sigma_t/E$$

$$S_r' = \sigma_r'/E$$

$$S_t' = \sigma_t'/E$$

$$S_p = p/E \tag{19.68}$$

$$\theta = \frac{dw}{dr} = \frac{dW}{dX} \tag{19.69}$$

or

$$W = \int \theta dX \tag{19.70}$$

Now from standard circular plate theory,

$$\sigma_r^{/} = \frac{6D}{t^2}\left(\frac{d\theta}{dr} + \frac{v\theta}{r}\right)$$

and

$$\sigma_t^{/} = \frac{6D}{t^2}\left(\frac{\theta}{r} + v\frac{d\theta}{dr}\right)$$

Hence,

$$S_r^{/} = \frac{1}{2\left(1 - v^2\right)}\left(\frac{d\theta}{dX} + \frac{v\theta}{X}\right) \tag{19.71}$$

and

$$S_t^{/} = \frac{1}{2\left(1 - v^2\right)}\left(\frac{\theta}{X} + v\frac{d\theta}{dX}\right) \tag{19.72}$$

Now from elementary two-dimensional stress theory,

$$\frac{uE}{r} = \sigma_t - v\sigma_r$$

or

$$U = X\left(S_t - vS_r\right) \tag{19.73}$$

where u is the in-plane radial deflection at r.

Substituting equations (19.68) to (19.73) into equations (19.65) to (19.67), the fundamental equations take the form of equations (19.74) to (19.76):

$$\frac{1}{12(1-\nu^2)}\frac{d}{dX}\left\{\frac{1}{X}\frac{d(X\theta)}{dX}\right\} = S_p\frac{X}{2} + S_r\theta \tag{19.74}$$

$$\frac{d(XS_r)}{dX} - S_t = 0 \tag{19.75}$$

$$X\frac{d}{dX}(S_r + S_t) + \frac{\theta^2}{2} = 0 \tag{19.76}$$

Solution of equations (19.74) to (19.76) can be achieved through a power series solution.

Now S_r is a symmetrical function, i.e. $S_r(X) = S_r(-X)$, so that it can be approximated in an even series powers of X.

Furthermore, as θ is antisymmetrical, i.e. $\theta(X) = -\theta(X)$, it can be expanded in an odd series power of X. Let

$$S_r = B_1 + B_2X^2 + B_3X^4 + \ldots$$

and

$$\theta = C_1X + C_2X^3 + C_3X^5 + \ldots$$

or

$$S_r = \sum_{i=1}^{\infty} B_i X^{2i-2} \tag{19.77}$$

and

$$\theta = \sum_{i=1}^{\infty} C_i X^{2i-1} \tag{19.78}$$

Now from equation (19.75)

$$S_t = \frac{d(XS_r)}{dX} = \sum_{i=1}^{\infty} (2i-1)\,B_i X^{2i-2} \tag{19.79}$$

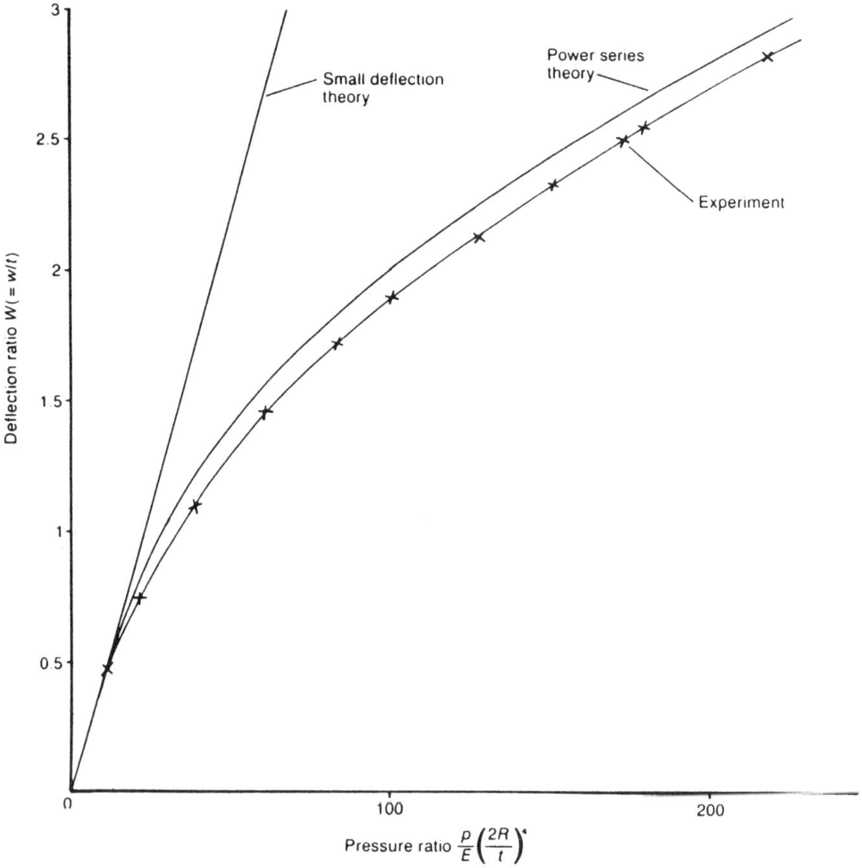

Figure 19.8 Central deflection versus pressure for a simply-supported plate.

$$W = \int \theta \, dX = \sum_{i=1}^{\infty} \left(\frac{1}{2i}\right) C_i X^{2i} \tag{19.80}$$

Hence

$$S_r' = \sum_{i=1}^{\infty} \frac{(2i + \nu - 1)}{2(1 - \nu^2)} C_i X^{2i-2} \tag{19.81}$$

$$S_t' = \sum_{i=1}^{\infty} \frac{\{1 + \nu(2i - 1)\}}{2(1 - \nu^2)} C_i X^{2i-2} \tag{19.82}$$

Now

$$U = X(S_t - vS_r)$$
$$= \sum_{i=1}^{\infty} (2i - 1 - v)B_i X^{2i-1} \tag{19.83}$$

for $i = 1, 2, 3, 4 \rightarrow \infty$.

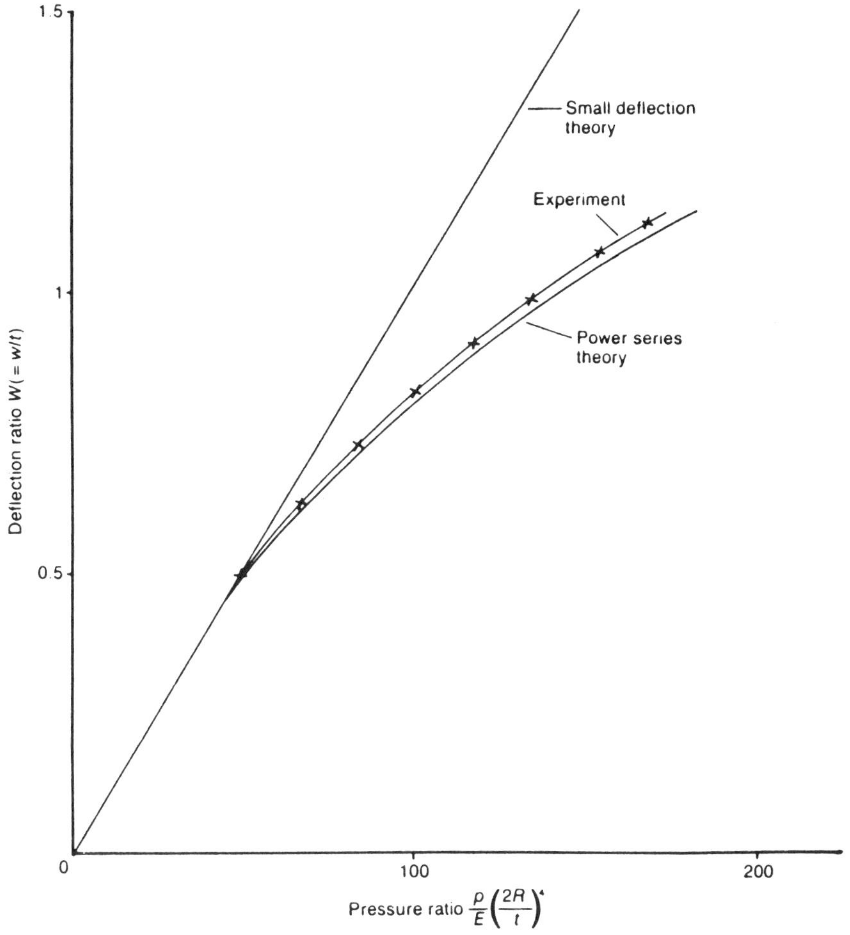

Figure 19.9 Central deflection versus pressure for an encastré plate.

From equations (19.77) to (19.83), it can be seen that if B_1 and C_1 are known all quantities of interest can readily be determined.

Way has shown that

$$B_k = \frac{-\sum_{m=1}^{k-1} C_m C_{k-m}}{8k(k-1)}$$

for $k = 2, 3, 4$ etc. and

$$C_k = \frac{3(1-v^2)}{k(k-1)} \sum_{m=1}^{k-1} B_m C_{k-m}$$

for $k = 3, 4, 5$ etc. and

$$C_2 \quad \frac{3(1-v^2)}{2}\left(\frac{S_p}{2} + B_1 C_1\right)$$

Once B_1 and C_1 are known, the other constants can be found. In fact, using this approach, Hewitt and Tannent[6] have produced a set of curves which under uniform lateral pressure, as shown in Figures 19.8 to 19.12. Hewitt and Tannent have also compared experiment and small deflection theory with these curves.

19.4 Shear deflections of very thick plates

If a plate is very thick, so that membrane effects are insignificant, then it is possible that shear deflections can become important.

For such cases, the bending effects and shear effects must be added together, as shown by equation (19.84), which is rather similar to the method used for beams in Chapter 13,

$$\delta = \delta_{bending} + \delta_{shear}$$

which for a plate under uniform pressure p is

$$\delta = pR\left\{k_1\left(\frac{R}{t}\right)^3 + k_2\left(\frac{t}{R}\right)^2\right\} \qquad (19.84)$$

where k_1 and k_2 are constants.

From equations (19.84), it can be seen that δ_{shear} becomes important for large values of (t/R).

[6]Hewitt D A, Tannent J O, *Large deflections of circular plates*, Portsmouth Polytechnic Report M195, 1973-74.

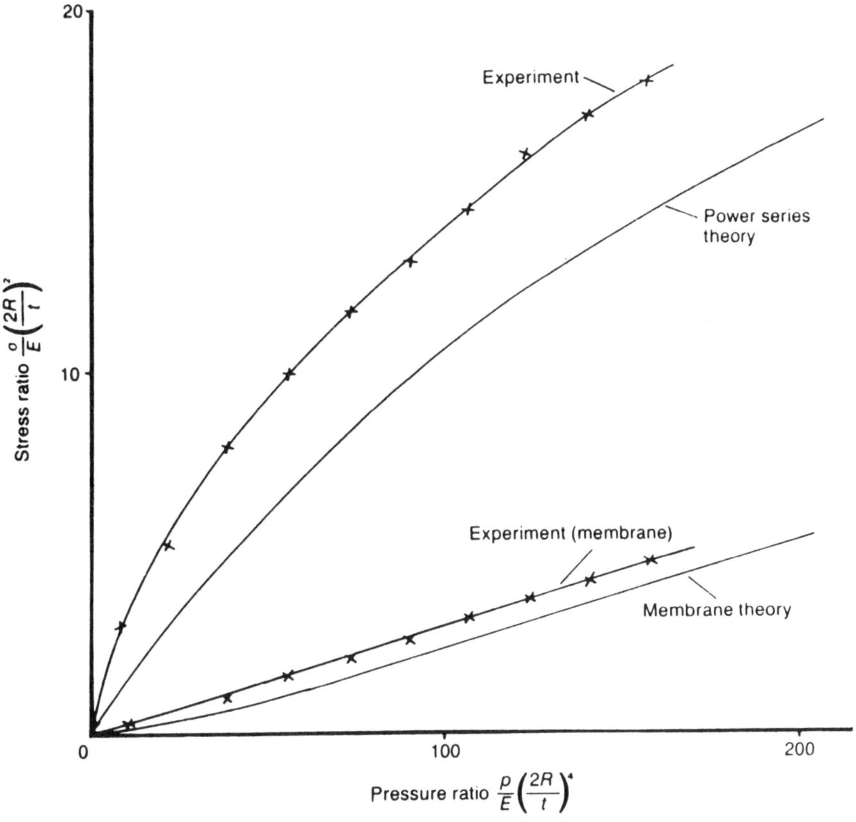

Figure 19.10 Central stress versus pressure for an encastré plate.

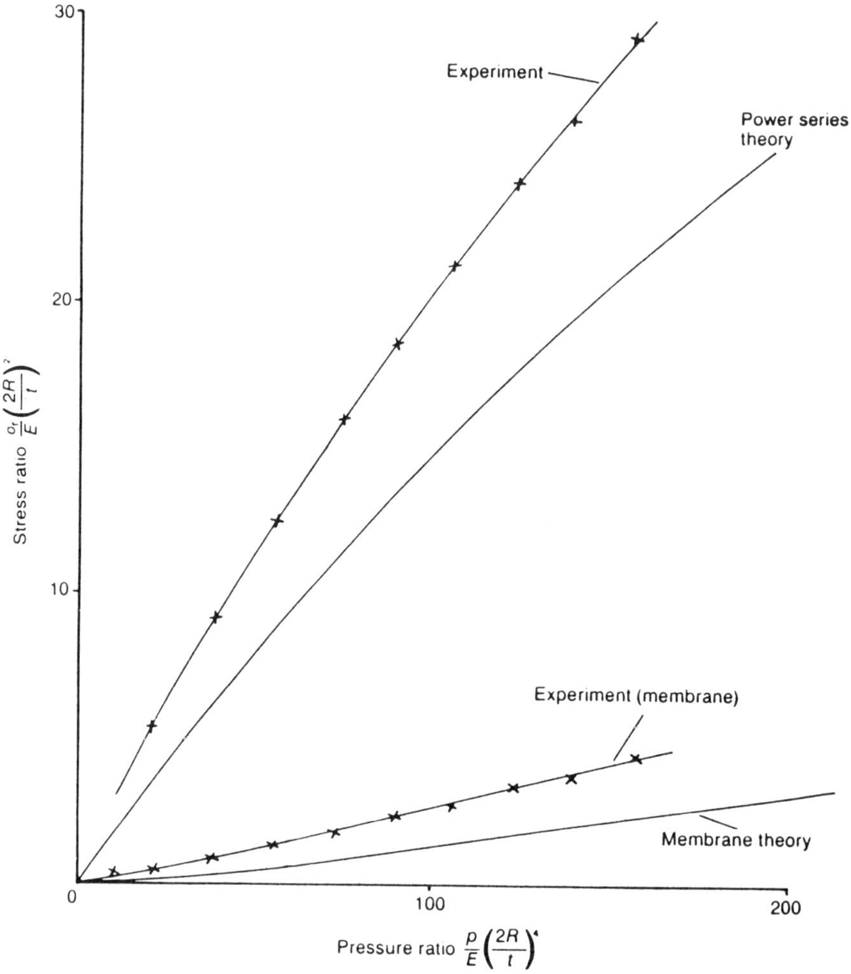

Figure 19.11 Radial stresses near edge versus pressure for an encastré plate.

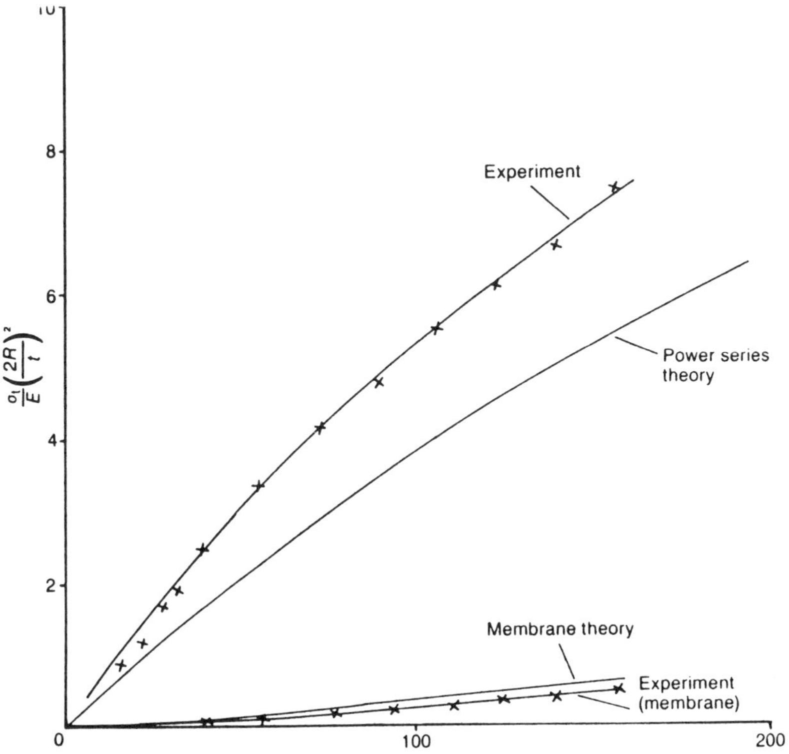

Figure 19.12 Circumferential stresses versus pressure near edge for an encastré plate.

Further problems *(answers on page 694)*

19.6 Determine an expression for the deflection of a circular plate of radius R, simply-supported around its edges, and subjected to a centrally placed concentrated load $\boldsymbol{W}$.

19.7 Determine expressions for the deflection and circumferential bending moments for a circular plate of radius R, simply-supported around its edges and subjected to a uniform pressure p.

19.8 Determine an expression for the maximum deflection of a simply-supported circular plate, subjected to the loading shown in Figure 19.13.

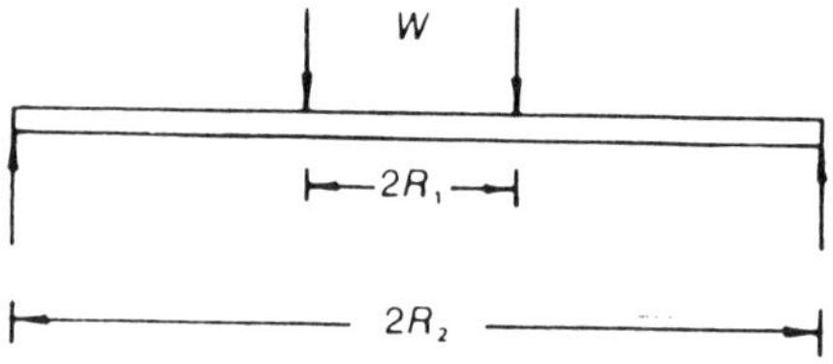

Figure 19.13 Simply-supported plate.

19.9 Determine expressions for the maximum deflection and bending moments for the concentrically loaded circular plates of Figure 19.14(a) and (b).

(a) Simply supported.

(b) Clamped.

Figure 19.14 Problem 19.9

19.10 A flat circular plate of radius R is firmly clamped around its boundary. The plate has stepped variation in its thickness, where the thickness inside a radius of $(R/5)$ is so large that its flexural stiffness may be considered to approach infinity. When the plate is subjected to a pressure p over its entire surface, determine the maximum central deflection and the maximum surface stress at any radius r. $\nu = 0.3$.

19.11 If the loading of Example 19.9 were replaced by a centrally applied concentrated load W, determine expressions for the central deflection and the maximum surface stress at any radius r.

20 Torsion of non-circular sections

20.1 Introduction

The torsional theory of circular sections (Chapter 16) cannot be applied to the torsion of non-circular sections, as the shear stresses for non-circular sections are no longer circumferential. Furthermore, plane cross-sections do not remain plane and undistorted on the application of torque, and in fact, warping of the cross-section takes place.

As a result of this behaviour, the polar second moment of area of the section is no longer applicable for static stress analysis, and it has to be replaced by a torsional constant, whose magnitude is very often a small fraction of the magnitude of the polar second moment of area.

20.2 To determine the torsional equation

Consider a prismatic bar of uniform non-circular section, subjected to twisting action, as shown in Figure 20.1.

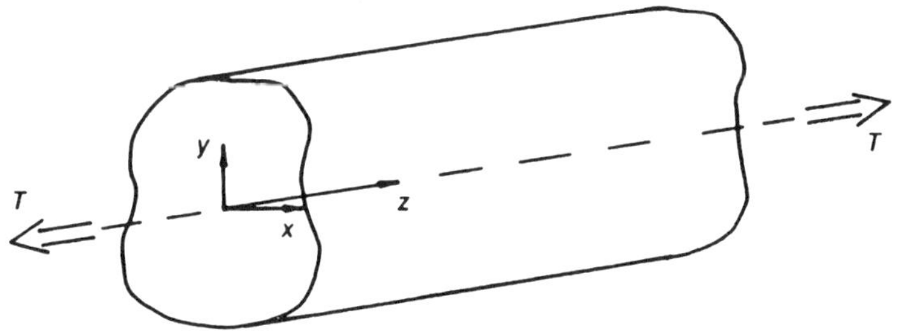

Figure 20.1 Non-circular section under twist.

Let,

T	=	torque
u	=	displacement in the x direction
v	=	displacement in the y direction
w	=	displacement in the z direction
	=	the warping function
θ	=	rotation / unit length
x, y, z	=	Cartesian co-ordinates

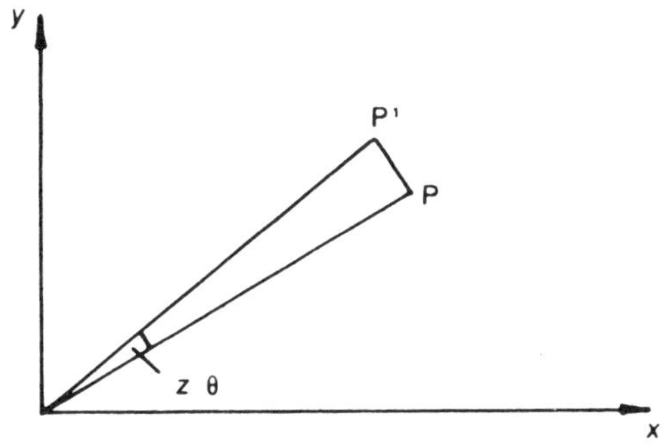

Figure 20.2 Displacement of *P*.

Consider any point *P* in the section, which, owing to the application of *T*, will rotate and warp, as shown in Figure 20.2:

$$u = -yz\theta$$
$$v = xz\theta \qquad (20.1)$$

due to rotation, and

$$w = \theta \times \psi(x, y))$$
$$= \theta \times \psi \qquad (20.2)$$

due to warping. The theory assumes that,

$$\varepsilon_x = \varepsilon_y = \varepsilon_z = \gamma_{xy} = 0 \qquad (20.3)$$

and therefore the only shearing strains that exist are γ_{xz} and γ_{yz}, which are defined as follows:

γ_{xz} = shear strain in the $x-z$ plane

$$= \frac{\partial w}{\partial x} + \frac{\partial u}{\partial z} = \theta\left(\frac{\partial \psi}{\partial x} - y\right) \qquad (20.4)$$

$$\gamma_{yz} = \text{shear strain in the } y\text{-}z \text{ plane}$$

$$= \frac{\partial w}{\partial y} + \frac{\partial v}{\partial z} = \theta\left(\frac{\partial \psi}{\partial y} + x\right) \tag{20.5}$$

The equations of equilibrium of an infinitesimal element of dimensions $dx \times dy \times dz$ can be obtained with the aid of Figure 20.3, where,

$$\tau_{xz} = \tau_{zx}$$

and

$$\tau_{yz} = \tau_{zy}$$

Resolving in the z-direction

$$\frac{\partial \tau_{yz}}{\partial y} \times dy \times dx \times dz + \frac{\partial \tau_{xz}}{\partial x} \times dx \times dy \times dz = 0$$

or

$$\frac{\partial \tau_{xz}}{\partial x} + \frac{\partial \tau_{yz}}{\partial y} = 0 \tag{20.6}$$

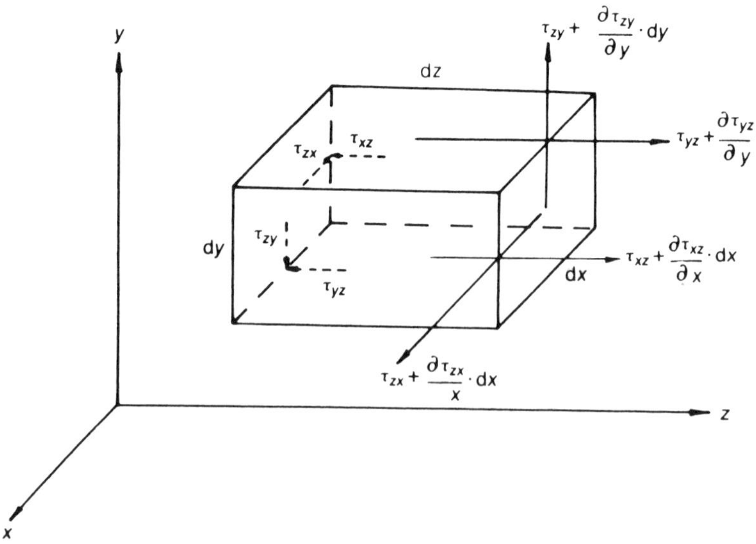

Figure 20.3 Shearing stresses acting on an element.

However, from equations (20.4) and (20.5):

$$\tau_{xz} = G\gamma_{xz} = G\theta\left(\frac{\partial\psi}{\partial x} - y\right) \tag{20.7}$$

and

$$\tau_{yz} = G\gamma_{yz} = G\theta\left(\frac{\partial\psi}{\partial y} + x\right) \tag{20.8}$$

Let,

$$\frac{\partial\chi}{\partial y} = \frac{\partial\psi}{\partial x} - y \tag{20.9}$$

and,

$$-\frac{\partial\chi}{\partial x} = \frac{\partial\psi}{\partial y} + x \tag{20.10}$$

where χ is a shear stress function.

By differentiating equations (20.9) and (20.10) with respect to y and x, respectively, the following is obtained:

$$\frac{\partial_1^2\chi}{\partial x^2} + \frac{\partial_1^2\chi}{\partial y^2} = \frac{\partial^2\Psi}{\partial x\,.\,\partial y} - 1 - \frac{\partial^2\Psi}{\partial x\,.\,\partial y} - 1$$

or,

$$\frac{\partial_1^2\chi}{\partial x^2} + \frac{\partial_1^2\chi}{\partial y^2} = -2 \tag{20.11}$$

Equation (20.11) can be described as the *torsion equation for non-circular sections*.

From equations (20.7) and (20.8):

$$\tau_{xz} = G\theta\frac{\partial\chi}{\partial y} \tag{20.12}$$

and

$$\tau_{yz} = -G\theta\frac{\partial\chi}{\partial x} \tag{20.13}$$

Equation (20.11), which is known as Poisson's equation, can be put into the alternative form of equation (20.14), which is known as Laplace's equation.

$$\frac{\partial^2 \psi}{\partial x^2} + \frac{\partial^2 \psi}{\partial y^2} = 0 \tag{20.14}$$

20.3 To determine expressions for the shear stress τ and the torque *T*

Consider the non-circular cross-section of Figure 20.4.

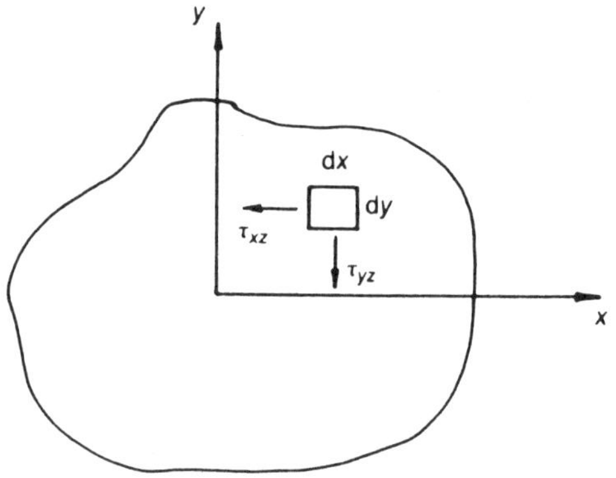

Figure 20.4 Shearing stresses acting on an element.

From Pythagoras' theorem

τ = shearing stress at any point (x, y) on the cross-section

$$= \sqrt{\left(\tau_{xz}^2 + \tau_{yz}^2\right)} \tag{20.15}$$

From Figure 20.4, the torque is

$$T = \iint \left(\tau_{xz} \times y - \tau_{yz} \times x\right) dx \, . \, dy \tag{20.16}$$

To determine the *boundary value for χ*, consider an element on the boundary of the section, as shown in Figure 20.5, where the shear stress acts tangentially. Now, as the shear stress perpendicular to the boundary is zero,

$$\tau_{yz} \sin\varphi + \tau_{xz} \cos\varphi = 0$$

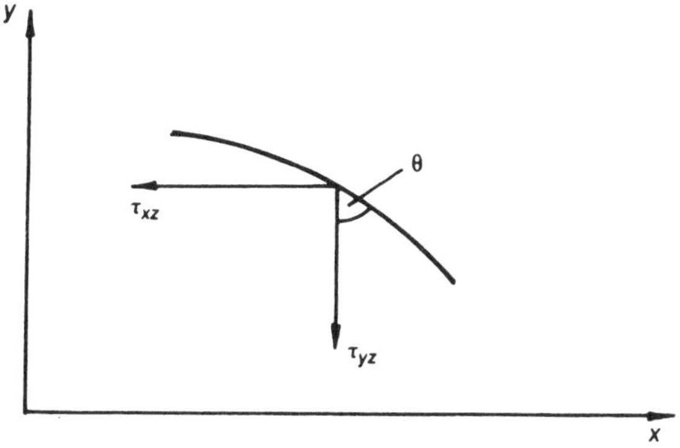

Figure 20.5 Shearing stresses on boundary.

or

$$-G\theta \times \frac{\partial \chi}{\partial x}\left(-\frac{dx}{ds}\right) + G\theta \times \frac{\partial \chi}{\partial y}\left(\frac{dy}{ds}\right) = 0$$

or

$$G\theta \frac{d\chi}{ds} = 0$$

where s is any distance along the boundary, i.e. χ is a constant along the boundary.

Problem 20.1 Determine the shear stress function χ for an elliptical section, and hence, or otherwise, determine expressions for the torque T, the warping function w and the torsional constant J.

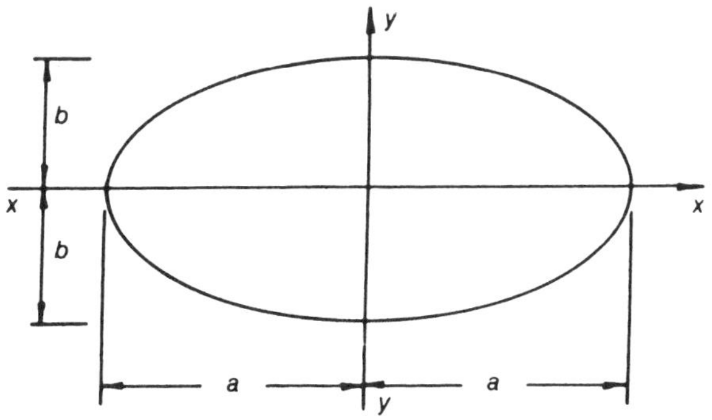

Figure 20.6 Elliptical section.

<u>*Solution*</u>

The equation for the ellipse of Figure 20.6 is given by

$$\frac{x^2}{a^2} + \frac{y^2}{b^2} = 1 \tag{20.17}$$

and this equation can be used for determining the shear stress function χ as follows:

$$\chi = C\left(\frac{x^2}{a^2} + \frac{y^2}{b^2} - 1\right) \tag{20.18}$$

where C is a constant, to be determined.

Equation (20.18) ensures that χ is constant along the boundary, as required. The constant C can be determined by substituting equation (20.18) into (20.11), i.e.

$$C\left(\frac{2}{a^2} + \frac{2}{b^2}\right) = -2$$

therefore

$$C = \frac{-a^2 b^2}{a^2 + b^2}$$

and

$$\chi = \frac{a^2 b^2}{(a^2 + b^2)}\left(1 - \frac{x^2}{a^2} - \frac{y^2}{b^2}\right) \tag{20.19}$$

where χ is the required stress function for the elliptical section.

Now,

$$\tau_{xz} = G\theta\frac{\partial\chi}{\partial y} = -G\theta\frac{2ya^2}{a^2 + b^2}$$

$$\tau_{yz} = -G\theta\frac{\partial\chi}{\partial x} = \frac{G\theta \;\; 2xb^2}{a^2 + b^2}$$

and

$$T = \int (\tau_{xz} y - \tau_{yz} x)\, dA$$

$$= -G\theta \int \left(\frac{2x^2 b^2}{a^2 + b^2} + \frac{2y^2 a^2}{a^2 + b^2} \right) dA$$

$$= -2G\theta \frac{a^2 b^2}{a^2 + b^2} \left[\int \frac{x^2}{a^2}\, dA + \int \frac{y^2}{b^2}\, dA \right]$$

but

$$\int y^2\, dA = I_{xx} = \frac{\pi a b^3}{4} = \text{second moment of area about } x\text{–}x$$

and,

$$\int x^2\, dA = I_{yy} = \frac{\pi a^3 b}{4} = \text{second moment of area about } y\text{–}y$$

therefore

$$T = -2G\theta \frac{a^2 b^2}{a^2 + b^2} \left(\frac{\pi a b}{4} + \frac{\pi a b}{4} \right)$$

$$T = \frac{-G\theta \pi a^3 b^3}{a^2 + b^2} \tag{20.20}$$

therefore

$$\tau_{xz} = \frac{-2a^2 y}{(a^2 + b^2)} \cdot \frac{-(a^2 + b^2)T}{\pi a^3 b^3}$$

$$\tau_{xz} = \frac{2Ty}{\pi a b^3} \tag{20.21}$$

$$T_{yz} = \frac{-2Tx}{\pi a^3 b} \tag{20.22}$$

By inspection, it can be seen that $\hat{\tau}$ is obtained by substituting $y = b$ into (20.21), provided $a > b$.

$$\hat{\tau} = \text{maximum shear stress}$$

$$= \frac{2T}{\pi a b^2} \tag{20.23}$$

and occurs at the extremities of the minor axis.

The warping function can be obtained from equation (20.2). Now,

$$\frac{\partial \chi}{\partial y} = \frac{\partial \psi}{\partial x} - y$$

or

$$\frac{2ya^2b^2}{(a^2 + b^2)b^2} = \frac{\partial \psi}{\partial x} - y$$

i.e.

$$\frac{\partial \psi}{\partial x} = \frac{(-2a^2 + a^2 + b^2)}{(a^2 + b^2)}y$$

therefore

$$\psi = \left(\frac{b^2 - a^2}{a^2 + b^2}\right) xy \tag{20.24}$$

Similarly, from the expression

$$-\frac{\partial \chi}{\partial x} = \frac{\partial \psi}{\partial y} + x$$

the same equation for ψ, namely equation (20.24), can be obtained. Now,

$$w = \text{warping function}$$

$$= \theta \times \psi$$

therefore

$$w = \frac{(b^2 - a^2)}{(a^2 + b^2)} \theta xy \tag{20.25}$$

From simple torsion theory,

$$\frac{T}{J} = G\theta \tag{20.26}$$

or

$$T = G\theta J \tag{20.27}$$

Equating (20.20) and (20.27), and ignoring the negative sign in (20.20),

$$G\theta J = \frac{G\theta\pi a^3 b^3}{(a^2 + b^2)}$$

therefore

J = torsional constant for an elliptical section

$$J = \frac{\pi a^3 b^3}{(a^2 + b^2)} \tag{20.28}$$

Problem 20.2 Determine the shear stress function χ and the value of the maximum shear stress $\hat{\tau}$ for the equilateral triangle of Figure 20.7.

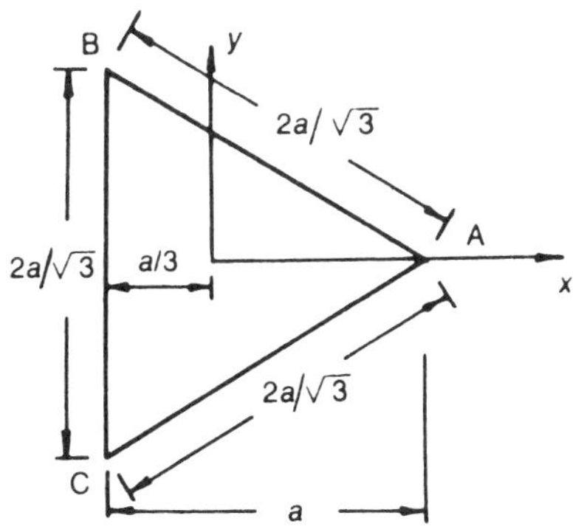

Figure 20.7 Equilateral triangle.

Solution

The equations of the three straight lines representing the boundary can be used for determining χ, as it is necessary for χ to be a constant along the boundary.

Side BC
This side can be represented by the expression

$$x = -\frac{a}{3} \quad \text{or} \quad x + \frac{a}{3} = 0 \tag{20.29}$$

Side AC
This side can be represented by the expression

$$x - \sqrt{3}\, y - \frac{2a}{3} = 0 \tag{20.30}$$

Side AB
This side can be represented by the expression

$$x + \sqrt{3}\, y - \frac{2a}{3} = 0 \tag{20.31}$$

The stress function χ can be obtained by multiplying together equations (20.29) to (20.31):

$$\begin{aligned} \chi &= C(x + a/3) \times \left(x - \sqrt{3}\, y - 2a/3\right) \times \left(x + \sqrt{3}\, y - 2a/3\right) \\ &= C\left\{\left(x^3 - 3xy^2\right) - a\left(x^2 + y^2\right) + 4a^3/27\right\} \end{aligned} \tag{20.32}$$

From equation (20.32), it can be seen that $\chi = 0$ (i.e. constant) along the external boundary, so that the boundary condition is satisfied.

Substituting χ into equation (20.11),

$$C(6x - 2a) + C(-6x - 2a) = -2$$

$$-4aC = -2$$

$$C = 1/(2a)$$

therefore

$$\chi \; \frac{1}{2a}\left(x^3 - 3xy^2\right) - \frac{1}{2}\left(x^2 + y^2\right) + \frac{2a^2}{27} \tag{20.33}$$

Now

$$\tau_{xz} = G\theta \frac{\partial \chi}{\partial y}$$

$$= G\theta \left\{ \frac{1}{2a} (-6xy) - \frac{1}{2} \times 2y \right\}$$

$$\tau_{xz} = -G\theta \left(\frac{3xy}{2a} + y \right) \tag{20.34}$$

Along

$y = 0, \ \tau_{xz} = 0.$

Now

$$\tau_{yz} = -G\theta \frac{\partial \chi}{\partial x} = -G\theta \left\{ \frac{1}{2a} (3x^2 - 3y^2) - \frac{1}{2} \times 2x \right\}$$

therefore

$$\tau_{yz} = -\frac{3G\theta}{2a} \left\{ (x^2 - y^2) - \frac{2ax}{3} \right\} \tag{20.35}$$

As the triangle is equilateral, the maximum shear stress $\hat{\tau}$ can be obtained by considering the variation of τ_{yz} along any edge. Consider the edge *BC* (i.e. $x = -a/3$):

$$\tau_{yz} \text{ (edge } BC) = -\frac{3G\theta}{2a} \left(\frac{a^2}{9} - y^2 + \frac{2a^2}{9} \right)$$

$$= -\frac{3G\theta}{2a} \left(\frac{a^2}{3} - y^2 \right) \tag{20.36}$$

where it can be seen from (20.36) that $\hat{\tau}$ occurs at $y = 0$. Therefore

$$\hat{\tau} = -G\theta a/2 \tag{20.37}$$

20.4 Numerical solution of the torsional equation

Equation (20.11) lends itself to satisfactory solution by either the finite element method or the finite difference method and Figure 20.8 shows the variation of χ for a rectangular section, as obtained by the computer program LAPLACE. (The solution was carried out on an Apple II + microcomputer, and the screen was then photographed.) As the rectangular section had two axes of symmetry, it was only necessary to consider the top right-hand quadrant of the rectangle.

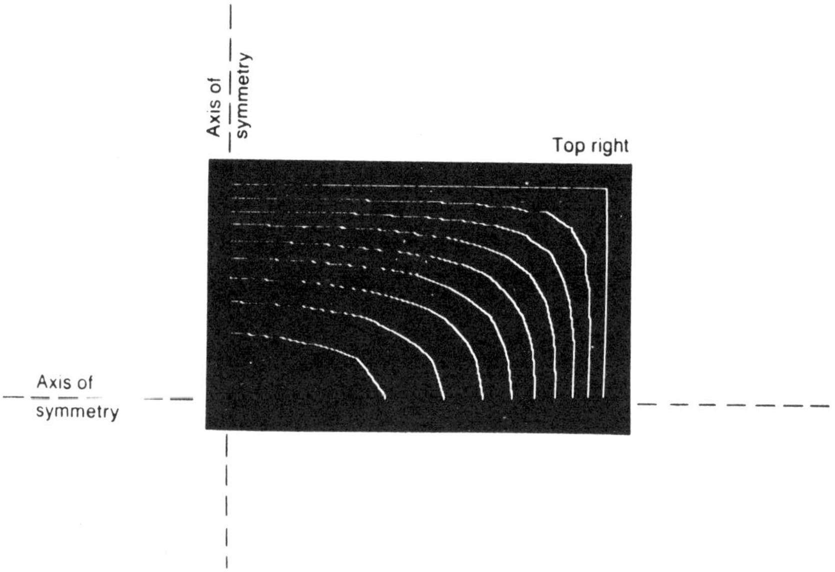

Figure 20.8 Shear stress contours.

20.5 Prandtl's membrane analogy

Prandtl noticed that the equations describing the deformation of a thin weightless membrane were similar to the torsion equation. Furthermore, he realised that as the behaviour of a thin weightless membrane under lateral pressure was more readily understood than that of the torsion of a non-circular section, the application of a membrane analogy to the torsion of non-circular sections considerably simplified the stress analysis of the latter.

Prior to using the membrane analogy, it will be necessary to develop the differential equation of a thin weightless membrane under lateral pressure. This can be done by considering the equilibrium of the element AA ′ BB ′in Figure 20.9.

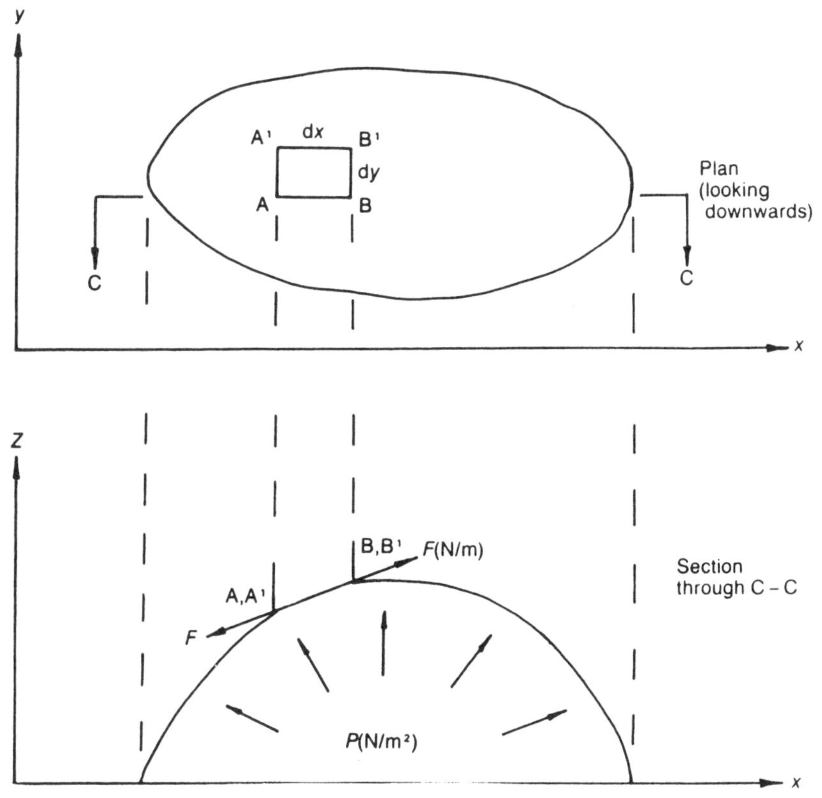

Figure 20.9 Membrane deformation.

Let,

F = membrane tension per unit length (N/m)

Z = deflection of membrane (m)

P = pressure (N/m^2)

Component of force on AA ′ in the z-direction is $F \times \dfrac{\partial Z}{\partial x} \times dy$ ↓

Component of force on BB ′in the z-direction is $F\left(\dfrac{\partial Z}{\partial x} + \dfrac{\partial^2 Z}{\partial x^2} \times dx\right) dy$ ↑

Component of force on AB in the z-direction is $F \times \frac{\partial Z}{\partial y} \times dx$ ↓

Component of force on A′ B′ in the z-direction is $F \times \left(\frac{\partial Z}{\partial y} + \frac{\partial^2 Z}{\partial y^2} \times dy \right) dx$ ↑

Resolving vertically

$$F \left(\frac{\partial^2 Z}{\partial x^2} + \frac{\partial^2 Z}{\partial y^2} \right) dz \times dy = -P \times dx \times dy$$

therefore

$$\frac{\partial^2 Z}{\partial x^2} + \frac{\partial^2 Z}{\partial y^2} = -\frac{P}{F} \tag{20.38}$$

If $Z = \chi$ in equation (20.38), and the pressure is so adjusted that $P/F = 2$, then it can be seen that equation (20.38) can be used as an analogy to equation (20.11).

From equations (20.12) and (20.13), it can be seen that

$$\begin{aligned} \tau_{xz} &= G\,\theta \times \text{slope of the membrane in the } y \text{ direction} \\ \tau_{yz} &= G\,\theta \times \text{slope of the membrane in the } x \text{ direction} \end{aligned} \tag{20.39}$$

Now, the torque is

$$\begin{aligned} T &= \iint \left(\tau_{xz} \times y - \tau_{yz} \times x \right) dx\,dy \\ &= G\theta \iint \left(\frac{\partial Z}{\partial y} \times y + \frac{\partial Z}{\partial x} \times x \right) dxdy \end{aligned} \tag{20.40}$$

Consider the integral

$$\iint \frac{\partial Z}{\partial y} \times y \times dxdy = \iint \partial Z \times y \times dx$$

Now y and dx are as shown in Figure 20.10, where it can be seen that $\iint y \times dx$ is the area of section. Therefore the

$$\iint \frac{\partial Z}{\partial y} \times y \times dx \times dy = \text{volume under membrane} \tag{20.41}$$

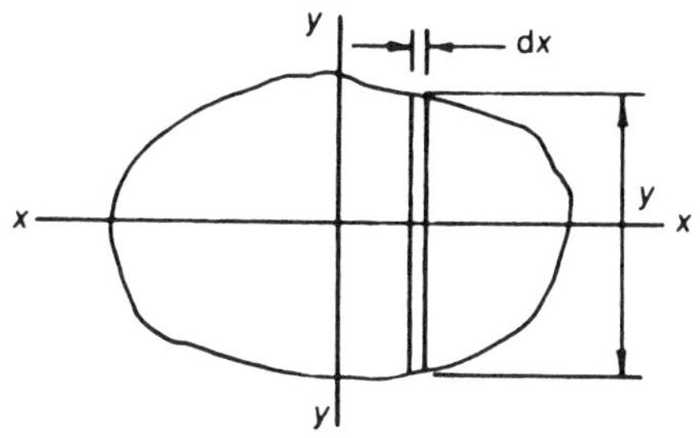

Figure 20.10

Similarly, it can be shown that the volume under membrane is

$$\iint \frac{\partial Z}{\partial x} \times x \times dx \times dy \qquad (20.42)$$

Substituting equations (20.41) and (20.42) into equation (20.40):

$$T = 2G\theta \times \text{volume under membrane} \qquad (20.43)$$

Now

$$\frac{T}{J} = G\theta$$

which, on comparison with equation (20.43), gives

$$J = \text{torsional constant}$$

$$= 2 \times \text{volume under membrane} \qquad (20.44)$$

20.6 Varying circular cross-section

Consider the varying circular section shaft of Figure 20.11, and assume that,

$$u = w = 0$$

where,

u = radial deflection
v = circumferential deflection
w = axial deflection

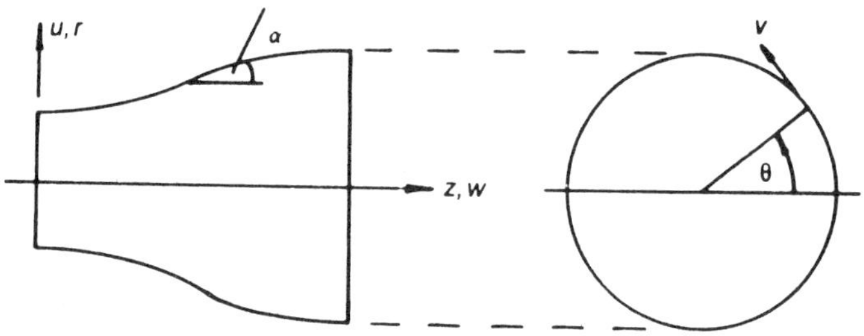

Figure 20.11 Varying section shaft.

As the section is circular, it is convenient to use polar co-ordinates. Let,

ε_r = radial strain = 0

ε_θ = hoop strain = 0

ε_z = axial strain = 0

γ_{rz} = shear strain in a longitudinal radial plane = 0

r = any radius on the cross-section

Thus, there are only two shear strains, $\gamma_{r\theta}$ and $\gamma_{\theta z}$, which are defined as follows:

$\gamma_{r\theta}$ = shear strain in the r–θ plane $= \dfrac{\partial v}{\partial r} - \dfrac{v}{r}$

$\gamma_{\theta z}$ = shear strain in the θ–z plane $= \dfrac{\partial v}{\partial z}$

But

$$\tau_{r\theta} = G\gamma_{r\theta} = G\left(\frac{\partial v}{\partial r} - \frac{v}{r}\right) \tag{20.45}$$

and

$$\tau_{\theta z} = G\gamma_{\theta z} = G\frac{\partial v}{\partial z} \tag{20.46}$$

From equilibrium considerations,

$$\frac{\partial \tau_{r\theta}}{\partial r} + \frac{\partial \tau_{\theta z}}{\partial z} + \frac{2\tau_{r\theta}}{r} = 0$$

which, when rearranged, becomes

$$\frac{\partial}{\partial r}\left(r^2 \tau_{r\theta}\right) + \frac{\partial}{\partial z}\left(r^2 \tau_{\theta z}\right) = 0 \tag{20.47}$$

Let κ be the shear stress function

where

$$\frac{\partial \kappa}{\partial r} = r^2 \tau_{\theta z} \tag{20.48}$$

and

$$\frac{\partial \kappa}{\partial z} = -r^2 \tau_{r\theta} \tag{20.49}$$

which satisfies equation (20.47).

From compatibility considerations

$$\frac{\partial \gamma_{r\theta}}{\partial z} = \frac{\partial \gamma_{\theta z}}{\partial r} - \frac{\gamma_{\theta z}}{r}$$

or

$$\frac{\partial \tau_{r\theta}}{\partial z} = \frac{\partial \tau_{\theta z}}{\partial r} - \frac{\tau_{\theta z}}{r} \tag{20.50}$$

From equation (20.49)

$$\frac{\partial \tau_{r\theta}}{\partial z} = -\frac{1}{r^2}\frac{\partial^2 \kappa}{\partial z^2} \tag{20.51}$$

From equation (20.48)

$$\frac{\partial \tau_{\theta z}}{\partial r} = \frac{1}{r^2}\frac{\partial^2 \kappa}{\partial r^2} - \frac{2}{r^3}\frac{\partial \kappa}{\partial r} \tag{20.52}$$

Substituting equations (20.59) and (20.52) into equation (20.50) gives

$$-\frac{1}{r^2}\frac{\partial^2 \kappa}{\partial z^2} = \frac{1}{r^2}\frac{\partial^2 \kappa}{\partial r^2} - \frac{2}{r^3}\frac{\partial \kappa}{\partial r} - \frac{1}{r^3}\frac{\partial \kappa}{\partial r}$$

or

$$\frac{\partial^2 \kappa}{\partial r^2} - \frac{3}{r}\frac{\partial \kappa}{\partial r} + \frac{\partial^2 \kappa}{\partial z^2} = 0 \qquad (20.53)$$

From considerations of equilibrium on the boundary,

$$\tau_{r\theta} \cos\alpha - \tau_{\theta z} \sin\alpha = 0 \qquad (20.54)$$

where

$$\cos\alpha = \frac{dz}{ds}$$
$$\qquad (20.55)$$
$$\sin\alpha = \frac{dr}{ds}$$

Substituting equations (20.48), (20.49) and (20.55) into equation (20.54),

$$-\frac{1}{r^2}\frac{\partial \kappa}{\partial z}\frac{dz}{ds} - \frac{1}{r^2}\frac{\partial \kappa}{\partial r}\frac{dr}{ds} = 0$$

or

$$\frac{2}{r^2}\frac{d\kappa}{ds} = 0$$

i.e. κ is a constant on the boundary, as required.

Equation (20.53) is the torsion equation for a tapered circular section, which is of similar form to equation (20.11).

20.7 Plastic torsion

The assumption made in this section is that the material is ideally elastic-plastic, as described in Chapter 15, so that the shear stress is everywhere equal to τ_{yp}, the yield shear stress. As the shear stress is constant, the slope of the membrane must be constant, and for this reason, the membrane analogy is now referred to as a sand-hill analogy.

Consider a *circular section*, where the sand-hill is shown in Figure 20.12.

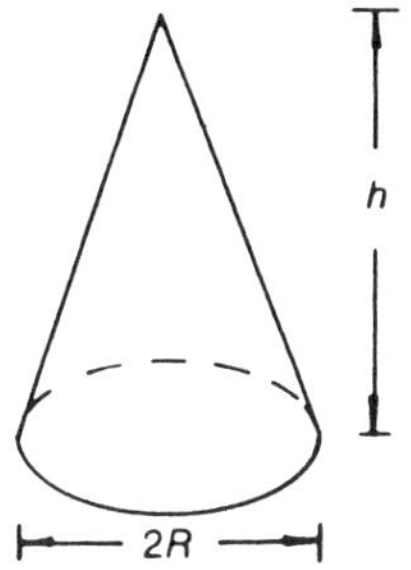

Figure 20.12 Sand-hill for a circular section.

From Figure 20.12, it can be seen that the volume (*Vol*) of the sand-hill is

$$Vol = \frac{1}{3}\pi R^2 h$$

but

$$\tau_{yp} = G\theta \times \text{slope of the sand-hill}$$

where

$$\theta = \text{twist/unit length} \rightarrow \infty$$

$$G = \text{modulus of rigidity} \rightarrow 0$$

$$\therefore \tau_{yp} = G\theta\,\frac{h}{R}$$

or

$$h = R\tau_{yp}/G\theta$$

and

$$Vol = \frac{\pi R^3 \tau_{yp}}{3G\theta}$$

Now

$$J = 2 \times Vol = 2\pi R^3 \tau_{yp}/(3G\theta)$$

and

$$T_p = G\theta J = G\theta \times 2\pi R^3 \tau_{yp}/(3G\theta)$$

therefore

$$T_p = 2\pi R^3 \tau_{yp}/3$$

where T_p is the fully plastic torsional moment of resistance of the section, which agrees with the value obtained in Chapter 4.

Consider a *rectangular section,* where the sand-hill is shown in Figure 20.13.

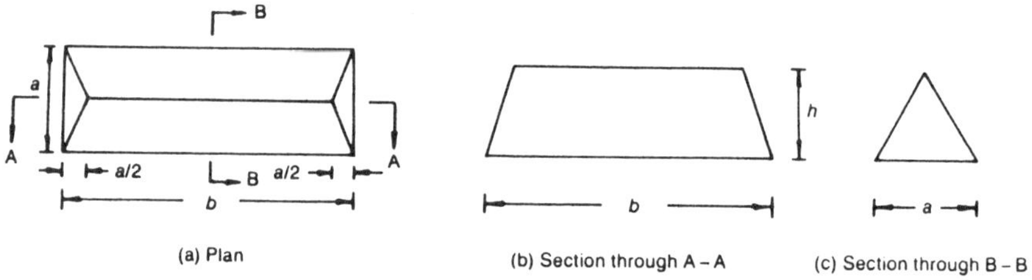

Figure 20.13 Sand-hill for rectangular section.

The volume under sand-hill is

$$Vol = \frac{1}{2}abh - \frac{1}{3}\left(\frac{1}{2}a \times \frac{a}{2}\right) \times h \times 2$$

$$= \frac{1}{2}abh - \frac{a^2 h}{6}$$

$$= \frac{ah}{6}(3b - a)$$

and $\tau_{yp} = G\theta \times \text{slope of sand-hill} = G\theta \times 2h/a$

or

$$h = \frac{a\tau_{yp}}{2G\theta}$$

therefore

$$Vol = \frac{a\,(3b - a)a\tau_{yp}}{12G\theta}$$

Now

$$J = 2 \times \text{vol} = a^2(3b - a)\tau_{yp}/(6G\theta)$$

and

$$T_p = G\theta J$$

therefore

$$T_p = a^2(3b - a)\tau_{yp}/6$$

where T_p is the fully plastic moment of resistance of the rectangular section.

Consider an equilateral triangular section, where the sand-hill is shown in Figure 20.14.

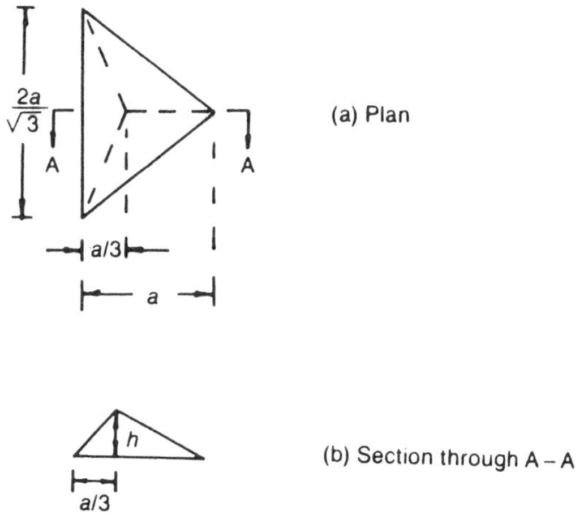

Figure 20.14 Sand-hill for triangular section.

Now

$$\tau_{yp} = G\theta \times \text{slope of sand-hill}$$

or

$$\tau_{yp} = G\theta \times \frac{3h}{a}$$

and

$$h = \frac{a\tau_{yp}}{3G\theta}$$

therefore, the volume of the sand-hill is

$$Vol = \frac{1}{3}\left(\frac{1}{2} \times \frac{2a}{\sqrt{3}} \times a\right) \times h$$

$$= \frac{a^2}{3\sqrt{3}} \times \frac{a\tau_{yp}}{3G\theta}$$

$$= \frac{a^3\tau_{yp}}{9\sqrt{3}G\theta}$$

and

$$T_p = 2G\theta \times \frac{a^3\tau_{yp}}{9\sqrt{3}G\theta}$$

$$T_p = \frac{2a^3\tau_{yp}}{9\sqrt{3}}$$

where T_p is the fully plastic torsional resistance of the triangular section.

21 Thick circular cylinders, discs and spheres

21.1 Introduction

Thin shell theory is satisfactory when the thickness of the shell divided by its radius is less than 1/30. When the thickness: radius ratio of the shell is greater than this, errors start to occur and thick shell theory should be used. Thick shells appear in the form of gun barrels, nuclear reactor pressure vessels, and deep diving submersibles.

21.2 Derivation of the hoop and radial stress equations for a thick-walled circular cylinder

The following convention will be used, where all the stresses and strains are assumed to be tensile and positive. At any radius, r

σ_θ = hoop stress

σ_r = radial stress

σ_z = longitudinal stress

ε_θ = hoop strain

ε_r = radial strain

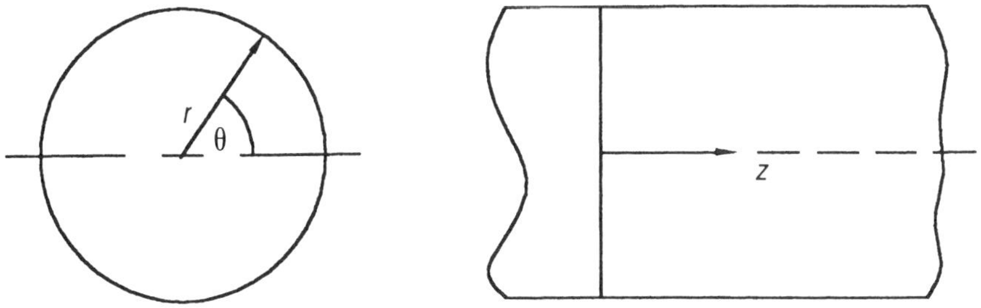

Figure 21.1 Thick cylinder.

ε_z = longitudinal strain (assumed to be constant)

w = radial deflection

From Figure 21.2, it can be seen that at any radius r,

$$\varepsilon_\theta = \frac{2\pi(r + w) - 2\pi r}{2\pi r}$$

or

$$\varepsilon_\theta = w/r \tag{21.1}$$

Similarly,

$$\varepsilon_r = \frac{\delta w}{\delta r} = \frac{dw}{dr} \tag{21.2}$$

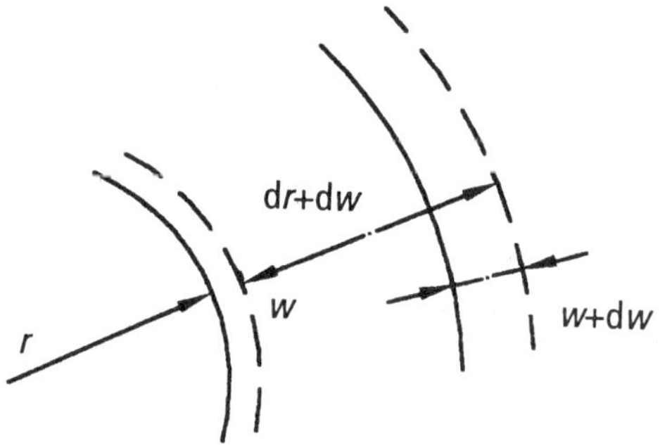

Figure 21.2. Deformation at any radius r.

From the standard stress–strain relationships,

$$E\varepsilon_z = \sigma_z - \nu\,\sigma_\theta - \nu\,\sigma_r = \text{a constant}$$

$$E\varepsilon_\theta = \frac{Ew}{r} = \sigma_\theta - \nu\sigma_z - \nu\sigma_r \tag{21.3}$$

$$E\varepsilon_r = E\frac{dw}{dr} = \sigma_r - \nu\sigma_\theta - \nu\sigma_z \tag{21.4}$$

Multiplying equation (21.3) by r,

$$Ew = \sigma_\theta \times r - \nu\sigma_z \times r - \nu\sigma_r \times r \tag{21.5}$$

and differentiating equation (21.5) with respect to r, we get

$$E\frac{dw}{dr} = \sigma_\theta - \nu\sigma_z - \nu\sigma_r + r\left(\frac{d\sigma_\theta}{dr} - \nu\frac{d\sigma_z}{dr} - \nu\frac{d\sigma_r}{dr}\right) \tag{21.6}$$

Subtracting equation (21.4) from equation (21.6),

$$(\sigma_\theta - \sigma_r)(1+\nu) + r\frac{d\sigma_\theta}{dr} - \nu r\frac{d\sigma_z}{dr} - \nu r\frac{d\sigma_r}{dr} = 0 \tag{21.7}$$

As ε_z is constant

$$\sigma_z - \nu\sigma_\theta - \nu\sigma_r = \text{constant} \tag{21.8}$$

Differentiating equation (21.8)with respect to r,

$$\frac{d\sigma_z}{dr} - \nu\frac{d\sigma_\theta}{dr} - \nu\frac{d\sigma_r}{dr} = 0$$

or

$$\frac{d\sigma_z}{dr} = \nu\left(\frac{d\sigma_\theta}{dr} + \frac{d\sigma_r}{dr}\right) \tag{21.9}$$

Substituting equation (21.9) into equation (21.7),

$$(\sigma_\theta - \sigma_r)(1+\nu) + r(1 - \nu^2)\frac{d\sigma_\theta}{dr} - \nu r(1+\nu)\frac{d\sigma_r}{dr} = 0 \tag{21.10}$$

and dividing equation (21.10) by $(1 + \nu)$, we get

$$\sigma_\theta - \sigma_r + r(1+\nu)\frac{d\sigma_\theta}{dr} - \nu r\frac{d\sigma_r}{dr} = 0 \tag{21.11}$$

Considering now the radial equilibrium of the shell element, shown in Figure 21.3,

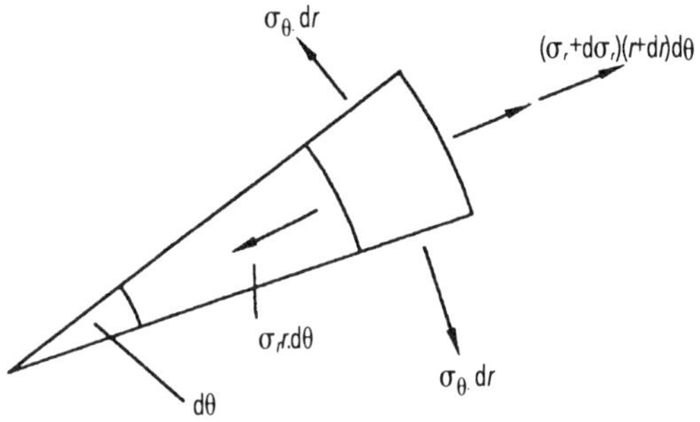

Figure 21.3 Shell element.

$$2\sigma_\theta \quad \delta_r \sin\left(\frac{\delta\theta}{2}\right) + \sigma_r \quad r \quad \delta_\theta - \left(\sigma_r + \delta\sigma_r\right)\left(r + \delta r\right)\delta\theta = 0 \tag{21.12}$$

Neglecting higher order terms in the above, we get

$$\sigma_\theta - \sigma_r - r\frac{\mathrm{d}\sigma_r}{\mathrm{d}r} = 0 \tag{21.13}$$

Subtracting equation (21.11) from equation (21.12)

$$\frac{\mathrm{d}\sigma_\theta}{\mathrm{d}r} + \frac{\mathrm{d}\sigma_r}{\mathrm{d}r} = 0 \tag{21.14}$$

$$\therefore \sigma_\theta + \sigma_r = \text{constant} = 2A \tag{21.15}$$

Subtracting equation (21.13) from equation (21.15),

$$2\sigma_r + r\frac{\mathrm{d}\sigma_r}{\mathrm{d}r} = 2A$$

or

$$\frac{1}{r}\,\frac{\mathrm{d}\left(\sigma_r r^2\right)}{\mathrm{d}r} = 2A$$

$$\frac{\mathrm{d}\left(\sigma_r \quad r^2\right)}{\mathrm{d}r} = 2Ar$$

Integrating the above,

$$\sigma_r \, r^2 = Ar^2 - B$$

$$\sigma_r = A - \frac{B}{r^2} \tag{21.16}$$

From equation (21.15),

$$\sigma_\theta = A + \frac{B}{r^2} \tag{21.17}$$

21.3 Lamé line

If equations (21.16) and (21.17) are plotted with respect to a horizontal axis, where $1/r^2$ is the horizontal axis, the two equations appear as a single straight line, where σ_r lies to the left and σ_θ to the right, as shown by Figure 21.4. For the case shown in Figure 21.4, σ_r is compressive and σ_θ tensile, where

$\sigma_{\theta 1}$ = internal hoop stress, which can be seen to be the maximum stress

$\sigma_{\theta 2}$ = external hoop stress

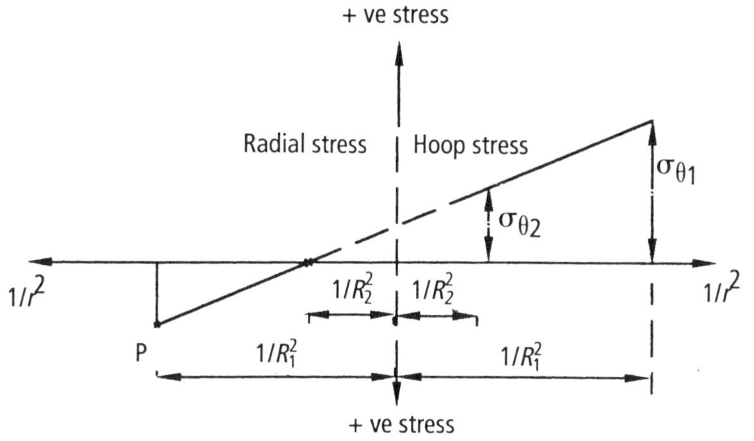

Figure 21.4 Lamé line for the case of internal pressure.

To calculate $\sigma_{\theta 1}$ and $\sigma_{\theta 2}$, equate similar triangles in Figure 21.4,

$$\frac{\sigma_{\theta 1}}{\left(\frac{1}{R_1^2} + \frac{1}{R_2^2}\right)} = \frac{P}{\left(\frac{1}{R_1^2} - \frac{1}{R_2^2}\right)}$$

or

$$\sigma_{\theta 1} = \frac{P\left(\frac{1}{R_1^2} + \frac{1}{R_2^2}\right)}{\left(\frac{1}{R_1^2} - \frac{1}{R_2^2}\right)} \times \frac{R_1^2 R_2^2}{R_1^2 R_2^2} \tag{21.18}$$

$$\sigma_{\theta 1} = \frac{P\left(R_1^2 + R_2^2\right)}{\left(R_2^2 - R_1^2\right)}$$

Similarly, from Figure 21.4

$$\frac{\sigma_{\theta 2}}{\left(\frac{1}{R_2^2} + \frac{1}{R_2^2}\right)} = \frac{P}{\left(\frac{1}{R_1^2} - \frac{1}{R_2^2}\right)}$$

or,

$$\sigma_{\theta 2} = \frac{P\left(\frac{1}{R_2^2}\right) \times 2}{\left(\frac{1}{R_1^2} - \frac{1}{R_2^2}\right)} \times \frac{R_1^2 R_2^2}{R_1^2 R_2^2} \tag{21.19}$$

$$\sigma_{\theta 2} = \frac{2PR_1^2}{\left(R_2^2 - R_1^2\right)}$$

Problem 21.1 A thick-walled circular cylinder of internal diameter 0.2 m is subjected to an internal pressure of 100 MPa. If the maximum permissible stress in the cylinder is limited to 150 MPa, determine the maximum possible external diameter d_2.

Solution

$$\frac{100}{\left(\frac{1}{0.2^2} - \frac{1}{d_2^2}\right)} = \frac{150}{\left(\frac{1}{0.2^2} + \frac{1}{d_2^2}\right)}$$

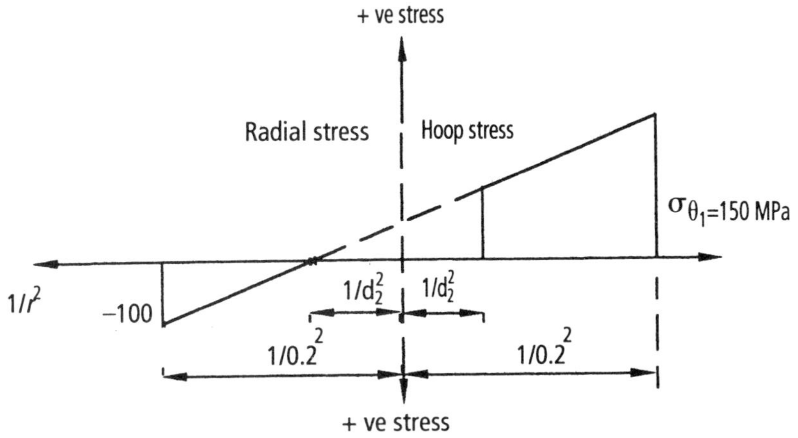

Figure 21.5 Lamé line for thick cylinder.

or

$$\frac{\left(\frac{1}{0.2^2} + \frac{1}{d_2^2}\right)}{\left(\frac{1}{0.2^2} - \frac{1}{d_2^2}\right)} \times \left(\frac{0.2^2\ d_2^2}{0.2^2\ d_2^2}\right) = 1.5$$

$$\left(\frac{d_2^2 + 0.2^2}{d_2^2 - 0.2^2}\right) = 1.5$$

or $$d_2^2 + 0.2^2 = 1.5\left(d_2^2 - 0.2^2\right)$$

or $$0.2^2\left(1+1.5\right) = d_2^2\left(1.5-1\right)$$

$$d_2^2 = 0.2\,\mathrm{m}^2$$

$$d_2 = 0.447\,\mathrm{m}$$

Problem 21.2 If the cylinder in the previous problem were subjected to an external pressure of 100 MPa and an internal pressure of zero, what would be the maximum magnitude of stress.

<u>*Solution*</u>

Now $\dfrac{1}{d_1^2} = 25$ and $\dfrac{1}{d_2^2} = 5,$

hence the Lamé line would take the form of Figure 21.6.

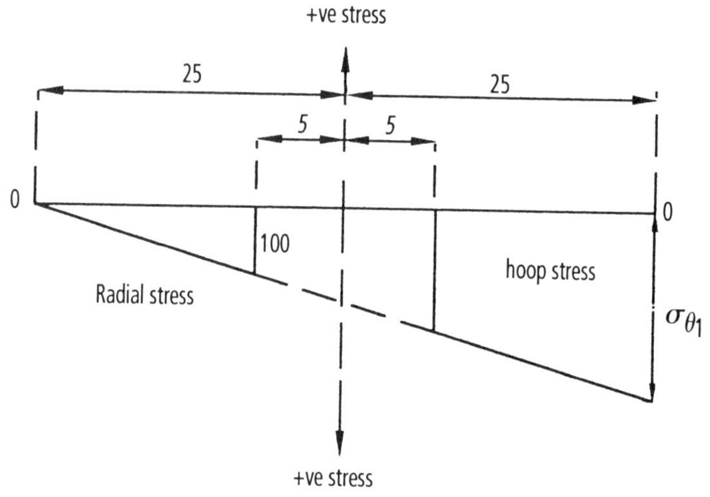

Figure 21.6 Lamé line for external pressure case.

By equating similar triangles,

$$\frac{-100}{(25 - 5)} = \frac{\sigma_{\theta i}}{25 + 25}$$

where $\sigma_{\theta i}$ is the internal stress which has the maximum magnitude

$$\therefore\ \sigma_{\theta i} = \frac{-50 \times 100}{20} = -250\ \text{MPa}$$

Problem 21.3 A steel disc of external diameter 0.2 m and internal diameter 0.1 m is shrunk onto a solid steel shaft of external diameter 0.1 m, where all the dimensions are nominal. If the interference fit, based on diameters, between the shaft and the disc at the common surface is 0.2 mm, determine the maximum stress.

For steel, $E = 2 \times 10^{11}$ N/m², $\nu = 0.3$

Solution

Consider the steel disc. In this case the radial stress on the internal surfaces is the unknown P_c. Hence, the Lamé line will take the form shown in Figure 21.7.

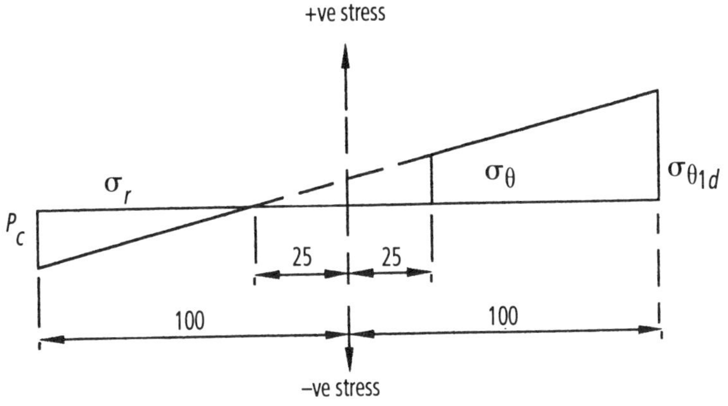

Figure 21.7 Lamé line for steel ring.

Let,

$\sigma_{\theta 1d}$ = hoop stress (maximum stress) on the internal surface of the disc

σ_{r1d} = radial stress on the internal surface of the disc

Equating similar triangles, in Figure 21.7

$$\frac{P_c}{(100 - 25)} = \frac{\sigma_{\sigma 1d}}{100 + 25}$$

$$\therefore\ \sigma_{\theta 1d} = \frac{125\ P_c}{75} = 1.667\ P_c$$

Consider now the solid shaft. In this case, the internal diameter of the shaft is zero and as $1/0^2 \rightarrow \infty$, the Lamé line must be horizontal or the shaft's hoop stress will be infinity, which is impossible; see Figure 21.8.

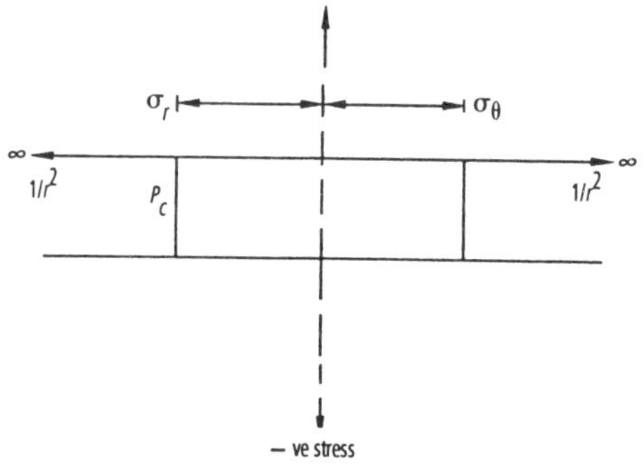

Figure 21.8 Lamé line for a solid shaft.

Let

$$P_c = \text{external pressure on the shaft}$$

$$\therefore \sigma_r = \sigma_\theta = -P_c \text{ (everywhere)} \qquad (21.20)$$

Let,

$$w_d = \text{increase in the radius of the disc at its inner surface}$$

$$w_s = \text{increase in the radius of the shaft at its outer surface}$$

Now, applying the expression

$$E\varepsilon_\theta = \frac{w}{r} = \sigma_\theta - \nu\sigma_r - \nu\sigma_x$$

to the inner surface of the disc

$$\frac{EW_d}{5 \times 10^{-2}} = \sigma_{\theta 1d} - \nu\sigma_{r1d}$$

but,

$$\sigma_{r1d} = -P_c$$

therefore

$$\frac{2 \times 10^{11} \times w_d}{5 \times 10^{-2}} = 1.667\ P_c + 0.3\ P_c \tag{21.21}$$

$$w_d = 4.918 \times 10^{-13}\ P_c$$

Similarly, for the shaft

$$\frac{Ew_s}{5 \times 10^{-2}} = \sigma_{\theta s} - \nu\sigma_{rs}$$

but $\sigma_{\theta s} = \sigma_{rs} = P_c$

$$\therefore \frac{2 \times 10^{11}\ w_s}{5 \times 10^{-2}} = -P_c\ (1 - \nu)$$

$$w_s = -1.75 \times 10^{-13}\ P_c \tag{21.22}$$

but $w_d - w_s = 2 \times 10^{-3}/2$

$$(4.918 \times 10^{-13} + 1.75 \times 10^{-13})\ P_c = 1 \times 10^{-4}$$

$$\therefore P_c = 150\ \text{MPa}$$

Maximum stress is

$$\sigma_{\theta 1d} = 1.667\ P_c = 250\ \text{MPa}$$

21.4 Compound tubes

A compound tube is usually made from two cylinders of different materials where one is shrunk onto the other.

Problem 21.4 A circular steel cylinder of external diameter 0.2 m and internal diameter 0.1 m is shrunk onto a circular aluminium alloy cylinder of external diameter 0.1 m and internal diameter 0.05 m, where the dimensions are nominal.

Determine the radial pressure at the common surface due to shrinkage alone, so that when there is an internal pressure of 300 MPa, the maximum hoop stress in the inner cylinders is 150 Mpa. Sketch the hoop stress distributions.

For steel, $E_s = 2 \times 10^{11}$ N/m², $\nu_s = 0.3$

For aluminium alloy, $E_a = 6.7 \times 10^{10}$ N/m², $\nu_a = 0.32$

Solution

σ_θ^P = the hoop stress due to pressure alone

σ_θ^s = the hoop stress due to shrinkage alone

$\sigma_{\theta,2s}$ = hoop stress in the steel on the 0.2 m diameter

$\sigma_{\theta,1s}$ = hoop stress in the steel on the 0.1 m diameter

$\sigma_{r,2s}$ = radial stress in the steel on the 0.2 m diameter

$\sigma_{r,1s}$ = radial stress in the steel on the 0.1 m diameter

$\sigma_{\theta,1a}$ = hoop stress in the aluminium on the 0.1 m diameter

$\sigma_{r,1a}$ = radial stress in the aluminium on the 0.1 m diameter

$\sigma_{\theta,5a}$ = hoop stress in the aluminum on the 0.05 m diameter

$\sigma_{r,5a}$ = radial stress in the aluminium on the 0.05 m diameter

Consider first the stress due to shrinkage alone, as shown in Figures 21.9 and 21.10.

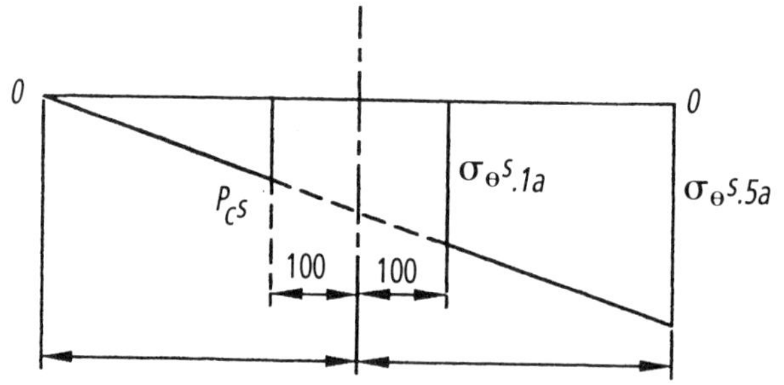

Figure 21.9 Lamé line for aluminium alloy tube.

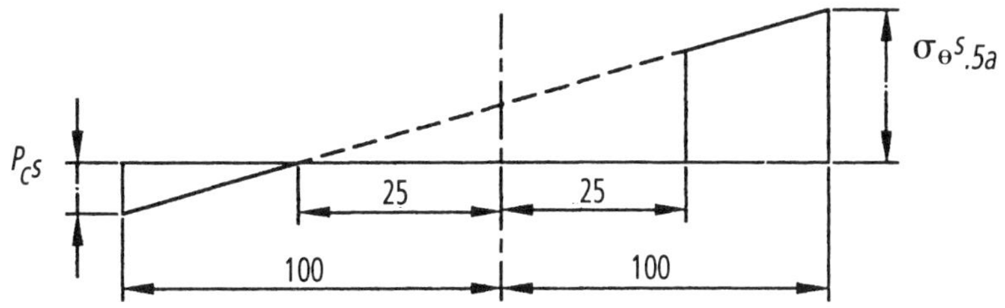

Figure 21.10 Lamé line for steel tube, due to shrinkage with respect to e.

Equating similar triangles in Figure 21.9.

$$\frac{\sigma^s_{\theta,5a}}{400 + 400} = \frac{-P^s_c}{400 - 100} \tag{21.23}$$

$$\sigma^s_{\theta,5a} = -2.667\ P^s_c$$

Similarly, from figure 21.9,

$$\frac{\sigma^s_{\theta,1a}}{400 + 100} = \frac{-P^s_c}{400 - 100} \tag{21.24}$$

$$\sigma^s_{\theta,1a} = -1.667\ P^s_c$$

Equating similar triangles in Figure 21.10.

$$\frac{\sigma^s_{\theta,1s}}{100 + 25} = \frac{P^s_c}{100 - 25} \tag{21.25}$$

$$\sigma^s_{\theta,1s} = 1.667\ P^s_c$$

Consider the stresses due to pressure alone

P_c = internal pressure

P^P_c = pressure at the common surface due to pressure alone

The Lamé lines will be as shown in Figures 21.11 and 21.12.

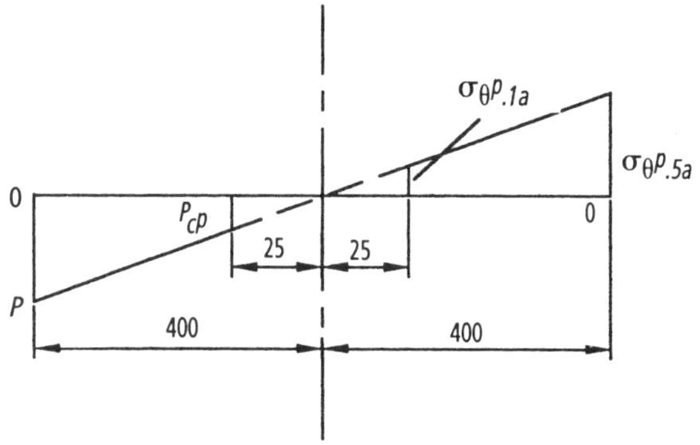

Figure 21.11 Lamé line in aluminium alloy, due to pressure alone.

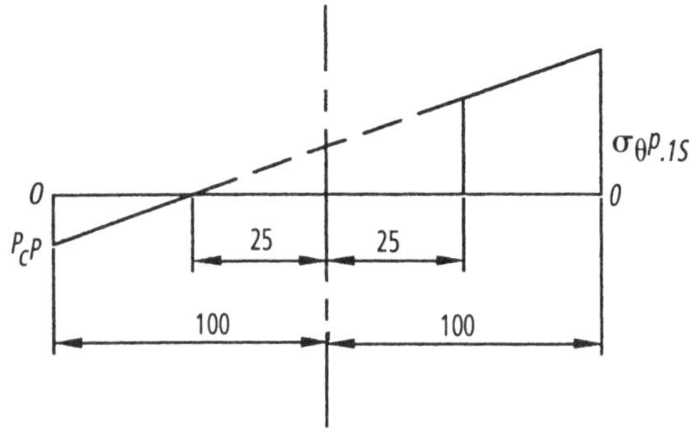

Figure 21.12 Lamé line for steel, due to pressure alone.

Equating similar triangles in Figure 21.11.

$$\frac{P - P_c^P}{400 - 100} = \frac{\sigma_{\theta,1a}^P + P}{400 + 100}$$

or
$$\frac{300 - P_c^P}{300} = \frac{\sigma_{\theta,1a}^P + 300}{500} \tag{21.26}$$

or
$$\sigma_{\theta.1a}^P = 200 - 1.667\,P_c^P$$

Similarly, from Figure 21.11,

$$\frac{P - P_c^P}{300} = \frac{\sigma_{\theta,5a}^P + P}{800}$$

$$\frac{300 - P_c^P}{300} = \frac{\sigma_{\theta,5a}^P + 300}{800} \tag{21.27}$$

or
$$\sigma_{\theta,5a}^P = \frac{8}{3}\left(300 - P_c^P\right) - 300$$

$$\sigma_{\theta,5a}^P = 500 - 2.667\,P_c^P$$

Similarly , from Figure 21.12,

$$\frac{\sigma_{\theta,1s}^P}{100+25} = \frac{P_c^P}{100-25} \tag{21.28}$$

$$\sigma_{\theta,1s}^P = 1.667\,P_c^P$$

Owing to pressure alone, there is *no interference fit*, so that

$$w_a^P = w_s^P$$

Now

$$\frac{E_s\,w_s^P}{0.05} = \sigma_{\theta,1s}^P + \nu_s\,P_c^P$$

or
$$w_s^P = \frac{0.05}{2 \times 10^{11}}\left(1.667\,P_c^P + 0.3\,P_c^P\right)$$

or
$$w_s = 4.917 \times 10^{-13}\,P_c^P \tag{21.29}$$

Similarly

$$\frac{E_a\,w_a^P}{0.05} = \sigma_{\theta,1a}^P + \nu_a\,P_c^P$$

or
$$w_a^P = \frac{0.05}{6.7 \times 10^{10}}\left(\sigma_{\theta,1a}^P + 0.32\,P_c^P\right)$$

$$= \frac{0.05}{6.7 \times 10^{10}}\left(200 - 1.667\,P_c^P + 0.32\,P_c^P\right) \tag{21.30}$$

$$w_a^P = 1.493 \times 10^{-10} - 1.0 \times 10^{-12}\,P_c^P$$

Equating (21.29) and (21.30)

$$4.917 \times 10^{-13}\,P_c^P = 1.493 \times 10^{-10} - 1.0 \times 10^{-12} \tag{21.31}$$

$$\therefore\ P_c^P = 100\,MPa$$

Substituting equation (21.31) into equations (21.26) and (21.27)

$$\sigma_{\theta,5a} = 500 - 2.667 \times 100 = 233.3\ \text{MPa} \tag{21.32}$$

$$\sigma_{\theta,1a}^P = 200 - 1.667 \times 100 = 33.3\ \text{MPa} \tag{21.33}$$

Now the maximum hoop stress in the inner tube lies either on its internal surface or its external surface, so that either

$$\sigma_{\theta,1a}^P + \sigma_{\theta,1a}^s = 150 \tag{21.34}$$

or

$$\sigma_{\theta,5a}^P + \sigma_{\theta,5a}^s = 150 \tag{21.35}$$

Substituting equations (21.32) and (21.24) into equation (21.34), we get

$$33.3 - 1.667\ P_c^s = 150$$

$$\text{or } P_c^s = -70\ \text{MPa}$$

Substituting equations (21.33) and (21.23) into equation (21.35), we get

$$233.3 - 2.667\ P_c^s = 150$$

$$\therefore\ P_c^s = 31.2\ \text{MPa}$$

i.e. P_c^s = 31.2 MPa, as P_c^s cannot be negative!

$$P_c = P_c^s + P_c^P = 31.2 + 100 = 131.2 \text{ MPa} \tag{21.36}$$

$$\frac{\sigma_{\theta,2s}}{25 + 25} = \frac{P_c^s + P_c^P}{100 - 25}$$

$$\sigma_{\theta,2s} = 87.5 \text{ MPa}$$

$$\sigma_{\theta,1s} = 1.667\left(P_c^s + P_c^P\right) = 218.7 \text{ MPa}$$

$$\sigma_{\theta,1a} = 200 - 1.667\left(P_c^s + P_c^P\right) = -18.7 \text{ MPa}$$

$$\sigma_{\theta,5a} = 500 - 2.667\left(P_c^s + P_c^P\right) = 150 \text{ MPa}$$

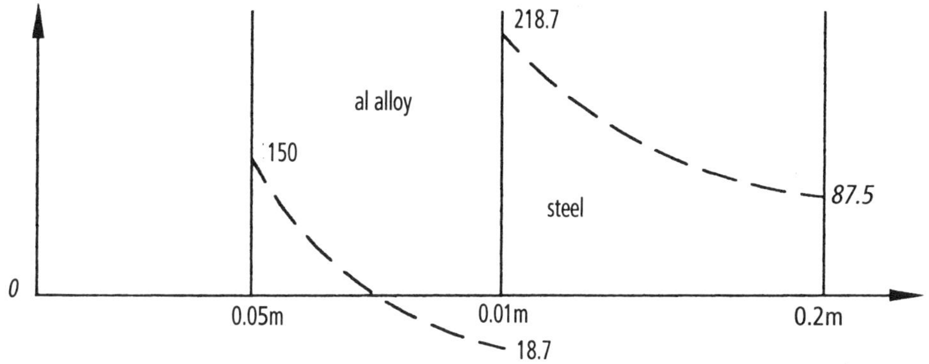

Figure 21.13 Hoop stress distribution.

21.5 Plastic deformation of thick tubes

The following assumptions will be made in this theory:

1. Yielding will take place according to the maximum shear stress theory, (Tresca).
2. The material of construction will behave in an ideally elastic-plastic manner.
3. The longitudinal stress will be the 'minimax' stress in the three-dimensional system of stress.

For this case, the equilibrium considerations of equation (21.13) apply, so that

$$\sigma_\theta - \sigma_r - r\frac{d\sigma_r}{dr} = 0 \tag{21.37}$$

Now, according to the maximum shear stress criterion of yield,

$$\begin{aligned}\sigma_\theta - \sigma_r &= \sigma_{yp} \\ \sigma_\theta &= \sigma_{yp} + \sigma_r\end{aligned} \tag{21.38}$$

Substituting equation (21.38) into equation (21.37),

$$\sigma_{yp} + \sigma_r - \sigma_r - r\frac{d\sigma_r}{dr} = 0$$

$$d\sigma_r = \sigma_{yp}\frac{dr}{r} \tag{21.39}$$

$$\sigma_r = \sigma_{yp}\ln r + C$$

For the case of the partially plastic cylinder shown in Figure 21.14,

at $r = R_2$, $\sigma_r = -P_2$

Substituting this boundary condition into equation (21.39), we get

$$-P_2 = \sigma_{yp}\ln R_2 + C$$

therefore

$$C = -\sigma_{yp}\ln R_2 - P_2$$

and,

$$\sigma_r = \sigma_{yp}\ln\left(\frac{r}{R_2}\right) - P_2 \tag{21.40}$$

Similarly, from equation (21.38),

$$\sigma_\theta = \sigma_{yp}\left\{1 + \ln\left(\frac{r}{R_2}\right)\right\} - P_2 \tag{21.41}$$

where,

R_1 = internal radius

R_2 = outer radius of plastic section of cylinder

R_3 = external radius

P_1 = internal pressure

P_2 = external pressure

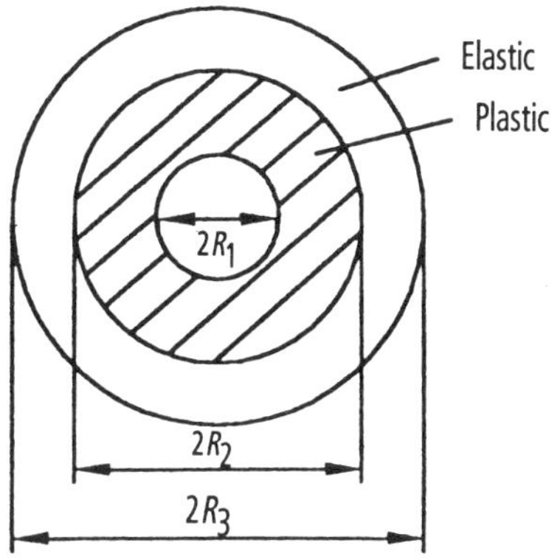

Figure 21.14 Partially plastic cylinder.

The vessel can be assumed to behave as a compound cylinder, with the internal portion behaving plastically, and the external portion elastically. The Lamé line for the elastic portion of the cylinder is shown in Figure 21.15.

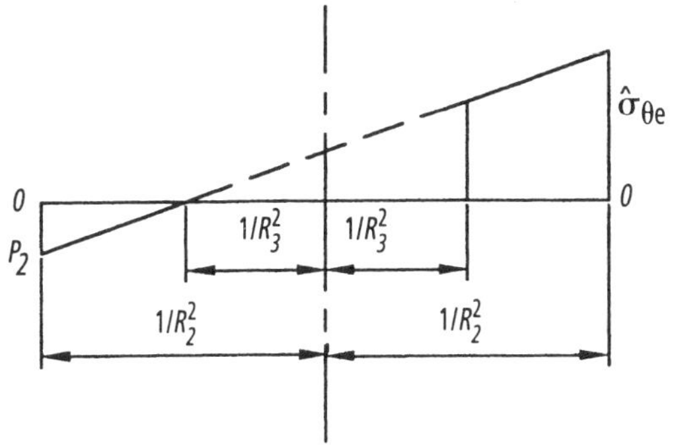

Figure 21.15 Lamé line for elastic zone.

In Figure 21.15,

$$\hat{\sigma}_{\theta e} = \text{elastic hoop stress at } r = R_2$$

so that according to the maximum shear stress criterion of yield on this radius,

$$\sigma_{yp} = \hat{\sigma}_{\theta e} + P_2 \tag{21.42}$$

From Figure 21.15

$$\frac{P_2}{\left(\dfrac{1}{R_2^2} - \dfrac{1}{R_3^2}\right)} = \frac{\hat{\sigma}_{\theta e}}{\left(\dfrac{1}{R_2^2} + \dfrac{1}{R_3^2}\right)}$$

therefore

$$\hat{\sigma}_{\theta e} = \frac{P_2\left(R_3^2 + R_2^2\right)}{\left(R_3^2 - R_2^2\right)} \tag{21.43}$$

Substituting equation (21.43) into equation (21.42),

$$P_2 = \sigma_{yp}\left(R_3^2 - R_2^2\right) / \left(2R_3^2\right) \tag{21.44}$$

Consider now the portion of the cylinder that is plastic. Substituting equation (21.44) into equation (21.41), the stress distributions in the plastic zone are given by:

$$\sigma_r = -\sigma_{yp}\left\{\ln\left(\frac{R_2}{r}\right) + \frac{\left(R_3^2 - R_2^2\right)}{2R_3^2}\right\} \tag{21.45}$$

$$\sigma_\theta = \sigma_{yp}\left\{\frac{\left(R_3^2 + R_2^2\right)}{2R_3^2} - \ln\left(\frac{R_2}{r}\right)\right\} \tag{21.46}$$

To find the *pressure to just cause yield,* put

$$\sigma_r = -P_1 \quad \text{when} \quad r = R_1$$

where P_l is the internal pressure that causes the onset of yield. Therefore,

$$P_1 = \sigma_{yp} \left\{ \ln \left(\frac{R_2}{R_1} \right) + \left(\frac{R_3^2 - R_2^2}{2R_3^2} \right) \right\} \tag{21.47}$$

but, if yield is only on the inside surface,

$$R_1 = R_2$$

in (21.61), so that,

$$P_1 = \sigma_{yp} \left\{ \left(R_3^2 - R_1^2 \right) / \left(2R_3^2 \right) \right\} \tag{21.48}$$

To determine the *plastic collapse pressure* P_p, put $R_2 = R_3$ in equation (21.47), to give

$$P_P = \sigma_{yp} \ln \left(\frac{R_3}{R_1} \right) \tag{21.49}$$

To determine the hoop stress distribution in the plastic zone, $\sigma_{\theta p}$, it must be remembered that

$$\sigma_{yp} = \sigma_\theta - \sigma_r$$

therefore

$$\sigma_{\theta P} = \sigma_{yp} \left\{ 1 + \ln \left(R_3 / R_1 \right) \right\} \tag{21.50}$$

Plots of the stress distributions in a partially plastic cylinder, under internal pressure, are shown in Figure 21.16.

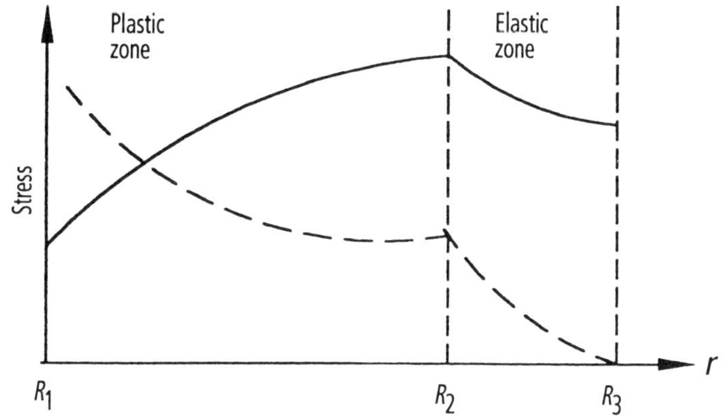

Figure 21.16 Stress distribution plots.

Problem 21.5 A circular cylinder of 0.2 m external diameter and of 0.1 m internal diameter is shrunk onto another circular cylinder of external diameter 0.1 m and of bore 0.05 m, where the dimensions are nominal. If the interference fit is such that when an internal pressure of 10 MPa is applied to the inner face of the inner cylinder, the inner face of the inner cylinder is on the point of yielding. What internal pressure will cause plastic penetration through half the thickness of the inner cylinder. It may be assumed that the Young's modulus and Poisson's ratio for both cylinders is the same, but that the outer cylinder is made of a higher grade steel which will not yield under these conditions. The yield stress of the inner cylinder may be assumed to be 160 MPa.

Solution

The Lamé line for the compound cylinder at the onset of yield is shown in Figure 21.17.

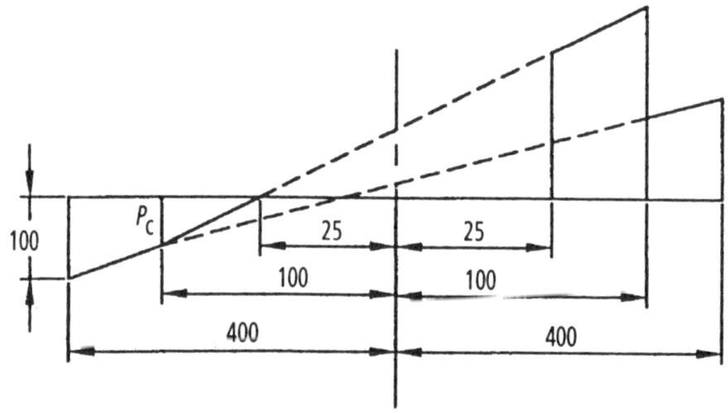

Figure 21.17 Lamé line for compound cylinder.

In Figure 21.17,

σ_1 = hoop stress on inner surface of inner cylinder.

σ_2 = hoop stress on outer surface of inner cylinder.

σ_3 = hoop stress on inner surface of outer cylinder.

As yield occurs on the inner surface of the inner surface when an internal pressure of 50 MPa is applied,

$$\sigma_1 - (-100) = 160$$

$$\therefore \sigma_1 = 60 \text{ MPa}$$

Equating similar triangles in Figure 21.17, we get

$$\frac{\sigma_1 + 100}{400 + 400} = \frac{100 - P_c}{400 - 100}$$

$$\frac{160 \times 300}{800} = 100 - P_c \tag{21.51}$$

$$\therefore P_c = 40 \text{ MPa}$$

Similarly from Figure 21.17

$$\frac{\sigma_2 + 100}{400 + 100} = \frac{100 - P_c}{400 - 100} \tag{21.52}$$

$$\sigma_2 = 0$$

Also from Figure 21.17,

$$\frac{\sigma_3}{100 + 25} = \frac{P_c}{100 - 25} \tag{21.53}$$

$$\therefore \sigma_3 = \frac{400 \times 125}{75} = 66.7 \text{ MPa}$$

Consider, now, plastic penetration of the inner cylinder to a diameter 0.075. The Lamé line in the elastic zones will be as shown in Figure 21.17. From Figure 21.18,

$$\sigma_6 + P_3 = 160$$

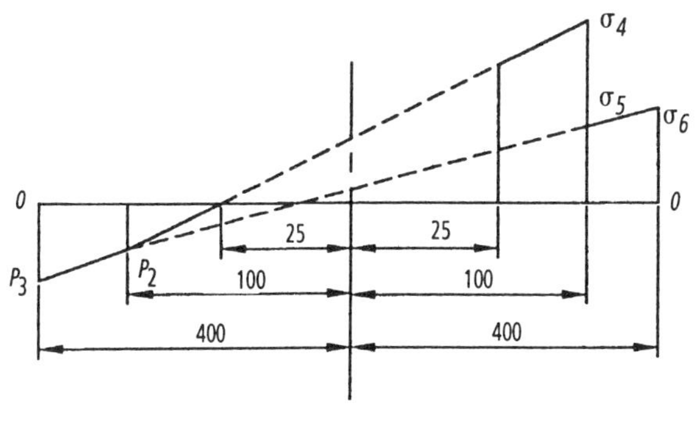

Figure 21.18 Lamé line in elastic zones.

therefore

$$\therefore \sigma_6 = 160 - P_3 \tag{21.54}$$

Similarly

$$\frac{P_3 - P_2}{400 - 100} = \frac{\sigma_6 + P_3}{400 + 400} = \frac{160}{800} \tag{21.55}$$

$$\therefore P_3 = 60 + P_2 \tag{21.56}$$

Also from Figure 21.18

$$\frac{\sigma_4}{100 + 25} = \frac{P_2}{100 - 25} \tag{21.57}$$

or $$\sigma_4 = 1.667\, P_2 \tag{21.58}$$

Substituting equation (21.56) into equation (21.58), we get

$$\sigma_4 = 1.667\,(P_3 - 60)$$

or $$\sigma_4 = 1.667\, P_3 - 100$$

Also from equation (21.55)

$$\frac{\sigma_5 + P_3}{100 + 400} = \frac{P_3 - P_2}{400 - 100} = \frac{160}{800}$$

$$\therefore \sigma_5 = 100 - P_3 \tag{21.59}$$

Now,

$$w = \frac{r}{E}\,(\sigma_\theta - v\sigma_r)$$

which will be the same for both cylinders at the common surface, i.e.,

$$\frac{1}{E}\,\{(\sigma_5 - \sigma_2) - v(P_2 - P_c)\} = \left\{\frac{1}{E}\,(\sigma_4 - \sigma_3) - v(P_2 - P_c)\right\}$$

or

$$\sigma_5 - \sigma_2 = \sigma_4 - \sigma_3$$

Substituting equations (21.52), (21.53), (21.58) and (21.59) into the above, we get

$$100 - P_3 - 0 = 1.667\ P_3 - 100 - 66.7$$

or

$$2.667\ P_3 = 100 + 100 + 66.7$$

$$P_3 = 100$$

Consider now the yielded portion

$$\sigma_r = \sigma_{yp} \ln r + c$$

$$\sigma_{yp} = 160$$

$$\text{at } r = 0.0375 \text{ m},$$

$$\sigma_r = -P_3 = -100$$

or

$$-100 = 160 \ln (0.0375) + C$$

$$C = -100 + 525.3$$

$$\therefore C = 425.3$$

Now, at $r = 0.025$ m,

$$-P = 160 \ln (0.025) + 425.3$$

$$= -590.2 + 425.3$$

$$P = 164.9 \text{ MPa}$$

which is the pressure to cause plastic penetration.

Problem 21.6 Determine the internal pressure that will cause complete plastic collapse of the compound cylinder given that the yield stress for the material of the outer cylinder is 700 MPa.

Solution

Now,

$$P_p = \sigma_{yp} \ln\left(\frac{R_3}{R_1}\right) \tag{21.60}$$

$$= \sigma_{yp2} \ln\left(\frac{R_3}{R_2}\right) + \sigma_{yp1} \ln\left(\frac{R_2}{R_1}\right)$$

$$= 700 \ln\left(\frac{0.1}{0.05}\right) + 160 \ln\left(\frac{0.05}{0.0375}\right)$$

$$= 485 + 46$$

$$P_n = 531 \text{ MPa}$$

which is the plastic collapse pressure of the compound cylinder.

21.6 Thick spherical shells

Consider a thick hemispherical shell element of radius r, under a compressive radial stress P, as shown in Figure 21.19.

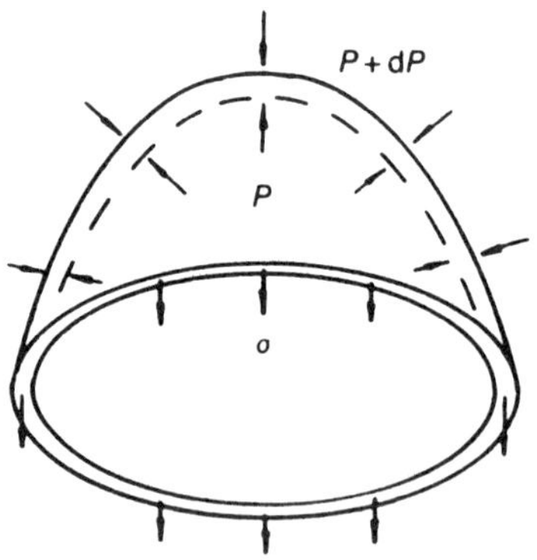

Figure 21.19 Thick hemispherical shell element.

Let w be the radial deflection at any radius r,

so that

$$\text{hoop strain} = w/r$$

and

$$\text{radial strain} = \frac{dw}{dr}$$

From three-dimensional stress–strain relationships,

$$E\,\frac{w}{r} = \sigma - \nu\sigma + \nu P \tag{21.61}$$

and

$$E\,\frac{\mathrm{d}w}{\mathrm{d}r} = -P - \nu\sigma - \nu\sigma \tag{21.62}$$

$$= -P - 2\nu\sigma$$

Now

$$Ew = \sigma\; r - \nu\sigma\; r + \nu P\; r$$

which, on differentiating with respect to r, gives

$$\begin{aligned} E\frac{\mathrm{d}w}{\mathrm{d}r} &= \sigma + r\frac{\mathrm{d}\sigma}{\mathrm{d}r} - \nu\sigma - \nu r\frac{\mathrm{d}\sigma}{\mathrm{d}r} + \nu P + \nu r\frac{\mathrm{d}P}{\mathrm{d}r} \\ &= (1-\nu)\left(\sigma - r\frac{\mathrm{d}\sigma}{\mathrm{d}r}\right) + \nu\left(P + r\frac{\mathrm{d}P}{\mathrm{d}r}\right) \end{aligned} \tag{21.63}$$

Equating (21.62) and (21.63),

$$-P - 2\nu\sigma = (1-\nu)\left(\sigma - r\,\frac{\mathrm{d}\sigma}{\mathrm{d}r}\right) + \nu\left(P + r\,\frac{\mathrm{d}P}{\mathrm{d}r}\right)$$

or

$$(1+\nu)(\sigma + P) + r(1-\nu)\,\frac{\mathrm{d}\sigma}{\mathrm{d}r} + \nu r\,\frac{\mathrm{d}P}{\mathrm{d}r} = 0 \tag{21.64}$$

Considering now the equilibrium of the hemispherical shell element,

$$\sigma \times 2\pi r \times dr = P \times \pi r^2 - (P + dP) \times \pi \times (r + dr)^2 \tag{21.65}$$

Neglecting higher order terms, equation 21.65 becomes

$$\sigma + P = (-r/2)\frac{dP}{dr} \tag{21.66}$$

Substituting equation (21.66) into equation (21.64),

$$-(r/2)(dP/dr)(1 + v) + r(1 - v)(d\sigma/dr) + vr(dP/dr) = 0$$

or

$$\frac{d\sigma}{dr} - \frac{1}{2}\frac{dP}{dr} = 0 \tag{21.67}$$

which on integrating becomes,

$$\sigma - P/2 = A \tag{21.68}$$

Substituting equation (21.68) into equation (21.66)

$$3P/2 + A = (-r/2)(dP/dr)$$

or

$$-\frac{1}{r^2}\frac{d(P \times r^3)}{dr} = 2A$$

or

$$\frac{d(P \times r^3)}{dr} = -2Ar^2$$

which on integrating becomes,

$$P \times r^3 = -2Ar^3/3 + B$$

or

$$P = -2A/3 + B/r^3 \tag{21.69}$$

and $\sigma = 2A/3 + B/(2r^3)$ (21.70)

21.7 Rotating discs

These are of much importance in engineering components that rotate at high speeds. If the speed is high enough, such components can shatter when the centrifugal stresses become too large. The theory for thick circular cylinders can be extended to deal with problems in this category.

Consider a uniform thickness disc, of density ρ, rotating at a constant angular velocity ω.

From

$$E\frac{dw}{dr} = \sigma_r - v\sigma_\theta \tag{21.71}$$

and,

$$E\frac{w}{r} = \sigma_\theta - v\sigma_r \tag{21.72}$$

or,

$$Ew = \sigma_\theta \times r - v\sigma_r \times r \tag{21.73}$$

Differentiating equation (21.73) with respect to r,

$$E\frac{dw}{dr} = \sigma_\theta + r\frac{d\sigma_\theta}{dr} - v\sigma_r - vr\frac{d\sigma_r}{dr} \tag{21.74}$$

Equating (21.71) and (21.74),

$$(\sigma_\theta - \sigma_r)(1 + v) + r\frac{d\sigma_\theta}{dr} - vr\frac{d\sigma_r}{dr} = 0 \tag{21.75}$$

Considering radial equilibrium of an element of the disc, as shown in Figure 21.20,

$$2\sigma_\theta \times dr \times \sin\left(\frac{d\theta}{2}\right) + \sigma_r \times r \times d\theta$$

$$- (\sigma_r + d\sigma_r)(r + dr)d\theta = \rho \times \omega^2 \times r^2 \times dr \times d\theta$$

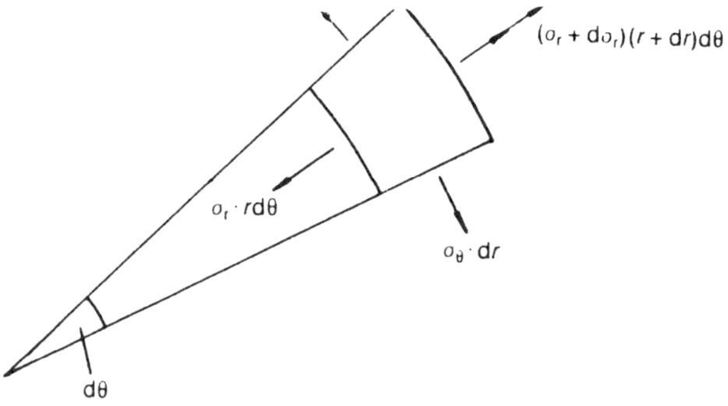

Figure 21.20 Element of disc.

In the limit, this reduces to

$$\sigma_\theta - \sigma_r - r\frac{d\sigma_r}{dr} = \rho\omega^2 r^2 \tag{21.76}$$

Substituting equation (21.76) into equation (21.75),

$$\left(r\frac{d\sigma_r}{dr} + \rho\omega^2 r^2\right)(1 + v) + r\frac{d\sigma_\theta}{dr} - vr\frac{d\sigma_r}{dr} = 0$$

or,

$$\frac{d\sigma_\theta}{dr} + \frac{d\sigma_r}{dr} = -\rho\omega^2 r^2 (1 + v)$$

which on integrating becomes,

$$\sigma_\theta + \sigma_r = -(\rho\omega^2 r^2/2)(1 + v) + 2A \tag{21.77}$$

Subtracting equation (21.76) from equation (21.77),

$$2\sigma_r + r\frac{d\sigma_r}{dr} = -(\rho\omega^2 r^2/2)(3 + v) + 2A$$

or,

$$\frac{1}{r}\frac{d(\sigma_r \times r^2)}{dr} = -\frac{\rho\omega^2 r^2 (3 + v)}{2} + 2A$$

which on integrating becomes,

$$\sigma_r\, r^2 = -\left(\rho\omega^2 r^4/8\right)(3+\nu) + Ar^2 - B \tag{21.78}$$

or

$$\sigma_r = A - B/r^2 - (3+\nu)\left(\rho\omega^2 r^2/8\right)$$

and,

$$\sigma_\theta = A + B/r^2 - (1+3\nu)\left(\rho\omega^2 r^2/8\right) \tag{21.79}$$

Problem 21.7 Obtain an expression for the variation in the thickness of a disc, in its radial direction, so that it will be of constant strength when it is rotated at an angular velocity ω.

Solution

Let,

t_0 = thickness at centre

t = thickness at a radius r

$t + \mathrm{d}t$ = thickness at a radius $r + \mathrm{d}r$

σ = stress = constant (everywhere)

Consider the radial equilibrium of an element of this disc at any radius r as shown in Figure 21.21.

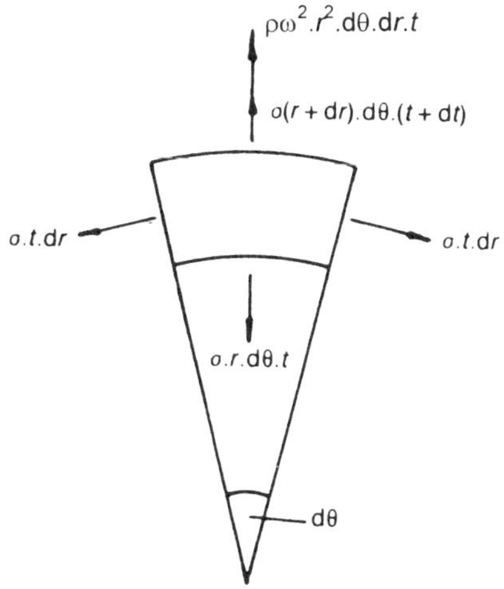

Figure 21.21 Element of constant strength disc.

Resolving forces radially

$$2\sigma \times t \times dr \sin\left(\frac{d\theta}{2}\right) + \sigma tr\, d\theta = \sigma(r + dr)(t + dt)\, d\theta + \rho\omega^2 r^2 t\, d\theta\, dr$$

Neglecting higher order terms, this equation becomes

$$\sigma t\, dt = \sigma r\, dt + \sigma t\, dr + \rho\omega^2 rt\, dr$$

or

$$\frac{dt}{dr} = -\rho\omega^2 rt/\sigma$$

which on integrating becomes,

$$\ln t = -\rho\omega^2 r^2 t/(2\sigma) + \ln C$$

or

$$t = Ce^{(-\rho\omega^2 r^2/2\sigma)}$$

Now, at $r = 0,\ t = t_0 \ \therefore C = t_0$

Hence,

$$t = t_0 e^{(-\rho\omega^2 r^2/2\sigma)} \tag{21.80}$$

21.7.1 Plastic collapse of rotating discs

Assume that $\sigma_\theta > \sigma_r$, and that plastic collapse occurs when

$$\sigma_\theta = \sigma_{yp}$$

where σ_{yp} is the yield stress.

Let R be the external radius of the disc. Then,

from *equilibrium considerations*,

$$\sigma_{yp} - \sigma_r - r\frac{d\sigma_r}{dr} = \rho\omega^2 r^2$$

or,

$$\int r \mathrm{d}\sigma_r = \int \left\{\sigma_{yp} - \sigma_r - \rho\omega^2 r^2\right\} \mathrm{d}r$$

Integrating the left-hand side of the above equation by parts,

$$r\ \sigma_r - \int \sigma_r\ \mathrm{d}r = \sigma_{yp}\ r - \int \sigma_r\ \mathrm{d}r\ \ \rho\omega^2 r^3/3 + A$$

therefore

$$\sigma_r = \sigma_{yp} - \rho\omega^2 r^2/3 + A/r \qquad (21.81)$$

For a solid disc, at $r = 0$, $\sigma_r \neq \infty$, or the disc will collapse at small values of ω. Therefore

$$A = 0$$

and

$$\sigma_r = \sigma_{yp} - \rho\omega^2 r^2/3$$

at $r = R$, $\sigma_r = 0$; therefore

$$0 = \sigma_{yp} - \rho\omega^2 R^2/3$$

$$\therefore\ \omega = \frac{1}{R}\sqrt{\frac{3\sigma_{yp}}{\rho}} \qquad (21.82)$$

where, ω is the angular velocity of the disc, which causes plastic collapse of the disc.

For an *annular disc*, of internal radius R_1 and external radius R_2, suitable boundary conditions for equation (21.81) are:

at $r = R_2$, $\sigma_r = 0$; therefore

$$A = \left(\rho\omega^2 R_1^2/3 - \sigma_{yp}\right)R_1$$

$$\therefore\ \sigma_r = \sigma_{yp} - \rho\omega^2 r^2/3 + \left(\rho\omega^2 R_1^2/3 - \sigma_{yp}\right)\left(R_1/r\right) \tag{21.83}$$

at $r = R_2$, $\sigma_r = 0$; therefore

$$0 = \sigma_{yp} - \rho\omega^2 R_2^2/3 + \left(\rho\omega^2 R_1^2/3 - \sigma_{yp}\right)\left(R_1/R_2\right)$$

$$\text{Hence,}\quad \omega = \sqrt{\left\{\left(\frac{3\sigma_{yp}}{\rho}\right)\frac{\left(R_2 - R_1\right)}{\left(R_2^3 - R_1^3\right)}\right\}} \tag{21.84}$$

21.8 Collapse of rotating rings

Consider the radial equilibrium of the thin semicircular ring element shown in Figure 21.21.

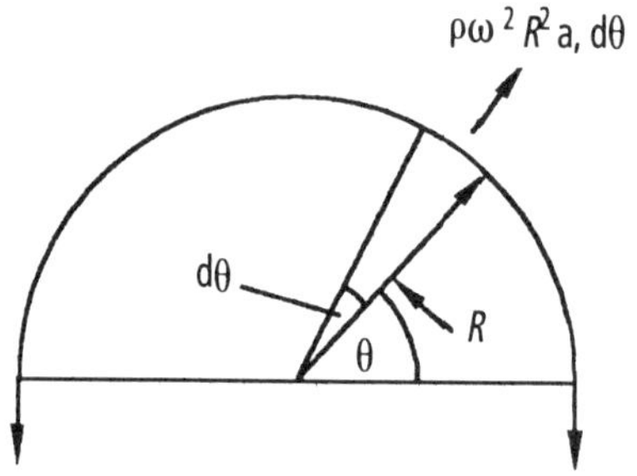

Figure 21.21 Ring element.

Let,

a = cross-sectional area of ring

R = mean radius of ring

Resolving forces vertically

$$\sigma_\theta \times a \times 2 = \int_0^\pi \rho\omega^2 R^2 a \, d\theta \quad \sin\theta$$

$$= \rho\omega^2 R^2 a \left[-\cos\theta\right]_0^\pi$$

$$= 2\rho\omega^2 R^2 a$$

$$\therefore \sigma_\theta = \rho\omega^2 R^2$$

at collapse,

$$\sigma_\theta = \sigma_{yp}$$

$$\therefore \omega = \frac{1}{R}\sqrt{\left(\frac{\sigma_{yp}}{\rho}\right)} \tag{21.85}$$

where ω is the angular velocity required to fracture the ring.

22 Introduction to matrix algebra

22.1 Introduction

Since the advent of the digital computer with its own memory, the importance of matrix algebra has continued to grow along with the developments in computers. This is partly because matrices allow themselves to be readily manipulated through skilful computer programming, and partly because many physical laws lend themselves to be readily represented by matrices.

The present chapter will describe the laws of matrix algebra by a methodological approach, rather than by rigorous mathematical theories. This is believed to be the most suitable approach for engineers, who will use matrix algebra as a tool.

22.2 Definitions

A rectangular matrix can be described as a table or array of quantities, where the quantities usually take the form of numbers, as shown be equations (22.1) and (22.2):

$$[\mathbf{A}] = \begin{bmatrix} a_{11} & a_{12} & a_{13} & \cdots & a_{1n} \\ a_{21} & a_{22} & a_{23} & \cdots & a_{2n} \\ a_{31} & a_{32} & a_{33} & \cdots & a_{3n} \\ \cdot & & & & \\ \cdot & & & & \\ \cdot & & & & \\ a_{m1} & a_{m2} & a_{m3} & \cdots & a_{mn} \end{bmatrix} \tag{22.1}$$

$$[\mathbf{B}] = \begin{bmatrix} 2 & -1 & 0 \\ 3 & 4 & -2 \\ -3 & 5 & 6 \\ -4 & -5 & 7 \end{bmatrix} \tag{22.2}$$

The matrix [**A**] of equation (22.1) is said to be of order $m \times n$, where

m = number of rows

n = number of columns

A row can be described as a horizontal line of quantities, and a column can be described as a vertical line of quantities, so that the matrix [**B**] of equation (22.2) is of order 4 × 3.

The quantities contained in the third row of [**B**] are −3, 5 and 6, and the quantities contained in the second column of [**B**] are −1, 4, 5 and −5.

A square matrix has the same number of rows as columns, as shown by equation (22.3), which is said to be of order n:

$$[\mathbf{A}] = \begin{bmatrix} a_{11} & a_{12} & a_{13} & \cdots & a_{1n} \\ a_{21} & a_{22} & a_{23} & \cdots & a_{2n} \\ a_{31} & a_{32} & a_{33} & \cdots & a_{3n} \\ \cdot & & & & \\ \cdot & & & & \\ \cdot & & & & \\ a_{nl} & a_{n2} & a_{n3} & \cdots & a_{nn} \end{bmatrix} \tag{22.3}$$

A column matrix contains a single column of quantities, as shown by equation (22.4), where it can be seen that the matrix is represented by braces:

$$\{\mathbf{A}\} = \begin{Bmatrix} a_{11} \\ a_{21} \\ a_{31} \\ \cdot \\ \cdot \\ \cdot \\ a_{nl} \end{Bmatrix} \tag{22.4}$$

A row matrix contains a single row of quantities, as shown by equation (22.5), where it can be seen that the matrix is represented by the special brackets:

$$[\mathbf{A}] = \lfloor a_{11} \quad a_{12} \quad a_{13} \quad \cdots \quad a_{1n} \rfloor \tag{22.5}$$

The transpose of a matrix is obtained by exchanging its columns with its rows, as shown by equation (22.6):

$$[\mathbf{A}]^{\mathrm{T}} = \begin{bmatrix} 1 & 0 \\ 4 & -3 \\ -5 & 6 \end{bmatrix}^{T} = [\mathbf{B}]$$

$$= \begin{bmatrix} 1 & 4 & -5 \\ 0 & -3 & 6 \end{bmatrix} \tag{22.6}$$

In equation (22.6), the first row of [**A**], when transposed, becomes the first column of [**B**]; the second row of [**A**] becomes the second column of [**B**] and the third row of [**A**] becomes the third column of [**B**], respectively.

22.3 Matrix addition and subtraction

Matrices can be added together in the manner shown below. If

$$[\mathbf{A}] = \begin{bmatrix} 1 & 0 \\ 4 & -3 \\ -5 & 6 \end{bmatrix}$$

and

$$[\mathbf{B}] = \begin{bmatrix} 2 & 9 \\ -7 & 8 \\ -1 & -2 \end{bmatrix}$$

$$[\mathbf{A}] + [\mathbf{B}] = \begin{bmatrix} (1+2) & (0+9) \\ (4-7) & (-3+8) \\ (-5-1) & (6-2) \end{bmatrix} \tag{22.7}$$

$$\begin{bmatrix} 3 & 9 \\ -3 & 5 \\ -6 & 4 \end{bmatrix}$$

Similarly, matrices can be subtracted in the manner shown below:

$$[\mathbf{A}] - [\mathbf{B}] = \begin{bmatrix} (1 - 2) & (0 - 9) \\ (4 + 7) & (-3 - 8) \\ (-5 + 1) & (6 + 2) \end{bmatrix} = \begin{bmatrix} -1 & -9 \\ 11 & -11 \\ -4 & 8 \end{bmatrix} \quad (22.8)$$

Thus, in general, for two $m \times n$ matrices:

$$[\mathbf{A}] + [\mathbf{B}] = \begin{bmatrix} (a_{11} + b_{11})(a_{12} + b_{12}) \ldots (a_{1n} + b_{1n}) \\ (a_{21} + b_{21})(a_{22} + b_{22}) \ldots (a_{2n} + b_{2n}) \\ \cdot \\ \cdot \\ \cdot \\ (a_{m1} + b_{m1})(a_{m2} + b_{m2}) \ldots (a_{mn} + b_{mn}) \end{bmatrix} \quad (22.9)$$

and

$$[\mathbf{A}] - [\mathbf{B}] = \begin{bmatrix} (a_{11} - b_{11})(a_{12} - b_{12}) \ldots (a_{1n} - b_{1n}) \\ (a_{21} - b_{21})(a_{22} - b_{22}) \ldots (a_{2n} - b_{2n}) \\ \cdot \\ \cdot \\ \cdot \\ (a_{m1} - b_{m1})(a_{m2} - b_{m2}) \ldots (a_{mn} - b_{mn}) \end{bmatrix} \quad (22.10)$$

22.4 Matrix multiplication

Matrices can be multiplied together, by multiplying the rows of the premultiplier into the columns of the postmultiplier, as shown by equations (22.11) and (22.12).

If

$$[\mathbf{A}] = \begin{bmatrix} 1 & 0 \\ 4 & -3 \\ -5 & 6 \end{bmatrix}$$

and

$$[\mathbf{B}] = \begin{bmatrix} 7 & 2 & -2 \\ -1 & 3 & -4 \end{bmatrix}$$

$$[\mathbf{A}] \times [\mathbf{B}] = [\mathbf{C}]$$

$$= \begin{bmatrix} (1\times 7+0\times(-1)) & (1\times 2+0\times 3) & (1\times(-2)+0\times(-4)) \\ (4\times 7+(-3)\times(-1)) & (4\times 2+(-3)\times 3) & (4\times(-2)+(-3)\times(-4)) \\ (-5\times 7+6\times(-1)) & (-5\times 2+6\times 3) & (-5\times(-2)+6\times(-4)) \end{bmatrix} \tag{22.11}$$

$$= \begin{bmatrix} (7+0) & (2+0) & (-2+0) \\ (28+3) & (8-9) & (-8+12) \\ (-35-6) & (-10+18) & (10-24) \end{bmatrix}$$

$$[\mathbf{C}] = \begin{bmatrix} 7 & 2 & -2 \\ 31 & -1 & 4 \\ -41 & 8 & -14 \end{bmatrix} \tag{22.12}$$

i.e. to obtain an element of the matrix [**C**], namely C_{ij}, the ith row of the premultiplier [**A**] must be premultiplied into the jth column of the postmultiplier [**B**] to give

$$C_{ij} = \sum_{k=1}^{p} a_{ik} \times b_{kj}$$

where

P = the number of columns of the premultiplier and also, the number of rows of the postmultiplier.

NB The premultiplying matrix [**A**] must have the same number of columns as the rows in the postmultiplying matrix [**B**].

In other words, if [**A**] is of order ($m \times P$) and [**B**] is of order ($P \times n$), then the product [**C**] is of order ($m \times n$).

22.5 Some special types of square matrix

A *diagonal matrix* is a square matrix which contains all its non-zero elements in a diagonal from the top left corner of the matrix to its bottom right corner, as shown by equation (22.13). This diagonal is usually called the main or leading diagonal.

$$[\mathbf{A}] = \begin{bmatrix} a_{11} & 0 & 0 & 0 \\ 0 & a_{22} & & 0 \\ 0 & 0 & a_{33} & 0 \\ & 0 & & \\ & & & 0 \\ 0 & 0 & 0 & a_{nn} \end{bmatrix} \quad (22.13)$$

A special case of diagonal matrix is where all the non-zero elements are equal to unity, as shown by equation (22.14). This matrix is called a *unit matrix*, as it is the matrix equivalent of unity.

$$[\mathbf{I}] = \begin{bmatrix} 1 & 0 & 0 & 0 \\ 0 & 1 & 0 & 0 \\ 0 & 0 & 1 & \\ & & & 0 \\ 0 & 0 & 0 & 1 \end{bmatrix} \quad (22.14)$$

A *symmetrical matrix* is shown in equation (22.15), where it can be seen that the matrix is symmetrical about its leading diagonal:

$$[\mathbf{A}] = \begin{bmatrix} 8 & 2 & -3 & 1 \\ 2 & 5 & 0 & 6 \\ -3 & 0 & 9 & -7 \\ 1 & 6 & -7 & 4 \end{bmatrix} \tag{22.15}$$

i.e. for a symmetrical matrix, all

$$a_{ij} = a_{ji}$$

22.6 Determinants

The determinant of the 2×2 matrix of equation (22.16) can be evaluated, as follows:

$$[\mathbf{A}] = \begin{vmatrix} 4 & 2 \\ -1 & 6 \end{vmatrix} \tag{22.16}$$

Determinant of $[\mathbf{A}| = 4 \times 6 - 2 \times (-1) = 24 + 2 = 26$

so that, in general, the determinant of a 2 × 2 matrix, namely **det**[**A**], is given by:

$$\mathbf{det}\ [\mathbf{A}] = a_{11} \times a_{22} - a_{12} \times a_{21} \tag{22.17}$$

where

$$|\mathbf{A}| = \begin{vmatrix} a_{11} & a_{12} \\ a_{21} & a_{22} \end{vmatrix} \tag{22.18}$$

Similarly, the determinant of the 3×3 matrix of equation (22.19) can be evaluated, as shown by equation (22.20):

$$\mathbf{det}\ |\mathbf{A}| = \begin{vmatrix} a_{11} & a_{12} & a_{13} \\ a_{21} & a_{22} & a_{23} \\ a_{31} & a_{32} & a_{33} \end{vmatrix} \tag{22.19}$$

$$= a_{11}\begin{vmatrix} a_{22} & a_{23} \\ a_{32} & a_{33} \end{vmatrix} - a_{12}\begin{vmatrix} a_{21} & a_{23} \\ a_{31} & a_{33} \end{vmatrix} + a_{13}\begin{vmatrix} a_{21} & a_{22} \\ a_{31} & a_{32} \end{vmatrix} \qquad (22.20)$$

For example, the determinant of equation (22.21) can be evaluated, as follows:

$$\mathbf{det}\ |\mathbf{A}| = \begin{vmatrix} 8 & 2 & -3 \\ 2 & 5 & 0 \\ -3 & 0 & 9 \end{vmatrix} \qquad (22.21)$$

$$= 8\begin{vmatrix} 5 & 0 \\ 0 & 9 \end{vmatrix} -2\begin{vmatrix} 2 & 0 \\ -3 & 9 \end{vmatrix} + (-3)\begin{vmatrix} 2 & 5 \\ -3 & 0 \end{vmatrix}$$

$$= 8\ (45 - 0) -2(18 - 0) -3\ (0 + 15)$$

or

$$\mathbf{det}\ |\mathbf{A}| = 279$$

For a determinant of large order, this method of evaluation is unsatisfactory, and readers are advised to consult Ross, C T F, *Advanced Applied Finite Element Methods* (Horwood 1998), or Collar, A R, and Simpson, A, *Matrices and Engineering Dynamics* (Ellis Horwood, 1987) which give more suitable methods for expanding larger order determinants.

22.7 Cofactor and adjoint matrices

The cofactor of a third order matrix is obtained by removing the appropriate columns and rows of the cofactor, and evaluating the resulting determinants, as shown below.

If

$$[\mathbf{A}] = \begin{bmatrix} a_{11} & a_{12} & a_{13} \\ a_{21} & a_{22} & a_{23} \\ a_{31} & a_{32} & a_{33} \end{bmatrix}$$

then $[\mathbf{A}]^c$ = the cofactor matrix of [**A**], where,

$$[\mathbf{A}]^{\mathbf{c}} = \begin{bmatrix} a_{11}^c & a_{12}^c & a_{13}^c \\ a_{21}^c & a_{22}^c & a_{23}^c \\ a_{31}^c & a_{32}^c & a_{33}^c \end{bmatrix} \tag{22.22}$$

and the cofactors are evaluated, as follows:

$$a_{11}^c = \begin{vmatrix} a_{22} & a_{23} \\ a_{32} & a_{33} \end{vmatrix}$$

$$a_{12}^c = -\begin{vmatrix} a_{21} & a_{23} \\ a_{31} & a_{33} \end{vmatrix}$$

$$a_{13}^c = \begin{vmatrix} a_{21} & a_{22} \\ a_{31} & a_{32} \end{vmatrix}$$

$$a_{21}^c = -\begin{vmatrix} a_{12} & a_{13} \\ a_{32} & a_{33} \end{vmatrix}$$

$$a_{22}^c = \begin{vmatrix} a_{11} & a_{13} \\ a_{31} & a_{33} \end{vmatrix}$$

$$a_{23}^c = -\begin{vmatrix} a_{11} & a_{12} \\ a_{31} & a_{32} \end{vmatrix}$$

$$a_{31}^c = \begin{vmatrix} a_{12} & a_{13} \\ a_{22} & a_{23} \end{vmatrix}$$

$$a_{32}^c = -\begin{vmatrix} a_{11} & a_{13} \\ a_{21} & a_{23} \end{vmatrix}$$

$$a_{33}^c = \begin{vmatrix} a_{11} & a_{12} \\ a_{21} & a_{22} \end{vmatrix}$$

The adjoint or adjugate matrix, $[\mathbf{A}]^a$ is obtained by transposing the cofactor matrix, as follows:

ie $$[\mathbf{A}]^{\mathbf{a}} = [\mathbf{A}^{\mathbf{c}}]^{\mathbf{T}} \qquad (22.23)$$

22.8 Inverse of a matrix $[\mathbf{A}]^{-1}$

The inverse or reciprocal matrix is required in matrix algebra, as it is the matrix equivalent of a scalar reciprocal, and it is used for division.

The inverse of the matrix [**A**] is given by equation (22.24):

$$[\mathbf{A}]^{-1} = \frac{[\mathbf{A}]^{\mathbf{a}}}{\det |\mathbf{A}|} \qquad (22.24)$$

For the 2 × 2 matrix of equation (22.25),

$$\begin{bmatrix} a_{11} & a_{12} \\ a_{21} & a_{22} \end{bmatrix} \qquad (22.25)$$

the cofactors are given by

$$a_{11}^c = a_{22}$$

$$a_{12}^c = -a_{21}$$

$$a_{21}^c = -a_{12}$$

$$a_{22}^c = a_{11}$$

and the determinant is given by:

$$\textbf{det}\ |\mathbf{A}| = a_{11} \times a_{22} - a_{12} \times a_{21}$$

so that

$$\begin{bmatrix} a_{11} & a_{12} \\ a_{21} & a_{22} \end{bmatrix}^{-1} = \frac{\begin{bmatrix} a_{22} & -a_{12} \\ -a_{21} & a_{11} \end{bmatrix}}{(a_{11} \times a_{22} - a_{12} \times a_{21})} \tag{22.26}$$

In general, inverting large matrices through the use of equation (22.24) is unsatisfactory, and for large matrices, the reader is advised to refer to Ross, C T F, *Advanced Applied Finite Element Methods* (Horwood 1998), where a computer program is presented for solving *n*th order matrices on a microcomputer.

The inverse of a unit matrix is another unit matrix of the same order, and the inverse of a diagonal matrix is obtained by finding the reciprocals of its leading diagonal.

The inverse of an orthogonal matrix is equal to its transpose. A typical orthogonal matrix is shown in equation (22.27):

$$[\mathbf{A}] = \begin{bmatrix} c & s \\ -s & c \end{bmatrix} \tag{22.27}$$

where

$$c = \cos\alpha$$

$$s = \sin\alpha$$

The cofactors of [**A**] are:

$$a_{11}^{c} = c$$

$$a_{12}^{c} = s$$

$$a_{21}^{c} = -s$$

$$a_{22}^{c} = c$$

and

$$\textbf{det}\ |\mathbf{A}| = c^2 + s^2 = 1$$

so that

$$[\mathbf{A}]^{-1} = \begin{bmatrix} c & -s \\ s & c \end{bmatrix}$$

i.e. for an orthogonal matrix

$$[\mathbf{A}]^{-1} = [\mathbf{A}]^{\mathrm{T}} \tag{22.28}$$

22.9 Solution of simultaneous equations

The inverse of a matrix can be used for solving the set of linear simultaneous equations shown in equation (22.29). If,

$$[\mathbf{A}]\,\{x\} = \{c\} \tag{22.29}$$

where [**A**] and {c} are known and $\{x\}$ is a vector of unknowns, then $\{x\}$ can be obtained from equation (22.30), where $[\mathbf{A}]^{-1}$ has been pre-multiplied on both sides of this equation:

$$\{x\} = [\mathbf{A}]^{-1}\,\{c\} \tag{22.30}$$

Another method of solving simultaneous equations, which is usually superior to inverting the matrix, is by triangulation. For this case, the elements of the matrix below the leading diagonal are eliminated, so that the last unknown can readily be determined, and the remaining unknowns obtained by back-substitution.

Further problems *(answers on page 695)*

If

$$[\mathbf{A}] = \begin{bmatrix} 4 & 1 \\ 2 & 3 \end{bmatrix} \quad \text{and} \quad [\mathbf{B}] = \begin{bmatrix} -1 & 0 \\ 2 & -4 \end{bmatrix}$$

Determine:

22.1 [**A**] + [**B**]

22.2 [**A**] − [**B**]

22.3 $[\mathbf{A}]^T$

22.4 $[\mathbf{B}]^T$

22.5 $[\mathbf{A}] \times [\mathbf{B}]$

22.6 $[\mathbf{B}] \times [\mathbf{A}]$

22.7 det $[\mathbf{A}]$

22.8 det $[\mathbf{B}]$

22.9 $[\mathbf{A}]^{-1}$

22.10 $[\mathbf{B}]^{-1}$

If

$$[\mathbf{C}] = \begin{bmatrix} 1 & -2 & 0 \\ -2 & 1 & -2 \\ 0 & -2 & 1 \end{bmatrix}$$

and

$$[\mathbf{D}] = \begin{bmatrix} 9 & 1 & -2 \\ -1 & 8 & 3 \\ -4 & 0 & 6 \end{bmatrix}$$

determine:

22.11 $[\mathbf{C}] + [\mathbf{D}]$

22.12 $[\mathbf{C}] - [\mathbf{D}]$

22.13 $[\mathbf{C}]^{T}$

22.14 $[\mathbf{D}]^{T}$

22.15 $[\mathbf{C}] \times [\mathbf{D}]$

22.16 $[\mathbf{D}] \times [\mathbf{C}]$

22.17 det $[\mathbf{C}]$

22.18 det $[\mathbf{D}]$

22.19 $[\mathbf{C}]^{-1}$

22.20 $[\mathbf{D}]^{-1}$

If

$$[\mathbf{E}] = \begin{bmatrix} 2 & 4 \\ -3 & 1 \\ 5 & 6 \end{bmatrix}$$

and

$$[\mathbf{F}] = \begin{bmatrix} 0 & 7 & -1 \\ 8 & -4 & -5 \end{bmatrix}$$

determine:

22.21 $[\mathbf{E}]^{T}$

22.22 $[\mathbf{F}]^{T}$

22.23 $[\mathbf{E}] \times [\mathbf{F}]$

22.24 $[\mathbf{F}]^T \times [\mathbf{E}]^T$

22.25 If

$$
\begin{aligned}
x_1 - 2x_2 + 0 &= -2 \\
-x_1 + x_2 - 2x_3 &= 1 \\
0 - 2x_2 + x_3 &= 3
\end{aligned}
$$

determine

x_1, x_2 and x_3

23 Matrix methods of structural analysis

23.1 Introduction

This chapter describes and applies the matrix displacement method to various problems in structural analysis. The matrix displacement method first appeared in the aircraft industry in the 1940s[7], where it was used to improve the strength-to-weight ratio of aircraft structures.

In today's terms, the structures that were analysed then were relatively simple, but despite this, teams of operators of mechanical, and later electromechanical, calculators were required to implement it. Even in the 1950s, the inversion of a matrix of modest size, often took a few weeks to determine. Nevertheless, engineers realised the importance of the method, and it led to the invention of the finite element method in 1956[8], which is based on the matrix displacement method. Today, of course, with the progress made in digital computers, the matrix displacement method, together with the finite element method, is one of the most important forms of analysis in engineering science.

The method is based on the elastic theory, where it can be assumed that most structures behave like complex elastic springs, the load–displacement relationship of which is linear. Obviously, the analysis of such complex springs is extremely difficult, but if the complex spring is subdivided into a number of simpler springs, which can readily be analysed, then by considering equilibrium and compatibility at the boundaries, or nodes, of these simpler elastic springs, the entire structure can be represented by a large number of simultaneous equations. Solution of the simultaneous equations results in the displacements at these nodes, whence the stresses in each individual spring element can be determined through Hookean elasticity.

In this chapter, the method will first be applied to pin-jointed trusses, and then to continuous beams and rigid-jointed plane frames.

23.2 Elemental stiffness matrix for a rod

A pin-jointed truss can be assumed to be a structure composed of line elements, called rods, which possess only axial stiffness. The joints connecting the rods together are assumed to be in the form of smooth, frictionless hinges. Thus these rod elements in fact behave like simple elastic springs, as described in Chapter 1.

Consider now the rod element of Figure 23.1, which is described by two nodes at its ends, namely, node 1 and node 2.

---

[7]Levy, S., Computation of Influence Coefficients for Aircraft Structures with Discontinuities and Sweepback, *J. Aero. Sci.*, **14**, 547–560, October 1947.

[8]Turner, M.J., Clough, R.W., Martin, H.C. and Topp, L.J., Stiffness and Deflection Analysis of Complex Structures, *J. Aero. Sci.*, **23**, 805–823, 1956.

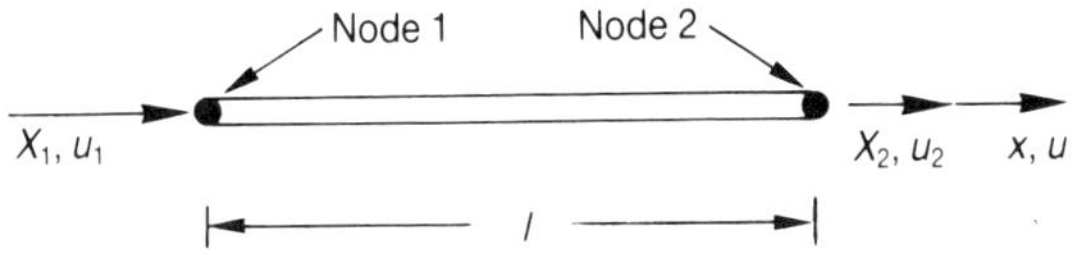

Figure 23.1 Simple rod element.

Let

X_1 = axial force at node 1

X_2 = axial force at node 2

u_1 = axial deflection at node 1

u_2 = axial deflection at node 2

A = cross-sectional area of the rod element

l = elemental length

E = Young's modulus of elasticity

Applying Hooke's law to node 1,

$$\frac{\sigma}{\varepsilon} = E$$

but

$$\sigma = X_1/A$$

and

$$\varepsilon = (u_1 - u_2)/l$$

so that

$$X_1 = AE\,(u_1 - u_2)/l \tag{23.1}$$

From equilibrium considerations

$$X_2 = -X_1 = AE\,(u_2 - u_1)/l \tag{23.2}$$

Rewriting equations (23.1) and (23.2), into matrix form, the following relationship is obtained:

$$\begin{Bmatrix} X_1 \\ X_2 \end{Bmatrix} = \frac{AE}{l}\begin{bmatrix} 1 & -1 \\ -1 & 1 \end{bmatrix}\begin{Bmatrix} u_1 \\ u_2 \end{Bmatrix} \tag{23.3}$$

or in short form, equation (23.3) can be written

$$\{P_i\} = [\mathbf{k}]\{u_i\} \tag{23.4}$$

where,

$$\{P_i\} = \begin{Bmatrix} X_1 \\ X_2 \end{Bmatrix} = \text{a vector of loads}$$

$$\{u_i\} = \begin{Bmatrix} u_1 \\ u_2 \end{Bmatrix} = \text{a vector of nodal displacements}$$

Now, as Force = stiffness × displacement

$$[\mathbf{k}] = \frac{AE}{l}\begin{bmatrix} 1 & -1 \\ -1 & 1 \end{bmatrix} \tag{23.5}$$

= the stiffness matrix for a rod element

23.3 System stiffness matrix [K]

A structure such as pin-jointed truss consists of several rod elements; so to demonstrate how to form the system or structural stiffness matrix, consider the structure of Figure 23.2, which is composed of two in-line rod elements.

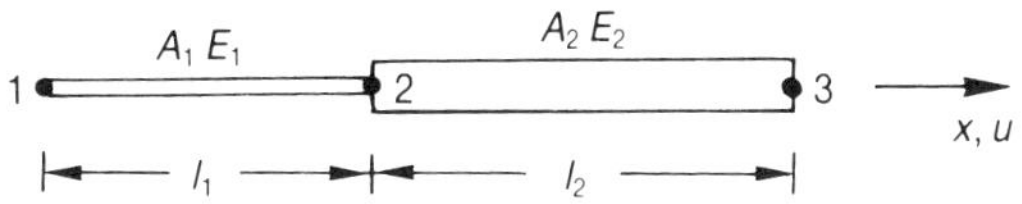

Figure 23.2 Two-element structure.

Consider element 1–2. Then from equation (23.5), the stiffness matrix for the rod element 1–2 is

$$[\mathbf{k}_{1-2}] = \frac{A_1 E_1}{l_1} \begin{bmatrix} 1 & -1 \\ -1 & 1 \end{bmatrix} \begin{matrix} u_1 \\ u_2 \end{matrix} \qquad (23.6)$$

The element is described as 1–2, which means it points from node 1 to node 2, so that its start node is 1 and its finish node is 2. The displacements u_1 and u_2 are not part of the stiffness matrix, but are used to describe the coefficients of stiffness that correspond to those displacements.

Consider element 2–3. Substituting the values A_2, E_2 and l_2 into equation (23.5), the elemental stiffness matrix for element 2–3 is given by

$$[\mathbf{k}_{2-3}] = \frac{A_2 E_2}{l_2} \begin{matrix} u_2 \quad u_3 \\ \begin{bmatrix} 1 & -1 \\ -1 & 1 \end{bmatrix} \end{matrix} \begin{matrix} u_2 \\ u_3 \end{matrix} \qquad (23.7)$$

Here again, the displacements u_2 and u_3 are not part of the stiffness matrix, but are used to describe the components of stiffness corresponding to these displacements.

The system stiffness matrix **[K]** is obtained by superimposing the coefficients of stiffness of the elemental stiffness matrices of equations (23.6) and (23.7), into a system stiffness matrix of pigeon holes, as shown by equation (23.8):

$$[\mathbf{K}] = \begin{bmatrix} u_1 & u_2 & u_3 \\ A_1E_1/l_1 & -A_1E_1/l_1 & 0 \\ -A_1E_1/l_1 & A_1E_1/l_1 + A_2E_2/l_2 & -A_2E_2/l_2 \\ 0 & -A_2E_2/l_2 & A_2E_2/l_2 \end{bmatrix} \begin{matrix} \\ u_1 \\ u_2 \\ u_3 \end{matrix} \qquad (23.8)$$

It can be seen from equation (23.8), that the components of stiffness are added together with reference to the displacements u_1, u_2 and u_3. This process, effectively mathematically joins together the two springs at their common node, namely node 2.

Let

$$\{q\} = \begin{Bmatrix} P_1 \\ P_2 \\ P_3 \end{Bmatrix} \tag{23.9}$$

= a vector of known externally applied loads at the nodes, 1, 2 and 3, respectively

$$\{u_i\} = \begin{Bmatrix} u_1 \\ u_2 \\ u_3 \end{Bmatrix} \tag{23.10}$$

= a vector of unknown nodal displacements, due to $\{q\}$, at nodes 1, 2 and 3 respectively

Now for the entire structure,

force = stiffness × displacement, or

$$\{q\} = [\mathbf{K}]\,\{u_i\} \tag{23.11}$$

where [**K**] is the system or structural stiffness matrix.

Solution of equation (23.11) cannot be carried out, as [**K**] is singular, i.e. the structure is floating in space and has not been constrained. To constrain the structure of Figure 23.2, let us assume that it is firmly fixed at (say) node 3, so that $u_3 = 0$.

Equation (23.11) can now be partitioned with respect to the free displacements, namely u_1 and u_2, and the constrained displacement, namely u_3, as shown by equation (23.12):

$$\left\{\begin{array}{c} q_F \\ \hline R \end{array}\right\} = \left[\begin{array}{c|c} K_{11} & K_{12} \\ \hline K_{21} & K_{22} \end{array}\right] \left\{\begin{array}{c} u_F \\ \hline u_3 = 0 \end{array}\right\} \tag{23.12}$$

where

$$\{q_F\} = \begin{Bmatrix} P_1 \\ P_2 \end{Bmatrix} \tag{23.13}$$

= a vector of known nodal forces, corresponding to the free displacements, namely u_1 and u_2

$$\{u_F\} = \begin{Bmatrix} u_1 \\ u_2 \end{Bmatrix} \tag{23.14}$$

= a vector of free displacements, which have to be determined

$$[\mathbf{K}_{11}] = \begin{bmatrix} A_1 E_1/l_1 & -AE_1/l_1 \\ -AE_1/l_1 & (A_1E_1/l_1 + A_2E_2/l_2) \end{bmatrix} \tag{23.15}$$

= that part of the system stiffness matrix that corresponds to the free displacements, which in this case is u_1 and u_2

$\{R\}$ = a vector of reactions corresponding to the constrained displacements, which in this case is u_3

$$[\mathbf{K}_{22}] = [A_2 E_2 / l_2]$$

$$[\mathbf{K}_{21}] = [0 - A_2 E_2 / l_2]$$

in this case

$$[\mathbf{K}_{12}] = \begin{bmatrix} 0 \\ -A_2 E_2 / l_2 \end{bmatrix}$$

Expanding the top part of equation (23.12):

$$\{q_F\} = [\mathbf{K}_{11}]\{u_F\}$$

$$\therefore \{u_F\} = [\mathbf{K}_{11}]^{-1}\{q_F\} \tag{23.16}$$

Once $\{u_F\}$ is determined, the initial stresses can be determined through Hookean elasticity.

For some cases u_3 may not be zero but may have a known value, say u_c. For these cases, equation (23.12) becomes

$$\begin{Bmatrix} q_F \\ \hline R \end{Bmatrix} = \left[\begin{array}{c|c} K_{11} & K_{12} \\ \hline K_{21} & K_{22} \end{array}\right] \begin{Bmatrix} u_F \\ \hline u_c \end{Bmatrix} \tag{23.17}$$

so that

$$\{u_F\} = [\mathbf{K}_{11}]^{-1}(\{q_F\} - [\mathbf{K}_{12}]\{u_c\}) \tag{23.18}$$

and

$$\{R\} = [K_{21}]\{u_F\} + [K_{22}]\{u_c\} \tag{23.19}$$

23.4 Relationship between local and global co-ordinates

The rod element of Figure 23.1 is not very useful element because it lies horizontally, when in fact a typical rod element may lie at some angle to the horizontal, as shown in Figures 23.3 and 23.4, where the x–$y°$ axes are the global axes and the x–y axes are the local axes.

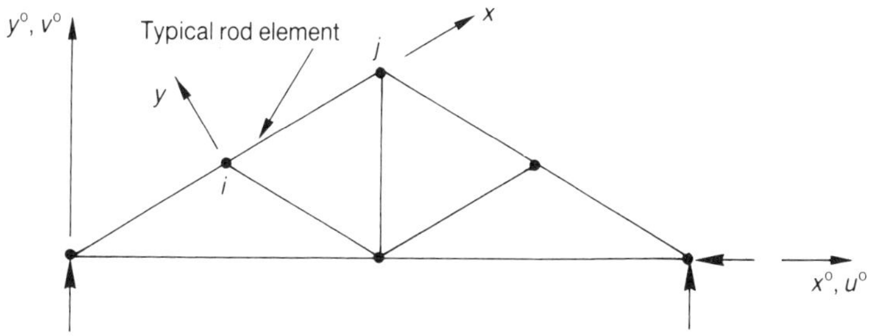

Figure 23.3 Plane pin-jointed truss.

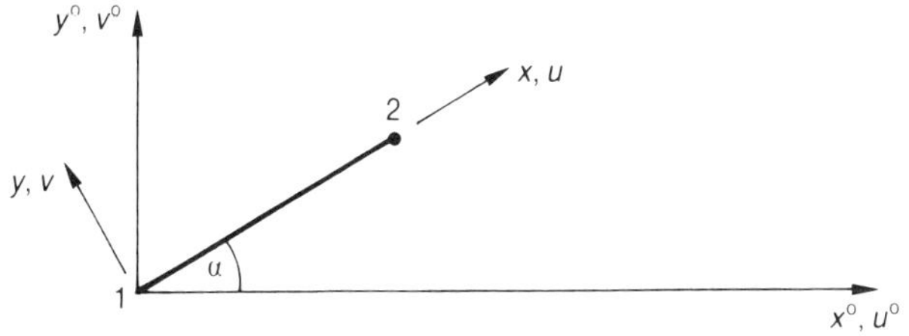

Figure 23.4 Rod element, shown in local and global systems.

From Figure 23.4, it can be seen that the relationships between the local displacements u and v, and the global displacements $u°$ and $v°$, are given by equation (23.20):

$$\begin{aligned} u &= u^\circ \cos\alpha + v^\circ \sin\alpha \\ v &= -u^\circ \sin\alpha + v^\circ \cos\alpha \end{aligned} \tag{23.20}$$

which, when written in matrix form, becomes:

$$\left\{\frac{u}{v}\right\} = \begin{bmatrix} \cos\alpha & \sin\alpha \\ -\sin\alpha & \cos\alpha \end{bmatrix} \left\{\frac{u^\circ}{v^\circ}\right\} \tag{23.21}$$

For node 1,

$$\left\{\frac{u_1}{v_1}\right\} = \begin{bmatrix} c & s \\ -s & c \end{bmatrix} \begin{Bmatrix} u_1^\circ \\ v_1^\circ \end{Bmatrix} \tag{23.22}$$

where,

$c = \cos\alpha$

$s = \sin\alpha$

Similarly, for node 2

$$\left\{\frac{u_2}{v_2}\right\} = \begin{bmatrix} c & s \\ -s & c \end{bmatrix} \begin{Bmatrix} u_2^\circ \\ v_2^\circ \end{Bmatrix}$$

Or, for both nodes,

$$\begin{Bmatrix} u_1 \\ v_1 \\ u_2 \\ v_2 \end{Bmatrix} = \left[\begin{array}{c|c} \zeta & 0_2 \\ \hline 0_2 & \zeta \end{array}\right] \begin{Bmatrix} u_1^\circ \\ v_1^\circ \\ u_2^\circ \\ v_2^\circ \end{Bmatrix} \tag{23.23}$$

where,

$$[\zeta] = \begin{bmatrix} c & s \\ -s & c \end{bmatrix}$$

$$[\mathbf{0_2}] = \begin{bmatrix} 0 & 0 \\ 0 & 0 \end{bmatrix}$$

Equation (23.23) can be written in the form:

$$\{u_i\} = [\mathbf{DC}]\ \{u_i{}^\circ\} \tag{23.24}$$

where,

$$[\mathbf{DC}] = \left[\begin{array}{c|c} \zeta & \mathbf{0_2} \\ \hline \mathbf{0_2} & \zeta \end{array}\right] \tag{23.25}$$

= a matrix of directional cosines

$$\{u_i\} = \begin{Bmatrix} u_1 \\ v_1 \\ u_2 \\ v_2 \end{Bmatrix}$$

$$\{u_i{}^\circ\} = \begin{Bmatrix} u_1{}^\circ \\ v_1{}^\circ \\ u_2{}^\circ \\ v_2{}^\circ \end{Bmatrix}$$

From equation (23.25), it can be seen that [**DC**] is orthogonal, i.e.

$$[\mathbf{DC}]^{-1} = [\mathbf{DC}]^{\mathrm{T}}$$

$$\therefore \{u_i{}^\circ\} = [\mathbf{DC}]^{\mathrm{T}}\ \{u_i\} \tag{23.26}$$

Similarly, it can be shown that

$$\{P_i\} = [\mathbf{DC}]\,\{P_i{}^\circ\} \tag{23.27}$$

and $$\{P_i{}^\circ\} = [\mathbf{DC}]^{\mathrm{T}}\,\{P_i\}$$

where

$$\{P_i\} = \begin{Bmatrix} X_1 \\ Y_1 \\ X_2 \\ Y_2 \end{Bmatrix}$$

and

$$\{P_i{}^\circ\} = \begin{Bmatrix} X_1{}^\circ \\ Y_1{}^\circ \\ X_2{}^\circ \\ Y_2{}^\circ \end{Bmatrix}$$

23.5 Plane rod element in global co-ordinates

For this case, there are four degrees of freedom per element, namely $u_1{}^\circ$, $v_1{}^\circ$, $u_2{}^\circ$ and $v_2{}^\circ$. Thus, the elemental stiffness matrix for a rod in local co-ordinates must be written as a 4 × 4 matrix, as shown by equation (23.28):

$$\begin{matrix} & & u_1 & v_1 & u_2 & v_2 & \\ [\mathrm{k}] = & \dfrac{AE}{l} & \left[\begin{matrix} 1 \\ 0 \\ -1 \\ 0 \end{matrix}\right. & \begin{matrix} 0 \\ 0 \\ 0 \\ 0 \end{matrix} & \begin{matrix} -1 \\ 0 \\ 1 \\ 0 \end{matrix} & \left.\begin{matrix} 0 \\ 0 \\ 0 \\ 0 \end{matrix}\right] & \begin{matrix} u_1 \\ v_1 \\ u_2 \\ v_2 \end{matrix} \end{matrix} \tag{23.28}$$

The reason why the coefficients of the stiffness matrix under v_1 and v_2 are zero, is that the rod only possesses axial stiffness in the local x-direction, as shown in Figure 23.1.

For the inclined rod of Figure 23.4, although the rod only possesses stiffness in the x-direction, it has components of stiffness in the global $x°$- and $y°$-directions.

The elemental stiffness matrix for a rod in global co-ordinates is obtained, as follows. From equation (23.4):

$$\{P_i\} = [\mathbf{k}]\{u_i\} \tag{23.29}$$

but

$$\{P_i\} = [\mathbf{DC}]\{P_i°\} \tag{23.30}$$

and

$$\{u_i\} = [\mathbf{DC}]\{u_i°\} \tag{23.31}$$

Substituting equations (23.30) and (23.31) into equation (23.29), the following is obtained:

$$[\mathbf{DC}]\{P_i°\} = [\mathbf{k}][\mathbf{DC}]\{u_i°\} \tag{23.32}$$

Premultiplying both sides by $[\mathbf{DC}]^{-1}$,

$$\{P_i°\} = [\mathbf{DC}]^{-1}[\mathbf{k}][\mathbf{DC}]\{u_i°\}$$

but from equation (22.28),

$$[\mathbf{DC}]^{-1} = [\mathbf{DC}]^{\mathrm{T}}$$

$$\therefore \{P_i°\} = [\mathbf{DC}]^{\mathrm{T}}[\mathbf{k}][\mathbf{DC}]\{u_i°\} \tag{23.33}$$

Now,

$$\text{force} = \text{stiffness} \times \text{deflection}$$

$$\therefore \{P_i°\} = [\mathbf{k}°]\{u_i°\} \tag{23.34}$$

Comparing equation (23.34) with (23.33),

$$[\mathbf{k}^\circ] = [\mathbf{DC}]^{\mathrm{T}}\,[\mathbf{k}]\,[\mathbf{DC}] \tag{23.35}$$

= elemental stiffness matrix in global co-ordinates

$$[\mathbf{k}^\circ] = \frac{AE}{l}\begin{array}{c} \begin{array}{cccc} u_1^\circ & v_1^\circ & u_2^\circ & v_2^\circ \end{array} \\ \begin{bmatrix} c^2 & cs & -c^2 & -cs \\ cs & s^2 & -cs & -s^2 \\ -c^2 & -cs & c^2 & cs \\ -cs & -s^2 & cs & s^2 \end{bmatrix} \end{array}\begin{array}{c} u_1^\circ \\ v_1^\circ \\ u_2^\circ \\ v_2^\circ \end{array} \tag{23.36}$$

= the elemental stiffness matrix for a rod in global co-ordinates

Problem 23.1 The plane pin-jointed truss below may be assumed to be composed of uniform section members, with the same material properties. If the truss is subjected to the load shown, determine the forces in the members of the truss.

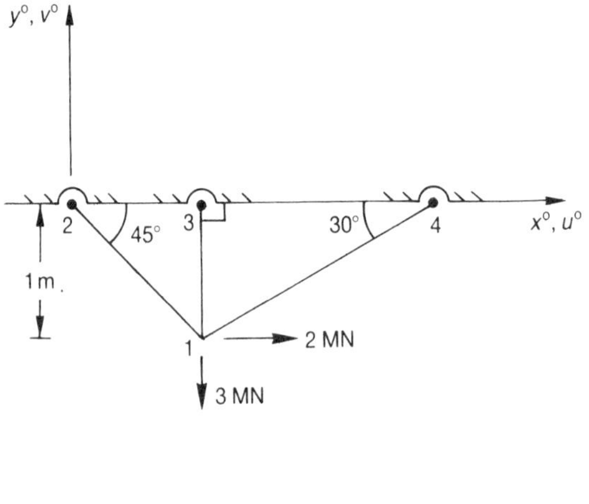

Solution

This truss has two free degrees of freedom, namely, the unknown displacements u_1° and v_1°.

Element 1–2

This element points from 1 to 2, so that its start node is 1 and its end node is 2, as shown:

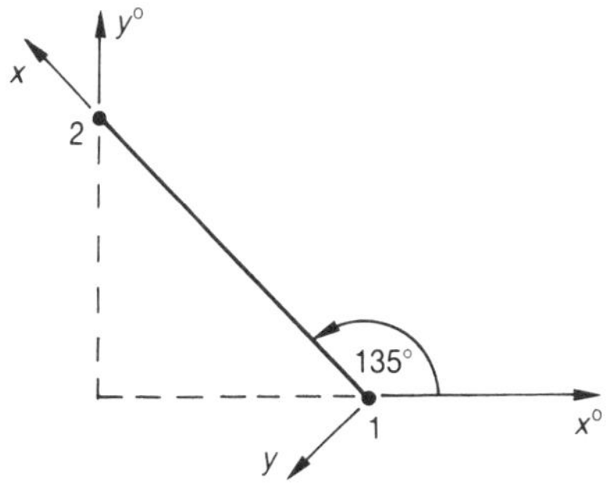

$\alpha = 135°$ $\therefore c = -0.707,$ $s = 0.707,$ $l = 1.414$ m

Substituting the above information into equation (23.36), and removing the rows and columns corresponding to the zero displacements, namely $u_2°$ and $v_2°$, the elemental stiffness matrix for element 1–2 is given by

$$\begin{array}{c} \\ [\mathbf{k}_{1\text{-}2}°] = \dfrac{AE}{1.414} \end{array} \begin{array}{c} \begin{matrix} u_1° & v_1° & u_2° & v_2° \end{matrix} \\ \begin{bmatrix} 0.5 & -0.5 & & \\ -0.5 & 0.5 & & \\ & & & \\ & & & \end{bmatrix} \end{array} \begin{matrix} u_1° \\ v_1° \\ u_2° \\ v_2° \end{matrix} \qquad (23.37)$$

Element 1–3

This member points from 1 to 3, so that its start node is 1 and its end node is 3, as shown below.

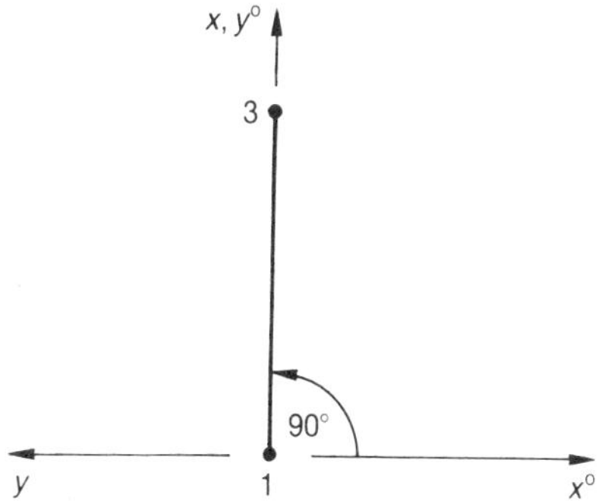

$\alpha = 90°,$ $c = 0,$ $s = 1,$ $l = 1$ m

Substituting the above values into equation (23.36) and removing the rows and columns corresponding to the zero displacements, namely u_3° and v_3°, the elemental stiffness matrix for element 1–3 is given by:

$$[\mathbf{k}_{1\text{-}3}^\circ] = \frac{AE}{1}\begin{array}{c} \begin{array}{cccc} u_1^\circ & v_1^\circ & u_3^\circ & v_3^\circ \end{array} \\ \left[\begin{array}{cccc} 0 & 0 & & \\ 0 & 1 & & \\ & & & \\ & & & \end{array}\right] \end{array} \begin{array}{c} u_1^\circ \\ v_1^\circ \\ u_3^\circ \\ v_3^\circ \end{array} \qquad (23.38)$$

Element 4 –1

This element points from 4 to 1, so that its start node is 4 and its end node is 1, as shown:

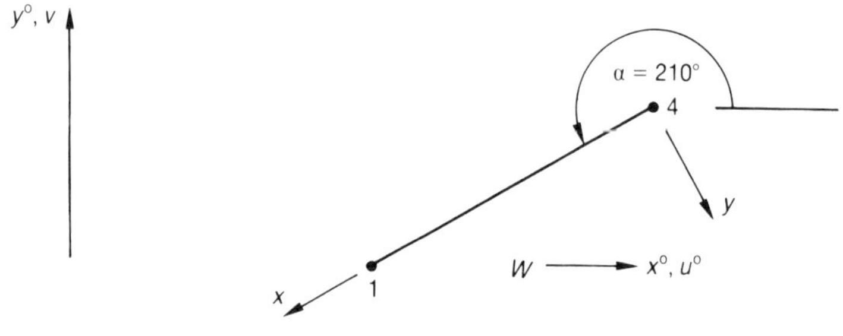

$\alpha = 210^\circ$

or $\alpha = -150^\circ$

$c = -0.866$

$s = -0.5$

$l = 2$

Substituting the above information into equation (23.36), and removing the rows and columns corresponding to the zero displacements, which in this case are u_4° and v_4°, the elemental stiffness matrix is given by

$$
[\mathbf{k}_{4\text{-}1}{}^{\circ}] = \frac{AE}{2}
\begin{array}{c} \begin{matrix} u_4{}^{\circ} & v_4{}^{\circ} & u_1{}^{\circ} & v_1{}^{\circ} \end{matrix} \\
\left[\begin{matrix} & & & \\ & & & \\ & & 0.75 & 0.433 \\ & & 0.433 & 0.25 \end{matrix}\right]
\begin{matrix} u_4{}^{\circ} \\ v_4{}^{\circ} \\ u_1{}^{\circ} \\ v_1{}^{\circ} \end{matrix}
\end{array}
\qquad (23.39)
$$

The system stiffness matrix corresponding to the free displacements, namely $u_1{}^{\circ}$ and $v_1{}^{\circ}$, is given by adding together the appropriate coefficients of equations (23.37) to (23.39), as shown by equation (23.40):

$$
[\mathbf{K}_{11}] = AE
\begin{array}{c} \begin{matrix} u_1{}^{\circ} & \qquad v_1{}^{\circ} \end{matrix} \\
\left[\begin{array}{c|c} 0.354 + 0 & -0.354 + 0 \\ +0.375 & +0.217 \\ \hline -0.354 + 0 & 0.354 + 1 \\ +0.217 & +0.125 \end{array}\right]
\begin{matrix} u_1{}^{\circ} \\ \\ v_1{}^{\circ} \\ \end{matrix}
\end{array}
\qquad (23.40)
$$

or

$$
[\mathbf{K}_{11}] = AE
\begin{array}{c} \begin{matrix} u_1{}^{\circ} & v_1{}^{\circ} \end{matrix} \\
\left[\begin{matrix} 0.729 & -0.137 \\ -0.137 & 1.479 \end{matrix}\right]
\begin{matrix} u_1{}^{\circ} \\ v_1{}^{\circ} \end{matrix}
\end{array}
\qquad (23.41)
$$

NB $[\mathbf{K}_{11}]$ is of order two, as it corresponds to the two free displacements $u_1{}^{\circ}$ and $v_1{}^{\circ}$, which are unknown.

The vector of external loads $\{q_F\}$, corresponds to the two free displacements $u_1{}^{\circ}$ and $v_1{}^{\circ}$, and can readily be shown to be given by equation (23.42), ie

$$
\{qF\} = \begin{Bmatrix} 2 \\ -3 \end{Bmatrix} \begin{matrix} u_1{}^{\circ} \\ v_1{}^{\circ} \end{matrix} \qquad (23.42)
$$

where the load value 2 is in the $u_1{}^{\circ}$ direction, and the load value -3 is in the $v_1{}^{\circ}$ direction.

Substituting equations (23.41) and (23.42) into equation (23.16)

$$\{u_F{}^\circ\} = \begin{Bmatrix} u_1{}^\circ \\ v_1{}^\circ \end{Bmatrix} = [\mathbf{K}_{11}]^{-1} \begin{Bmatrix} 2 \\ -3 \end{Bmatrix}$$

$$= \frac{\dfrac{1}{AE}\begin{bmatrix} 1.479 & 0.137 \\ 0.137 & 0.729 \end{bmatrix}\begin{Bmatrix} 2 \\ -3 \end{Bmatrix}}{(0.729 \times 1.479 - 0.137 \times 0.137)}$$

$$= \frac{1}{AE}\begin{bmatrix} 1.396 & 0.129 \\ 0.129 & 0.688 \end{bmatrix}\begin{Bmatrix} 2 \\ -3 \end{Bmatrix}$$

i.e.

$$\begin{Bmatrix} u_1{}^\circ \\ v_1{}^\circ \end{Bmatrix} = \frac{1}{AE}\begin{Bmatrix} 2.405 \\ -1.806 \end{Bmatrix} \tag{23.43}$$

These displacements are in global co-ordinates, so it will be necessary to resolve these displacements along the length of each rod element, to discover how much each rod extends or contacts along its length, and then through the use of Hookean elasticity to obtain the internal forces in each element.

Element 1– 2

Now,

$$c = -0.707, \quad s = 0.707 \quad \text{and} \quad l = 1.414 \text{ m}$$

Hence, from equation (23.23),

$$u_1 = [c \quad s]\begin{Bmatrix} u_1{}^\circ \\ v_1{}^\circ \end{Bmatrix}$$

$$= [-0.707 \quad 0.707]\,\frac{1}{AE}\begin{Bmatrix} 2.405 \\ -1.806 \end{Bmatrix}$$

$$u_1 = -2.977/AE$$

From Hooke's law,

$$F_{1-2} = \text{force in element 1–2}$$

$$= \frac{AE}{l}(u_2 - u_1)$$

$$= \frac{2.977}{1.414}$$

$$F_{1-2} = 2.106 \text{ MN (tension)}$$

Element 1–3

$$c = 0, \quad s = 1 \quad \text{and} \quad l = 1 \text{ m}$$

From equation (23.23),

$$u_1 = [c \quad s]\begin{Bmatrix} u_1^\circ \\ v_1^\circ \end{Bmatrix}$$

$$= [0 \quad 1]\frac{1}{AE}\begin{Bmatrix} 2.405 \\ -1.806 \end{Bmatrix}$$

$$u_1 = -1.806/AE$$

From Hooke's law,

$$F_{1-3} = \text{force in element 1–3}$$

$$= \frac{AE}{l}(u_3 - u_1)$$

$$F_{1-3} = 1.806 \text{ MN (tension)}$$

Element 4–1

$$c = -0.866, \quad s = 0.5 \quad \text{and} \quad l = 2 \text{ m}$$

From equation (23.23),

$$u_1 = [c \quad s] \begin{Bmatrix} u_1{}^{\circ} \\ v_1{}^{\circ} \end{Bmatrix}$$

$$= [-0.866 \quad 0.5] \frac{1}{AE} \begin{Bmatrix} 2.405 \\ -1.806 \end{Bmatrix}$$

$$u_1 = -1.1797 \,/\, AE$$

From Hooke's law,

$$F_{4-1} = \text{force in element 1–4}$$

$$= \frac{AE}{l}(u_1 - u_4)$$

$$= \frac{AE}{2} \frac{(-1.1797 - 0)}{AE}$$

$$F_{4-1} = -0.59 \text{ MN (compression)}$$

Problem 23.2 Using the matrix displacement method, determine the forces in the members of the plane pin-jointed truss below, which is free to move horizontally at node 3, but not vertically. It may also be assumed that the truss is firmly pinned at node 1, and that the material and geometrical properties of its members are given in the table below.

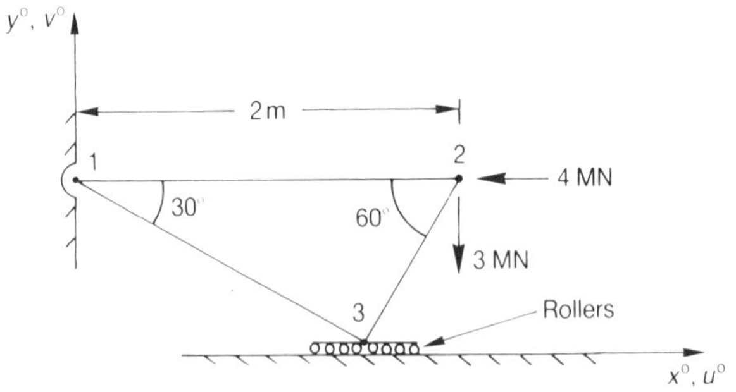

Member	***A***	***E***
1–2	$2A$	E
1–3	A	$3E$
2–3	$3A$	$2E$

<u>*Solution*</u>

Element 1–2

$$\alpha = 0, \qquad c = 1, \qquad s = 0 \quad \text{and} \quad l = 2\text{ m}$$

Substituting the above values into equation (23.36),

$$\left[\mathbf{k}_{1\text{-}2}^{\circ}\right] = \frac{2AE}{2}\begin{array}{c} \begin{matrix} u_1^{\circ} & v_1^{\circ} & u_2^{\circ} & v_2^{\circ} \end{matrix} \\ \left[\begin{matrix} & & & \\ & & & \\ & & 1 & 0 \\ & & 0 & 0 \end{matrix}\right]\begin{matrix} u_1^{\circ} \\ v_1^{\circ} \\ u_2^{\circ} \\ v_2^{\circ} \end{matrix} \end{array} \tag{23.44}$$

Element 2–3

$$\alpha = 240^{\circ}, \quad c = -0.5, \quad s = -0.866 \quad \text{and} \quad l = 1\text{ m}$$

$$\left[\mathrm{k}_{1-3}^{\circ}\right] = \frac{3A \times 2E}{1}\begin{array}{c} \begin{matrix} u_2^{\circ} & v_2^{\circ} & u_3^{\circ} & v_3^{\circ} \end{matrix} \\ \left[\begin{matrix} 0.25 & 0.433 & -0.25 & \\ 0.433 & 0.75 & -0.433 & \\ -0.25 & -0.433 & 0.25 & \\ & & & \end{matrix}\right]\begin{matrix} u_2^{\circ} \\ v_2^{\circ} \\ u_3^{\circ} \\ v_3^{\circ} \end{matrix} \end{array}$$

$$= AE \begin{array}{c} \begin{matrix} u_2^{\circ} & v_2^{\circ} & u_3^{\circ} \end{matrix} \\ \left[\begin{matrix} 1.5 & 2.6 & -1.5 \\ 2.6 & 4.5 & -2.6 \\ -1.5 & -2.6 & 1.5 \end{matrix}\right]\begin{matrix} u_2^{\circ} \\ v_2^{\circ} \\ u_3^{\circ} \end{matrix} \end{array} \tag{23.45}$$

Element 3–1

$$\alpha = 150°, \quad c = -0.866, \quad s = 0.5 \quad \text{and } l = 1.732 \text{ m}$$

$$[\mathbf{k}_{3-1}°] = \frac{A \times 3E}{1.732} \begin{bmatrix} 0.75 & & & \\ & & & \\ & & & \\ & & & \end{bmatrix} \begin{matrix} u_3° \\ v_3° \\ u_1° \\ v_1° \end{matrix}$$

(columns: $u_3°$, $v_3°$, $u_1°$, $v_1°$)

$$= \begin{matrix} u_3° \\ [1.3] \end{matrix} u_3° \tag{23.46}$$

The system stiffness matrix $[\mathbf{K}_{11}]$ is obtained by adding together the appropriate components of stiffness, from the elemental stiffness matrices of equations (23.44) to (23.46), with reference to the free degrees of freedom, namely, $u_2°$, $v_2°$ and $u_3°$, as shown by equation (23.47):

$$[\mathbf{K}_{11}] = AE \begin{bmatrix} 1 + 1.5 & 0 + 2.6 & -1.5 \\ 0 + 2.6 & 0 + 4.5 & -2.6 \\ -1.5 & -2.6 & 1.5 + 1.3 \end{bmatrix} \begin{matrix} u_2° \\ v_2° \\ u_3° \end{matrix} \tag{23.47}$$

(columns: $u_2°$, $v_2°$, $u_3°$)

$$= AE \begin{bmatrix} 2.5 & 2.6 & -1.5 \\ 2.6 & 4.5 & -2.6 \\ -1.5 & -2.6 & 2.8 \end{bmatrix} \begin{matrix} u_2° \\ v_2° \\ u_3° \end{matrix} \tag{23.48}$$

(columns: $u_2°$, $v_2°$, $u_3°$)

The vector of loads $\{q_F\}$, corresponding to the free degrees of freedom, namely, $u_2°$, $v_2°$ and $u_3°$ is given by:

$$\{q_F\} = \begin{Bmatrix} -4 \\ -3 \\ 0 \end{Bmatrix} \begin{matrix} u_2^\circ \\ v_2^\circ \\ u_3^\circ \end{matrix} \qquad (23.49)$$

Substituting equations (23.48) and (23.49) into equation (23.16) and solving, the vector of free displacements $\{u_F\}$ is given by

$$\begin{Bmatrix} u_2^\circ \\ v_2^\circ \\ u_3^\circ \end{Bmatrix} = \frac{1}{AE} \begin{Bmatrix} -2.27 \\ -0.125 \\ -1.332 \end{Bmatrix} \qquad (23.50)$$

The member forces will be obtained by resolving these displacements along the length of each rod element, and then by finding the amount that each rod extends or contracts, to determine the force in each member through Hookean elasticity.

Element 1–2

$$c = 1, \quad s = 0 \quad \text{and} \quad l = 2 \text{ m}$$

From equation (23.23),

$$u_2 = [c \quad s] \begin{Bmatrix} u_2^\circ \\ v_2^\circ \end{Bmatrix}$$

$$= [1 \quad 0] \frac{1}{AE} \begin{Bmatrix} -2.27 \\ -0.125 \end{Bmatrix}$$

$$u_2 = -2.27/AE$$

From Hooke's law,

F_{1-2} = force in element 1–2

$$= \frac{2AE}{2}\left(-\frac{2.27}{AE} - 0\right)$$

$$F_{1-2} = -2.27 \text{ MN (compression)}$$

Element 2–3

$$c = -0.5, \quad s = -0.866 \quad \text{and} \quad l \quad 1\ \text{m}$$

From equation (23.23),

$$u_2 = [c \quad s] \begin{Bmatrix} u_2^\circ \\ v_2^\circ \end{Bmatrix}$$

$$= [-0.5 \quad -0.866] \frac{1}{AE} \begin{Bmatrix} -2.27 \\ -0.125 \end{Bmatrix}$$

$$u_2 = 1.243/AE$$

Similarly, from equation (23.23),

$$u_3 = [c \quad s] \begin{Bmatrix} u_3^\circ \\ v_3^\circ \end{Bmatrix}$$

$$= [-0.5 \quad -0.866] \frac{1}{AE} \begin{Bmatrix} -1.332 \\ 0 \end{Bmatrix}$$

$$u_3 = 0.666/AE$$

From Hooke's law,

$$F_{2-3} = \text{force in element 2–3}$$

$$= \frac{3A \times 2E}{1} (u_3 - u_2)$$

$$= 6AE \times \frac{(-0.577)}{AE}$$

$$F_{2-3} = -3.46\ \text{MN (compression)}$$

Element 3–1

$$c = -0.866, \quad s = 0.5 \quad \text{and} \quad l = 1.732 \text{ m}$$

$$u_3 = \begin{bmatrix} c & s \end{bmatrix} \begin{Bmatrix} u_3^{\circ} \\ v_3^{\circ} \end{Bmatrix}$$

$$= \begin{bmatrix} -0.866 & 0.5 \end{bmatrix} \frac{1}{AE} \begin{Bmatrix} -1.332 \\ 0 \end{Bmatrix}$$

$$u_3 = 1.154/AE$$

From Hooke's law,

$$F_{3-1} = \text{force in element 1–3}$$

$$= \frac{A \times 3E}{1.732} \left(0 - \frac{1.154}{AE} \right)$$

$$F_{3-1} = -2 \text{ MN (compression)}$$

23.6 Pin-jointed space trusses

In three dimensions, the relationships between forces and displacements for the rod element of Figure 23.5 are given by equation (23.51):

$$\begin{Bmatrix} X_1 \\ Y_1 \\ Z_1 \\ X_2 \\ Y_2 \\ Z_2 \end{Bmatrix} = \frac{AE}{l} \begin{bmatrix} 1 & 0 & 0 & -1 & 0 & 0 \\ 0 & 0 & 0 & 0 & 0 & 0 \\ 0 & 0 & 0 & 0 & 0 & 0 \\ -1 & 0 & 0 & 1 & 0 & 0 \\ 0 & 0 & 0 & 0 & 0 & 0 \end{bmatrix} \begin{Bmatrix} u_1 \\ v_1 \\ w_1 \\ u_2 \\ v_2 \\ w_2 \end{Bmatrix} \qquad (23.51)$$

where,

X_1 = load in the x direction at node 1

= $AE\,(u_1 - u_2)/l$

Y_1 = load in the y direction at node 1

= 0

Z_1 = load in the z direction at node 1

= 0

X_2 = load in the x direction at node 2

= $AE(u_2 - u_1)/l$

Y_2 = load in the y direction at node 2

= 0

Z_2 = load in the z direction at node 2

= 0

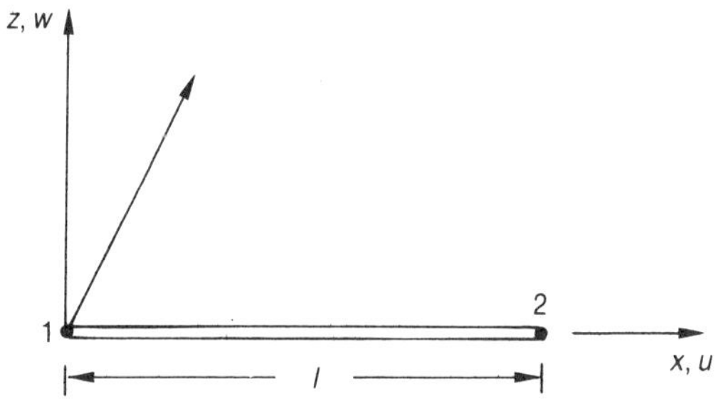

Figure 23.5 Three-dimensional rod in local co-ordinates.

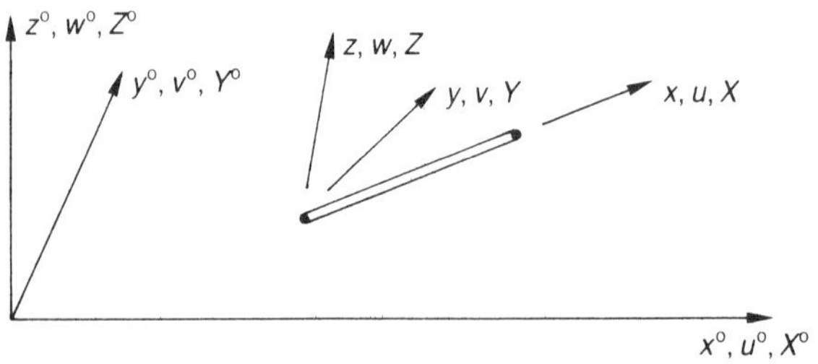

Figure 23.6 Rod in three dimensions.

For the case of the three dimensional rod in the global co-ordinate system of Figure 23.6, it can be shown through resolution that the relationship between local loads and global loads is given by:

$$\begin{Bmatrix} X_1 \\ Y_1 \\ Z_1 \\ X_2 \\ Y_2 \\ Z_2 \end{Bmatrix} = \begin{bmatrix} \zeta & 0_3 \\ 0_3 & \zeta \end{bmatrix} \begin{Bmatrix} X_1{}^{\circ} \\ Y_1{}^{\circ} \\ Z_1{}^{\circ} \\ X_2{}^{\circ} \\ Y_2{}^{\circ} \\ Z_2{}^{\circ} \end{Bmatrix} \tag{23.52}$$

where

$$[\zeta] = \begin{bmatrix} C_{x,x^{\circ}} & C_{x,y^{\circ}} & C_{x,z^{\circ}} \\ C_{y,x^{\circ}} & C_{y,y^{\circ}} & C_{y,z^{\circ}} \\ C_{z,x^{\circ}} & C_{z,y^{\circ}} & C_{z,z} \end{bmatrix} \tag{23.53}$$

x, y, z = local axes

$x^{\circ}, y^{\circ}, z^{\circ}$ = global axes

$C_{x,x}$, $C_{x,y}$, $C_{x,z}$, etc = the directional cosines of x with x°, x with y°, x with z°, respectively, etc.

$X_1{}^{\circ}$ = force in x° direction at node 1

$Y_1{}^{\circ}$ = force in y° direction at node 1

$Z_1{}^{\circ}$ = force in z° direction at node 1

$X_2{}^{\circ}$ = force in x° direction at node 2

$Y_2{}^{\circ}$ = force in y° direction at node 2

$Z_2{}^{\circ}$ = force in z° direction at node 2

Now from equation (23.35) the elemental stiffness matrix for a rod in global co-ordinates is given by:

$$[\mathbf{k}^\circ] = [\mathbf{DC}]^\mathbf{T}\ [\mathbf{k}]\ [\mathbf{DC}]$$

$$= \begin{bmatrix} \zeta & 0_3 \\ 0_3 & \zeta \end{bmatrix}^T [\mathbf{k}] \begin{bmatrix} \zeta & 0_3 \\ 0_3 & \zeta \end{bmatrix} \qquad (23.54)$$

$$[\mathbf{k}^\circ] = \begin{bmatrix} a & -a \\ -a & a \end{bmatrix}$$

where

$$[\mathbf{a}] = \frac{AE}{l}\begin{bmatrix} C^2_{x,x^\circ} & C_{x,x^\circ}C_{x,y^\circ} & C_{x,x^\circ}C_{x,z^\circ} \\ C_{x,x^\circ}C_{x,y^\circ} & C^2_{x,y^\circ} & C_{x,y^\circ}C_{x,z^\circ} \\ C_{x,x^\circ}C_{x,z^\circ} & C_{x,y^\circ}C_{x,z^\circ} & C^2_{x,z^\circ} \end{bmatrix} \qquad (23.55)$$

By Pythagoras' theorem in three dimensions:

$$l = \left[(x_2{}^\circ - x_1{}^\circ)^2 + (y_2{}^\circ - y_1{}^\circ)^2 + (z_2{}^\circ - z_1{}^\circ)^2\right]^{\frac{1}{2}} \qquad (23.56)$$

The directional cosines[9] can readily be shown to be given by equation (23.57):

$$C_{x,x}{}^\circ = (x_2{}^\circ - x_1{}^\circ)/l$$

$$C_{x,y}{}^\circ = (y_2{}^\circ - y_1{}^\circ)/l \qquad (23.57)$$

$$C_{x,z}{}^\circ = (z_2{}^\circ - z_1{}^\circ)/l$$

Problem 23.3 A tripod, with pinned joints, is constructed from three uniform section members, made from the same material. If the tripod is firmly secured to the ground at nodes 1 to 3, and loaded at node 4, as shown below, determine the forces in the members of the tripod, using the matrix displacement method.

---

[9] Ross, C T F, *Advanced Applied Element Methods*, Horwood, 1998.

<u>Solution</u>

Element 1–4

The element points from 1 to 4, so that the start node is 1 and the finish node is 4. From the figure below it can readily be seen that:

$x_1^\circ = 0,$ $\qquad y_1^\circ = 0.$ $\qquad z_1^\circ = 0,$

$z_4^\circ = 5$ m, $\qquad y_4^\circ = 5$ m, $\qquad z_4^\circ = 7.07$ m

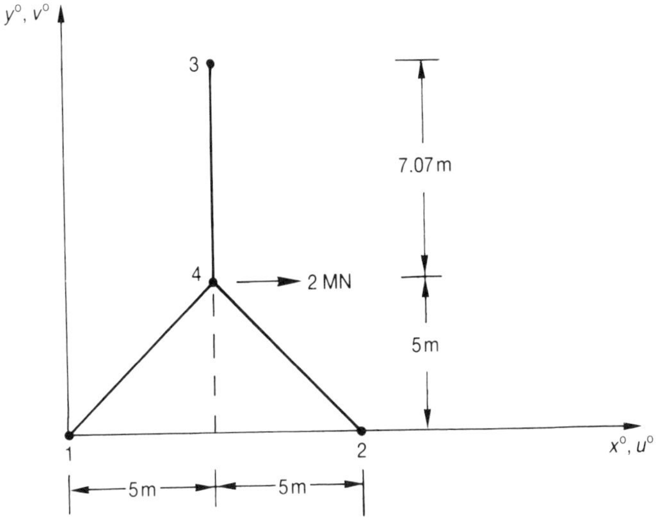

(a) Plan view of the tripod.

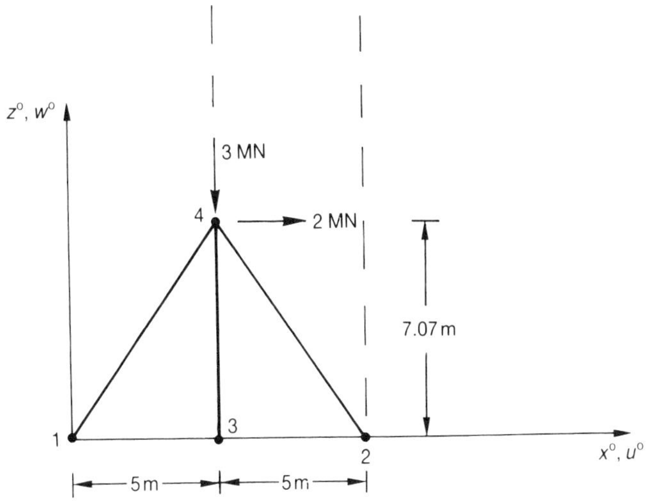

(b) Front view of tripod.

Substituting the above into equation (23.56),

$$l = \left[(5-0)^2 + (5-0)^2 + (7.07-0)^2\right]^{\frac{1}{2}}$$

$$l = 10 \text{ m}$$

Substituting the above into equation (23.57),

$$C_{x,x}^{\circ} = \frac{x_4^{\circ} - x_1^{\circ}}{l} = \frac{5-0}{10} = 0.5$$

$$C_{x,y}^{\circ} = \frac{y_4^{\circ} - y_1^{\circ}}{l} = \frac{5-0}{10} = 0.5$$

$$C_{x,z}^{\circ} = \frac{z_4^{\circ} - z_1^{\circ}}{l} = \frac{7.07-0}{10} = 0.707$$

Substituting the above values into equation (23.54), and removing the coefficients of the stiffness matrix corresponding to the zero displacements, which in this case are u_1°, v_1° and w_1°, the stiffness matrix for element 1–4 is given by equation 23.58):

$$\left[\mathbf{k}_{1-4}^{\circ}\right] = \frac{AE}{10} \begin{array}{c} \begin{array}{cccccc} u_1^{\circ} & v_1^{\circ} & w_1^{\circ} & u_4^{\circ} & v_4^{\circ} & w_4^{\circ} \end{array} \\ \left[\begin{array}{cccccc} & & & & & \\ & & & & & \\ & & & & & \\ & & & 0.25 & & \\ & & & 0.25 & 0.25 & \\ & & & 0.354 & 0.354 & 0.5 \end{array}\right] \end{array} \begin{array}{c} u_1^{\circ} \\ v_1^{\circ} \\ w_1^{\circ} \\ u_4^{\circ} \\ v_4^{\circ} \\ w_4^{\circ} \end{array} \qquad (23.58)$$

Element 2–4

The member points from 2 to 4, so that the start node is 2 and the finish node is 4. From the above figure,

$$x_2^{\circ} = 10, \qquad y_2^{\circ} = 0, \qquad z_2^{\circ} = 0$$

Substituting the above and x_4°, y_4° and z_4° into equation (23.56),

$$l = \left[(5 - 10)^2 + (5 - 0)^2 + (7.07 - 0)^2\right]^{\frac{1}{2}}$$

$$l = 10 \text{ m}$$

From equation (23.57),

$$C_{x,x}^\circ = \frac{x_4^\circ - x_2^\circ}{l} = \frac{5 - 10}{10} = -0.5$$

$$C_{x,y}^\circ = \frac{y_4^\circ - y_2^\circ}{l} = \frac{5 - 0}{10} = 0.5$$

$$C_{x,z}^\circ = \frac{z_4^\circ - z_2^\circ}{l} = \frac{7.07 - 0}{10} = 0.707$$

Substituting the above values into equation (23.54), and removing the rows and columns corresponding to the zero displacements, which in this case are u_2°, v_2° and w_2°, the stiffness matrix for element 2–4 is given by equation (23.59):

$$\left[\mathbf{k}_{2-4}^\circ\right] = \frac{AE}{10} \begin{array}{c} \begin{array}{cccccc} u_2^\circ & v_2^\circ & w_2^\circ & u_4^\circ & v_4^\circ & w_4^\circ \end{array} \\ \left[\begin{array}{cccccc} & & & & & \\ & & & & & \\ & & & & & \\ & & & 0.25 & & \\ & & & -0.25 & 0.25 & \\ & & & -0.354 & 0.354 & 0.5 \end{array}\right] \end{array} \begin{array}{c} u_2^\circ \\ v_2^\circ \\ w_2^\circ \\ u_4^\circ \\ v_4^\circ \\ w_4^\circ \end{array} \qquad (23.59)$$

Element 4–3

The member points from 4 to 3, so that the start node is 4 and the finish node is 3. From the figure at the start of this problem,

$$x_3^\circ = 5 \qquad y_3^\circ = 12.07 \qquad z_3^\circ = 0$$

Substituting the above and x_4°, y_4° and z_4° into equation (23.56),

$$l = \left[(5 - 5)^2 + (12.07 - 5)^2 + (0 - 7.07)^2\right]^{\frac{1}{2}}$$

$$l = 10 \text{ m}$$

From equation (23.57),

$$C_{x,x}^\circ = \frac{x_3^\circ - x_4^\circ}{l} = \frac{5 - 5}{10} = 0$$

$$C_{x,y}^\circ = \frac{y_3^\circ - y_4^\circ}{l} = \frac{12.07 - 5}{10} = 0.707$$

$$C_{x,z}^\circ = \frac{z_3^\circ - z_4^\circ}{l} = \frac{0 - 7.07}{10} = -0.707$$

Substituting the above into equation (23.54), and removing the rows and columns corresponding to the zero displacements, which in this case are u_3°, v_3° and w_3°, the stiffness matrix for element 4–3 is given by equation (23.60):

$$\left[\mathbf{k}_{4-3}^\circ\right] = \frac{AE}{10}\begin{bmatrix} 0 & & & & & \\ 0 & 0.5 & & & & \\ 0 & -0.5 & 0.5 & & & \\ & & & & & \\ & & & & & \\ & & & & & \end{bmatrix} \begin{matrix} u_4^\circ \\ v_4^\circ \\ w_4^\circ \\ u_3^\circ \\ v_3^\circ \\ w_3^\circ \end{matrix} \tag{23.60}$$

(columns: u_4°, v_4°, w_4°, u_3°, v_3°, w_3°)

To obtain $[\mathbf{K}_{11}]$, the system stiffness matrix corresponding to the free displacements, namely u_4°, v_4° and w_4°, the appropriate coefficients of the elemental stiffness matrices of equations (23.58) to (23.60) are added together, with reference to these free displacements, as shown by equation (23.61):

$$
[\mathbf{K}_{11}{}^{\circ}] = \frac{AE}{10}
\begin{array}{c}
\begin{array}{ccc} u_4{}^{\circ} \qquad & \qquad v_4{}^{\circ} \qquad & \qquad w_4{}^{\circ} \end{array} \\
\left[\begin{array}{c|c|c}
\begin{matrix}0.25\\+\,0.25\\+\,0\end{matrix} & & \\
\hline
\begin{matrix}0.25\\-\,0.25\\+\,0\end{matrix} & \begin{matrix}0.25\\+\,0.25\\+\,0.5\end{matrix} & \\
\hline
\begin{matrix}0.354\\-\,0.354\\+\,0\end{matrix} & \begin{matrix}0.354\\+\,0.354\\-\,0.5\end{matrix} & \begin{matrix}0.5\\+\,0.5\\+\,0.5\end{matrix}
\end{array}\right]
\end{array}
\begin{array}{c} \\ u_4{}^{\circ} \\ \\ \\ v_4{}^{\circ} \\ \\ \\ w_4{}^{\circ} \end{array}
\qquad (23.61)
$$

$$
= \frac{AE}{10}
\begin{array}{c}
\begin{array}{ccc} u_4{}^{\circ} \qquad & \qquad v_4{}^{\circ} \qquad & \qquad w_4{}^{\circ} \end{array} \\
\left[\begin{array}{ccc}
0.5 & 0 & 0 \\
0 & 1.0 & 0.208 \\
0 & 0.208 & 1.5
\end{array}\right]
\end{array}
\begin{array}{c} u_4{}^{\circ} \\ v_4{}^{\circ} \\ w_4{}^{\circ} \end{array}
\qquad (23.62)
$$

The vector of loads is obtained by considering the loads in the directions of the free displacements, namely $u_4{}^{\circ}$, $v_4{}^{\circ}$ and $w_4{}^{\circ}$, as shown by equation (23.63):

$$
\{q_F\} = \begin{Bmatrix} 2 \\ 0 \\ -3 \end{Bmatrix} \begin{array}{c} u_4{}^{\circ} \\ v_4{}^{\circ} \\ w_4{}^{\circ} \end{array} \qquad (23.63)
$$

Substituting equations (23.62) and (23.63) into (23.16), the following three simultaneous equations are obtained:

$$
2 = \left(\frac{AE}{10}\right) \times 0.5\ u_4{}^{\circ} \qquad (23.64a)
$$

$$
0 = \left(\frac{AE}{10}\right) \left(v_4{}^{\circ} + 0.208\ w_4{}^{\circ}\right) \qquad (23.64b)
$$

$$
-3 = \left(\frac{AE}{10}\right) \left(0.208\ v_4{}^{\circ} + 1.5\ w_4{}^{\circ}\right) \qquad (23.64c)
$$

From (23.64a)

$$u_4^\circ = 40/AE$$

Hence, from (20.64b) and (23.64c),

$$v_4^\circ = 4.284/AE$$

$$w_4^\circ = -20.594/AE$$

so that,

$$\{u_F\} = \begin{Bmatrix} u_4^\circ \\ v_4^\circ \\ w_4^\circ \end{Bmatrix} = \frac{1}{AE}\begin{Bmatrix} 40 \\ 4.284 \\ -20.594 \end{Bmatrix} \tag{23.65}$$

To determine the forces in the members, the displacements of equation (23.65) must be resolved along the length of each rod, so that the amount the rod contracts or extends can be determined. Then through the use of Hookean elasticity, the internal forces in each member can be obtained.

Element 1–4

$$C_{x,x}^\circ = 0.5, \quad C_{x,y}^\circ = 0.5, \quad C_{x,z}^\circ = 0.707, \quad l = 10\text{ m}$$

From equation (23.52):

$$u_4 = \begin{bmatrix} C_{x,x}^\circ & C_{x,y}^\circ & C_{x,z}^\circ \end{bmatrix}\begin{Bmatrix} u_4^\circ \\ v_4^\circ \\ w_4^\circ \end{Bmatrix}$$

$$= \begin{bmatrix} 0.5 & 0.5 & 0.707 \end{bmatrix}\frac{1}{AE}\begin{Bmatrix} 40 \\ 4.28 \\ -20.59 \end{Bmatrix}$$

$$u_4 = 7.568/AE$$

From Hooke's law,

$$F_{1\text{–}4} = \text{force in member 1–4}$$

$$= \frac{AE}{10}\left(u_4 - u_1\right) = \frac{AE}{10} \times \frac{7.568}{AE}$$

$$F_{1\text{–}4} = 0.757 \text{ MN (tension)}$$

Element 2–4

$$C_{x,x}{}^\circ = -0.5, \quad C_{x,y}{}^\circ = 0.5, \quad C_{x,z}{}^\circ = 0.707, \quad l = 10 \text{ m}$$

From equation (23.52):

$$u_4 = \begin{bmatrix} C_{x,x}{}^\circ & C_{x,y}{}^\circ & C_{x,z}{}^\circ \end{bmatrix} \begin{Bmatrix} u_4{}^\circ \\ v_4{}^\circ \\ w_4{}^\circ \end{Bmatrix}$$

$$= \begin{bmatrix} -0.5 & 0.5 & 0.707 \end{bmatrix} \frac{1}{AE} \begin{Bmatrix} 40 \\ 4.28 \\ -20.59 \end{Bmatrix}$$

$$u_4 = -32.417/AE$$

From Hooke's law,

$$F_{2\text{–}4} = \text{force in member 2–4}$$

$$= \frac{AE}{10}\left(u_4 - u_2\right) = \frac{AE}{10} \times (-32.417/AE)$$

$$F_{2\text{–}4} = 3.242 \text{ MN (tension)}$$

Element 4–3

$$C_{x,x}^{\circ} = 0, \qquad C_{x,y}^{\circ} = 0.707, \qquad C_{x,z}^{\circ} = -0.707, \qquad l = 10\text{ m}$$

$$u_4 = \begin{bmatrix} C_{x,x}^{\circ} & C_{x,y}^{\circ} & C_{x,z}^{\circ} \end{bmatrix} \begin{Bmatrix} u_4^{\circ} \\ v_4^{\circ} \\ w_4^{\circ} \end{Bmatrix}$$

$$u_4 = \begin{bmatrix} 0 & 0.707 & -0.707 \end{bmatrix} \frac{1}{AE} \begin{Bmatrix} 40 \\ 4.28 \\ -20.59 \end{Bmatrix}$$

$$u_4 = 17.58/AE$$

From Hooke's law,

$$F_{4-3} = \text{force in member 4–3}$$

$$= \frac{AE}{l}\left(u_3 - u_4\right)$$

$$= \frac{AE}{10}\left(0 - 17.58/AE\right)$$

$$F_{4-3} = -1.758 \text{ MN (compression)}$$

23.7 Beam element

The stiffness matrix for a beam element can be obtained by considering the beam element of Figure 23.7.

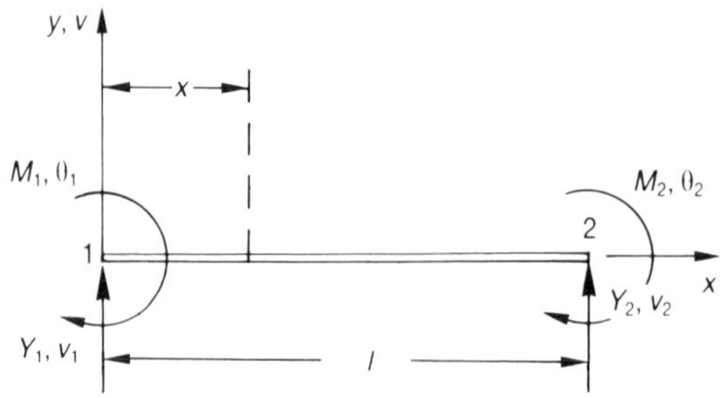

Figure 23.7 Beam element.

From equation (13.4),

$$EI\frac{d^2v}{dx^2} = M = Y_1x + M_1 \tag{23.66}$$

$$EI\frac{dv}{dx} = \frac{Y_1x^2}{2} + M_1x + A \tag{23.67}$$

$$EIv = \frac{Y_1x^3}{6} + \frac{M_1x^2}{2} + Ax + B \tag{23.68}$$

where

Y_1 = vertical reaction at node 1

Y_2 = vertical reaction at node 2

M_1 = clockwise couple at node 1

M_2 = clockwise couple at node 2

v_1 = vertical deflection at node 1

v_2 = vertical deflection at node 2

θ_1 = rotational displacement (clockwise) at node 1

θ_2 = rotational displacement (clockwise) at node 2

There are four unknowns in equation (23.68), namcly Y_1, M_1, A and B; therefore, four boundary values will have to be substituted into equations (23.67) and (23.68) to determine these four unknowns, through the solution of four linear simultaneous equations.

These four boundary values are as follows:

$$\text{At } x = 0, \quad v = v_1 \quad \text{and} \quad \theta_1 = -\left(\frac{dv}{dx}\right)_{x=0}$$

$$\text{At } x = l, \quad v = v_2 \quad \text{and} \quad \theta_2 = -\left(\frac{dv}{dx}\right)_{x=l}$$

Substituting these four boundary conditions into equations (23.67) and (23.68), the following are obtained:

$$Y_1 = -\frac{6EI}{l^2}(\theta_1 + \theta_2) + \frac{12EI}{l^3}(v_1 - v_2) \tag{23.69}$$

$$M_1 = \frac{6EI}{l^2}(v_2 - v_1) + \frac{EI}{l}(4\theta_1 + 2\theta_2) \tag{23.70}$$

$$Y_2 = \frac{6EI}{l^2}(\theta_1 + \theta_2) - \frac{12EI}{l^3}(v_1 - v_2) \tag{23.71}$$

$$M_2 = \frac{2EI}{l}\theta_1 + \frac{4EI}{l}\theta_2 - \frac{6EI}{l^2}(v_1 - v_2) \tag{23.72}$$

Equations (23.69) to (23.72) can be put in the form:

$$\{P_i\} = [\mathbf{k}]\{u_i\}$$

where,

$$[\mathbf{k}] = EI \begin{bmatrix} 12/l^3 & -6/l^2 & -12/l^3 & -6/l^2 \\ -6/l^2 & 4/l & 6/l^2 & 2/l \\ -12/l^3 & 6/l^2 & 12/l^3 & 6/l^2 \\ -6/l^2 & 2/l & 6/l^2 & 4/l \end{bmatrix} \begin{matrix} v_1 \\ \theta_1 \\ v_2 \\ \theta_2 \end{matrix} \tag{23.73}$$

(columns: v_1, θ_1, v_2, θ_2)

= the elemental stiffness matrix for a beam

$$\{P_i\} = \begin{Bmatrix} Y_1 \\ M_1 \\ Y_2 \\ M_2 \end{Bmatrix} = \text{a vector of generalised loads} \tag{23.74}$$

$$\{u_i\} = \begin{Bmatrix} v_1 \\ \theta_1 \\ v_2 \\ \theta_2 \end{Bmatrix} = \text{a vector of generalised displacements} \tag{23.75}$$

Problem 23.4 Determine the nodal displacements and bending moments for the uniform section beam below, which can be assumed to be fully fixed at its ends.

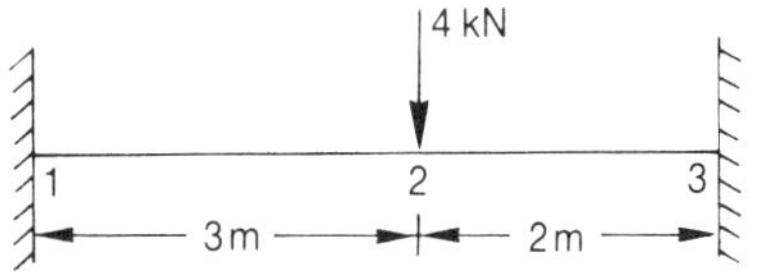

Solution

Prior to solving this problem, it must be emphasised that the nodes must be numbered in ascending order from left to right, because the beam element has been developed with the assumption that the start node is on the left and the finish node is on the right.

Element 1–2

$$l = 3 \text{ m}$$

Substituting this value of l into equation (23.73), and removing the components of the stiffness matrix corresponding to the zero displacements, namely v_1 and θ_1, the stiffness matrix for element 1–2 is given by equation (23.76):

$$\begin{array}{cccccc} & v_1 & \theta_1 & v_2 & \theta_2 & \\ [\mathbf{k}_{1\text{-}2}] = EI & \left[\begin{array}{c} \\ \\ \\ \\ \end{array} \right. & & \begin{array}{c} \\ \\ 0.444 \\ 0.667 \end{array} & \left. \begin{array}{c} \\ \\ 0.667 \\ 1.333 \end{array} \right] & \begin{array}{c} v_1 \\ \theta_1 \\ v_2 \\ \theta_2 \end{array} \end{array} \qquad (23.76)$$

Element 2–3

$$l = 2 \text{ m}$$

Substituting this value of l into equation (23.73), and removing the components of the stiffness matrix corresponding to the zero displacements, namely v_3 and θ_3, the following is obtained for the elemental stiffness matrix 2–3:

$$[\mathbf{k}_{2\text{-}3}] = EI \begin{bmatrix} 1.5 & -1.5 & & \\ -1.5 & 1.5 & & \\ & & & \\ & & & \end{bmatrix} \begin{matrix} v_2 \\ \theta_2 \\ v_3 \\ \theta_3 \end{matrix} \qquad (23.77)$$

with columns v_2, θ_2, v_3, θ_3.

The system stiffness matrix, which corresponds to the free displacements v_2 and θ_2, is obtained by adding together the appropriate components of the elemental stiffness matrices of equations (23.76) and (23.77), as shown by equation (23.78):

$$[\mathbf{K}_{11}] = EI \left[\begin{array}{c|c} \begin{matrix} 0.444 \\ +1.5 \end{matrix} & \begin{matrix} 0.667 \\ -1.5 \end{matrix} \\ \hline \begin{matrix} 0.667 \\ -1.5 \end{matrix} & \begin{matrix} 1.332 \\ +2.0 \end{matrix} \end{array}\right] \begin{matrix} v_4^\circ \\ \theta_2^\circ \end{matrix} \qquad (23.78)$$

with columns v_2°, θ_2°.

$$= EI \begin{bmatrix} 1.944 & -0.833 \\ -0.833 & 3.333 \end{bmatrix} \begin{matrix} v_2 \\ \theta_2 \end{matrix} \qquad (23.79)$$

with columns v_2, θ_2.

The vector of generalised loads is obtained by considering the loads in the directions of the free displacements v_2 and θ_2, as follows:

$$\{q_F\} = \begin{Bmatrix} -4 \\ 0 \end{Bmatrix} \begin{matrix} v_2 \\ \theta_2 \end{matrix}$$

From equation (23.11),

$$\begin{Bmatrix} -4 \\ 0 \end{Bmatrix} = EI \begin{bmatrix} 1.944 & -0.833 \\ -0.833 & 3.333 \end{bmatrix} \begin{Bmatrix} v_2 \\ \theta_2 \end{Bmatrix}$$

or,

$$\begin{Bmatrix} v_2 \\ \theta_2 \end{Bmatrix} = \frac{\dfrac{1}{EI} \begin{bmatrix} 3.333 & 0.833 \\ 0.833 & 1.944 \end{bmatrix} \begin{Bmatrix} -4 \\ 0 \end{Bmatrix}}{(1.944 \times 3.333 - 0.833^2)}$$

$$= \frac{1}{EI}\begin{bmatrix} 0.576 & 0.144 \\ 0.144 & 0.336 \end{bmatrix}\begin{Bmatrix} -4 \\ 0 \end{Bmatrix} \tag{23.80}$$

$$\begin{Bmatrix} v_2 \\ \theta_2 \end{Bmatrix} = \frac{1}{EI}\begin{Bmatrix} -2.304 \\ -0.576 \end{Bmatrix} \tag{23.81}$$

NB $\quad v_1 = \theta_1 = v_2 = \theta_2 = 0$

To obtain the nodal bending moments, these values of displacement must be substituted into the slope-deflection equations (23.70) and (23.72), as follows.

Element 1–2

Substituting v_1, θ_1, v_2 and θ_2 into equations (23.70) and (23.72):

$$M_1 = \frac{6EI}{9}\left(\frac{-2.304}{EI} - 0\right) + \frac{EI}{3}\left(4 \times 0 - \frac{2 \times 0.576}{EI}\right)$$

$$= -1.536 - 0.384$$

$$M_1 = -1.92 \text{ kNm}$$

and,

$$M_2 = \frac{2EI}{3} \times 0 + \frac{4EI}{3} \times \left(\frac{-0.576}{EI}\right) - \frac{6EI}{9}\left(0 + \frac{2.304}{EI}\right)$$

$$= -0.768 - 1.536$$

$$M_2 = -2.304 \text{ kNm}$$

Element 2–3

Substituting v_2, θ_2, v_3 and θ_3 into equations (23.70) and (23.72), and remembering that the first node is node 2 and the second node is node 3, the following is obtained for M_2 and M_3:

$$M_2 = \frac{4EI}{2} \times \left(\frac{-0.576}{EI}\right) + 0 - \frac{6EI}{4}\left(\frac{-2.304}{EI}\right)$$

$$= -1.152 + 3.456$$

$$M_2 = 2.304 \text{ kNm}$$

and,

$$M_3 = \frac{2EI}{2}\left(\frac{-0.576}{EI}\right) + 0 - \frac{6EI}{4}\left(\frac{-2.304}{EI} - 0\right)$$

$$= -0.576 + 3.456$$

$$M_3 = 2.88 \text{ kNm}$$

Problem 23.5 Determine the nodal displacements and bending moments for the encastré beam:

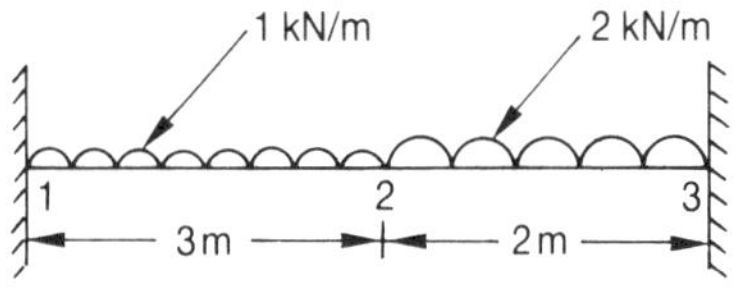

Solution

Now the matrix displacement method is based on applying the loads at the nodes, but for the above beam, the loading on each element is between the nodes. It will therefore be necessary to adopt the following process, which is based on the principle of superposition:

1. Fix the beam at its nodes and determine the end fixing forces, as shown in the following figure at (a) and (b) and as calculated below.

2. The beam in condition (1) is not in equilibrium at node 2, hence, it will be necessary to subject the beam to the negative resultants of the end fixing forces at node 2 to achieve equilibrium, as shown in the figure at (c). It should be noted that, as the beam is firmly fixed at nodes 1 and 3, any load or couple applied to these ends will in fact be absorbed by these walls.

3. Using the matrix displacement method, determine the nodal displacements due to the loads of the figure at (c) and, hence, the resulting bending moments.

4. To obtain the final values of nodal bending moments, the bending moments of condition (1) must be superimposed with those of condition (3).

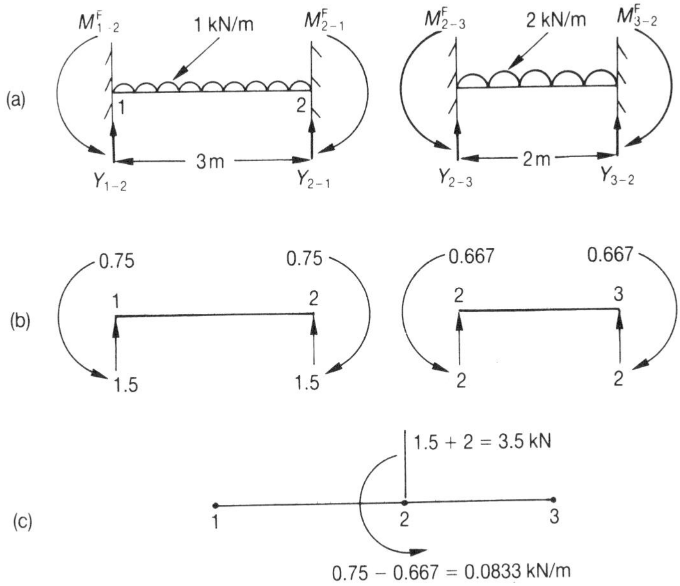

End-fixing forces

Element 1–2

$$M^F_{1-2} = -\frac{wl^2}{12} = -\frac{1 \times 3^2}{12} = -0.75 \text{ kNm}$$

$$M^F_{2-1} = \frac{wl^2}{12} = 0.75 \text{ kNm}$$

$$Y_{1-2} = Y_{2-1} = \frac{1 \times 3}{2} = 1.5 \text{ kN}$$

Element 2–3

$$M^F_{2-3} = -\frac{wl^2}{12} = -\frac{2 \times 2^2}{12} = -0.667 \text{ kNm}$$

$$M^F_{3-2} = \frac{wl^2}{12} = 0.667 \text{ kNm}$$

$$Y_{2-3} = Y_{3-2} = \frac{wl}{2} = \frac{2 \times 2}{2} = 2 \text{ kN}$$

From the figure above, at (c), the vector of generalised loads is obtained by considering the free degrees of freedom, which in this case, are v_2 and θ_2.

$$\{q_F\} = \begin{Bmatrix} -3.5 \\ -0.0833 \end{Bmatrix} \begin{matrix} v_2 \\ \theta_2 \end{matrix} \tag{23.82}$$

From equation (23.80),

$$[\mathbf{K}_{11}]^{-1} = \frac{1}{EI}\begin{bmatrix} 0.576 & 0.144 \\ 0.144 & 0.336 \end{bmatrix}$$

and from equation (23.16),

$$\{u_F\} = \begin{Bmatrix} v_2 \\ \theta_2 \end{Bmatrix} = \frac{1}{EI}\begin{bmatrix} 0.576 & 0.144 \\ 0.144 & 0.336 \end{bmatrix}\begin{Bmatrix} -3.5 \\ -0.0833 \end{Bmatrix}$$

or,

$$\begin{Bmatrix} v_2 \\ \theta_2 \end{Bmatrix} = \frac{1}{EI}\begin{Bmatrix} -2.028 \\ -0.532 \end{Bmatrix} \tag{23.83}$$

NB $v_1 = \theta_1 = v_3 = \theta_3 = 0$

To determine the nodal bending moments, the nodal bending moments obtained from the equations (23.70) and (20.72) must be superimposed with the end-fixing bending moment of the figure above, as follows.

Element 1–2

Substituting equation (23.83) into equation (23.70) and adding the end-fixing bending moment from the figure above (b),

$$M_1 = \frac{6EI}{9}\left(\frac{-2.028}{EI} - 0\right) + \frac{EI}{3}\left(4 \times 0 - \frac{2 \times 0.532}{EI}\right) - 0.75$$

$$= -1.352 - 0.355 - 0.75$$

$$M_1 = -2.457 \text{ kNm}$$

Similarly, substituting equation (23.83) into equation (23.72) and adding the end-fixing bending moment of the above figure at (b),

$$M_2 = -\frac{6EI}{3^2}\left(0 + \frac{2.028}{EI}\right) + \frac{EI}{3}\left(0 - \frac{4 \times 0.532}{EI}\right) + 0.75$$

$$= -1.352 - 0.709 + 0.75$$

$$M_2 = 1.311 \text{ kN/m}$$

Element 2–3

Substituting equation (23.83) into equations (23.70) and (23.72) and remembering that node 2 is the first node and node 3 is the second node, and adding the end fixing moments from the above figure at (b),

$$M_2 = \frac{6EI}{4}\left(\frac{2.028}{EI} + 0\right) + \frac{EI}{2}\left(-\frac{4 \times 0.532}{EI}\right) - 0.667$$

$$= 3.042 - 1.064 - 0.667$$

$$M_2 = 1.311 \text{ kNm}$$

$$M_3 = \frac{6EI}{4}\left(\frac{2.028}{EI} + 0\right) + \frac{EI}{2}\left(-\frac{2 \times 0.532}{EI}\right) + 0.667$$

$$= 3.042 - 0.532 + 0.667$$

$$M_3 = 3.177 \text{ kNm}$$

23.8 Rigid-jointed plane frames

The elemental stiffness matrix for a rigid-jointed plane frame element in local co-ordinates, can be obtained by superimposing the elemental stiffness matrix for the rod element of equation (23.28) with that of the beam element of equation (23.73), as shown by equation (23.84):

$$[\mathbf{k}] = EI\begin{bmatrix} (A/lI) & 0 & 0 & (-A/lI) & 0 & 0 \\ 0 & 12/l^3 & -6/l^2 & 0 & -12/l^3 & -6/l^2 \\ 0 & -6/l^2 & 4/l & 0 & 6/l^2 & 2/l \\ (-A/lI) & 0 & 0 & (A/lI) & 0 & 0 \\ 0 & -12/l^3 & 6/l^2 & 0 & 12/l^3 & 6/l^2 \\ 0 & -6/l^2 & 2/l & 0 & 6/l^2 & 4/l \end{bmatrix} \tag{23.84}$$

= the elemental stiffness matrix for a rigid-jointed plane frame element, in local co-ordinates

Now the stiffness matrix of equation (23.84) is of little use in that form, as most elements for a rigid-jointed plane frame will be inclined at some angle to the horizontal, as shown by Figure 23.8.

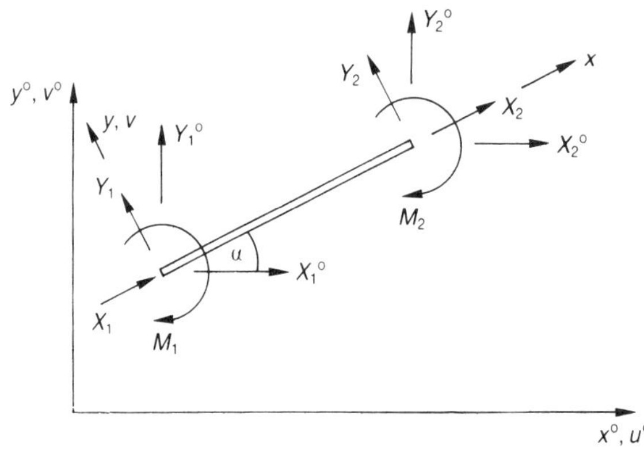

Figure 23.8 Rigid-jointed plane frame element.

It can readily be shown that the relationships between the local and global forces for the element are:

$$\begin{Bmatrix} X_1 \\ Y_1 \\ M_1 \\ X_2 \\ Y_2 \\ M_2 \end{Bmatrix} = \begin{bmatrix} c & s & 0 & & & \\ -s & c & 0 & & 0_3 & \\ 0 & 0 & 1 & & & \\ & & & c & s & 0 \\ & 0_3 & & -s & c & 0 \\ & & & 0 & 0 & 1 \end{bmatrix} \begin{Bmatrix} X_1{}^{\circ} \\ Y_1{}^{\circ} \\ M_1{}^{\circ} \\ X_2{}^{\circ} \\ Y_2{}^{\circ} \\ M_2{}^{\circ} \end{Bmatrix} \tag{23.85}$$

or,

$$\{P_i\} = [\mathbf{DC}]\{P_i^{\circ}\}$$

where,

$$[\mathbf{DC}] = \begin{bmatrix} \zeta & 0_3 \\ 0_3 & \zeta \end{bmatrix}$$

$$[\zeta] = \begin{bmatrix} c & s & 0 \\ -s & c & 0 \\ 0 & 0 & 1 \end{bmatrix}$$

Now, from equation (23.35):

$$\begin{aligned} [\mathbf{k}^\circ] &= [\mathbf{DC}]^{\mathbf{T}}\ [\mathbf{k}]\ [\mathbf{DC}] \\ &= [\mathbf{k_r}^\circ] + [\mathbf{k_b}^\circ] \end{aligned} \tag{23.86}$$

where,

$$[\mathbf{k_r}^\circ] = \frac{AE}{l} \begin{array}{c} \begin{array}{cccccc} u_1^\circ & v_1^\circ & \theta_1 & u_2^\circ & v_2^\circ & \theta_2 \end{array} \\ \left[\begin{array}{cccccc} c^2 & cs & 0 & -c^2 & -cs & 0 \\ cs & s^2 & 0 & -cs & -s^2 & 0 \\ 0 & 0 & 0 & 0 & 0 & 0 \\ -c^2 & -cs & 0 & c^2 & cs & 0 \\ -cs & -s^2 & 0 & cs & s^2 & 0 \\ 0 & 0 & 0 & 0 & 0 & 0 \end{array}\right] \end{array} \begin{array}{c} u_1^\circ \\ v_1^\circ \\ \theta_1 \\ u_2^\circ \\ v_2^\circ \\ \theta_2 \end{array} \tag{23.87}$$

$$[\mathbf{k}_b^\circ] = EI \begin{array}{c} \begin{array}{cccccc} u_1^\circ & v_1^\circ & \theta_1 & u_2^\circ & v_2^\circ & \theta_2 \end{array} \\ \left[\begin{array}{cccccc} \frac{12}{l^3}s^2 & & & & & \\ \frac{12}{l^3}cs^2 & \frac{12}{l^3}c^2 & & & & \\ \frac{6}{l^2}s & -\frac{6}{l^2}c & \frac{4}{l} & & & \\ -\frac{12}{l^3}s^2 & \frac{12}{l^3}cs & -\frac{6}{l^2}s & \frac{12}{l^3}s^2 & & \\ \frac{12}{l^3}cs & -\frac{12}{l^3}c^2 & \frac{6}{l^2}c & -\frac{12}{l^3}cs & \frac{12}{l^3}c^2 & \\ \frac{6}{l^2}s & -\frac{6}{l^2}c & \frac{2}{l} & -\frac{6}{l^2}s & \frac{6}{l^2}c & \frac{4}{l} \end{array}\right] \end{array} \begin{array}{c} u_1^\circ \\ v_1^\circ \\ \theta_1^\circ \\ u_2^\circ \\ v_2^\circ \\ \theta_2 \end{array} \tag{23.88}$$

c = cos α

s = sin α

A = cross-sectional area

I = second moment of area of the element's cross-section

l = elemental length

E = Young's modulus of elasticity

Problem 23.6 Using the matrix displacement method, determine the nodal bending moments for the rigid-jointed plane frame shown in the figure below. It may be assumed that the axial stiffness of each element is very large compared to the flexural stiffness, so that $v_2^\circ = v_3^\circ = 0$, and $u_2^\circ = u_3^\circ$.

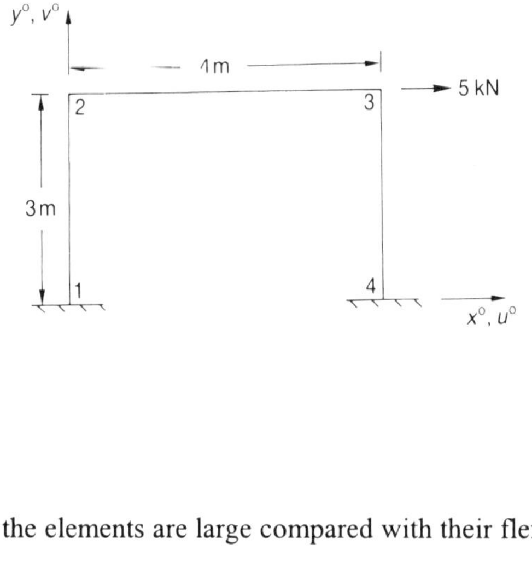

Solution

As the axial stiffness of the elements are large compared with their flexural stiffness, the effects of $[\mathbf{k}_r^\circ]$ can be ignored.

Element 1–2

$\alpha = 90°$ $\qquad c = 0$ $\qquad s = 1$ $\qquad l = 3$ m

Substituting the above into equation (23.88), and removing the rows and columns corresponding to the zero displacements, which in this case are u_1°, v_1°, θ_1 and v_2°, the elemental stiffness matrix for member 1–2 becomes

$$
\begin{array}{c}
\quad u_1^\circ \quad v_1^\circ \quad \theta_1 \quad u_2^\circ \quad v_2^\circ \quad \theta_2 \\
\left[\mathbf{k}_{1-2}^\circ\right] = EI \begin{bmatrix}
 & & & & & \\
 & & & & & \\
 & & & & & \\
 & & & 0.444 & & \\
 & & & -0.667 & & 1.333 \\
 & & & & &
\end{bmatrix}
\begin{matrix} u_1^\circ \\ v_1^\circ \\ \theta_1 \\ u_2^\circ \\ v_2^\circ \\ \theta_2 \end{matrix}
\end{array}
\qquad (23.89)
$$

Element 2–3

$$\alpha = 0, \qquad c = 1, \qquad s = 0, \qquad l = 4 \text{ m}$$

Substituting the above into equation (23.88), and removing the columns and rows corresponding to zero displacements, which in this case are v_2° and v_3°, the elemental stiffness matrix for member 2–3 is given by

$$
\begin{array}{c}
\quad u_2^\circ \quad v_2^\circ \quad \theta_2 \quad u_3^\circ \quad v_3^\circ \quad \theta_3 \\
\left[\mathbf{k}_{2-3}^\circ\right] = EI \begin{bmatrix}
0 & & & & & \\
 & & & & & \\
0 & & 1 & & & \\
0 & & 0 & & & \\
 & & & & & \\
0 & & 0.5 & 0 & & 1
\end{bmatrix}
\begin{matrix} u_2^\circ \\ v_2^\circ \\ \theta_2 \\ u_3^\circ \\ v_3^\circ \\ \theta_3 \end{matrix}
\end{array}
\qquad (23.90)
$$

Element 3–4

$$\alpha = -90^\circ, \qquad c = 0, \qquad s = -1, l = 3 \text{ m}$$

Substituting the above into equation (23.88), and removing the columns and rows corresponding to zero displacements, namely v_3°, u_4°, v_4° and θ_4, the elemental stiffness matrix for member 3–4 is given by

$$
\left[\mathbf{k}_{3-4}{}^{\circ}\right] = EI
\begin{array}{c}
\begin{array}{cccccc} u_3{}^{\circ} & v_3{}^{\circ} & \theta_3 & u_4{}^{\circ} & v_4{}^{\circ} & \theta_4 \end{array} \\
\left[\begin{array}{cccccc}
0.444 & & & & & \\
 & & & & & \\
-0.667 & & 1.333 & & & \\
 & & & & & \\
 & & & & & \\
 & & & & &
\end{array}\right]
\begin{array}{c} u_3{}^{\circ} \\ v_3{}^{\circ} \\ \theta_3 \\ u_4{}^{\circ} \\ v_4{}^{\circ} \\ \theta_4 \end{array}
\end{array}
\tag{23.91}
$$

Superimposing the stiffness influence coefficients, corresponding to the free displacements, $u_2{}^{\circ}$, θ_2, $u_3{}^{\circ}$ and θ_3, the system stiffness matrix $[\mathbf{K}_{11}]$ is obtained, as shown by equation (23.92):

$$
[\mathbf{K}_{11}{}^{\circ}] = EI
\begin{array}{c}
\begin{array}{cccc} u_2{}^{\circ} & \theta_2 & u_3{}^{\circ} & \theta_3 \end{array} \\
\left[\begin{array}{c|c|c|c}
\begin{array}{c} 0.444 \\ +0 \end{array} & & & \\
\hline
\begin{array}{c} -0.667 \\ +0 \end{array} & \begin{array}{c} 1.333 \\ +1 \end{array} & & \\
\hline
 & & 0.444 & \\
 & 0.5 & -0.667 & 1+1.333
\end{array}\right]
\begin{array}{c} u_2{}^{\circ} \\ \theta_2 \\ u_3{}^{\circ} \\ \theta_3 \end{array}
\end{array}
\tag{23.92}
$$

$$
\left[\mathbf{K}_{11}{}^{\circ}\right] = EI
\begin{array}{c}
\begin{array}{cccc} u_2{}^{\circ} & \theta_2 & u_3{}^{\circ} & \theta_3 \end{array} \\
\left[\begin{array}{cccc}
0.444 & -0.667 & 0 & 0 \\
-0.667 & 2.333 & 0 & 0.5 \\
0 & 0 & 0.444 & -0.667 \\
0 & 0.5 & -0.667 & 2.333
\end{array}\right]
\begin{array}{c} u_2{}^{\circ} \\ \theta_2 \\ u_3{}^{\circ} \\ \theta_3 \end{array}
\end{array}
\tag{23.93}
$$

The vector of loads corresponding to these free displacements is given by

$$\{q_F\} = \begin{Bmatrix} 0 \\ 0 \\ 5 \\ 0 \end{Bmatrix} \begin{matrix} u_2^\circ \\ \theta_2 \\ u_3^\circ \\ \theta_3 \end{matrix} \tag{23.94}$$

Rewriting equations (23.93) and (23.94) in the form of four linear simultaneous equations, and noting that the 5 kN load is shared between members 1–2 and 3–4, the following is obtained:

$$\begin{aligned} 2.5 &= EI(0.444\,u_2^\circ - 0.667\,\theta_2) \\ 0 &= EI(-0.667\,u_2^\circ + 2.333\,\theta_2 + 0.5\,\theta_3) \\ 2.5 &= EI(0.444\,u_3^\circ - 0.667\,\theta_3) \\ 0 &= EI(0.5\,\theta_2 - 0.667\,u_3^\circ + 2.333\,\theta_3) \end{aligned} \tag{23.95}$$

Now for this case

$$\theta_2 = \theta_3 \tag{23.96}$$

and

$$u_2^\circ = u_3^\circ$$

Hence, equation (23.95) can be reduced to the form shown in equation (23.97):

$$\begin{aligned} 2.5 &= 0.444\ EIu_2^\circ - 0.667\ EI\theta_2 \\ 0 &= -0.667\ EIu_2^\circ + 2.833\ EI\theta_2 \end{aligned} \tag{23.97}$$

Solving the above

$$u_2^\circ = u_3^\circ = 8.707/EI$$

and

$$\theta_2 = \theta_3 = 2.049/EI \tag{23.98}$$

To determine the nodal bending moments, the displacements in the local v and θ directions will have to be calculated, prior to using equations (23.70) and (23.72).

Element 1–2

$$c = 0, \qquad s = 1, \qquad l = 3 \text{ m}$$

From equation (23.23):

$$v_2 = \begin{bmatrix} -s & c \end{bmatrix} \begin{Bmatrix} u_2^\circ \\ v_2^\circ \end{Bmatrix}$$

$$= \begin{bmatrix} -1 & 0 \end{bmatrix} \frac{1}{EI} \begin{Bmatrix} 8.707 \\ 0 \end{Bmatrix}$$

$$v_2 = -8.707/EI$$

By inspection,

$$v_1 = \theta_1 = 0 \quad \text{and} \quad \theta_2 = 2.049/EI$$

Substituting the above values into the slope–deflection equations (23.70) and (23.72)

$$M_{1\text{-}2} = 0 + \frac{2EI}{3} \times \frac{2.049}{EI} - \frac{6EI}{9}\left(0 + \frac{8.707}{EI}\right)$$

$$= 1.366 - 5.805$$

$$M_{1\text{-}2} = -4.43 \text{ kNm}$$

$$M_{2\text{-}1} = 0 + \frac{4EI}{3} \times \frac{2.049}{EI} - \frac{6EI}{9}\left(0 + \frac{8.707}{EI}\right)$$

$$= 2.732 - 5.805$$

$$M_{2\text{-}1} = -3.07 \text{ kNm}$$

Element 2–3

$$l = 4 \text{ m}$$

By inspection,

$$v_2 = v_3 = 0$$

and

$$\theta_2 = \theta_3 = 2.049/EI$$

Substituting the above values into the slope–deflection equations (23.70) an (23.72):

$$M_{2\text{-}3} = \frac{4EI}{4} \times \frac{2.049}{EI} + \frac{2EI}{4} \times \frac{2.049}{EI}$$

$$M_{2\text{-}3} = 3.07 \text{ kNm}$$

Element 3–4

$$c = 0, \qquad s = -1, \qquad l = 3 \text{ m}$$

From equation (23.23):

$$v_3 = \begin{bmatrix} -s & c \end{bmatrix} \begin{Bmatrix} u_3^\circ \\ v_3^\circ \end{Bmatrix}$$

$$= \begin{bmatrix} 1 & 0 \end{bmatrix} \frac{1}{EI} \begin{Bmatrix} 8.707 \\ 0 \end{Bmatrix}$$

$$v_3 = 8.707/EI$$

By inspection,

$$v_4 = \theta_4 = 0 \qquad \text{and } \theta_3 = 2.049/EI$$

Substituting the above values into equations (23.70) and (23.72),

$$M_{3\text{-}4} = \frac{4EI}{3} \times \frac{2.049}{EI} + 0 - \frac{6EI}{9}\left(\frac{8.707}{EI} - 0\right)$$

$$= 2.732 - 5.805$$

$$M_{3\text{-}4} = -3.07 \text{ kNm}$$

$$M_{4\text{-}3} = \frac{2EI}{3} \times 2.049 + 0 - \frac{6EI}{9}\left(\frac{8.707}{EI} - 0\right)$$

$$= 1.366 - 5.805$$

$$M_{4\text{-}3} = -4.44 \text{ kNm}$$

Problem 23.7 Using the matrix displacement method, determine the nodal bending moments for the rigid-jointed plane frame shown below.

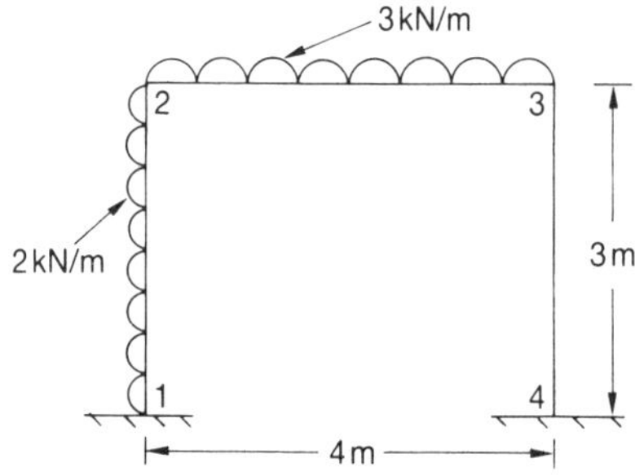

Solution

As this frame has distributed loading between some of the nodes, it will be necessary to treat the problem in a manner similar to that described in the solution of Problem 23.5.

There are four degrees of freedom for this structure, namely, u_2°, θ_2, u_3° and θ_3, hence $\{q_F\}$ will be of order 4×1.

To determine $\{q_F\}$, it will be necessary to fix the structure at its nodes, and calculate the end fixing forces, as shown and calculated below.

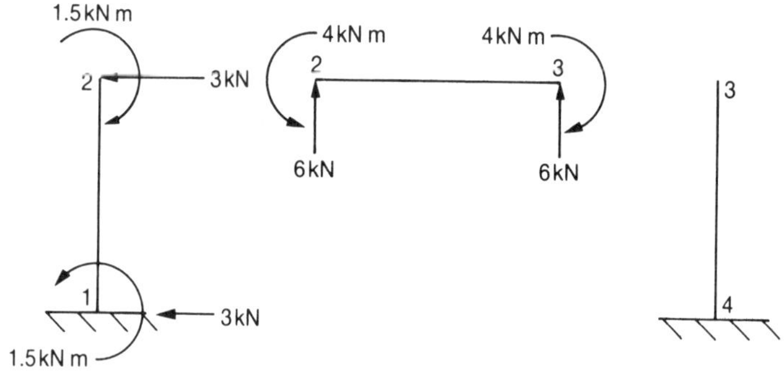

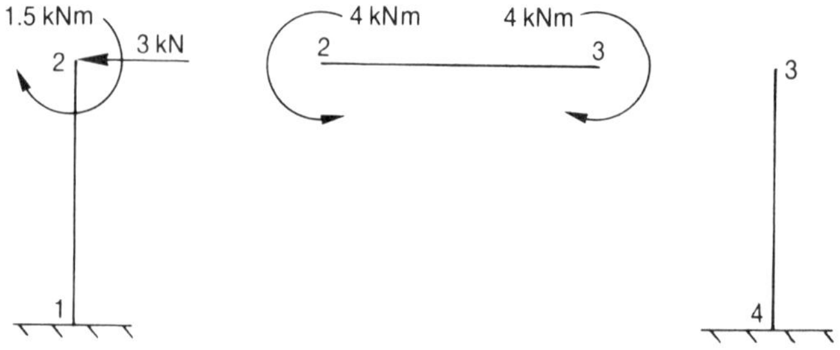

End fixing forces

$$M_{1-2}^F = -\frac{wl^2}{12} = -\frac{2 \times 3^2}{12} = -1.5 \text{ kNm}$$

$$M_{2-1}^F = \frac{wl^2}{12} = 1.5 \text{ kNm}$$

$$\text{Horizontal reaction at node 1} = \frac{wl}{2} = \frac{2 \times 3}{2} = 3 \text{kN}$$

$$\text{Horizontal reaction at node 2} = \frac{wl}{2} = \frac{2 \times 3}{2}$$

$$= 3 \text{ kN}$$

$$M_{2-3}^F = -\frac{wl^2}{12} = -\frac{3 \times 4^2}{12} = -4 \text{ kNm}$$

$$M_{3-2}^F = -M_{2-3}^F = 4 \text{ kNm}$$

$$\text{Vertical reaction at node 2} = \frac{wl}{2} = \frac{3 \times 4}{2} = 6 \text{ kN}$$

$$\text{Vertical reaction at node 3} = \frac{wl}{2} = \frac{3 \times 4}{2}$$

$$= 6 \text{ kN}$$

Now, for this problem, as

$$u_1^\circ = v_1^\circ = \theta_1 = v_2^\circ = v_3^\circ = u_4^\circ = v_4^\circ = \theta_4 = 0$$

the only components of the end-fixing forces required for calculating $\{q_F\}$ are shown below:

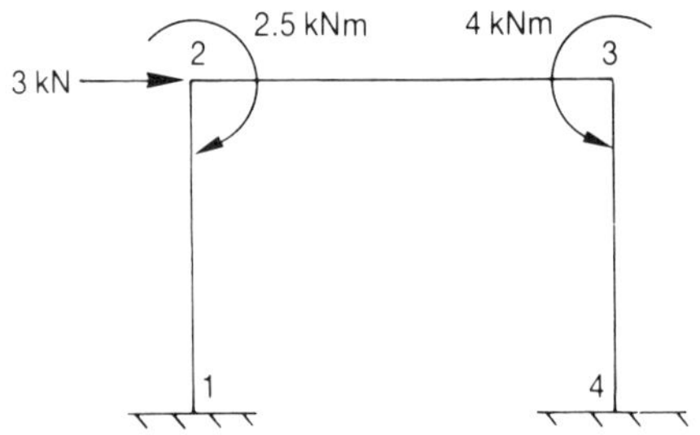

The negative resultants of the end-fixing forces are shown below, where it can be seen that

$$q_F = \begin{Bmatrix} 3 \\ 2.5 \\ 0 \\ -4 \end{Bmatrix} \begin{matrix} u_2^\circ \\ \theta_2 \\ u_3^\circ \\ \theta_3 \end{matrix} \tag{23.99}$$

From equation (23.93),

$$[K_{11}] = EI \begin{bmatrix} 0.444 & -0.667 & 0 & 0 \\ -0.667 & 2.333 & 0 & 0.5 \\ 0 & 0 & 0.444 & -0.667 \\ 0 & 0.5 & -0.667 & 2.333 \end{bmatrix} \begin{matrix} u_2^\circ \\ \theta_2 \\ u_3^\circ \\ \theta_3 \end{matrix} \tag{23.100}$$

(columns: u_2°, θ_2, u_3°, θ_3)

Rewriting equations (23.99) and (23.100) in the form of four simultaneous equations,

$$3 = 0.444u_2^\circ / EI - 0.667\theta_2 / EI \tag{23.101a}$$

$$2.5 = -0.667u_2^\circ / EI + 2.333\theta_2 / EI + 0.5\theta_3 / EI \tag{23.101b}$$

$$0 = 0.444u_3^\circ / EI - 0.667\theta_3 / EI \tag{23.101c}$$

$$-4 = 0.5\theta_2 / EI - 0.667u_3^\circ / EI + 2.333\theta_3 / EI \tag{23.101d}$$

Now, as the 2.5 kN load is shared between elements 1–2 and 3–4, equation (23.101a) *must be added* to equation (23.101c), as shown by equation (23.102):

$$3 = 0.888u_2^\circ / EI - 0.667\theta_2 / EI - 0.667\theta_3 / EI \tag{23.102}$$

Putting $u_2^\circ = u_3^\circ$, the simultaneous equations (23.101) now become:

$$\begin{aligned} 3 &= 0.888\,u_2^\circ/EI - 0.667\,\theta_2/EI - 0.667\,\theta_3/EI \\ 2.5 &= -0.667\,u_2^\circ/EI + 2.333\,\theta_2/EI + 0.5\,\theta_3/EI \\ -4 &= -0.667\,u_2^\circ/EI + 0.5\,\theta_2/EI + 2.333\,\theta_3/EI \end{aligned} \tag{23.103}$$

Solving the above,

$$u_2^\circ = u_3^\circ = 4.61/EI$$

$$\theta_2 = 2.593/EI$$

$$\theta_3 = -0.953/EI$$

To determine the nodal bending moments, the end fixing moments will have to be added to the moments obtained from the slope–deflection equations.

Element 1–2

$$c = 0 \qquad s = 1 \qquad l = 3\text{ m}$$

From equation (23.23)

$$v_2 = \begin{bmatrix} -s & c \end{bmatrix} \begin{Bmatrix} u_2^\circ \\ v_2^\circ \end{Bmatrix}$$

$$= \begin{bmatrix} -1 & 0 \end{bmatrix} \begin{Bmatrix} 4.61/EI \\ 0 \end{Bmatrix}$$

$$v_2 = -4.61/EI$$

By inspection,

$$v_1 = \theta_1 = 0 \qquad \text{and} \qquad \theta_2 = -0.953/\text{EI}$$

Substituting the above into the slope–deflection equations (23.70) and (23.72), and adding the end fixing moments,

$$M_{1\text{-}2} = 0 + \frac{2EI}{3} \times (2.593/EI) - \frac{6EI}{9}(0 + 4.61/EI) - 1.5$$

$$= 1.729 - 3.07 - 1.5$$

$$M_{1\text{-}2} = -2.84 \text{ kNm}$$

and

$$M_{2\text{-}1} = \frac{4EI}{3} \times \frac{2.593}{EI} - 3.07 + 1.5$$

$$M_{2\text{-}1} = 1.89 \text{ kNm}$$

Element 2–3

By inspection,

$$v_2 = v_3 = 0$$

and

$$\theta_2 = 2.593/\text{EI}, \qquad \theta_3 = -0.953/\text{EI}$$

Substituting the above into equations (23.70) and (23.72), adding the end-fixing moments for this element, and remembering that node 2 is the first node and node 3 the second node,

$$M_{2\text{-}3} = \frac{4EI}{4} \times \frac{2.593}{EI} + \frac{2EI}{4} \times \left(\frac{-0.953}{EI}\right) - 4$$

$$M_{2\text{-}3} = -1.88 \text{ kNm}$$

$$M_{3\text{-}2} = \frac{2EI}{4} \times \frac{2.593}{EI} + \frac{4EI}{4} \times \left(\frac{-0.953}{EI}\right) + 4$$

$$M_{3\text{-}2} = 4.34 \text{ kNm}$$

Element 3–4

$$c = 0, \qquad s = 1, \qquad l = 3\text{ m}$$

From equation (23.23),

$$v_3 = \begin{bmatrix} -s & c \end{bmatrix} \begin{Bmatrix} u_3^{\circ} \\ v_3^{\circ} \end{Bmatrix}$$

$$= \begin{bmatrix} 1 & 0 \end{bmatrix} \begin{Bmatrix} 4.61/EI \\ 0 \end{Bmatrix}$$

$$v_3 = 4.61/EI$$

By inspection,

$$u_3 = u_4 = v_4 = \theta_4 = 0$$

and

$$\theta_3 = -0.953/EI$$

$$M_{3-4} = \frac{4EI}{3} \times \left(\frac{-0.953}{EI}\right) + 0 - \frac{6EI}{9}(4.61/EI)$$

$$M_{3-4} = -4.34\text{ kNm}$$

$$M_{4-3} = \frac{2EI}{3} \times \left(\frac{-0.953}{EI}\right) + 0 - \frac{6EI}{9}(4.61/EI)$$

$$M_{4-3} = -3.71\text{ kNm}$$

Further problems *(answers on page 697)*

23.8 Determine the forces in the members of the framework of the figure below, under the following conditions:

(a) all joints are pinned;
(b) all joints are rigid (i.e. welded).

The following may be assumed:

AE = 100 EI
A = cross-sectional area
I = second moment of area
E = Young's modulus
$[\mathbf{k}]^\circ$ = the elemental stiffness matrix
= $[\mathbf{k}_b{}^\circ] + [\mathbf{k}_r{}^\circ]$

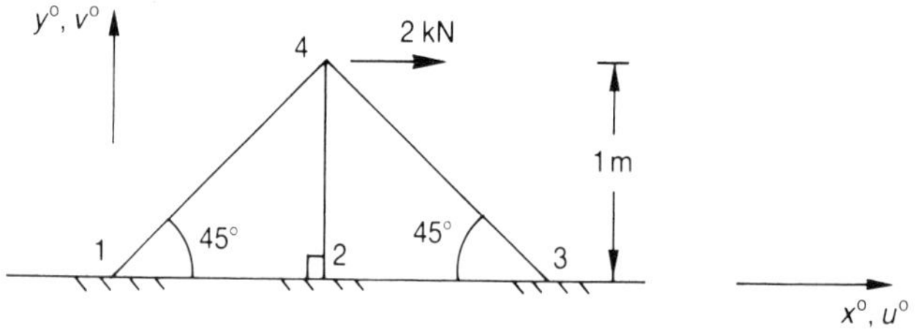

(Portsmouth, 1987, Standard level)

23.9 Determine the displacements at node 5 for the framework shown below under the following conditions:

(a) all joints are pinned;
(b) all joints are rigid (i.e. welded).

It may be assumed, for all members of the framework,

A = 100 EI

where

A = cross-sectional area
I = second moment of area
E = Young's modulus
$[\mathbf{k}]^\circ$ = the stiffness matrix
= $[\mathbf{k}_b{}^\circ] + [\mathbf{k}_r{}^\circ]$

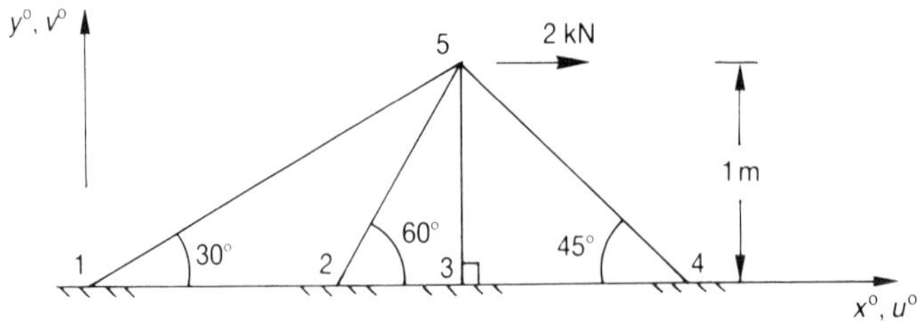

(Portsmouth, 1987, Honours level)

23.10 Determine the nodal displacements and moments for the beams shown below, using the *matrix displacement* method.

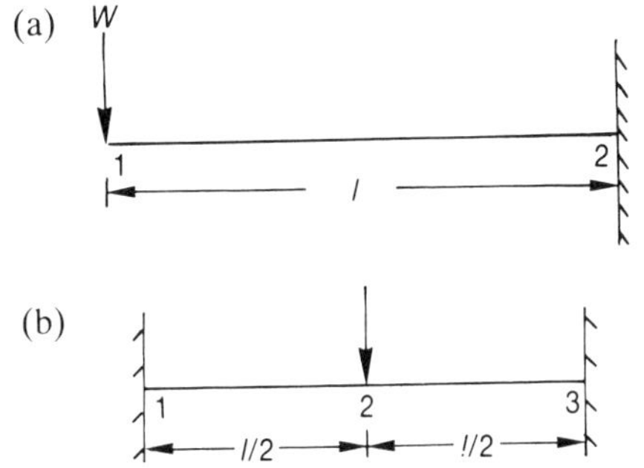

23.11 Determine the nodal bending moments in the continuous beam below, using the *matrix displacement* method.

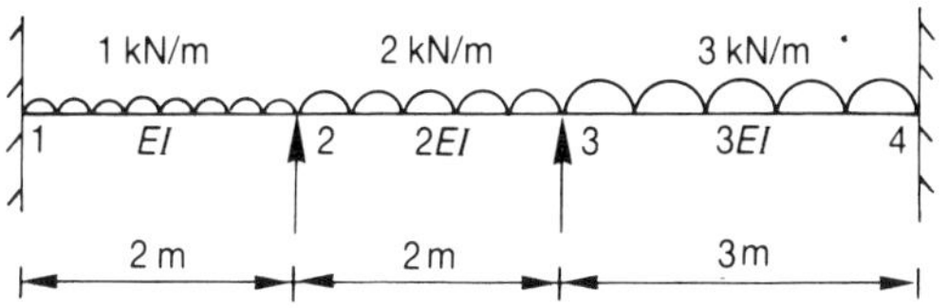

23.12 A ship's bulkhead stiffener is subjected to the hydrostatic loading shown below. If the stiffener is firmly supported at nodes 2 and 3, and fixed at nodes 1 and 4, determine the nodal displacements and moments.

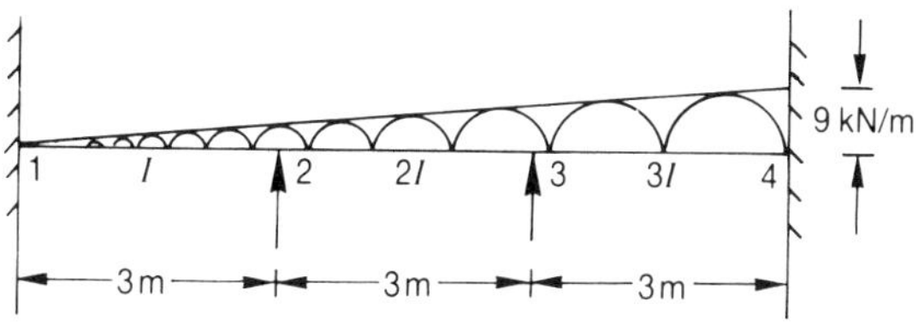

23.13 Using the *matrix displacement* method, determine the forces in the pin-jointed space trusses shown in the following figures. It may be assumed that AE = a constant.

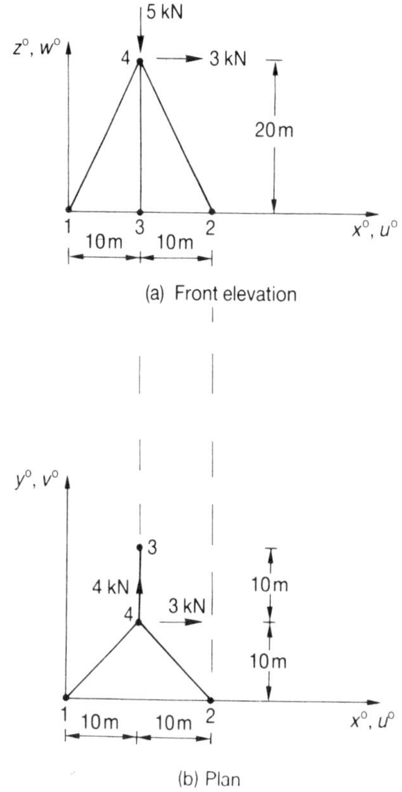

(Portsmouth, 1989)

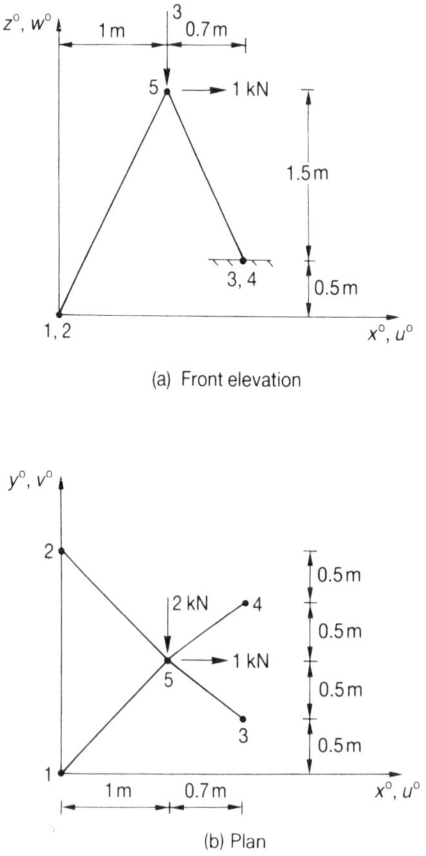

(Portsmouth, 1983)

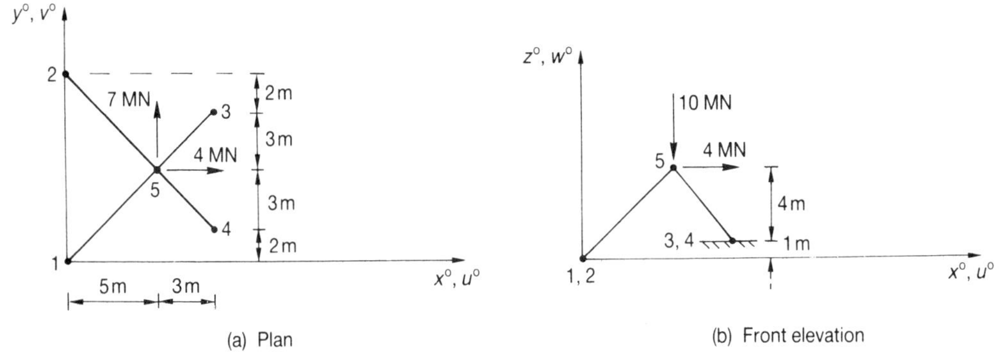

(Portsmouth, 1989)

23.14 Determine the nodal displacements and moments for the uniform section rigid-jointed plane frames shown in the two figures below.

It may be assumed that the axial stiffness of each member is large compared with its flexural stiffness, so that,

$$v_2^{\circ} = v_3^{\circ} = 0$$

and

$$u_2^{\circ} = u_3^{\circ}$$

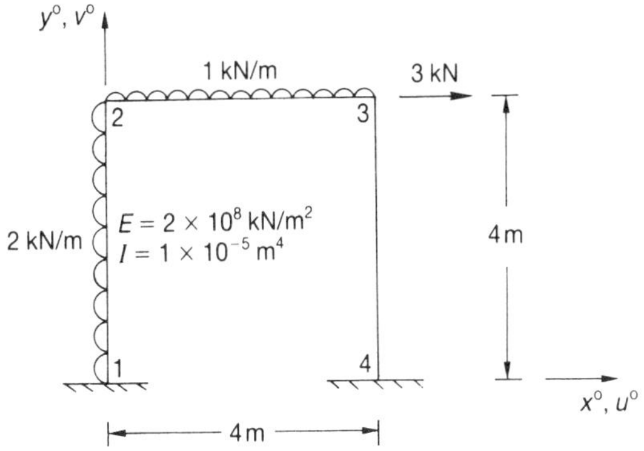

(Portsmouth, 1984)

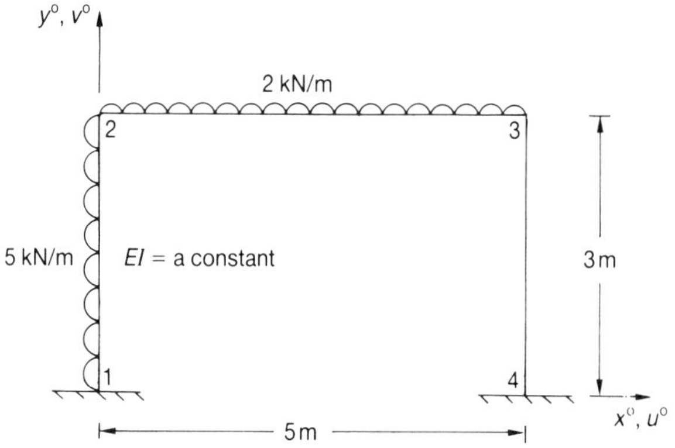
y°, v°
2 kN/m
2
3
5 kN/m
EI = a constant
3m
1
4
5m
x°, u°

24 The finite element method

24.1 Introduction

In this chapter the finite element method proper[10] will be described with the aid of worked examples.

The finite element method is based on the matrix displacement method described in Chapter 23, but its description is separated from that chapter because it can be used for analysing much more complex structures, such as those varying from the legs of an integrated circuit to the legs of an offshore drilling rig, or from a gravity dam to a doubly curved shell roof. Additionally, the method can be used for problems in structural dynamics, fluid flow, heat transfer, acoustics, magnetostatics, electrostatics, medicine, weather forecasting, etc.

The method is based on representing a complex shape by a series of simpler shapes, as shown in Figure 24.1, where the simpler shapes are called finite elements.

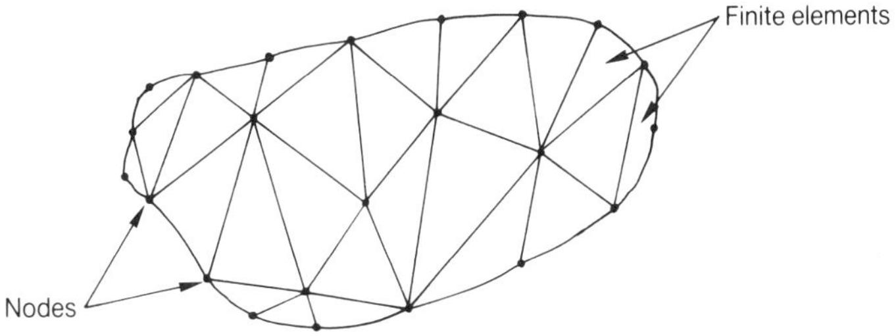

Figure 24.1 Complex shape, represented by finite elements.

Using the energy methods described in Chapter 17, the stiffness and other properties of the finite element can be obtained, and then by considering equilibrium and compatibility along the inter-element boundaries, the stiffness and other properties of the entire domain can be obtained.

[10] Turner M J, Clough R W, Martin H C and Topp L J, Stiffness and Deflection Analysis of Complex Structures, *J Aero. Sci*, 23, 805–23, 1956.

This process leads to a large number of simultaneous equations, which can readily be solved on a high-speed digital computer. It must be emphasised, however, that the finite element method is virtually useless without the aid of a computer, and this is the reason why the finite element method has been developed alongside the advances made with digital computers. Today, it is possible to solve massive problems on most computers, including microcomputers, laptop and notepad computers; and in the near future, it will be possible to use the finite element method with the aid of hand-held computers.

Finite elements appear in many forms, from triangles and quadrilaterals for two-dimensional domains to tetrahedrons and bricks for three-dimensional domains, where, in general, the finite element is used as a 'space' filler.

Each finite element is described by nodes, and the nodes are also used to describe the domain, as shown in Figure 24.1, where corner nodes have been used.

If, however, mid-side nodes are used in addition to corner nodes, it is possible to develop curved finite elements, as shown in Figure 24.2, where it is also shown how ring nodes can be used for axisymmetric structures, such as conical shells.

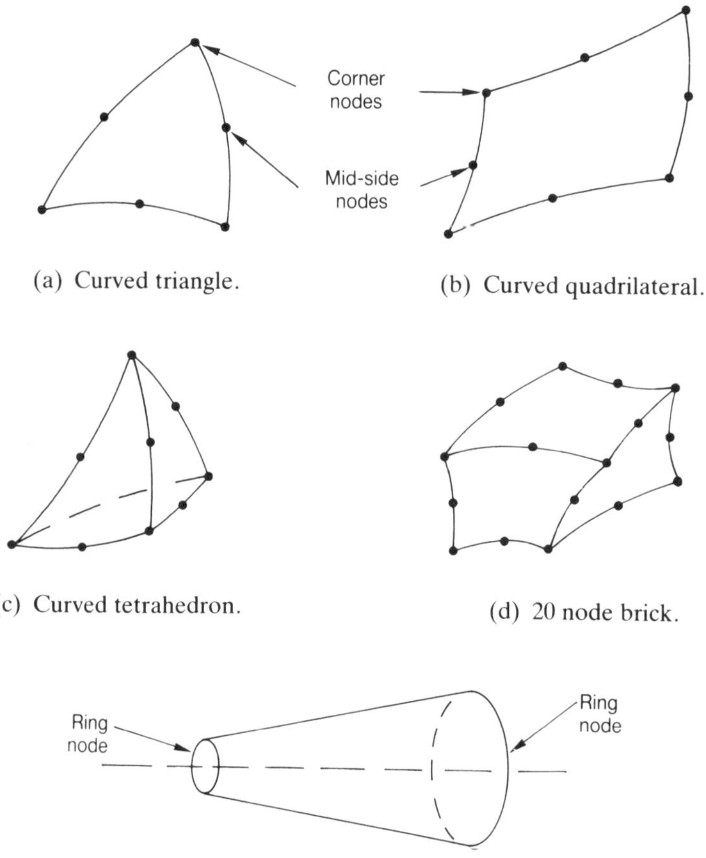

(a) Curved triangle.

(b) Curved quadrilateral.

(c) Curved tetrahedron.

(d) 20 node brick.

Figure 24.2 Some typical finite elements.

The finite element was invented in 1956 by Turner *et al.* where the important three node in-plane triangular finite element was first presented.

The derivation of the stiffness matrix for this element will now be described.

24.2 Stiffness matrices for some typical finite elements

The in-plane triangular element of Turner *et al.* is shown in Figure 24.3. From this figure, it can be seen that the element has six degrees of freedom, namely, $u_1°$, $u_2°$, $u_3°$, $v_1°$, $v_2°$ and $v_3°$, and because of this, the assumptions for the displacement polynomial distributions $u°$ and $v°$ will involve six arbitrary constants. It is evident that with six degrees of freedom, a total of six simultaneous equations will be obtained for the element, so that expressions for the six arbitrary constants can be defined in terms of the nodal displacements, or boundary values.

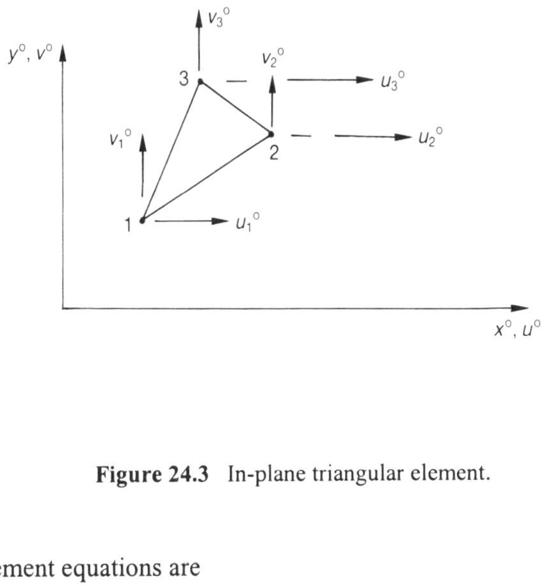

Figure 24.3 In-plane triangular element.

Convenient displacement equations are

$$u° = \alpha_1 + \alpha_2 x° + \alpha_3 y° \tag{24.1}$$

and

$$v° = \alpha_4 + \alpha_5 x° + \alpha_6 y° \tag{24.2}$$

where α_1 to α_6 are the six arbitrary constants, and $u°$ and $v°$ are the displacement equations.

Suitable boundary conditions, or boundary values, at node 1 are:

$$\text{at } x° = x_1° \text{ and } y° = y_1°, \quad u° = u_1° \text{ and } v° = v_1°$$

Substituting these boundary values into equations (24.1) and (24.2),

$$u_1^\circ = \alpha_1 + \alpha_2 x_1^\circ + \alpha_3 y_1^\circ \tag{24.3}$$

and
$$v_1^\circ = \alpha_4 + \alpha_5 x_1^\circ + \alpha_6 y_1^\circ \tag{24.4}$$

Similarly, at node 2,

at $x^\circ = x_2^\circ$ and $y^\circ = y_2^\circ$, $u^\circ = u_2^\circ$ and $v^\circ = v_2^\circ$

When substituted into equations (24.1) and (24.2), these give

$$u_2^\circ = \alpha_1 + \alpha_2 x_2^\circ + \alpha_3 y_2^\circ \tag{24.5}$$

and
$$v_2^\circ = \alpha_4 + \alpha_5 x_2^\circ + \alpha_6 y_2^\circ \tag{24.6}$$

Likewise, at node 3,

at $x^\circ = x_3^\circ$ and $y^\circ = y_3^\circ$, $u^\circ = u_3^\circ$ and $v^\circ = v_3^\circ$

which, when substituted into equation (24.1) and (24.2), yield

$$u_3^\circ = \alpha_1 + \alpha_2 x_3^\circ + \alpha_3 y_3^\circ \tag{24.7}$$

and
$$v_3^\circ = \alpha_4 + \alpha_5 x_3^\circ + \alpha_6 y_3^\circ \tag{24.8}$$

Rewriting equations (24.3) to (24.8) in matrix form, the following equation is obtained:

$$\begin{Bmatrix} u_1^\circ \\ u_2^\circ \\ u_3^\circ \\ v_1^\circ \\ v_2^\circ \\ v_3^\circ \end{Bmatrix} = \begin{bmatrix} 1 & x_1^\circ & y_1^\circ & & & \\ 1 & x_2^\circ & y_2^\circ & & 0_3 & \\ 1 & x_3^\circ & y_3^\circ & & & \\ & & & 1 & x_1^\circ & y_1^\circ \\ & 0_3 & & 1 & x_2^\circ & y_2^\circ \\ & & & 1 & x_3^\circ & y_3^\circ \end{bmatrix} \begin{Bmatrix} \alpha_1 \\ \alpha_2 \\ \alpha_3 \\ \alpha_4 \\ \alpha_5 \\ \alpha_6 \end{Bmatrix} \tag{24.9}$$

or

$$\{u_1^\circ\} = \begin{bmatrix} A & 0_3 \\ 0_3 & A \end{bmatrix} \{a_i\} \tag{24.10}$$

and

$$\{a_i\} = \begin{bmatrix} A^{-1} & 0_3 \\ 0_3 & A^{-1} \end{bmatrix} \begin{Bmatrix} u_1^\circ \\ u_2^\circ \\ u_3^\circ \\ v_1^\circ \\ v_2^\circ \\ v_3^\circ \end{Bmatrix} \qquad (24.11)$$

where

$$[\mathbf{A}]^{-1} = \begin{bmatrix} a_1 & a_2 & a_3 \\ b_1 & b_2 & b_3 \\ c_1 & c_2 & c_3 \end{bmatrix} / \ \mathbf{det}|\mathbf{A}| \qquad (24.12)$$

$$\begin{aligned} a_1 &= x_2^\circ y_3^\circ - x_3^\circ y_2^\circ \\ a_2 &= x_3^\circ y_1^\circ - x_1^\circ y_3^\circ \\ a_3 &= x_1^\circ y_2^\circ - x_2^\circ y_1^\circ \\ b_1 &= y_2^\circ - y_3^\circ \\ b_2 &= y_3^\circ - y_1^\circ \\ b_3 &= y_1^\circ - y_2^\circ \\ c_1 &= x_3^\circ - x_2^\circ \\ c_2 &= x_1^\circ - x_3^\circ \\ c_3 &= x_2^\circ - x_1^\circ \end{aligned} \qquad (24.13)$$

$$\mathbf{det}\ |A| = x_2^\circ y_3^\circ - y_2^\circ x_3^\circ - x_1^\circ (y_3^\circ - y_2^\circ) + y_1^\circ (x_3^\circ - x_2^\circ) = 2\Delta$$

Δ = area of triangle

Substituting equations (24.13) and (24.12) into equations (24.1) and (24.2)

$$\begin{Bmatrix} u^\circ \\ v^\circ \end{Bmatrix} = \begin{bmatrix} N_1 & N_2 & N_3 & 0 & 0 & 0 \\ 0 & 0 & 0 & N_1 & N_2 & N_3 \end{bmatrix} \begin{Bmatrix} u_1^\circ \\ u_2^\circ \\ u_3^\circ \\ v_1^\circ \\ v_2^\circ \\ v_3^\circ \end{Bmatrix} \tag{24.14}$$

or

$$\{u\} = [\mathbf{N}]\,\{u_i\} \tag{24.15}$$

where $[\mathbf{N}]$ = a matrix of shape functions:

$$\begin{aligned} N_1 &= \frac{1}{2\Delta}\left(a_1 + b_1 x^\circ + c_1 y^\circ\right) \\ N_2 &= \frac{1}{2\Delta}\left(a_2 + b_2 x^\circ + c_2 y^\circ\right) \\ N_3 &= \frac{1}{2\Delta}\left(a_3 + b_3 x^\circ + c_3 y^\circ\right) \end{aligned} \tag{24.16}$$

For a two-dimensional system of strain, the expressions for strain[11] are given by

$$\begin{aligned} \varepsilon_x &= \text{strain in the } x^\circ \text{ direction} = \partial u^\circ/\partial x^\circ \\ \varepsilon_y &= \text{strain in the } y^\circ \text{ direction} = \partial v^\circ/\partial y^\circ \\ \gamma_{xy} &= \text{shear strain in the } x^\circ\text{–}y^\circ \text{ plane} \\ &= \partial u^\circ/\partial y^\circ + \partial v^\circ/\partial x^\circ \end{aligned} \tag{24.17}$$

which when applied to equation (24.14) becomes

---

[11] Fenner R T, *Engineering Elasticity*, Ellis Horwood, 1986.

$$2\Delta\varepsilon_x = b_1 u_1^\circ + b_2 u_2^\circ + b_3 u_3^\circ$$

$$2\Delta\varepsilon_y = c_1 v_1^\circ + c_2 v_2^\circ + c_3 v_3^\circ$$

$$2\Delta\gamma_{xy} = c_1 u_1^\circ + c_2 u_2^\circ + c_3 u_3^\circ + b_1 v_1^\circ + b_2 v_2^\circ + b_3 v_3^\circ \tag{24.18}$$

Rewriting equation (24.18) in matrix form, the following is obtained:

$$\begin{Bmatrix} \varepsilon_x \\ \varepsilon_y \\ \gamma_{xy} \end{Bmatrix} = \frac{1}{2\Delta} \begin{bmatrix} b_1 & b_2 & b_2 & 0 & 0 & 0 \\ 0 & 0 & 0 & c_1 & c_2 & c_3 \\ c_1 & c_2 & c_3 & b_1 & b_2 & b_3 \end{bmatrix} \begin{Bmatrix} u_1^\circ \\ u_2^\circ \\ u_3^\circ \\ v_1^\circ \\ v_2^\circ \\ v_3^\circ \end{Bmatrix} \tag{24.19}$$

or

$$\{\varepsilon\} = [\mathbf{B}]\{u_i\} \tag{24.20}$$

where [**B**] is a matrix relating strains and nodal displacements

$$[\mathbf{B}] = \frac{1}{2\Delta} \begin{bmatrix} b_1 & b_2 & b_3 & 0 & 0 & 0 \\ 0 & 0 & 0 & c_1 & c_2 & c_3 \\ c_1 & c_2 & c_3 & b_1 & b_2 & b_3 \end{bmatrix} \tag{24.21}$$

Now, from Chapter 5, the relationship between stress and strain for plane stress is given by

$$\sigma_x = \frac{E}{(1 - \nu^2)} (\varepsilon_x + \nu\varepsilon_y)$$

$$\sigma_y = \frac{E}{(1 - \nu^2)} (\nu\varepsilon_x + \varepsilon_y) \tag{24.22}$$

$$\tau_{xy} = \frac{E}{2(1 + \nu)} \gamma_{xy}$$

where

σ_x = direct stress in the $x°$-direction

σ_y = direct stress in the $y°$-direction

τ_{xy} = shear stress in the $x°$–$y°$ plane

E = Young's modulus of elasticity

ν = Poisson's ratio

ε_x = direct strain in the $x°$-direction

ε_y = direct strain in the $y°$-direction

γ_{xy} = shear strain in the $x°$–$y°$ plane

G = shear modulus = $\dfrac{E}{2(1+\nu)}$

Rewriting equation (24.22) in matrix form,

$$\begin{Bmatrix} \sigma_x \\ \sigma_y \\ \tau_{xy} \end{Bmatrix} = \frac{E}{(1-\nu^2)} \begin{bmatrix} 1 & \nu & 0 \\ \nu & 1 & 0 \\ 0 & 0 & (1-\nu)/2 \end{bmatrix} \begin{Bmatrix} \varepsilon_x \\ \varepsilon_y \\ \gamma_{xy} \end{Bmatrix} \tag{24.23}$$

or

$$\{\sigma\} = [\mathbf{D}]\{\varepsilon\} \tag{24.24}$$

where, for plane stress,

$$[\mathbf{D}] = \frac{E}{(1-\nu^2)} \begin{bmatrix} 1 & \nu & 0 \\ \nu & 1 & 0 \\ 0 & 0 & (1-\nu)/2 \end{bmatrix} \tag{24.25}$$

= a matrix of material constants

and for plane strain,[12]

$$[\mathrm{D}] = \frac{E}{(1+v)(1-2v)}\begin{bmatrix}(1-v) & v & 0\\ v & (1-v) & 0\\ 0 & 0 & (1-2v)/2\end{bmatrix} \tag{24.26}$$

or, in general,

$$[\mathbf{D}] = E^1\begin{bmatrix}1 & \mu & 0\\ \mu & 1 & 0\\ 0 & 0 & \gamma\end{bmatrix} \tag{24.27}$$

where, for plane stress,

$$E^1 = E/(1 - v^2)$$

$$\mu = v$$

$$\gamma = (1 - v)/2$$

and for plane strain,

$$E^1 = E(1 - v)/[(1 + v)(1 - 2v)]$$

$$\mu = v/(1 - v)$$

$$\gamma = (1 - 2v)/[2(1 - v)]$$

Now from Section 1.13, it can be seen that the general expression for the strain energy of an elastic system, U_e, is given by

$$U_e = \int \frac{\sigma^2}{2E}\, d(\text{vol})$$

but

$$\sigma = E\varepsilon$$

$$\therefore\; U_e = \frac{1}{2}\int E\varepsilon^2\, d(\text{vol})$$

---

[12] Ross, C T F, *Mechanics of Solids*, Prentice Hall, 1996.

which, in matrix form, becomes

$$U_e = \frac{1}{2}\int \{\varepsilon\}^T [\mathbf{D}] \{\varepsilon\} d\,(\text{vol}) \tag{24.28}$$

where,

$\{\varepsilon\}$ = a vector of strains, which for this problem is

$$\{\varepsilon\} = \begin{Bmatrix} \varepsilon_x \\ \varepsilon_y \\ \gamma_{xy} \end{Bmatrix} \tag{24.29}$$

$[\mathbf{D}]$ = a matrix of material constants

It must be remembered that U_e is a scalar and, for this reason, the vector and matrix multiplication of equation (24.28) must be carried out in the manner shown.

Now, the work done by the nodal forces is

$$\text{WD} = -\{u_i^\circ\}^T \{P_i^\circ\} \tag{24.30}$$

where $\{P_i^\circ\}$ is a vector of nodal forces

and the total potential is

$$\pi_p = U_e + \text{WD}$$

$$= \frac{1}{2}\int \{\varepsilon\}^T [\mathbf{D}] \{\varepsilon\}\ d(\text{vol}) - \{u_i^\circ\} \{P_i^\circ\} \tag{24.31}$$

It must be remembered that WD is a scalar and, for this reason, the premultiplying vector must be a row vector, and the postmultiplying vector must be a column vector.

Substituting equation (24.20) into (24.31):

$$\pi_p = \frac{1}{2}\{u_1^\circ\}^T \int [\mathrm{B}]^T [\mathrm{D}][\mathrm{B}] d(\text{vol}) \{u_i^\circ\} - \{u_i^\circ\}^T \{P_i^\circ\} \tag{24.32}$$

but according to the method of minimum potential (see Chapter 17),

$$\frac{\partial \pi_p}{\partial \{u_i^\circ\}} = 0$$

or

$$0 = \int [\mathbf{B}]^T [\mathbf{D}] [\mathbf{B}]\, d(\text{vol}) \{u_i\} - \{P_i\}$$

i.e.

$$\{P_i\} = \int [\mathbf{B}]^T [\mathbf{D}] [\mathbf{B}]\, d(\text{vol}) \{u_i\} \tag{24.33}$$

but,

$$\{P_i\} = [\mathbf{k}°] \{u_i\}$$

or,

$$[\mathbf{k}°] = \int [\mathbf{B}]^T [\mathbf{D}] [\mathbf{B}]\, d(\text{vol}) \tag{24.34}$$

Substituting equations (24.21) and (24.27) into equation (24.34):

$$[\mathbf{k}°] = t \begin{bmatrix} P_{ij} & Q_{ij} \\ Q_{ji} & R_{ij} \end{bmatrix} \tag{24.35}$$

$$P_{ij} = 0.25\, E^1\, (b_i b_j + \gamma c_i c_j)/\Delta$$

$$Q_{ij} = 0.25\, E^1\, (\mu b_i c_j + \gamma c_i b_j)/\Delta$$

$$Q_{ji} = 0.25\, E^1\, (\mu b_j c_i + \gamma c_j b_i)/\Delta$$

$$R_{ij} = 0.25\, E^1\, (c_i c_j + \gamma b_i b_j)/\Delta \tag{24.36}$$

where i and j vary from 1 to 3 and t is the plate thickness

Problem 24.1 Working from first principles, determine the elemental stiffness matrix for a rod element, whose cross-sectional area varies linearly with length. The element is described by three nodes, one at each end and one at mid-length, as shown below. The cross-sectional area at node 1 is A and the cross-sectional area at node 3 is $2A$.

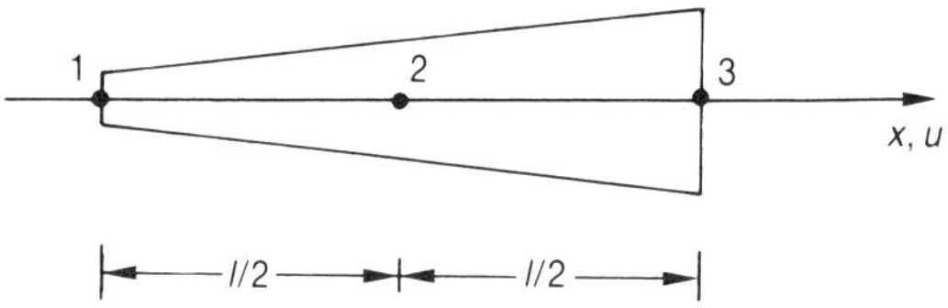

Solution

As there are three degrees of freedom, namely u_1, u_2 and u_3, it will be convenient to assume a polynomial involving three arbitrary constants, as shown by equation (24.37):

$$u = \alpha_1 + \alpha_2 x + \alpha_3 x^2 \tag{24.37}$$

To obtain the three simultaneous equations, it will be necessary to assume the following three boundary conditions or boundary values:

$$\begin{aligned} &\text{At } x = 0,\ u = u_1 \\ &\text{At } x = l/2,\ u = u_2 \\ &\text{At } x = l,\ u = u_3 \end{aligned} \tag{24.38}$$

Substituting equations (24.38) into equation (24.37), the following three simultaneous equations will be obtained:

$$u_1 = \alpha_1 \tag{24.39a}$$

$$u_2 = \alpha_1 + \alpha_2 l/2 + \alpha_3 l^2/4 \tag{24.39b}$$

$$u_3 = \alpha_1 + \alpha_2 l + \alpha_3 l^2 \tag{24.39c}$$

From (24.39a)

$$\alpha_1 = u_1 \tag{24.40}$$

Dividing (24.39c) by 2 gives

$$u_3/2 = u_1/2 + \alpha_2 l/2 + \alpha_3 l^2/2 \tag{24.41}$$

Taking (24.41) from (24.39b),

$$u_2 - u_3/2 = u_1 - u_1/2 - \alpha_3 l^2/4$$

or

$$\alpha_3 \frac{l^2}{4} = u_1/2 - u_2 + u_3/2$$

$$\alpha_3 = \frac{1}{l^2}(2u_1 - 4u_2 + 2u_3) \tag{24.42}$$

Substituting equations (24.40) and (24.42) into equation (24.39c),

$$\alpha_2 l = u_3 - u_1 - 2u_1 + 4u_2 - 2u_3$$

or

$$\alpha_2 = \frac{1}{l}(-3u_1 + 4u_2 - u_3) \tag{24.43}$$

Substituting equations (24.40), (24.42) and (24.43) into equation (24.37),

$$u = u_1 + (-3u_1 + 4u_2 - u_3)\xi + (2u_1 - 4u_2 + 2u_3)\xi^2$$

$$= u_1(1 - 3\xi + 2\xi^2) + u_2(4\xi - 4\xi^2) + u_3(-\xi + 2\xi^2)$$

$$u = \left[(1 - 3\xi + 2\xi^2)(4\xi - 4\xi^2)(-\xi + 2\xi^2)\right]\begin{Bmatrix} u_1 \\ u_2 \\ u_3 \end{Bmatrix}$$

$$= [\mathbf{N}]\{u_i\}$$

where

$$\xi = x/l$$

$$[\mathbf{N}] = \left[(1 - 3\xi + 2\xi^2)\ (4\xi - 4\xi^2)\ (-\xi + 2\xi^2)\right] \tag{24.44}$$

Now,

$$\varepsilon = \frac{du}{dx} = \frac{du}{l\,d\xi}$$

$$= \frac{1}{l}\left[(-3 + 4\xi)\ (4 - 8\xi)\ (-1 + 4\xi)\right]\begin{Bmatrix} u_1 \\ u_2 \\ u_3 \end{Bmatrix}$$

$$= [\mathbf{B}]\{u_i\}$$

or

$$[\mathbf{B}] = \frac{1}{l}\left[(-3 + 4\xi)\ (4 - 8\xi)\ (-1 + 4\xi)\right] \tag{24.45}$$

Now, for a rod,

$$\frac{\sigma}{\varepsilon} = E$$

or

$$\sigma = E\varepsilon$$

$$\therefore [\mathbf{D}] = E$$

Now.

$$[\mathbf{k}] = \int [\mathbf{B}]^{\mathrm{T}}\,[\mathbf{D}]\,[\mathbf{B}]\,d(\mathrm{vol})$$

$$= \int [\mathbf{B}]^{\mathrm{T}}\,[\mathbf{D}]\,[\mathbf{B}]\,al\ d\xi$$

where

$$a = \text{area at } \xi = A(1 + \xi)$$

$$\therefore [\mathbf{k}] = \int_0^{\xi} \frac{1}{l^2}\begin{bmatrix} \begin{pmatrix} -3 + 4\xi \\ 4 - 8\xi \\ -1 + 4\xi \end{pmatrix} \end{bmatrix}$$

$$\times\ E\left[(-3 + 4\xi)(4 - 8\xi)(-1 + 4\xi)\right]\ A(1 + \xi)\ l\,d\xi$$

$$= \begin{bmatrix} k_{11} & k_{12} & k_{13} \\ k_{21} & k_{22} & k_{23} \\ k_{31} & k_{32} & k_{33} \end{bmatrix}$$

where,

$$k_{11} = \frac{AE}{l} \int_0^1 (-3 + 4\xi)^2 (1 + \xi)\, d\xi$$

$$= 2.8333\ AE/l$$

$$k_{22} = \frac{AE}{l} \int_0^1 (4 - 8\xi)^2 (1 + \xi) d\xi$$

$$= 8\ AE/l$$

$$k_{33} = \frac{AE}{l} \int_0^1 (-1 + 4\xi)^2 (1 + \xi)\, d\xi$$

$$= 4.167\ AE/l$$

$$k_{12} = k_{21} = \frac{AE}{l} \int_0^1 (-3 + 4\xi)\ (4 - 8\xi)\ (1 + \xi) d\xi$$

$$= -3.33 AE/l$$

$$k_{13} = k_{31} = \frac{AE}{l} \int_0^1 (-3 + 4\xi)\ (-1 + 4\xi)\ (1 + \xi) d\xi$$

$$= AE/2l$$

$$k_{23} = k_{32} = \frac{AE}{l} \int_0^1 (4 - 8\xi)\ (-1 + 4\xi)\ (1 + \xi)\, d\xi$$

$$= -4.667\ AE/l$$

In this chapter, it has only been possible to introduce the finite element method, and for more advanced work on this topic, the reader is referred to Ross, C T F, *Advanced Applied Finite Element Methods*, Ellis Horword; Zienkiewicz, O C, and Taylor, R L. *The Finite Element Method*, McGraw-Hill, Vol 1, 1989, Vol 2, 1991.

Further problems *(answers on page 698)*

24.2 Using equation (24.34), determine the stiffness matrix for a uniform section rod element, with two degrees of freedom.

24.3 A rod element has a cross-sectional area which varies linearly from A_1 at node 1 to A_2 at node 2, where the nodes are at the ends of the rod. If the rod element has two degrees of freedom, determine its elemental stiffness matrix using equation (24.34).

24.4 Using equation (24.34), determine the stiffness matrix for a uniform section torque bar which has two degrees of freedom.

24.5 Using equation (24.34), determine the stiffness matrix for a two node uniform section beam, which has four degrees of freedom; two rotational and two translational.

25 Structural vibrations

25.1 Introduction

In this chapter, we will commence with discussing the free vibrations of a beam, which will be analysed by traditional methods. This fundamental approach will then be extended to forced vibrations and to damped oscillations, all on beams and by traditional methods.

The main snag with using traditional methods for vibration analysis, however, is that it is extremely difficult to analyse complex structures by this approach. For this reason, the finite element method discussed in the previous chapters will be extended to free vibration analysis, and applications will then be made to a number of simple structures.

Vibrations of structures usually occur due to pulsating or oscillating forces, such as those due to gusts of wind or from the motion of machinery, vehicles etc. If the pulsating load is oscillating at the same natural frequency of the structure, the structure can vibrate dangerously (i.e. resonate). If these vibrations continue for any length of time, the structure can suffer permanent damage.

25.2 Free vibrations of a mass on a beam

We can simplify the treatment of the free vibrations of a beam by considering its mass to be concentrated at the mid-length. Consider, for example, a uniform simply-supported beam of length L and flexural stiffness EI, Figure 25.1.

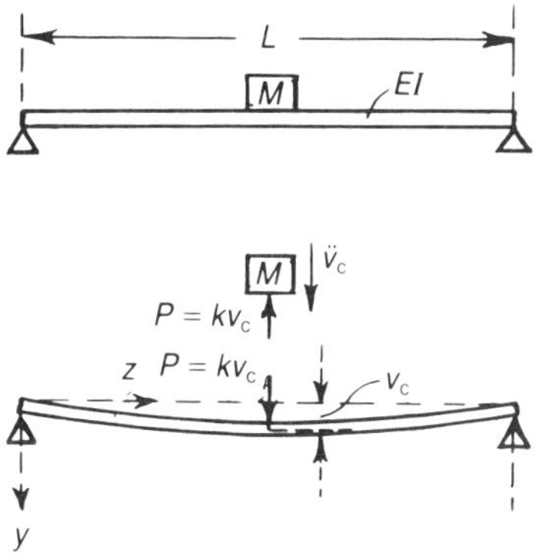

Figure 25.1 Vibrations of a concentrated mass on a beam.

Suppose the beam itself is mass-less, and that a concentrated mass M is held at the mid-span. If we ignore for the moment the effect of the gravitational field, the beam is undeflected when the mass is at rest. Now consider the motion of the mass when the beam is deflected laterally to some position and then released. Suppose, v_c is the lateral deflection of the beam at the mid-span at a time t; as the beam is mass-less the force P on the beam at the mid-span is

$$P = \frac{48\,EIv_c}{L^3}$$

If $k = 48\,EI/L^3$, then

$$P = kv_c$$

The mass-less beam behaves then as a simple elastic spring of stiffness k. In the deflected position there is an equal and opposite reaction P on the mass. The equation of vertical motion of the mass is

$$M\frac{d^2v_c}{dt^2} = -P = -kv_c$$

Thus

$$\frac{d^2v_c}{dt^2} + \frac{kv_c}{M} = 0$$

The general solution of this differential equation is

$$v_c = A\cos\sqrt{\frac{k}{M}}\,t + B\sin\sqrt{\frac{k}{M}}\,t$$

where A and B are arbitrary constants; this may also be written in the form

$$v_c = C\sin\left(\sqrt{\frac{k}{M}}\,t + \varepsilon\right)$$

where C and ε are also arbitrary constants. Obviously C is the amplitude of a simple-harmonic motion of the beam (Figure 25.2); v_c first assumes its peak value when

$$\sqrt{\frac{k}{M}}\,t_1 + \varepsilon = \frac{\pi}{2}$$

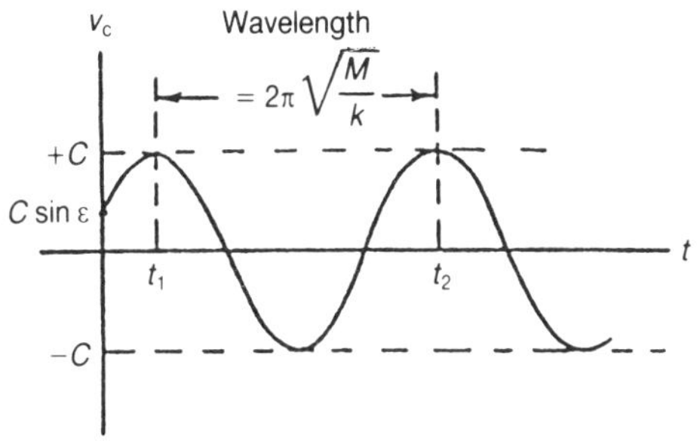

Figure 25.2 Variations of displacement of beam with time.

and again attains this value when

$$\sqrt{\frac{k}{M}}\,t_2 + \varepsilon = \frac{5\pi}{2}$$

This period T of one complete oscillation is then

$$T = t_1 - t_2 = 2\pi\sqrt{\frac{M}{k}} \qquad (25.1)$$

The number of complete oscillations occurring in unit time is the frequency of vibrations; this is denoted by n, and is given by

$$n = \frac{1}{T} = \frac{1}{2\pi}\sqrt{\frac{k}{M}} \qquad (25.2)$$

The behaviour of the system is therefore directly analogous to that of a simple mass–spring system. On substituting for the value of k we have

$$n = \frac{1}{T} = \frac{1}{2\pi}\sqrt{\frac{48EI}{ML^3}} \qquad (25.3)$$

Problem 25.1 A steel I-beam, simply supported at each end of a span of 10 m, has a second moment of area of 10^{-4} m^4. It carries a concentrated mass of 500 kg at the mid-span. Estimate the natural frequency of lateral vibrations.

Solution

In this case

$$EI = (200 \times 10^9)(10^{-4}) = 20 \times 10^6 \text{ Nm}^2$$

Then

$$k = \frac{48EI}{L^3} = \frac{48(20 \times 10^6)}{(10)^3} = 960 \times 10^3 \text{ N/m}$$

The natural frequency is

$$n = \frac{1}{2\pi}\sqrt{\frac{k}{M}} = \frac{1}{2\pi}\sqrt{\frac{960 \times 10^3}{500}} = 6.97 \text{ cycles/sec} = 6.97 \ Hz$$

25.3 Free vibrations of a beam with distributed mass

Consider a uniform beam of length L, flexural stiffness EI, and mass m per unit length (Figure 25.3); suppose the beam is simply-supported at each end, and is vibrating freely in the yz-plane, the displacement at any point parallel to the y-axis being v. We assume first that the beam vibrates in a sinusoidal form

$$v = a \sin\frac{\pi z}{L}\sin 2\pi n t \tag{25.4}$$

where a is the lateral displacement, or amplitude, at the mid-length, and n is the frequency of oscillation. The kinetic energy of an elemental length δz of the beam is

$$\frac{1}{2}m\,\delta z\left(\frac{dv}{dt}\right)^2 = \frac{1}{2}m\,\delta z\left[2\pi n a \sin\frac{\pi z}{L}\cos 2\pi n t\right]^2$$

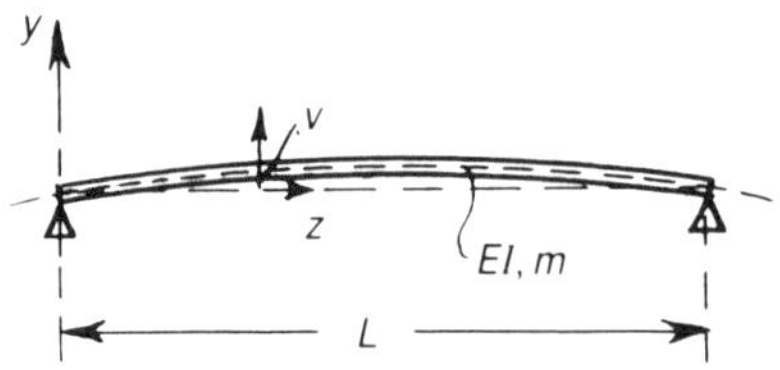

Figure 25.3 Vibrations of a beam having an intrinsic mass.

The bending strain energy in an elemental length is

$$\frac{1}{2} EI\left(\frac{d^2v}{dz^2}\right)^2 \delta z = \frac{1}{2} EI \left|\frac{a\pi^2}{L^2} \sin \frac{\pi z}{L} \sin 2\pi nt\right|^2 \delta z$$

The total kinetic energy at any time t is then

$$\frac{1}{2}m \left[4\pi^2 n^2 a^2 \cos^2 2\pi nt \int_0^L \sin^2 \frac{\pi z}{L}\, dz\right] \tag{25.5}$$

The total strain energy at time t is

$$\frac{1}{2}EI \frac{a^2\pi^4}{L^4} \sin^2 2\pi nt \int_0^L \sin^2 \frac{\pi z}{L}\, dz \tag{25.6}$$

For the free vibrations we must have the total energy, i.e. the sum of the kinetic and strain energies, is constant and independent of time. This is true if

$$\frac{1}{2}m \left(4\pi^2 n^2 a^2\right) \cos^2 2\pi nt + \frac{1}{2}EI\left(\frac{\pi^4 a^2}{L^4}\right) \sin^2 2\pi nt = \text{constant}$$

For this condition we must have

$$\frac{1}{2}m \left(4\pi^2 n^2 a^2\right) = \frac{1}{2}EI\left(\frac{\pi^4 a^2}{L^4}\right)$$

This gives

$$n^2 = \frac{\pi^2 EI}{4mL^4} \tag{25.7}$$

Now $mL = M$, say is the total mass of the beam, so that

$$n = \frac{\pi}{2}\sqrt{\frac{EI}{ML^3}} \tag{25.8}$$

This is the frequency of oscillation of a simply-supported beam in a single sinusoidal half-wave. If we consider the possibility of oscillations in the form

$$v = a \sin \frac{2\pi z}{L} \sin 2\pi n_2 t$$

then proceeding by the same analysis we find that

$$n_2 = 4n_1 = 2\pi \sqrt{\frac{EI}{ML^3}} \tag{25.9}$$

This is the frequency of oscillations of two sinusoidal half-waves along the length of the beam, Figure 25.4, and corresponds to the second mode of vibration. Other higher modes are found similarly.

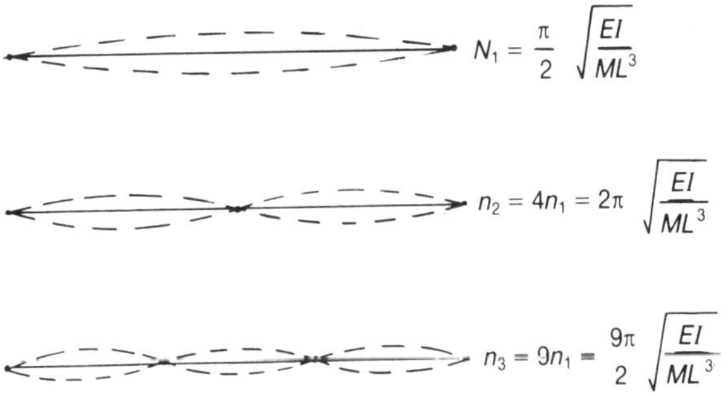

Figure 25.4 Modes of vibration of a simply-supported beam.

As in the case of the beam with a concentrated mass at the mid-length, we have ignored gravitation effects; when the weight of the beam causes initial deflections of the beam, oscillations take place about this deflected condition; otherwise the effects of gravity may be ignored.

The effect of distributing the mass uniformly along a beam, compared with the whole mass being concentrated at the mid-length, is to increase the frequency of oscillations from

$$\frac{1}{2\pi}\sqrt{\frac{48EI}{ML^3}} \quad \text{to} \quad \frac{\pi}{2}\sqrt{\frac{EI}{ML^3}}$$

If

$$n_1 = \frac{1}{2\pi}\sqrt{\frac{48EI}{ML^3}}, \quad \text{and} \quad n_2 = \frac{\pi}{2}\sqrt{\frac{EI}{ML^3}}$$

then

$$\frac{n_2}{n_1} = \left(\frac{\pi}{2}\right) \frac{2\pi}{\sqrt{48}} = \frac{\pi^2}{4\sqrt{3}} = 1.42 \qquad (25.10)$$

Problem 25.2 If the steel beam of the Problem 25.1 has a mass of 15 kg per metre run, estimate the lowest natural frequency of vibrations of the beam itself.

Solution

The lowest natural frequency of vibrations is

$$n_1 = \frac{\pi}{2} \sqrt{\frac{EI}{ML^3}}$$

Now

$$EI = 20 \times 10^6 \text{ Nm}^2$$

and

$$ML^3 = (15)\,(10)\,(10)^3 = 150 \times 10^3 \text{ kg.m}^3$$

Then

$$\frac{EI}{ML^3} = \frac{20 \times 10^6}{150 \times 10^3} = 133 \ s^{-2}$$

Thus

$$n_1 = \frac{\pi}{2} \sqrt{133} = 18.1 \text{ cycles per sec} = 18.1 \ Hz$$

25.4 Forced vibrations of a beam carrying a single mass

Consider a light beam, simply-supported at each end and carrying a mass M at mid-span, Figure 25.5. Suppose the mass is acted upon by an alternating lateral force

$$P \sin 2\pi Nt \qquad (25.11)$$

which is applied with a frequency N. If v_c is the central deflection of the beam, then the equation of motion of the mass is

$$M \frac{d^2 v_c}{dt^2} + k v_c = P \sin 2\pi N t$$

where $k = 48\, EI/L^3$. Then

$$\frac{d^2 v_c}{dt^2} + \frac{k}{M} v_c = \frac{P}{M} \sin 2\pi N t$$

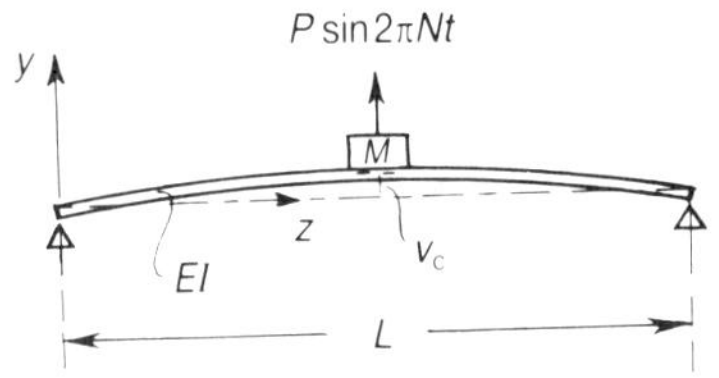

Figure 25.5 Alternating force applied to a beam.

The general solution is

$$v_c = A \cos \sqrt{\frac{k}{M}} t + B \sin \sqrt{\frac{k}{M}} t + \frac{\frac{P}{k} \sin 2\pi N t}{1 - 4\pi^2 N^2 \frac{M}{k}} \qquad (25.12)$$

in which A and B are arbitrary constants. Suppose initially, i.e. at time $t = 0$, both v_c and dv_c/dt are zero. Then $A = 0$ and

$$B = -\frac{2\pi N . \frac{P}{k}}{1 - 4\pi^2 N^2 \frac{M}{k}} \frac{1}{\sqrt{\frac{k}{M}}}$$

Then

$$v_c = \frac{P/k}{1 - 4\pi^2 N^2 \dfrac{M}{k}} \left[\sin 2\pi Nt - 2\pi N \sqrt{\frac{M}{k}} \sin \sqrt{\frac{k}{M}} t \right] \tag{25.13}$$

Now, the natural frequency of free vibrations of the system is

$$n = \frac{1}{2\pi} \sqrt{\frac{k}{M}}$$

Then

$$\sqrt{k/M} = 2\pi n$$

and

$$v_c = \frac{P/k}{1 - N^2/n^2} \left[\sin 2\pi Nt - \frac{N}{n} \sin 2\pi nt \right] \tag{25.14}$$

Now, the maximum value that the term

$$\left(\sin 2\pi Nt - \frac{N}{n} \sin 2\pi nt \right)$$

may assume is

$$\left[1 + \frac{N}{n} \right]$$

and occurs when $\sin 2\pi Nt = -\sin 2\pi nt = 1$. Then

$$v_{c\max} = \frac{P/k \left(1 + \dfrac{N}{n} \right)}{1 - \dfrac{N^2}{n^2}} = \frac{P/k}{1 - \dfrac{N}{n}} \tag{25.15}$$

Thus, if $N < n$, $v_{c\,max}$ is positive and in phase with the alternating load $P \sin 2\pi Nt$. As N approaches n, the values of $v_{c\,max}$ become very large. When $N > n$, $v_{c\,max}$ is negativc and out of phase with $P \sin 2\pi Nt$. When $N = n$, the beam is in a condition of resonance.

25.5 Damped free oscillations of a beam

The free oscillations of practical systems are inhibited by damping forces. One of the commonest forms of damping is known as velocity, or *viscous*, damping; the damping force on a particle or mass is proportional to its velocity.

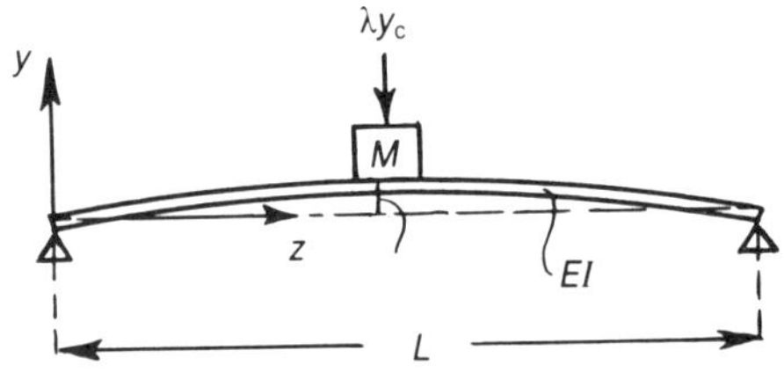

Figure 25.6 Effect of damping on free vibrations.

Suppose in the beam problem discussed in Section 25.2 we have as the damping force $\mu(dv_c/dt)$. Then the equation of motion of the mass is

$$M\frac{d^2v_c}{dt} = -kv_c - \mu\frac{dv_c}{dt}$$

Thus

$$M\frac{d^2v_c}{dt^2} + \mu\frac{dv_c}{dt} + kv_c = 0$$

Hence

$$\frac{d^2v_c}{dt^2} + \frac{\mu}{M}\frac{dv_c}{dt} + \frac{k}{M}v_c = 0$$

The general solution of this equation is

$$v_c = Ae^{\left\{-\mu/2M + \sqrt{(\mu/2M)^2 - k/M}\right\}t} + Be^{\left\{-\mu/2M - \sqrt{(\mu/2M)^2 - k/M}\right\}t} \qquad (25.16)$$

Now (k/M) is usually very much greater than $(\mu/2M)^2$, and so we may write

$$
\begin{aligned}
v_c &= Ae^{\left(-\mu/2M + i\sqrt{k/M}\right)t} + Be^{\left(-\mu/2M - i\sqrt{(k/M)}\right)t} \\
&= e^{-(\mu/2M)t}\left[AE^{i\sqrt{(k/M)}\,t} + Be^{-i\sqrt{(k/M)}\,t}\right] \qquad (25.17) \\
&= e^{-(\mu/2M)t}\left[C\cos\left\{\sqrt{\frac{k}{M}}\,t + \varepsilon\right\}\right]
\end{aligned}
$$

Thus, when damping is present, the free vibrations given by

$$C\cos\left(\sqrt{\frac{k}{M}}\,t + \varepsilon\right)$$

are damped out exponentially, Figure 25.7. The peak values on the curve of v_c correspond to points of zero velocity.

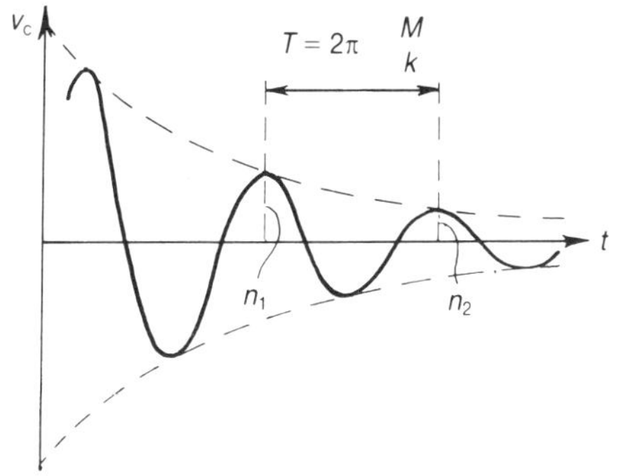

Figure 25.7 Form of damped oscillation of a beam.

These are given by

$$\frac{dv_c}{dt} = 0$$

or

$$\sqrt{\frac{k}{M}}\sin\left(\sqrt{\frac{k}{M}}\,t+\varepsilon\right) - \frac{\mu}{2M}\cos\left(\sqrt{\frac{k}{M}}\,t+\varepsilon\right) = 0$$

Obviously the higher peak values are separated in time by an amount

$$T = 2\pi\sqrt{\frac{M}{k}}$$

We note that successive peak values are in the ratio

$$\frac{v_{c1}}{v_{c2}} = \frac{e^{(-\mu/2M)t}\left[C\cos\left(\sqrt{(k/M)}\,t+\varepsilon\right)\right]}{e^{-(\mu/2M)\left(t+2\pi\sqrt{M/k}\right)}\left[C\cos\left(\sqrt{(k/M)}\,t+\varepsilon\right)\right]} = e^{(\mu/M)\pi\sqrt{M/k}} \tag{25.18}$$

Then

$$\log_e \frac{v_{c1}}{v_{c2}} = \frac{\pi\mu}{M}\sqrt{\frac{M}{k}} \tag{25.19}$$

Now

$$n = \frac{1}{2\pi}\sqrt{\frac{k}{M}}$$

Thus

$$\log_e \frac{v_{c1}}{v_{c2}} = \frac{\mu}{2Mn} \tag{25.20}$$

Hence

$$\mu = 2Mn\,\log_e \frac{v_{c1}}{v_{c2}} \tag{25.21}$$

25.6 Damped forced oscillations of a beam

We imagine that the mass on the beam discussed in Section 25.5 is excited by an alternating force $P \sin 2\pi Nt$. The equation of motion becomes

$$M \frac{d^2 v_c}{dt^2} + \mu \frac{dv_c}{dt} + kv_c = P \sin 2\pi Nt$$

The complementary function is the damped free oscillation; as this decreases rapidly in amplitude we may assume it to be negligible after a very long period. Then the particular integral is

$$v_c = \frac{P \sin 2\pi Nt}{MD^2 + \mu D + k}$$

This gives

$$v_c = \frac{P\left[\left(k - 4\pi^2 N^2 M\right)\sin 2\pi Nt - 2\pi N\mu \cos 2\pi Nt\right]}{\left(k - 4\pi^2 N^2 M\right)^2 + 4\pi^2 N^2 \mu^2} \tag{25.22}$$

If we write

$$n = \frac{1}{2\pi}\sqrt{\frac{k}{M}}$$

then

$$v_c = P\left[\frac{k\left(1 - \frac{N^2}{n^2}\right)\sin 2\pi Nt - 2\pi N\mu \cos 2\pi Nt}{k^2\left(1 - \frac{N^2}{n^2}\right)^2 + 4\pi N^2 \mu^2}\right] \tag{25.23}$$

The amplitude of this forced oscillation is

$$v_{c\max} = \frac{P}{\sqrt{k^2\left(1 - \frac{N^2}{n^2}\right)^2 + 4\pi^2 N^2 \mu^2}} \tag{25.24}$$

25.7 Vibrations of a beam with end thrust

In general, when a beam carries end thrust the period of free undamped vibrations is greater than when the beam carries no end thrust. Consider the uniform beam shown in Figure 25.8; suppose the beam is vibrating in the fundamental mode so that the lateral displacement at any section is given by

$$v = a \sin \frac{\pi z}{L} \sin 2\pi n t \tag{25.25}$$

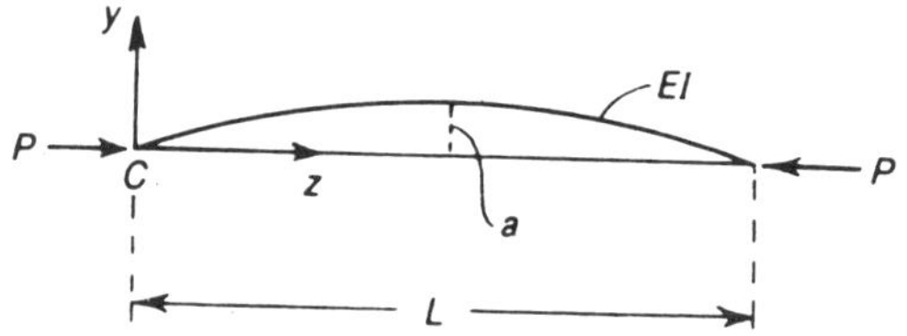

Figure 25.8 Vibrations of a beam carrying a constant end thrust.

If these displacements are small, the shortening of the beam from the straight configuration is approximately

$$\int_0^L \frac{1}{2} \left(\frac{dv}{dz} \right)^2 dz = \frac{a^2 \pi^2}{4L} \sin^2 2\pi n t \tag{25.26}$$

If m is the mass per unit length of the beam, the total kinetic energy at any instant is

$$\int_0^L \frac{1}{2} m \left(2\pi a \sin \frac{\pi z}{L} \cos 2\pi n t \right)^2 dz = m\pi^2 a^2 n^2 \, L \cos^2 2\pi n t \tag{25.27}$$

The total potential energy of the system is the strain energy stored in the strut together with the potential energy of the external loads; the total potential energy is then

$$\left[\frac{1}{2} EIL \left(\frac{a\pi^2}{L^2} \right)^2 - \frac{1}{4} P \left(\frac{a^2 \pi^2}{L} \right) \right] \sin^2 2\pi n t \tag{25.28}$$

If the total energy of the system is the same at all instants

$$m\pi^2 a^2 n^2 L = \frac{1}{4} EIL \left(\frac{a\pi^2}{L^2}\right)^2 - \frac{1}{4}P\left(\frac{a^2\pi^2}{L}\right)$$

This gives

$$n^2 = \frac{\pi^2 EI}{4mL^4}\left[1 - \frac{P}{P_e}\right] \tag{25.29}$$

where

$$P_e = \frac{\pi^2 EI}{L^2}$$

and is the Euler load of the column. If we write

$$n_1^2 = \frac{\pi^2 EI}{4mL^4} \tag{25.30}$$

then

$$n = n_1 \sqrt{1 - \frac{P}{P_e}}$$

Clearly, as P approaches P_e, the natural frequency of the column diminishes and approaches zero.

25.8 Derivation of expression for the mass matrix

Consider an infinitesimally small element of volume d(vol) and density ρ, oscillating at a certain time t, with a velocity $\dot{u}$.

The kinetic energy of this element (KE) is given by:

$$\text{KE} = \frac{1}{2}\rho \times d(\text{vol}) \times \dot{u}^2$$

and for the whole body,

$$\text{KE} = \frac{1}{2}\int \rho\, \dot{u}^2\, d(\text{vol}) \tag{25.31}$$

or in matrix form:

$$\mathrm{KE} = \frac{1}{2} \int_{\mathrm{vol}} \{\dot{u}\}^T \rho \, \{\dot{u}\} \, d(\mathrm{vol}) \tag{25.32}$$

NB The premultiplier of equation (25.32) must be a row and the postmultiplier of this equation must be a column, because KE is a scalar.

Assuming that the structure oscillates with simple harmonic motion, as described in Section 25.2,

$$\{u\} = \{C\} e^{jwt} \tag{25.33}$$

where

$\{C\}$ = a vector of amplitudes

ω = resonant frequency

j = $\sqrt{-1}$

Differentiating $\{u\}$ with respect to t,

$$\{\dot{u}\} = j\omega \, \{C\} \, e^{j\omega t} \tag{25.34}$$

$$= j\omega \, \{u\} \tag{25.35}$$

Substituting equation (25.35) into equation (25.32):

$$KE = -\frac{1}{2} \omega^2 \int_{\mathrm{vol}} \{u\}^T \rho \{u\} \, d(\mathrm{vol})$$

but,

$$\{u\} = [\mathbf{N}] \{u_i\}$$

$$\therefore KE = -\frac{1}{2} \omega^2 \{u_i\}^T \int_{\mathrm{vol}} [\mathbf{N}]^{\mathbf{T}} \rho \, [\mathbf{N}] \, d(\mathrm{vol}) \, \{u_i\} \tag{25.36}$$

but,

$$\text{KE} = \frac{M\dot{u}^2}{2}$$

or in matrix form:

$$\text{KE} = \frac{1}{2}\{\dot{u}_i\}^T \ [\mathbf{m}] \ \{\dot{u}_i\}$$

but,

$$\{\dot{u}_i\} = j\omega \ \{u_i\}$$

$$\therefore KE = -\frac{1}{2}\omega^2 \ \{u_i\}^T \ [\mathbf{m}] \ \{u_i\} \tag{25.37}$$

Comparing equation (25.37) with equation (25.36):

$$[\mathbf{m}] = \int_{\text{vol}} [N]^T \ \rho \ [\mathbf{N}] \ d(\text{vol}) \tag{25.38}$$

= elemental mass matrix

25.9 Mass matrix for a rod element

The one-dimensional rod element, which has two degree of freedom, is shown in Figure 23.1. As the rod element has two degrees of freedom, it will be convenient to assume a polynomial with two arbitrary constants, as shown in equation (25.39):

$$u = \alpha_1 + \alpha_2 x \tag{25.39}$$

The boundary conditions or boundary values are:

at $x = 0, \ u = u_1$

and

at $x = l, \ u = u_2$ (25.40)

Substituting equations (25.40) into equation (25.39),

$$\alpha_1 = u_1 \tag{25.41}$$

and

$$u_2 = u_1 + \alpha_2 l$$

or

$$\alpha_2 = (u_2 - u_1)/l \tag{25.42}$$

Substituting equations (25.41) and (25.42) into equation (25.39),

$$u = u_1 + (u_2 - u_1)x/l$$

or

$$u = u_1(1 - \xi) + u_2\xi \tag{25.43}$$

where,

$$\xi = x/l$$

Rewriting equation (25.43) in matrix form,

$$u = [(1 - \xi) \quad \xi] \begin{Bmatrix} u_1 \\ u_2 \end{Bmatrix}$$

$$= [\mathbf{N}] \{u_i\}$$

where

$$[\mathbf{N}] = [(1 - \xi) \quad \xi] \tag{25.44}$$

Substituting equation (25.44) into equation (24.38),

$$[\mathbf{m}] = \int [\mathbf{N}]^{\mathbf{T}} \rho [\mathbf{N}] \, d(\mathrm{vol})$$

$$= \rho \int_0^1 \begin{bmatrix} (1 - \xi) \\ \xi \end{bmatrix} [(1 - \xi) \quad \xi] \, Al \, d\xi$$

$$= \rho Al \int_0^1 \begin{bmatrix} (1 - 2\xi + \xi^2) & \xi - \xi^2 \\ \xi - \xi^2 & \xi^2 \end{bmatrix} d\xi$$

$$[\mathbf{m}] = \frac{\rho A l}{6}\begin{array}{c} \begin{array}{cc} u_1 & u_2 \end{array} \\ \begin{bmatrix} 2 & 1 \\ 1 & 2 \end{bmatrix} \end{array} \begin{array}{c} u_1 \\ u_2 \end{array} \qquad (25.45)$$

In two dimensions, it can readily be shown that the elemental mass matrix for a rod is

$$[\mathbf{m}] = \frac{\rho A l}{6}\begin{array}{c} \begin{array}{cccc} u_1 & v_1 & u_2 & v_2 \end{array} \\ \begin{bmatrix} 2 & 0 & 1 & 0 \\ 0 & 2 & 0 & 1 \\ 1 & 0 & 2 & 0 \\ 0 & 1 & 0 & 2 \end{bmatrix} \end{array} \begin{array}{c} u_1 \\ v_1 \\ u_2 \\ v_2 \end{array} \qquad (25.46)$$

The expression for the elemental mass matrix in global co-ordinates is given by an expression similar to that of equation (25.35), as shown by equation (25.47):

$$[\mathbf{m}^\circ] = [\mathbf{DC}]^{\mathbf{T}}\,[\mathbf{m}]\,[\mathbf{DC}] \qquad (25.47)$$

where,

$$[\mathbf{DC}] = \begin{bmatrix} \zeta & 0_2 \\ 0_2 & \zeta \end{bmatrix}$$

$$[\zeta] = \begin{bmatrix} c & s \\ -s & c \end{bmatrix} \qquad (25.48)$$

$c = \cos\alpha$

$s = \sin\alpha$

α is defined in Figure 23.4.

Substituting equations (23.25) and (25.46) into equation (25.47):

$$[\mathbf{m}^\circ] = \frac{\rho A l}{6} \begin{array}{c} \begin{array}{cccc} u_1^\circ & v_1^\circ & u_2^\circ & v_2^\circ \end{array} \\ \begin{bmatrix} 2 & 0 & 1 & 0 \\ 0 & 2 & 0 & 1 \\ 1 & 0 & 2 & 0 \\ 0 & 1 & 0 & 2 \end{bmatrix} \end{array} \begin{array}{c} \\ u_1^\circ \\ v_1^\circ \\ u_2^\circ \\ v_2^\circ \end{array} \tag{25.49}$$

= the elemental mass matrix for a rod in two dimensions, in global co-ordinates.

Similarly, in three dimensions, the elemental mass matrix for a rod in global co-ordinates, is given by:

$$[\mathbf{m}^\circ] = \frac{\rho A l}{6} \begin{array}{c} \begin{array}{cccccc} u_1^\circ & v_1^\circ & w_1^\circ & u_2^\circ & v_2^\circ & w_2^\circ \end{array} \\ \begin{bmatrix} 2 & & & & & \\ 0 & 2 & & & & \\ 0 & 0 & 2 & & & \\ 1 & 0 & 0 & 2 & & \\ 0 & 1 & 0 & 0 & 2 & \\ 0 & 0 & 1 & 0 & 0 & 2 \end{bmatrix} \end{array} \begin{array}{c} \\ u_1^\circ \\ v_1^\circ \\ w_1^\circ \\ u_2^\circ \\ v_2^\circ \\ w_2^\circ \end{array} \tag{25.50}$$

Equations (25.49) and (25.50) show the mass matrix for the self-mass of the structure, but if the effects of an additional concentrated mass are to be included at a particular node, this concentrated mass must be added to the mass matrix at the appropriate node, as follows:

$$M_a \begin{array}{c} \begin{array}{cc} u_i^\circ & v_i^\circ \end{array} \\ \begin{bmatrix} 1 & 0 \\ 0 & 1 \end{bmatrix} \end{array} \begin{array}{c} \\ u_i^\circ \\ v_1^\circ \end{array} \quad \text{(in two dimensions)} \tag{25.51}$$

and

$$M_a \begin{bmatrix} 1 & 0 & 0 \\ 0 & 1 & 0 \\ 0 & 0 & 1 \end{bmatrix} \begin{matrix} u_i^{\circ} \\ v_1^{\circ} \\ w_i^{\circ} \end{matrix} \quad \text{(in three dimensions)} \tag{25.52}$$

(columns: u_i° v_i° w_i°)

where

M_a = the value of the added mass

i = ith node

Problem 25.3 Determine the resonant frequencies and eigenmodes for the plane pin-jointed truss, below.

It may be assumed that the following apply:

$A = 1 \times 10^{-4}\ \text{m}^2$

$\rho = 7860\ \text{kg/m}^3$

$E = 2 \times 10^{11}\ \text{N/m}^2$

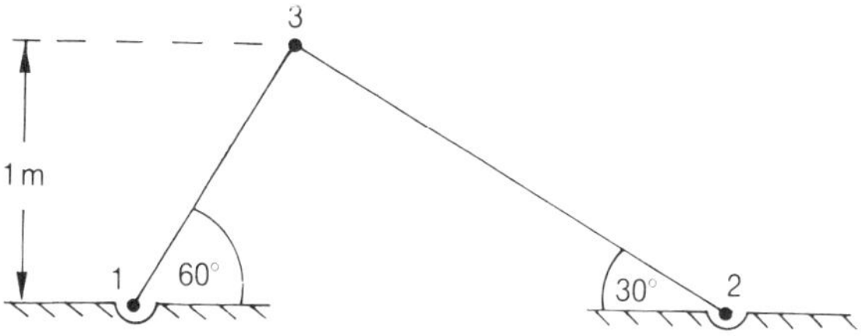

Solution

Element 1–3

$\alpha = 60°, \qquad c = 0.5, \qquad s = 0.866$

$$l_{1\text{-}3} = \frac{1\ \text{m}}{\sin 60} = 1.155\ \text{m} = \text{length of element 1–3}$$

Substituting the above values into equations (23.36) and (25.49), and removing the rows and columns corresponding to the zero displacements, namely $u_1°$ and $v_1°$, the stiffness and mass matrices for element 1–3 are given by:

$$\left[k_{1-3}°\right] = \frac{1\times10^{-4}\times2\times10^{11}}{1.155}\begin{bmatrix}0.25 & 0.433\\ 0.433 & 0.75\end{bmatrix}$$

$$\begin{matrix} & u_3° & v_3° & \\ & \begin{bmatrix}0.433\times10^7 & 0.75\times10^7\\ 0.75\times10^7 & 1.3\times10^7\end{bmatrix} & \begin{matrix}u_3°\\ v_3°\end{matrix}\end{matrix} \qquad (25.53)$$

$$\left[\mathbf{m}_{1-3}°\right] = \frac{7860\times1\times10^{-4}\times1.155}{6}\begin{bmatrix}2 & 0\\ 0 & 2\end{bmatrix}$$

$$\begin{matrix} & u_3° & v_3° & \\ = & \begin{bmatrix}0.303 & 0\\ 0 & 0.303\end{bmatrix} & \begin{matrix}u_3°\\ v_3°\end{matrix}\end{matrix} \qquad (25.54)$$

Element 2–3

$$\alpha = 150°, \qquad c = -0.866, \qquad s = 0.5$$

$$l_{2-3} = \frac{1\ \text{m}}{\sin 30} = 2\ \text{m} = \text{length of element 2–3}$$

Substituting the above values into equations (23.36) and (25.49), and removing the rows and columns corresponding to the zero displacements, namely $u_2°$ and $v_2°$, the stiffness and mass matrices for element 2–3 are given by:

$$[\mathbf{k}_{2-3}^\circ] = \frac{1 \times 10^{-4} \times 2 \times 10^{11}}{2}\begin{bmatrix} 0.75 & -0.433 \\ -0.433 & 0.25 \end{bmatrix}$$

$$= \begin{array}{c} \begin{matrix} u_3^\circ & \qquad & v_3^\circ \end{matrix} \\ \begin{bmatrix} 0.75 \times 10^7 & -0.433 \times 10^7 \\ -0.433 \times 10^7 & 0.25 \times 10^7 \end{bmatrix} \end{array} \begin{matrix} u_3^\circ \\ v_3^\circ \end{matrix} \tag{25.55}$$

$$[\mathbf{m}_{2-3}^\circ] = \frac{7860 \times 1 \times 10^{-4} \times 2}{6}\begin{bmatrix} 2 & 0 \\ 0 & 2 \end{bmatrix}$$

$$= \begin{array}{c} \begin{matrix} u_3^\circ & \quad & v_3^\circ \end{matrix} \\ \begin{bmatrix} 0.524 & 0 \\ 0 & 0.524 \end{bmatrix} \end{array} \begin{matrix} u_3^\circ \\ v_3^\circ \end{matrix} \tag{25.56}$$

The system stiffness matrix corresponding to the free displacements u_3° and v_3° is obtained by adding together equations (25.53) and (25.55), as shown by equation (25.57):

$$[\boldsymbol{K}_{11}] = \begin{array}{c} \begin{matrix} u_3^\circ & \qquad\qquad & v_3^\circ \end{matrix} \\ \left[\begin{array}{c|c} 0.433 \times 10^7 & 0.75 \times 10^7 \\ +0.75 \times 10^7 & -0.433 \times 10^7 \\ \hline 0.75 \times 10^7 & 1.3 \times 10^7 \\ -0.433 \times 10^7 & +0.25 \times 10^7 \end{array}\right] \end{array} \begin{matrix} u_3^\circ \\ \\ v_3^\circ \end{matrix} \tag{25.57}$$

$$[\mathbf{K}_{11}] = \begin{array}{c} \begin{matrix} u_3^\circ & \qquad & v_3^\circ \end{matrix} \\ \begin{bmatrix} 1.183 \times 10^7 & 0.317 \times 10^7 \\ 0.317 \times 10^7 & 1.55 \times 10^7 \end{bmatrix} \end{array} \begin{matrix} u_3^\circ \\ v_3^\circ \end{matrix} \tag{25.58}$$

The system mass matrix corresponding to the free displacements u_3° and v_3° is obtained by adding together equations (25.54) and (25.56), as shown by equation (25.59):

$$[\mathbf{M}_{11}] = \begin{array}{c|c} u_3^\circ & v_3^\circ \\ \hline \begin{matrix} 0.303 \\ +0.524 \end{matrix} & 0 \\ \hline 0 & \begin{matrix} 0.303 \\ +0.524 \end{matrix} \end{array} \begin{matrix} u_3^\circ \\ v_3^\circ \end{matrix} \tag{25.59}$$

$$= \begin{bmatrix} 0.827 & 0 \\ 0 & 0.827 \end{bmatrix} \begin{matrix} u_3^\circ \\ v_3^\circ \end{matrix} \quad (\text{columns } u_3^\circ,\ v_3^\circ) \tag{25.60}$$

Now, from Section 25.2,

$$\frac{d^2 v_c}{dt^2} + \frac{k v_c}{M} = 0 \tag{25.61}$$

If simple harmonic motion takes place, so that

$$v_c = C e^{j\omega t}$$

then,

$$\frac{d^2 v_c}{dt^2} = -\omega^2 \, C e^{j\omega t} = -\omega^2 v_c \tag{25.62}$$

Substituting equation (25.62) into equation (25.61),

$$-\omega^2 v_c + \frac{k v_c}{M} = 0 \tag{25.63}$$

In matrix form, equation (25.63) becomes

$$\left([\mathbf{K}] - \omega^2 [\mathbf{M}]\right) \{u_i\} = 0 \tag{25.64}$$

or, for a constrained structure,

$$\left(\left[\mathbf{K}_{11}\right] - \omega^2 \left[\mathbf{M}_{11}\right]\right) \{u_i\} = 0 \tag{25.65}$$

Now, in equation (25.65), the condition $\{u_i\} = \{0\}$ is not of practical interest, therefore the solution of equation (25.65) becomes equivalent to expanding the determinant of equation (25.66):

$$\left| \left[\mathbf{K}_{11}\right] - \omega^2 \left[\mathbf{M}_{11}\right] \right| = 0 \tag{25.66}$$

Substituting equations (25.58) and (25.60) into equation (25.66), the following is obtained:

$$\left| \begin{bmatrix} 1.183 \times 10^7 & 0.317 \times 10^7 \\ 0.317 \times 10^7 & 1.55 \times 10^7 \end{bmatrix} - \omega^2 \begin{bmatrix} 0.827 & 0 \\ 0 & 0.827 \end{bmatrix} \right| \tag{25.67}$$

Expanding equation (25.67), results in the quadratic equation (25.68):

$$(1.183 \times 10^7 - 0.827\omega^2)(1.55 \times 10^7 - 0.827\omega^2) - (0.317 \times 10^7)^2 = 0$$

or

$$1.834 \times 10^{14} - 2.26 \times 10^7 \omega^2 + 0.684 \omega^4 - 1 \times 10^{13} = 0$$

or

$$0.684 \omega^4 - 2.26 \times 10^7 \omega^2 + 1.734 \times 10^{14} = 0 \tag{25.68}$$

Solving the quadratic equation (25.68), the following are obtained for the roots ω_1^2 and ω_2^2:

$$\omega_1^2 = \frac{2.26 \times 10^7 - 6.028 \times 10^6}{1.368} = 1.211 \times 10^7$$

or

$$\omega_1 = 3480; \; n_1 = 533.9 \text{ Hz}$$

$$\omega_2^2 = \frac{2.26 \times 10^7 + 6.028 \times 10^6}{1.368} = 2.093 \times 10^7$$

or

$$\omega_2 = 4575; \; n_2 = 728 \text{ Hz}$$

To determine the eigenmodes, substitute ω_1^2 into the first row of equation (25.67) and substitute ω_2^2 into the second row of equation (25.67), as follows:

$$\left(1.183 \times 10^7 - 3480^2 \times 0.827\right)u_3^\circ + 0.317 \times 10^7 v_3^\circ = 0 \tag{25.69}$$

$$1.815 \times 10^6 u_3^\circ + 3.17 \times 10^6 v_3^\circ = 0$$

Let,

$$u_3^\circ = 1$$

$$\therefore v_3^\circ = -0.47$$

so that the first eigenmode is:

$[u_3^\circ \quad v_3^\circ] = [1 \quad -0.47]$ see the figure below at (a).

Similarly, to determine the second eigenmode, substitute ω_2^2 into the second row of equation (25.67), as follows:

$$0.317 \times 10^7 u_3^\circ + \left(1.55 \times 10^7 - 0.827 \times 4575^2\right) v_3^\circ = 0$$

or

$$0.317 \times 10^7 u_3^\circ - 1.81 \times 10^6 v_3^\circ = 0$$

Let,

$$v_3^\circ = 1$$

$$\therefore u_3 = 0.57$$

so that the second eigenmode is given by

$[u_3^\circ \quad v_3^\circ] = [0.57 \quad 1]$ see below at (b).

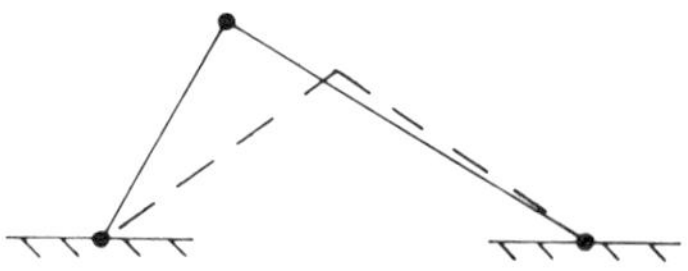

(a) First eigenmode.

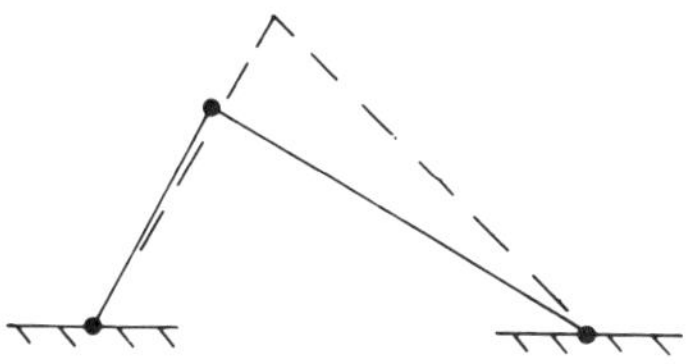

(b) Second eigenmode.

Problem 25.4 If the pin-jointed truss of Problem 25.3 had an additional mass of 0.75 kg attached to node 3, what would be the values of the resulting resonant frequencies?

Solution

From equation (25.58):

$$[K_{11}] = \begin{bmatrix} 1.183 \times 10^7 & 0.317 \times 10^7 \\ 0.317 \times 10^7 & 1.55 \times 10^7 \end{bmatrix} \begin{matrix} u_3^\circ \\ v_3^\circ \end{matrix} \quad \text{(columns: } u_3^\circ, v_3^\circ) \tag{25.70}$$

From equation (25.60)

$$[\mathbf{M}_{11}] = \begin{bmatrix} 0.827 & 0 \\ 0 & 0.827 \end{bmatrix} + \begin{bmatrix} 0.75 & 0 \\ 0 & 0.75 \end{bmatrix}$$

$$= \begin{bmatrix} 1.577 & 0 \\ 0 & 1.577 \end{bmatrix} \begin{matrix} u_3^\circ \\ v_3^\circ \end{matrix} \quad \text{(columns: } u_3^\circ, v_3^\circ) \tag{25.71}$$

Substituting equations (25.70) and (25.71) into equations (25.65), the following is obtained:

$$\left| \begin{bmatrix} 1.183 \times 10^7 & 0.317 \times 10^7 \\ 0.317 \times 10^7 & 1.55 \times 10^7 \end{bmatrix} - \omega^2 \begin{bmatrix} 1.577 & 0 \\ 0 & 1.577 \end{bmatrix} \right| = 0 \tag{25.72}$$

Expanding the determinant of equation (25.72), results in the quadratic equation (25.73):

$$\left(1.183 \times 10^7 - 1.577\,\omega^2\right)\left(1.55 \times 10^7 - 1.577\,\omega^2\right) - \left(0.317 \times 10^7\right)^2 = 0$$

or

$$1.834 \times 10^{14} - 4.31 \times 10^7\,\omega^2 + 2.487\,\omega^4 - 1 \times 10^{13} = 0$$

or

$$2.487\,\omega^4 - 4.31 \times 10^7\,\omega^2 + 1.734 \times 10^{14} = 0 \qquad (25.73)$$

The quadratic equation (25.73) has two roots, namely ω_1^2 and ω_2^2, which are obtained as follows:

$$\omega_1^2 = \frac{4.31 \times 10^7 - 1.178 \times 10^7}{4.974} = 6.297 \times 10^6$$

$$\omega_1 = 2509; \; n_1 = 399.3 \text{ Hz}$$

and

$$\omega_2^2 = \frac{4.31 \times 10^7 + 1.178 \times 10^7}{4.974} = 1.103 \times 10^7$$

$$\omega_2 = 3322; \; n_2 = 528.6 \text{ Hz}$$

Problem 25.5 Determine the resonant frequencies and eigenmodes for the pin-jointed space truss of Problem 23.3, given that,

$A = 2 \times 10^{-4}\ \text{m}^2$
$E = 2 \times 10^{11}\ \text{N/m}^2$
$\rho = 7860\ \text{kg/m}^3$

<u>*Solution*</u>

Element 1–4

From Problem 25.3,

$$l = 10 \text{ m}$$

Substituting this and other values into equation (25.50), and removing the rows and columns corresponding to the zero displacements, namely $u_1{}^\circ$, $v_1{}^\circ$ and $w_1{}^\circ$, the mass matrix for element 1–4 is given by

$$\left[\mathbf{m}_{1\text{-}4}{}^\circ\right] = \frac{7860 \times 2 \times 10^{-4} \times 10}{6}\begin{bmatrix} 2 & 0 & 0 \\ 0 & 2 & 0 \\ 0 & 0 & 2 \end{bmatrix} \tag{25.74}$$

$$= \begin{array}{c} \begin{array}{ccc} u_4{}^\circ & v_4{}^\circ & w_4{}^\circ \end{array} \\ \begin{bmatrix} 5.24 & 0 & 0 \\ 0 & 5.24 & 0 \\ 0 & 0 & 5.24 \end{bmatrix} \end{array} \begin{array}{c} u_4{}^\circ \\ v_4{}^\circ \\ w_4{}^\circ \end{array} \tag{25.75}$$

Element 2–4

From Problem 25.3,

$$l = 10 \text{ m}$$

Substituting this and other values into equation (25.50), and removing the rows and columns corresponding to the zero displacements, namely $u_2{}^\circ$, $v_2{}^\circ$ and $w_2{}^\circ$, the mass matrix for element 2–4 is given by

$$\left[\mathbf{m}_{2\text{-}4}{}^\circ\right] = \begin{array}{c} \begin{array}{ccc} u_4{}^\circ & v_4{}^\circ & w_4{}^\circ \end{array} \\ \begin{bmatrix} 5.24 & 0 & 0 \\ 0 & 5.24 & 0 \\ 0 & 0 & 5.24 \end{bmatrix} \end{array} \begin{array}{c} u_4{}^\circ \\ v_4{}^\circ \\ w_4{}^\circ \end{array} \tag{25.76}$$

Element 4–3

From Problem 25.3,

$$l = 10 \text{ m}$$

Substituting the above and other values into equation (25.50), and removing the rows and columns corresponding to the zero displacements, namely u_3°, v_3° and w_3°, the mass matrix for element 4–3 is given by

$$\left[\mathbf{m}_{4\text{-}3}^\circ\right] = \begin{array}{c} \begin{array}{ccc} u_4^\circ & v_4^\circ & w_4^\circ \end{array} \\ \begin{bmatrix} 5.24 & 0 & 0 \\ 0 & 5.24 & 0 \\ 0 & 0 & 5.24 \end{bmatrix} \end{array} \begin{array}{c} u_4^\circ \\ v_4^\circ \\ w_4^\circ \end{array} \tag{25.77}$$

To obtain $[\mathbf{M}_{11}]$, the system mass matrix corresponding to the free displacements u_4°, v_4° and w_4°, the elemental mass matrices of equations (25.75) to (25.77), are added together, as shown by equation (25.78):

$$\left[\mathbf{M}_{11}^\circ\right] = \begin{array}{c} \begin{array}{ccc} u_4^\circ & v_4^\circ & w_4^\circ \end{array} \\ \begin{bmatrix} 15.72 & 0 & 0 \\ 0 & 15.72 & 0 \\ 0 & 0 & 15.72 \end{bmatrix} \end{array} \begin{array}{c} u_4^\circ \\ v_4^\circ \\ w_4^\circ \end{array} \tag{25.78}$$

From equation (23.62),

$$\left[\mathbf{K}_{11}\right] = 1 \times 10^6 \begin{array}{c} \begin{array}{ccc} u_4^\circ & v_4^\circ & w_4^\circ \end{array} \\ \begin{bmatrix} 2 & 0 & 0 \\ 0 & 4 & 0.832 \\ 0 & 0.832 & 6 \end{bmatrix} \end{array} \begin{array}{c} u_4^\circ \\ v_4^\circ \\ w_4^\circ \end{array} \tag{25.79}$$

Substituting equations (25.78) and (25.79) into equation (25.65), the following determinant is obtained:

$$\left| 1 \times 10^6 \begin{bmatrix} 2 & 0 & 0 \\ 0 & 4 & 0.832 \\ 0 & 0.832 & 6 \end{bmatrix} - \omega^2 \begin{bmatrix} 15.72 & 0 & 0 \\ 0 & 15.72 & 0 \\ 0 & 0 & 15.72 \end{bmatrix} \right| \tag{25.80}$$

From the top line of equation (25.80):

$$2 \times 10^6 - 15.72\,\omega^2 = 0$$

or

$$\omega_1^2 = \frac{2 \times 10^6}{15.72} = 1.272 \times 10^5$$

$$\omega_1 = 356.7, \; n = 56.76 \text{ Hz}$$

As the first line of equation (25.80) is uncoupled, this equation can be reduced to the 2 × 2 determinant of equation (25.81):

$$\left| 1 \times 10^6 \begin{bmatrix} 4 & 0.832 \\ 0.832 & 6 \end{bmatrix} - \omega^2 \begin{bmatrix} 15.72 & 0 \\ 0 & 15.72 \end{bmatrix} \right| = 0 \quad (25.81)$$

Expanding equation (25.81), the quadratic equation (25.82) is obtained:

$$\left(4 \times 10^6 - 15.72\omega^2\right)\left(6 \times 10^6 - 15.72\omega^2\right) - \left(0.832 \times 10^6\right)^2 = 0$$

or

$$2.4 \times 10^{13} - 1.572 \times 10^8\omega^2 + 247.12\omega^4 - 6.922 \times 10^{11} = 0$$

or

$$247.12\omega^4 - 1.572 \times 10^8\omega^2 + 2.33 \times 10^{13} = 0 \quad (25.82)$$

Solving equation (25.82), the roots ω_2^2 and ω_3^2 are obtained, as follows:

$$\omega_2^2 = \frac{1.572 \times 10^8 - 0.41 \times 10^8}{492.24} = 2.361 \times 10^5$$

$$\omega_2 = 485.9; \; n_2 = 77.32 \text{ Hz}$$

$$\omega_3^2 = \frac{1.572 \times 10^8 + 0.41 \times 10^8}{492.24} = 4.026 \times 10^5$$

$$\omega_3 = 634.5; \; n_3 = 100.98 \text{ Hz}$$

To determine the eigenmodes

By inspection of the first line of equation (25.80),

$$u_4^\circ = 1, \quad v_4^\circ = 0 \quad \text{and} \quad w_4^\circ = 0$$

Therefore, the first eigenmode is

$$[u_4^\circ \quad v_4^\circ \quad w_4^\circ] = [1 \ 0 \ 0]$$

To obtain the second eigenmode, substitute ω_2^2 into the second line of equation (25.80) to give

$$0 \times u_4^\circ + \left[4 \times 10^6 - \left(485.9^2 \times 15.72\right)\right] v_4^\circ + 0.832 \times 10^6 w_4^\circ = 0$$

or

$$0.289 v_4^\circ + 0.832 w_4^\circ = 0 \tag{25.83}$$

Let,

$$v_4^\circ = 1$$

$$\therefore w_4^\circ = -0.347$$

Therefore, the second eigenmode is

$$[u_4^\circ \quad v_4^\circ \quad w_4^\circ] = [0 \ 1 \ -0.347]$$

To obtain the third eigenmode, substitute ω_3^2 into the third line of equation (25.80) to give

$$0 \times u_4^\circ + 0.832 \times 10^6 v_4^\circ + (6 \times 10^6 - 634.5^2 \times 15.72) w_4^\circ = 0$$

or

$$0.832 v_4^\circ - 0.329 w_4^\circ = 0 \tag{25.84}$$

Let,

$$w_4^\circ = 1$$

$$\therefore v_4^\circ = 0.395$$

Therefore, the third eigenmode is

$$[u_4^\circ \ v_4^\circ \ w_4^\circ] = [0 \ 0.395 \ 1]$$

Problem 25.6 Determine the resonant frequencies for the tripod of Problem 25.5, if this tripod has a mass of 10 kg added to node 4.

Solution

From equation (25.79),

$$[\mathbf{K}_{11}] = 1\times10^6 \begin{array}{c} \begin{array}{ccc} u_4^\circ & v_4^\circ & w_4^\circ \end{array} \\ \begin{bmatrix} 2 & 0 & 0 \\ 0 & 4 & 0.832 \\ 0 & 0.832 & 6 \end{bmatrix} \end{array} \begin{array}{c} u_4^\circ \\ v_4^\circ \\ w_4^\circ \end{array} \tag{25.85}$$

From equation (25.78):

$$[\mathbf{M}_{11}] = \begin{bmatrix} 15.72 & 0 & 0 \\ 0 & 15.72 & 0 \\ 0 & 0 & 15.72 \end{bmatrix} + \begin{bmatrix} 10 & 0 & 0 \\ 0 & 10 & 0 \\ 0 & 0 & 10 \end{bmatrix}$$

$$= \begin{array}{c} \begin{array}{ccc} u_4^\circ & v_4^\circ & w_4^\circ \end{array} \\ \begin{bmatrix} 25.72 & 0 & 0 \\ 0 & 25.72 & 0 \\ 0 & 0 & 25.72 \end{bmatrix} \end{array} \begin{array}{c} u_4^\circ \\ v_4^\circ \\ w_4^\circ \end{array} \tag{25.86}$$

Substituting equations (25.85) and (25.86) into equation (25.65), the following determinant is obtained:

$$\left| 1\times10^6 \begin{bmatrix} 2 & 0 & 0 \\ 0 & 4 & 0.832 \\ 0 & 0.832 & 6 \end{bmatrix} - \omega^2 \begin{bmatrix} 25.72 & 0 & 0 \\ 0 & 25.72 & 0 \\ 0 & 0 & 25.72 \end{bmatrix} \right| = 0 \tag{25.87}$$

From the first line of equation (25.65):

$$\omega_1^2 = \frac{2\times10^6}{25.72} = 7.776\times10^4$$

$$\omega_1 = 2789; \; n_1 = 44.1 \text{ Hz}$$

As first line is uncoupled, the determinant of equation (25.87) can be reduced to the 2 × 2 determinant of equation (25.88):

$$\left| 1 \times 10^6 \begin{bmatrix} 4 & 0.832 \\ 0.832 & 6 \end{bmatrix} - \omega^2 \begin{bmatrix} 25.72 & 0 \\ 0 & 25.72 \end{bmatrix} \right| = 0 \qquad (25.88)$$

Expanding the determinant of equation (25.88), the following quadratic is obtained:

$$(4 \times 10^6 - 25.72\,\omega^2)(6 \times 10^6 - 25.72\,\omega^2) - (0.832 \times 10^6)^2 = 0$$

or

$$2.4 \times 10^{13} - 2.572 \times 10^8\,\omega^2 + 661.5\,\omega^4 - 6.92 \times 10^{11} = 0$$

or

$$661.5\,\omega^4 - 2.572 \times 10^8\,\omega^2 + 2.33 \times 10^{13} = 0 \qquad (25.89)$$

Solving equation (25.89),

$$\omega_2^2 = \frac{2.572 \times 10^8 - 0.671 \times 10^8}{1323} = 1.437 \times 10^5$$

$$\omega_2 = 379.1; \; n_2 = 60.3 \text{ Hz}$$

$$\omega_3^2 = \frac{2.572 \times 10^8 + 0.671 \times 10^8}{1323} = 2.451 \times 10^5$$

$$\omega_3 = 495.1; \; n_3 = 78.8 \text{ Hz}$$

25.10 Mass matrix for a beam element

The beam element, which has four degrees of freedom, is shown in Figure 25.9.

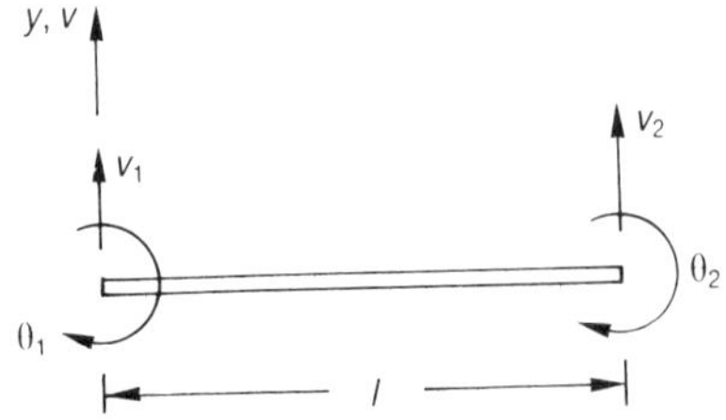

Figure 25.9 Beam element.

A convenient polynomial with which to describe the lateral deflection v is

$$v = \alpha_1 + \alpha_2 x + \alpha_3 x^2 + \alpha_4 x^3 \tag{25.90}$$

and

$$\frac{dv}{dx} = \alpha_2 + 2\alpha_3 x + 3\alpha_4 x^2 \tag{25.91}$$

In equation (25.90), it can be seen that the polynomial has four arbitrary constants, and this corresponds to the four degrees of freedom, namely, v_1, θ_1, v_2 and θ_2, i.e.

At $x = 0$, $\quad v = v_1 \quad$ and $\quad \theta_1 = -(dv/dx)_{x=0}$

At $x = l$, $\quad v = v_2 \quad$ and $\quad \theta_2 = -(dv/dx)_{x=l}$

Substituting the first two boundary conditions into equations (25.90) and (25.91):

$$\alpha_1 = v_1$$

and

$$\alpha_2 = -\theta_1$$

Substituting the remaining two boundary conditions into equations (25.90) and (25.91), the following two simultaneous equations are obtained:

$$v_2 = v_1 - \theta_1 l + \alpha_3 l^2 + \alpha_4 l^3 \tag{25.92}$$

and,

$$\theta_2 = \theta_1 - 2\alpha_3 l - 3\alpha_4 l^2 \tag{25.93}$$

Multiplying equation (25.92) by $2/l$, we get:

$$\frac{2}{l}\left(v_2 - v_1\right) = -2\theta_1 + 2\alpha_3 l + 2\alpha_4 l^2 \tag{25.94}$$

Adding equation (25.93) to equation (25.94):

$$\frac{2}{l}\left(v_2 - v_1\right) + \theta_2 = \theta_1 - 2\theta_1 - 3\alpha_4 l^2 + 2\alpha_4 l^2$$

or

$$-\alpha_4 l^2 = \frac{2}{l}(v_2 - v_1) + \theta_2 + \theta_1$$

$$\alpha_4 = -\frac{2}{l^3}(v_2 - v_1) - \frac{(\theta_2 + \theta_1)}{l^2} \quad (25.95)$$

Substituting equation (25.95) into equation (25.92):

$$v_2 - v_1 + \theta_1 l = \alpha_3 l^2 - 2(v_2 - v_1) - (\theta_2 + \theta_1)\, l$$

or

$$\alpha_3 = \frac{3}{l^2}(v_2 - v_1) + \frac{1}{l}(2\theta_1 + \theta_2) \quad (25.96)$$

Substituting the above values of α_1 to α_4 into equation (25.90)

$$v = v_1 - \theta_1 x + 3\xi^2 (v_2 - v_1) + \frac{x^2}{l}(2\theta_1 + \theta_2)$$

$$- 2\xi^2 (v_2 - v_1) + \frac{x^3}{l^2}(\theta_2 + \theta_1)$$

or

$$v = v_1 (1 - 3\xi^2 + 2\xi^3) + \theta_1 l\,(-\xi + 2\xi^2 - \xi^3)$$

$$+ v_2 (3\xi^2 - 2\xi^3) + \theta_2 l\,(\xi^2 - \xi^3) \quad (25.97)$$

where,

$$\xi = x/l$$

i.e.

$$v = \left| (1 - 3\xi^2 + 2\xi^3) \;\; l\,(-\xi + 2\xi^2 - \xi^3) \;\; (3\xi^2 - 2\xi^3) \;\; l\,(\xi^2 - \xi^3) \right| \begin{Bmatrix} v_1 \\ \theta_1 \\ v_2 \\ \theta_2 \end{Bmatrix}$$

$$= [N]\{u_i\} \quad (25.98)$$

where [**N**] is a matrix of shape functions for a beam element:

$$[\mathbf{N}] = \left[\left(1-3\xi^2+2\xi^3\right)\; l\left(-\xi+2\xi^2-\xi^3\right)\left(3\xi^2-2\xi^3\right) l\left(\xi^2-\xi^3\right)\right] \tag{25.99}$$

From equation (25.38):

$$[\mathbf{m}] = \int_0^1 [\mathbf{N}]^{\mathrm{T}}\ \rho[\mathbf{N}]\ Al\,\mathrm{d}\,\xi \tag{25.100}$$

Substituting equation (25.99) into equation (25.100), and integrating, the mass matrix for a beam element is given by

$$[\mathbf{m}] = \frac{\rho Al}{420}\begin{array}{c} \begin{array}{cccc} v_1 & \theta_1 & v_2 & \theta_2 \end{array} \\ \left[\begin{array}{cccc} 156 & & & \\ -22l & 4l^2 & & \\ 54 & -13l & 156 & \\ 13l & -3l^2 & 22l & 4l^2 \end{array}\right] \end{array} \begin{array}{c} \\ v_1 \\ \theta_1 \\ v_2 \\ \theta_2 \end{array} \tag{25.101}$$

Equation (25.101) is the mass matrix of a beam element due to the self-mass of the structure, but if an additional concentrated mass is added to node i, the following additional components of mass must be added to equation (25.102) at the appropriate node.

Added mass matrix at node i

$$= \begin{array}{c} \begin{array}{cc} v_i & \theta_i \end{array} \\ \left[\begin{array}{cc} M_a & 0 \\ 0 & MMI \end{array}\right] \end{array} \begin{array}{c} \\ v_i \\ \theta_i \end{array} \tag{25.102}$$

where MMI is the mass moment of inertia and M_a is the mass.

Problem 25.7 Determine the resonant frequencies for the beam of the figure in Problem 23.4, assuming that the 4 kN load is not present, and that

$E = 2\times10^{11}\ \mathrm{N/m^2}$, $\rho = 7860\ \mathrm{kg/m^3}$

$A = 1\times10^{-4}\ \mathrm{m^2}$, $I = 1\times10^{-7}\ \mathrm{m^4}$

Solution

Element 1–2

$$l = 3 \text{ m}$$

Substituting the above value of l into equation (25.101), together with the other properties of this element, and removing the columns and rows corresponding to the zero displacements v_1 and θ_1, the elemental mass matrix is given by

$$\left[\mathbf{m}_{1\text{-}2}\right] = \frac{7860 \times 1 \times 10^{-4} \times 3}{420} \begin{array}{c} \begin{matrix} v_2 & \theta_2 \end{matrix} \\ \begin{bmatrix} 156 & 66 \\ 66 & 36 \end{bmatrix} \end{array} \begin{array}{c} v_2 \\ \theta_2 \end{array}$$

$$= \begin{array}{c} \begin{matrix} v_2 & \theta_2 \end{matrix} \\ \begin{bmatrix} 0.876 & 0.371 \\ 0.371 & 0.202 \end{bmatrix} \end{array} \begin{array}{c} v_2 \\ \theta_2 \end{array} \tag{25.103}$$

Element 2–3

$$l = 2 \text{ m}$$

Substituting the above value of l into equation (25.101), together with the other properties of this element, and removing the columns and rows corresponding to the zero displacements v_3 and θ_3, the elemental mass matrix is given by:

$$\left[\mathbf{m}_{2\text{-}3}\right] = \begin{array}{c} \begin{matrix} v_2 & \theta_2 \end{matrix} \\ \begin{bmatrix} 0.584 & -0.165 \\ -0.165 & 0.0599 \end{bmatrix} \end{array} \begin{array}{c} v_2 \\ \theta_2 \end{array} \tag{25.104}$$

The system mass matrix $[\mathbf{M}_{11}]$ is obtained by adding together the elemental mass matrices of equations (25.103) and (25.104):

$$\left[\mathbf{M}_{11}\right] = \begin{array}{c} \begin{matrix} v_2 & \theta_2 \end{matrix} \\ \begin{bmatrix} 1.46 & 0.206 \\ 0.206 & 0.262 \end{bmatrix} \end{array} \begin{array}{c} v_2 \\ \theta_2 \end{array} \tag{25.105}$$

From equation (25.84),

$$
[\mathbf{K}_{11}] = \begin{bmatrix} 38\,880 & -16\,660 \\ -16\,660 & 66\,660 \end{bmatrix} \begin{matrix} v_2 \\ \theta_2 \end{matrix} \quad \text{(columns: } v_2, \theta_2\text{)} \tag{25.106}
$$

Substituting equations (25.105) and (25.106) into equation (25.65), the following determinant is obtained:

$$
\left| \begin{bmatrix} 38\,880 & -16\,660 \\ -16\,660 & 66\,660 \end{bmatrix} - \omega^2 \begin{bmatrix} 1.46 & 0.206 \\ 0.206 & 0.262 \end{bmatrix} \right| = 0 \tag{25.107}
$$

Expanding the determinant of equation (25.107), the following quadratic equation is obtained:

$$
(38\,880 - 1.46\omega^2)(66\,660 - 0.262\omega^2) - (-16\,660 - 0.206\omega^2)^2 = 0
$$

or,

$$
2592 \times 10^6 - 0.107 \times 10^6\,\omega^2 + 0.383\,\omega^4
$$

$$
- 278 \times 10^6 - 6864\,\omega^2 - 0.042\,\omega^4 = 0
$$

$$
0.341\,\omega^4 - 0.1139 \times 10^6\,\omega^2 + 2.314 \times 10^9 = 0 \tag{25.108}
$$

The roots of equation (25.108), namely, ω_1^2 and ω_2^2, can readily be shown to be:

$$
\omega_1^2 = \frac{0.1139 \times 10^6 - 99\,080}{0.682} = 2.173 \times 10^4
$$

or

$$
\omega_1 = 147.4; \; n_1 = 23.45 \text{ Hz}
$$

$$
\omega_2^2 = \frac{0.1139 \times 10^6 + 99\,080}{0.682} = 3.123 \times 10^5
$$

and,

$$
\omega_2 = 558.8; \; n_2 = 88.93 \text{ Hz}
$$

To obtain the first eigenmode, substitute ω_1^2 into the first line of equation (25.107), to give

$$(38\,880 - 1.46 \times 147.4^2)\, v_2 + (-\,16\,660 + 0.206 \times 147.4^2)\, \theta_2 = 0$$

or

$$7159\, v_2 - 21\,136\, \theta_2 = 0 \qquad (25.109)$$

i.e.

$[v_2 \;\; \theta_2] = [1 \;\; 0.339]$ – see the figure below at (a).

To obtain the second eigenmode, substitute $\omega_2^{\,2}$ into the second line of equation (25.107) to give:

$$(-\,16\,660 - 0.206 \times 558.8^2)\, v_2 + (66\,660 - 0.262 \times 558.8^2)\, \theta_2 = 0$$

or,

$$-\,80\,985\, v_2 - 15\,150\, \theta_2 = 0 \qquad (25.110)$$

i.e.

$[v_2 \qquad \theta_2] = [-\,0.187 \;\; 1]$ – see the figure below at (b).

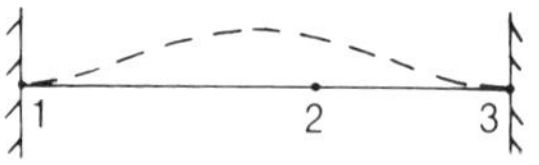

(a) First eigenmode

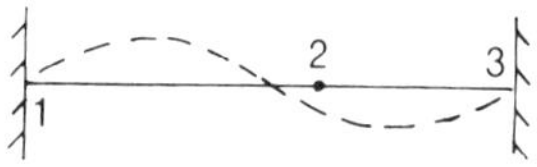

(b) Second eigenmode

Problem 25.8 If the beam of Problem 25.7 has a mass of 1 kg, with a mass moment of inertia of 0.1 kg m^2 added to node 2, determine the resonant frequencies of the beam.

<u>*Solution*</u>

From equation (25.105)

$$[M_{11}] = \begin{bmatrix} 1.46 & 0.206 \\ 0.206 & 0.262 \end{bmatrix} + \begin{bmatrix} 1 & 0 \\ 0 & 0.1 \end{bmatrix}$$

$$= \begin{bmatrix} 2.46 & 0.206 \\ 0.206 & 0.362 \end{bmatrix} \begin{matrix} v_2 \\ \theta_2 \end{matrix} \quad \text{(columns } v_2, \theta_2\text{)} \tag{25.111}$$

From equation (25.101),

$$[\mathbf{K}_{11}] = \begin{bmatrix} 38880 & -16660 \\ -16660 & 66660 \end{bmatrix} \tag{25.112}$$

Substituting equations (25.111) and (25.112) into equation (25.65),

$$\left| \begin{bmatrix} 38\,880 & -16\,660 \\ -16\,660 & 66\,660 \end{bmatrix} - \omega^2 \begin{bmatrix} 2.46 & 0.206 \\ 0.206 & 0.362 \end{bmatrix} \right| = 0 \tag{25.113}$$

$$(38\,880 - 2.46\,\omega^2)\,(66\,660 - 0.362\,\omega^2) - (16\,660 + 0.206\,\omega^2)^2 = 0$$

or

$$0.259 \times 10^{10} - 0.178 \times 10^6\,\omega^2 + 0.891\,\omega^4 - 2.776 \times 10^8 - 6864\,\omega^2 - 0.042\,\omega^4 = 0$$

or

$$0.849\,\omega^4 - 0.1849 \times 10^6\,\omega^2 + 0.231 \times 10^{10} = 0 \tag{25.114}$$

Solution of the quadratic equation (25.114) results in the roots ω_1^2 and ω_2^2, as follows:

$$\omega_1^2 = \frac{0.1849 \times 10^6 - 0.162 \times 10^6}{1.698} = 1.394 \times 10^4$$

or

$$\omega_1 = 116.1; \; n_1 = 18.48 \text{ Hz}$$

and,

$$\omega_2^2 = \frac{0.1849 \times 10^6 + 0.162 \times 10^6}{1.698} = 2.043 \times 10^5$$

or

$$\omega_2 = 452; \; n_2 = 71.93 \text{ Hz}$$

25.11 Mass matrix for a rigid-jointed plane frame element

Prior to obtaining the mass matrix for an element of a rigid-jointed plane frame, it will be necessary to obtain the mass matrix for the inclined beam of Figure 25.10.

The mass matrix for an inclined beam element in global co-ordinates is

$$[\mathbf{m}_b{}^\circ] = [\mathbf{DC}]^T [\mathbf{m}] [\mathbf{DC}] \tag{25.115}$$

where,

[**DC**] is given equation (25.85) and [**m**] is given by equation (25.101).

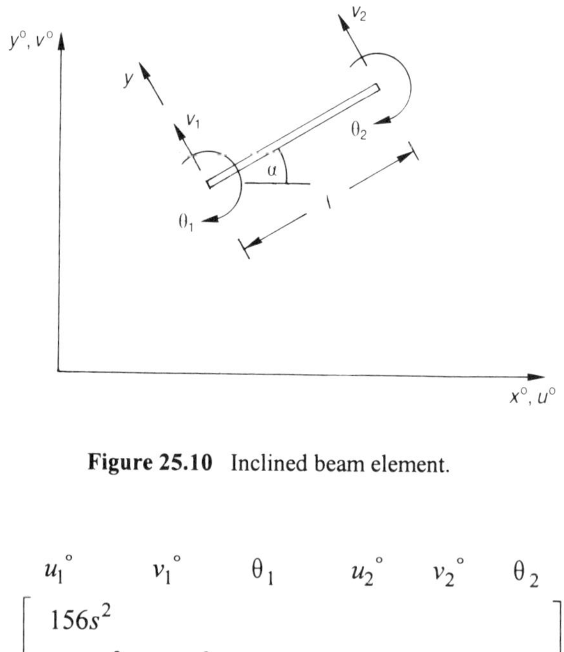

Figure 25.10 Inclined beam element.

$$\left[\mathrm{m}_b{}^\circ\right] = \frac{\rho A l}{420} \begin{array}{c} \begin{array}{cccccc} u_1{}^\circ & v_1{}^\circ & \theta_1 & u_2{}^\circ & v_2{}^\circ & \theta_2 \end{array} \\ \left[\begin{array}{cccccc} 156s^2 & & & & & \\ -156cs^2 & 156c^2 & & & & \\ 22ls & -22lc & 4l^2 & & & \\ 54s^2 & -54cs & 13ls & 156s^2 & & \\ -54cs & 54c^2 & -13cl & -156cs & 156c^2 & \\ -13ls & 13lc & -3l^2 & -22ls & 22lc & 4l^2 \end{array}\right] \end{array} \begin{array}{c} u_1{}^\circ \\ v_1{}^\circ \\ \theta_1{}^\circ \\ u_2{}^\circ \\ v_2{}^\circ \\ \theta_2 \end{array} \tag{25.116}$$

For the element of a rigid-jointed plane frame, the elemental mass matrix in global co-ordinates is given by

$$[\mathbf{m}^\circ] = [\mathbf{m}_b{}^\circ] + [\mathbf{m}_r{}^\circ] \tag{25.117}$$

where $[\mathbf{m}_r{}^\circ]$ is the axial part of the mass matrix of a rod element:

$$[\mathbf{m}_r{}^\circ] = [\mathbf{DC}]^T \frac{\rho A l}{6} \begin{bmatrix} 2 & 0 & 0 & 1 & 0 & 0 \\ 0 & 0 & 0 & 0 & 0 & 0 \\ 0 & 0 & 0 & 0 & 0 & 0 \\ 1 & 0 & 0 & 2 & 0 & 0 \\ 0 & 0 & 0 & 0 & 0 & 0 \\ 0 & 0 & 0 & 0 & 0 & 0 \end{bmatrix} [\mathbf{DC}] \tag{25.118}$$

where, in equation (25.118), the components of mass in the v displacement direction have been removed, because they have already been included in $[\mathbf{m}_b{}^\circ]$.

Substituting $[\mathbf{DC}]$ from equation (25.85) into equation (25.118):

$$\begin{array}{c} \quad\quad\quad u_1{}^\circ \quad v_1{}^\circ \quad \theta_1 \quad u_2{}^\circ \quad v_2{}^\circ \quad \theta_2 \\ [\mathbf{m}_r{}^\circ] = \dfrac{\rho A l}{6} \begin{bmatrix} 2c^2 & & & & & \\ 2cs & 2s^2 & & & & \\ 0 & 0 & 0 & & & \\ c^2 & cs & 0 & 2c^2 & & \\ cs & s^2 & 0 & 2cs & 2s^2 & \\ 0 & 0 & 0 & 0 & 0 & 0 \end{bmatrix} \begin{matrix} u_1{}^\circ \\ v_1{}^\circ \\ \theta_1 \\ u_2{}^\circ \\ v_2{}^\circ \\ \theta_2 \end{matrix} \end{array} \tag{25.119}$$

From equations (25.116) and (25.118), it can be seen that application of these elemental mass matrices, together with the elemental stiffness matrix of equation (25.85), to a realistic rigid-jointed plane frame will be extremely difficult without the aid of a computer.

Equation (25.117) shows the mass matrix for the self-mass of an element of a rigid-jointed plane frame, but if the effects of an additional concentrated mass are to be included at a particular node, the concentrated mass must be added to the appropriate node, as follows:

$$\begin{matrix} & u_1^\circ & v_1^\circ & \theta_1 & \\ \end{matrix}$$
$$\begin{bmatrix} M_a & 0 & 0 \\ 0 & M_a & 0 \\ 0 & 0 & \mathbf{MMI} \end{bmatrix} \begin{matrix} u_1^\circ \\ v_1^\circ \\ \theta_1 \end{matrix} \tag{25.120}$$

where

M_a = the value of the mass

MMI = the mass moment of inertia of this mass

25.12 Units in structural dynamics

Considerable care should be taken in choosing suitable units in structural dynamics. Recommended units are as follows:

(i) *Imperial*

Mass (lbf s^2/in); density (lbf s^2/in^4); E (lbf/in^2); time(s); length (in); Force(lbf); second moment of area (in^4); cross-sectional area (in^2).

(ii) *SI*

Mass (kg); density (kg/m^3); E (N/m^2); time (s); length (m); Force(N); second moment of area (m^4); cross-sectional area (m^2).

(iii) *Derived SI*

Mass (kg); density (kg/mm^3); E (mN/mm^2); time (s); length (mm); force(mN); second moment of area (mm^4); cross-sectional area (mm^2).

Further problems *(answers on page 698)*

25.9 A doubly symmetrical beam consists of a hollow rectangular steel section, having the cross-section shown, and of length 10 m. It is simply-supported in bending about both axes *Cx*, *Cy* at the ends. Estimate the lowest few natural frequencies of lateral vibrations of the beam about the axes *Cx* and *Cy*. Take E = 200 GN/m^2.

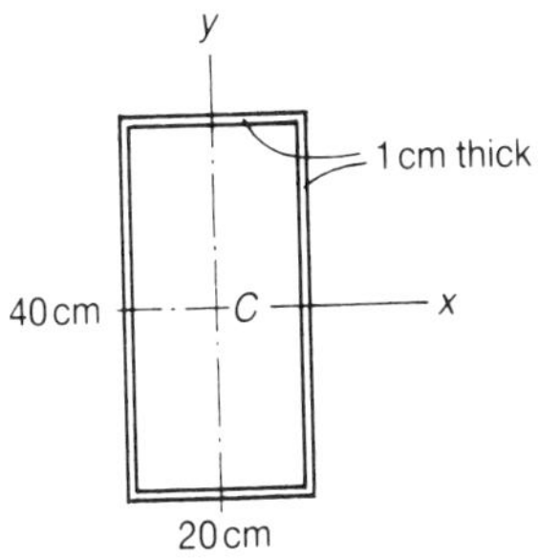

25.10 If the beam of Problem 25.7 carries an axial thrust of 10^3 kN, what is the lowest natural frequency of the beam?

25.11 A light, uniform cantilever, of length L and uniform flexural stiffness EI, carries a mass M at the free end. Estimate the natural frequency of vibrations.

25.12 Determine the resonant frequencies for the plane pin-jointed truss shown below, assuming that the truss is loaded with a mass of 1 kg at node 4, and that the following apply:

$A = 1 \times 10^{-4}\ \text{m}^2$

$E = 2 \times 10^{11}\ \text{N/m}^2$

$\rho = 7860\ \text{kg/m}^3$

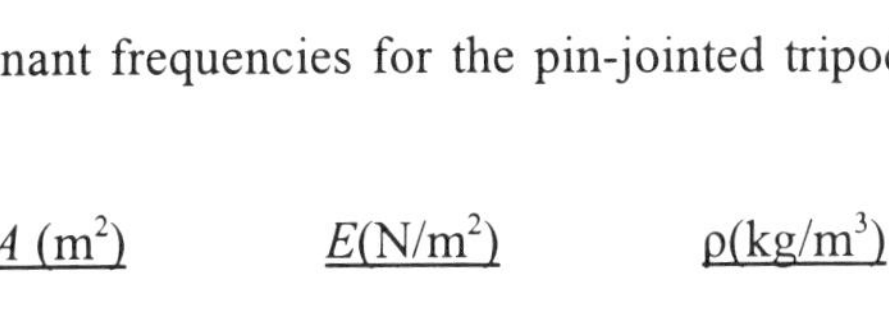

(*Portsmouth 1989*)

25.13 Determine the resonant frequencies for the pin-jointed tripod, below, given that the following apply:

Element	A (m^2)	E(N/m^2)	ρ(kg/m^3)
1–4	1×10^{-3}	2×10^{11}	7860
2–4	2×10^{-3}	2×10^{11}	7860
3–4	1×10^{-3}	2×10^{11}	7860

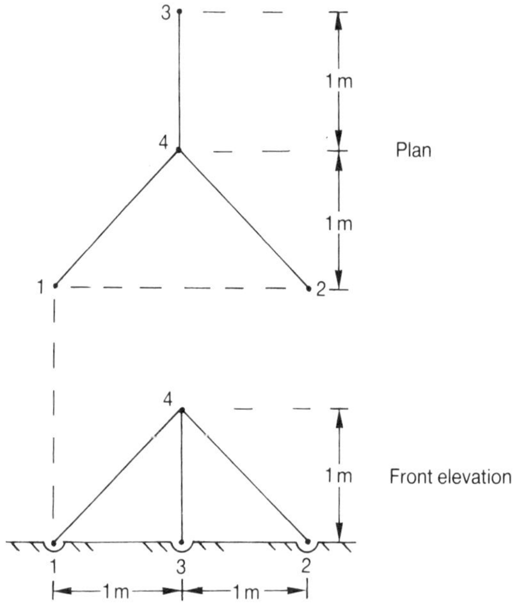

(*Portsmouth 1983*)

25.14 A continuous beam is fixed at the nodes 1 and 4, and simply-supported at the nodes 2 and 3, as shown in the figure below.

Determine the two lowest resonant frequencies of vibration, given the following:

$E = 2 \times 10^{11}\ \text{N/m}^2$

$\rho = 7860\ \text{kg/m}^3$

Element	A (m^2)	I(m^4)
1–2	1×10^{-4}	1×10^{-7}
2–3	2×10^{-4}	2×10^{-7}
3–4	1×10^{-4}	2×10^{-7}

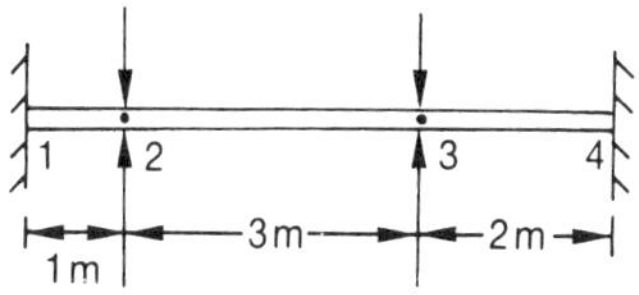

(*Portsmouth 1987*)

25.15 A continuous beam is fixed at the nodes 1 and 5, and simply-supported at the nodes 2, 3 and 4, as shown below.

Determine the two lowest resonant frequencies of vibration given the following:

$E = 2 \times 10^{11}$ N/m^2

$\rho = 7860$ kg/m^3

Element	A (m^2)	I (m^4)
1–2	1×10^{-4}	1×10^{-7}
2–3	2×10^{-4}	2×10^{-7}
3–4	2×10^{-4}	2×10^{-7}
4–5	1×10^{-4}	1×10^{-7}

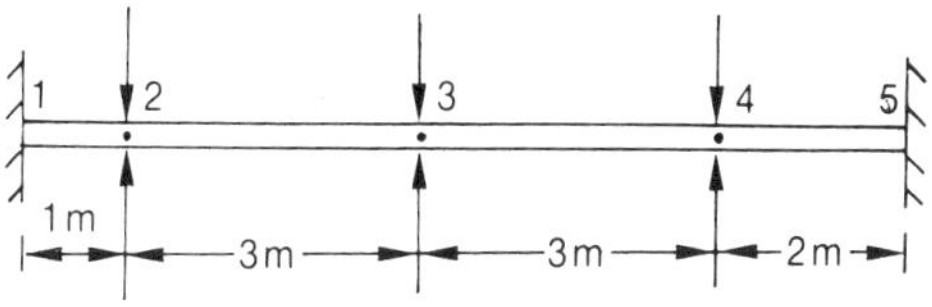

(*Portsmouth 1987, Honours*)

25.16 Calculate the three lowest natural frequencies of vibration for the continuous beam below, where

$A = 0.001$ m^2

$I = 1 \times 10^{-6}$ m^4

$E = 2 \times 10^{11}$ N/m^2

$\rho = 7860$ kg/m^3

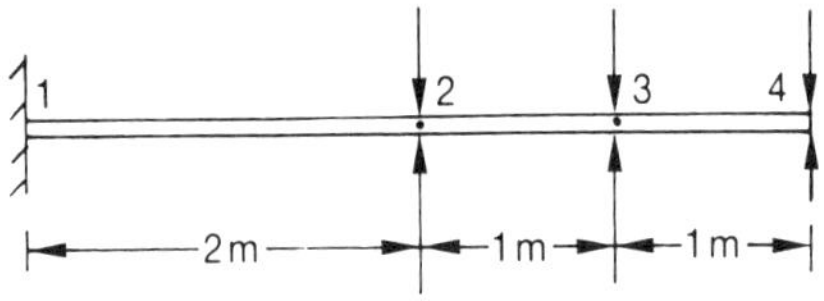
1
2
3
4
2 m
1 m
1 m

Answers to further problems

1.4	(a)	$R_A = 3.333$ kN	$R_B = 6.667$ kN		
	(b)	$R_A = 9.6$ kN	$R_B = 6.4$ kN		
	(c)	$R_A = 4.625$ kN	$R_B = 3.375$ kN		
1.5	(a)	$F_{ab} = 5$ kN	$F_{ac} = -8.66$ kN		
	(b)	$F_{ab} = -5$ kN	$F_{ac} = -8.66$ kN		
	(c)	$F_{ab} = 4.17$ kN	$F_{ac} = -3$ kN		
	(d)	$F_{ab} = -4.71$ kN	$F_{ac} = -7.454$ kN	$F_{bc} = 3.333$ kN	
	(e)	$F_{ab} = -4.71$ kN	$F_{ac} = 3.727$ kN	$F_{bc} = 3.333$ kN	

1.17 31.0 MN/m^2 (compressive); 0.098 cm.

1.18 0.902 cm.

1.19 0.865 cm.

1.21 51.0 kN.

1.22 65.5 MN/m^2 tensile in steel; 41.0 MN/m^2 compressive in copper; increase of length 0.611 cm; force to prevent expansion 135.5 kN.

2.2	$F_{ab} = -0.46$ W	$F_{ad} = 0.763$ W	$F_{bc} = -0.651$ W	
	$F_{cd} = 0.46$ W	$F_{de} = -0.008$ W	$F_{bd} = -0.54$ W	
2.3	$F_{ac} = 4$ kN	$F_{ad} = 2.829$ kN	$F_{bd} = -10$ kN	$F_{cd} = -4$ kN
	$F_{df} = -2$ kN	$F_{ce} = 5.66$ kN	$F_{ed} = -8$ kN	$F_{ef} = 5.66$ kN
2.4	$F_{ac} = 7.8$ kN	$F_{bc} = -1.45$ kN	$F_{bd} = -8.8$ kN	$F_{ce} = 6.8$ kN
	$F_{cd} = 1$ kN	$F_{ed} = -1.45$ kN	$F_{df} = -7.8$ kN	$F_{eg} = 5.8$ kN
	$F_{ef} = 1$ kN	$F_{fg} = -1.45$ kN	$F_{fh} = -6.8$ kN	$F_{gj} = 6.35$ kN
	$F_{gk} = 0.5$ kN	$F_{gh} = -3.9$ kN	$F_{hk} = -7.9$ kN	$F_{jk} = -4.5$ kN
	$F_{jl} = 4.5$ kN	$F_{kl} = -4$ kN		
2.5	$R_A = 1.333$ kN	$R_B = 1.667$ kN		
	$F_{ac} = -1.54$ kN	$F_{ac} = 0.77$ kN	$F_{ec} = 1.54$ kN	$F_{ef} = -1.54$ kN
	$F_{cf} = -0.38$ kN	$F_{cd} = 1.732$ kN	$F_{fd} = 0.38$ kN	$F_{fg} = -1.92$ kN
	$F_{dg} = 1.92$ kN	$F_{db} = 0.96$ kN	$F_{gb} = -1.92$ kN	

3.4 480 MN/m^2.

3.5 521 kW.

3.6 295 m/s^2.

3.7 188.5 Nm.

3.8 $d = 6.29$ cm; cotter thickness 1.57 cm; mean width of cotter 7.98 cm; distance of cotter hole from end of left-hand rod 2.97 cm; diameter of right-hand rod through cotter pin 8.28 cm; maximum diameter of right-hand rod 12.58 cm; distance of end of right-hand rod from cotter hole 2.97 cm.

3.9 8.93 cycles/sec.

3.10 0.6 MN/m^2.

4.7 376 kN/m.

5.7 50.0 MN/m^2 tensile; 28.9 MN/m^2 shearing.
5.8 greatest tensile stress 86.6 MN/m^2, on plane at 34° 44′ to cross-section; greatest shearing stress 64.0 MN/m^2, on planes at 10° 16′ and 79° 44′ to cross-section.
5.9 30 MN/m^2 tensile; 120 MN/m^2 compressive.
5.10 81.0 MN/m^2, inclined at 23° 27′ to horizontal.
5.11 90 MN/m^2 tensile; 60 MN/m^2 compressive; 5.40×10^{-4} tensile; 4.35×10^{-4} compressive.
5.12 7.5 MN/m^2 normal; 51.9 MN/m^2, shearing.

6.8 11.0 MN/m^2
6.9 1.03 kg/m.
6.10 0.114 per cent.

6.11 (a) copper: 38.2 MN/m^2; wire: 83.9 MN/m^2;
(b) copper: 28.6 MN/m^2 (compressive); wire: 230 MN/m^2.

6.12 1.19 MN/m^2.
6.13 171 MN/m^2.

7.10 489 kNM.
7.13 238 kNm; 0.75 m from *A*.

8.5 (a) 2.779×10^{-5} m^4.
(b) 10.83×10^{-6} m^4.
(c) 5.334×10^{-5} m^4.

8.6 (a) 1.419×10^{-5} m^4.
(b) 3.942×10^{-5} m^4.

8.7 H/3 BH^3 /36

9.12 40.6 kN.m.
9.13 69.9 MN/m^2.
9.14 15.9 MN/m^2.
9.15 86.0 MN/m^2.
9.16 6.17 cm.

10.4 1 cm thickness; 5 cm spacing of rivets, assuming one rivet at any cross-section.
10.5 maximum tensile stress of 124 MN/m^2 is greater than the allowable stress; maximum shearing stress of 18 MN/m^2 is less than the allowable stress.
10.6 96 per cent of shearing force carried by web; 88 per cent of bending moment carried by flanges.
10.7 web thickness 0.67 cm; weld throats 0.33 cm.
10.8 $\tau = 2450\,(RL/t)\sin\theta$, where θ is the angular position of any section from the vertical line through the centre of the tube.

10.9 bending is limiting, and gives an allowable superimposed load of 45 kN/m; required welds 0.26 cm throat thickness.

10.10 (a) 1.273 *R*.
(b) 1.72 *R*.

11.8 0.378×10^{-3} m^2; 13.02 kN/m.

11.9 114 kN.

11.10 wood 4.56 MN/m^2; steel 52.9 MN/m^2; glue 0.21 MN/m^2.

11.11 (i) 120 MN/m^2.
(ii) 1.00 MN/m^2.
(iii) 100 kN/m.
(iv) 0.75 cm.

12.3 tensile 155 MN/m^2, compressive 147 MN/m^2; neutral axis 0.365 m from outside of box-section.

12.4 17.68 kN; 11.8 MN/m^2 compressive.

12.6 maximum tensile 38.0 MN/m^2; maximum compressive 46.0 MN/m^2.

12.7 161 kN.

12.8 13.8 MN/m^2; 5.94 cm from tip of *T*.

13.2 1.80 cm and 2.48 cm.

13.3 3.06 cm.

14.2 maximum bending moment 105 kNm; points of inflexion at 1.75 m from each end.

14.3 169.7 kNm at left-hand end; 150.0 kNm at right-hand end; 1.52 m from left-hand end; 1.69 m from right-hand end.

15.2 217 kN.

15.3 62.4 kN/m.

15.4 required elastic section modulus 791 cm^3.

15.5 required elastic section modulus 2030 cm^3.

15.6 84.2 kN/m, with collapse in the end spans.

15.7 3.26 cm.

16.7 38.1 MN/m^2; 1.09°; 39.2 cm.

16.8 40.3 MN/m^2; 3.83°.

16.9 Shearing stress 37.7 MN/m^2; maximum tensile stress 37.7 MN/m^2; angle of twist 4.31°.

16.10 0.644: 1.

16.11 38.7 Nm.

16.12 147 MN/m^2 (tensile) at 34.8° to axis, 70.5 MN/m^2 (compressive) at 57.2° to axis.

17.15 No horizontal deflection.

17.16 609 kNm and 423 kNm.

17.17 $3WR/4\pi$, at the support, where W is the weight of the ring.

17.18 12.45 PR^3/EI.

17.19 2.89×10^{-4} m^3.

17.20 1.288×10^{-3} m^3.
17.21 6.324×10^{-4} m^3 (verticals); 9.486×10^{-4} m^3 (top left); 1.997×10^{-3} m^3 (top right).

18.2 970 N.
18.3 0.10 cm.
18.4 1.65 kN.
18.5 24.5 kN.

19.6
$$\left\{ w = \frac{W}{8\pi D}\left[\frac{(3+v)}{2(1+v)}\left(R^2 - r^2\right) + r^2 \ln\left(\frac{r}{R}\right)\right]\right\}$$

19.7
$$\left\{ \hat{w} = \frac{pR^4}{64D}\left[\frac{(5+v)}{(1+v)} - \frac{(6+2v)}{(1+v)}\left(\frac{r}{R}\right)^2 + \left(\frac{r}{R}\right)^4\right];\right.$$

$$M_r = -\frac{(3+v)}{16}pR^2\left[1 - \left(\frac{r}{R}\right)^2\right];$$

$$\left. M_t = \frac{pR^2}{16}\left[-(3+v) + (1+3v)\left(\frac{r}{R}\right)^2\right]\right\}$$

19.8
$$\left\{ \hat{w} = \frac{W}{16\pi D}\left[\frac{(3+v)}{(1+v)}\left(R_2^2 - R_1^2\right) + 2R_1^2 \ln\left(\frac{R_1}{R_2}\right)\right]\right\}$$

19.9 (a)
$$\left\{ \hat{w} = \frac{P}{16\pi D}\left[\frac{(3+v)}{(1+v)}R_2^2 - \frac{(7+3v)}{4(1+v)}R_1^2 + R_1^2 \ln\left(\frac{R_1}{R_2}\right)\right];\right.$$

$$\hat{M} = \frac{P}{4\pi}\left[1 - \frac{(1-v)}{4}\left(\frac{R_1}{R_2}\right)^2 - (1+v)\ln\left(\frac{R_1}{R_2}\right)\right];$$

(b)
$$\hat{w} = \frac{P}{16\pi D}\left[R_2^2 - 0.75R_1^2 + R_1^2 \ln\left(\frac{R_1}{R_2}\right)\right];$$

$$\hat{M} = \frac{P}{4\pi}\left[1 - 0.5\left(\frac{R_1}{R_2}\right)^2\right] \text{ for } R_1 / R_2 > 0.57$$

and $$\hat{M} = \frac{P}{4\pi}(1+v)\left[0.25\left(\frac{R_1}{R_2}\right)^2 - 1\text{n}\left(\frac{R_1}{R_2}\right)\right] \text{ for } R_1 \; R_2 < 0.57;$$

where $$P = p^* \pi R_1^2 \}$$

19.10 $$\left\{0.126pR^4 / \left(Et^3\right); p\left(\frac{R}{t}\right)^2\left[-1.238\left(\frac{r}{R}\right)^2 + 0.507 + 0.0105\left(\frac{R}{r}\right)^2\right]\right\}$$

19.11 $$\left\{0.115WR^2 / \left(Et^3\right); \frac{W}{t^2}\left[0.621 \ln\left(\frac{R}{r}\right) - 0.436 + 0.0224\left(\frac{R}{r}\right)^2\right]\right\}$$

22.1 $$\begin{bmatrix} 3 & 1 \\ 4 & -1 \end{bmatrix}$$

22.2 $$\begin{bmatrix} 5 & 1 \\ 0 & 7 \end{bmatrix}$$

22.3 $$\begin{bmatrix} 4 & 2 \\ 1 & 3 \end{bmatrix}$$

22.4 $$\begin{bmatrix} -1 & 2 \\ 0 & -4 \end{bmatrix}$$

22.5 $$\begin{bmatrix} -2 & -4 \\ 4 & -12 \end{bmatrix}$$

22.6 $$\begin{bmatrix} -4 & -1 \\ 0 & -10 \end{bmatrix}$$

22.7 10.

22.8 4.

22.9 $\begin{bmatrix} 0.3 & -0.1 \\ -0.2 & 0.4 \end{bmatrix}$

22.10 $\begin{bmatrix} -1 & 0 \\ -0.5 & -0.25 \end{bmatrix}$

22.11 $\begin{bmatrix} 10 & -1 & -2 \\ -3 & 9 & 1 \\ -4 & -2 & 7 \end{bmatrix}$

22.12 $\begin{bmatrix} -8 & -3 & 2 \\ -1 & -7 & -5 \\ 4 & -2 & -5 \end{bmatrix}$

22.13 $\begin{bmatrix} 1 & -2 & 0 \\ -2 & 1 & -2 \\ 0 & -2 & 1 \end{bmatrix}$

22.14 $\begin{bmatrix} 9 & -1 & -4 \\ 1 & 8 & 0 \\ -2 & 3 & 6 \end{bmatrix}$

22.15 $\begin{bmatrix} 11 & -15 & -8 \\ -11 & 6 & -5 \\ -2 & -16 & 0 \end{bmatrix}$

22.16 $\begin{bmatrix} 7 & -13 & -4 \\ -17 & 4 & -13 \\ -4 & -4 & 6 \end{bmatrix}$

22.17 -7.

22.18 362.

22.19 $\begin{bmatrix} 0.429 & -0.286 & -0.571 \\ -0.268 & -0.143 & -0.286 \\ -0.571 & -0.286 & 0.429 \end{bmatrix}$

22.20 $\begin{bmatrix} 0.133 & -1.66\times10^{-2} & -5.25\times10^{-2} \\ -1.66\times10^{-2} & 0.127 & -6.91\times10^{-2} \\ -8.84\times10^{-2} & -1.10\times10^{-2} & 0.202 \end{bmatrix}$

23.8 (a) $u_4^\circ = 2.828/AE$ $v_4^\circ = 0$
$F_{1\text{-}4} = 1.414$ kN $F_{2\text{-}4} = 0$ $F_{3\text{-}4} = -1.414$ kN
(b) $u_4^\circ = 2.626/AE$ $v_4^\circ = 0$
$F_{1\text{-}4} = 1.313$ kN $F_{2\text{-}4} = 0$ $F_{3\text{-}4} = 1.313$ kN

23.9 (a) $u_5^\circ = 2.178/AE$ $v_5^\circ = -0.243/AE$
(b) $u_5^\circ = 2.065/AE$ $v_5^\circ = -0.224/AE$ $\theta_5 = 2.528/AE$

23.10 (a) $v_1 = -Wl^3/3EI$ $\theta = -Wl^2/2EI$
$v_2 = \theta_2 = 0$
$M_1 = 0$ $M_2 = Wl$
(b) $v_1 = \theta_1 = v_3 = \theta_3 = 0;\ \theta_2 = 0$
$v_2 = -Wl^3/192EI$
$M_1 = -M_3 = -Wl/8$
$M_2 = \pm Wl/8$

23.11 $v_1 = \theta_1 = v_2 = v_3 = v_4 = \theta_4 = 0$
$\theta_2 = -1.136/EI$
$\theta_3 = 0.2/EI$
$M_1 = -0.345$ $M_2 = \pm 0.311$
$M_3 = \pm 1.45$ $M_4 = 2.65$

23.12 $v_1 = \theta_1 = v_2 = v_3 = v_4 = \theta_4 = 0$
$\theta_2 = 0.386/EI$ $\theta_3 = 0.193/EI$
$M_1 = -0.643$ $M_2 = \pm 1.864$
$M_3 = \pm 4.629$ $M_4 = 6.236$

23.13 (a) $u_4^\circ = 220.59/AE$ $v_4^\circ = 209.67/AE$ $w_4^\circ = -77.21/AE$
$F_{1\text{-}4} = 4.59$ kN $F_{2\text{-}4} = -2.75$ kN $F_{3\text{-}4} = -7.28$ kN
(b) $u_5^\circ = 2.287/AE$ $v_5^\circ = -8.591/AE$ $w_5^\circ = -1.904/AE$
$F_{1\text{-}5} = -1.685$ kN $F_{2\text{-}5} = 1.179$ kN $F_{3\text{-}5} = -2.927$ kN
$F_{4\text{-}5} = -0.054$ kN

23.13 (c) $u_5^\circ = 13.48/AE$ $v_5^\circ = 41.72/AE$ $w_5^\circ = -39.46/AE$
$F_{1\text{-}5} = 1.05$ MN $F_{2\text{-}5} = -4.51$ MN $F_{3\text{-}5} = -9.51$ MN
$F_{4\text{-}5} = -2.15$ MN

23.14 (a) $u_2^\circ = 1.257\times10^{-2} = u_3^\circ$
$\theta_2 = 0.162\times10^{-2}$ rads $\theta_3 = 0.162\times10^{-3}$ rads
$M_1 = -10.47$ kNm $M_2 = \pm 3.52$ kNm
$M_3 = \pm 6.188$ kNm $M_4 = -7.81$ kNm

(b) $u_2^\circ = 12.134/EI$ $\theta_2 = 3.777/EI$
$\theta_3 = 1.132/EI$
$M_1 = -9.32$ kNm $M_2 = \pm 0.7$ kNm
$M_3 = \pm 6.58$ kNm $M_4 = -7.33$ kNm

24.2 $$\frac{AE}{l}\begin{bmatrix} 1 & -1 \\ -1 & 1 \end{bmatrix}$$

24.3 $$\frac{(A_1 + A_2)E}{2l}\begin{bmatrix} 1 & -1 \\ -1 & 1 \end{bmatrix}$$

24.4 $$\frac{GJ}{l}\begin{bmatrix} 1 & -1 \\ -1 & 1 \end{bmatrix}$$

24.5 $$EI\begin{bmatrix} 12/l^3 & -6/l^2 & -12/l^3 & -6/l^2 \\ -6/l^2 & 4/l & 6/l^2 & 2/l \\ -12/l^3 & 6/l^2 & 12/l^3 & 6/l^2 \\ -6/l^2 & 2/l & 6/l^2 & 4/l \end{bmatrix}$$

25.7 7.00 Hz, 28Hz, etc., 11.85 Hz, 47.4 Hz etc.
25.8 4.73 Hz.
25.9 $(3EI/ML^3)^{1/2}/2\pi$.
25.10 404.1 Hz, 598.5 Hz.
25.11 294.8 Hz, 361.6 Hz, 485.4 Hz.
25.12 53.1 Hz, 164.1 Hz.
25.13 40.56 Hz.
25.14 191.8 Hz, 354.3 Hz, 907.8 Hz.

Index